Electric Power Systems Research

Special Issue Editor
Ying-Yi Hong

Special Issue Editor
Ying-Yi Hong
Chung Yuan Christian University
Taiwan

Editorial Office
MDPI AG
St. Alban-Anlage 66
Basel, Switzerland

This edition is a reprint of the Special Issue published online in the open access journal *Energies* (ISSN 1996-1073) from 2015–2016 (available at: http://www.mdpi.com/journal/energies/special_issues/electric-power-systems).

For citation purposes, cite each article independently as indicated on the article page online and as indicated below:

Author 1; Author 2; Author 3 etc. Article title. *Journal Name*. **Year**. Article number/page range.

ISBN 978-3-03842-404-8 (PDF)
ISBN 978-3-03842-405-5 (Pbk)

Table of Contents

About the Guest Editor

Ying-Yi Hong received his Ph.D. degree from the Department of Electrical Engineering in December 1990 from the National Tsing-Hua University, Taiwan. His main areas of interest are optimal power flow, power quality, power market, microgrid, AI application and FPGA chip design. Professor Hong is an IEEE senior member and an IET Fellow. He was the Chair of the IEEE PES Taipei Chapter from 2001–2002 and was the Vice Chair of IEEE Taipei Section from 2013–2014. He also received the honor of being the Outstanding Electrical Engineering Professor from the Chinese Institute of Electrical Engineering in 2006. Since 1991, Professor Hong has chaired/co-chaired more than 100 projects. To date, he has published 89 SCI journal papers and 185 conference papers. From 2006–2012, he was the Dean of College of EECS at Chung Yuan Christian University (CYCU). Presently, he is a Distinguished Professor and serves as a Secretary General at CYCU.

Article

Dynamic Equivalent Modeling for Small and Medium Hydropower Generator Group Based on Measurements

Bowei Hu [1], Jingtao Sun [1], Lijie Ding [2], Xinyu Liu [1] and Xiaoru Wang [1,*]

[1] School of Electrical Engineering, Southwest Jiaotong University, Chengdu 610031, China; hbw1115@163.com; (B.H.); jingtao_sun@126.com (J.S.); liuxinyu985@163.com (X.L.)
[2] Sichuan Electric Power Research Institute, Chengdu 610072, China; ding_lijie@163.com
*   Correspondence: xrwang@home.swjtu.edu.cn; Tel.: +86-28-6636-6009

Academic Editors: Gabriele Grandi and Ying-Yi Hong
Received: 19 December 2015; Accepted: 5 May 2016; Published: 12 May 2016

Abstract: At present, the common practice in the power system of China is to represent the small and medium hydropower generator group as a negative load. This paper presents a method to build a dynamic equivalent model of the hydropower generator group using a 3rd order generator model and a static characteristic load model. Based on phasor measurements in the tie line which connects to the modeled hydropower generator group, the dynamic multi-swarm particle swarm optimizer (DMS-PSO) algorithm is used to obtain parameters of the equivalent model. The proposed method is verified in the small and medium hydropower generator group of Sichuan power grid with both simulation and actual data. The results show that the dynamic responses and the transient stability are consistent before and after the equivalence. The proposed method can be used for modeling a group of small and medium hydropower generators whose structures and parameters are unknown.

Keywords: small and medium hydropower generator group; dynamic equivalence; estimation-based equivalent method; measurements; dynamic multi-swarm particle swarm optimizer (DMS-PSO) algorithm

1. Introduction

There are a large number of small and medium hydropower generators in abundant water resources areas because they are flexible, convenient and environmentally-friendly energy sources. If the region contains a lot of small and medium hydropower stations, the dynamic performance of the power system must be affected by the hydropower generator group. It is difficult to establish a detailed model because the parameters of each generating unit can barely be obtained if there is a small and medium hydropower generator group in the power system. The power flow calculation may not be converged and the dimension is too large for dynamic analysis, even if the typical parameters are used. At present, the common practice in the power system of China is to represent the small and medium hydropower generator group as a negative load, but the simulation results are not accurate. Dynamic equivalence is an accurate way to represent the characteristics of the equivalent system [1].

At present, the dynamic equivalence methods can be divided into three categories. The first one is the coherency-based equivalent method. In [2–4], the hydropower generator group is equivalent to a 3rd order generator and a load with a weighted average method, while a 4th order generator and a load are used as the equivalent model for the hydropower generator group in [5]. However, there are significant limitations when using the traditional coherency-based equivalent method to construct the equivalent model because the structures and parameters of the hydropower generator group must be known [6,7]. The second one is modal-based equivalent method. However, it is not appropriate to transient stability analysis because nonlinear models are linearized and state equations are used as the equivalent model [8–11]. The third one is estimation-based equivalent method, which functions only by making use of measurements. The estimation equivalent method is first proposed in paper [12]. Scholars have undertaken further study in the estimation-based equivalent method in two branches. A 16-machine test system is equivalent to using artificial neural networks in [13–15]. A 2nd order transfer function is proposed to equal the distributed network cell based on Prony analysis and nonlinear least square optimization method [16]. However, there is no physical meaning for these equivalent models. The other branch also has some disadvantages. A CEPRI 8-machine test system is equivalent to a 3rd order generator and a static characteristic load model with linearized approach during the process of calculating the equivalent model parameters so that the results are not accurate enough as a nonlinear method [17]. The eight generators are equivalent to a 3rd order generator based on PSASP (Power System Analysis Software Package) transient stability module and particle swarm optimizer (PSO) [18], but there is not much practical significance because the internal system model must be combined. Moreover, there is little research on dynamic equivalence for the small and medium hydropower generator group based on the estimation-based equivalent method. Therefore, it is significant to research the estimation-based equivalent method based on phasor measurements while the structures and parameters of the hydropower generator group are unknown.

In this paper, a measurement-based method for system dynamic equivalence is proposed to construct the dynamic equivalent model for a small and medium hydropower generator group whose structures and parameters are unknown. The proposed equivalent model is a parallel dynamic equivalent model composed of a 3rd order electromechanical transient generator model with damping and a static characteristic load model considering voltage. Based on the tie line phasor measurements, parameters of the equivalent model are obtained by combining nonlinear 4th order Runge-Kutta method and dynamic multi-swarm particle swarm optimizer (DMS-PSO) algorithm. The proposed method is verified in the small and medium hydropower generator group of Sichuan power grid with both simulation and actual data.

The rest of this paper is organized as follows. The equivalent model for the small and medium hydropower generator group is introduced in Section 2. In Section 3, methodology of the proposed equivalence is presented. The proposed equivalent model is validated by the case study in the Sichuan power grid with both simulation and actual data being presented in Section 4. Section 5 concludes this paper.

2. Equivalent Model for the Hydropower Generator Group

The small and medium hydropower generator group is always connected to the internal system through a tie line, and the electrical distance of each hydropower generator in the equivalent area is relatively close. Therefore, the hydropower generator group can be considered coherent when disturbances occur within the internal system [19]. Based on the above characteristics, the dynamic equivalent model is shown in Figure 1, which is in the form of a single generator and a parallel single load.

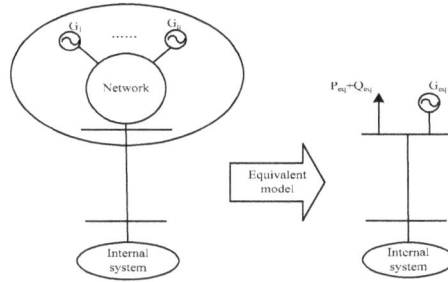

Figure 1. Dynamic equivalent model structure.

2.1. Equivalent Generator Model

The 2nd order generator model for system equivalence cannot accurately represent the dynamic behaviors of the original system since the 2nd order generator model does not consider the field winding and damping winding. The performances of different dynamic equivalent models are analyzed in [20], and the results show that the 3rd order generator model is good enough for accurate equivalence since the optimization time greatly increased with the 5th order generator model without improving accuracy significantly. In this paper, the 3rd order generator model is used by considering field windings and damping windings, and the proportional feedback model is used in an excitation system while the constant torque model is used in the speed control system. The equations of the generator model are:

$$\begin{cases} T_j \frac{d\omega}{dt} = P_m - \frac{E'V}{x'_d}\sin\delta - D\frac{d\delta}{dt} \\ \frac{d\delta}{dt} = (\omega - \omega_f)\,\omega_0,\, \omega_0 = 2\pi f \\ T'_{d0}\frac{dE'}{dt} = E_{f0} - K_V(V - V_0) - \frac{x_d}{x'_d}E' + (\frac{x_d}{x'_d} - 1)V\cos\delta \end{cases} \tag{1}$$

where T_j is the inertia time constant; P_m is the mechanical power; E' is the electromotive force after x'_d and instead of q-axis transient electromotive force; V is the generator voltage; x'_d is the d-axis transient reactance and the value is equal to q-axis synchronous reactance; δ is the power angle; D is the damping coefficient; t is time; ω is the rotor speed of generator; ω_f is the rotor speed of reference; T'_{d0} is the d-axis transient open circuit time constant; E_{f0} is the initial excitation voltage; K_V is the feedback coefficient of excitation voltage; V_0 is the initial generator voltage; x_d is the d-axis synchronous reactance.

2.2. Equivalent Load Model

Due to less influence to the system dynamic characteristics by the equivalent load, the static load model considering voltage is taken as the equivalent load model. The identifiability of a 3rd order generator and a static load is proved by the theory in [19]. The load model equations are:

$$\begin{cases} P_s = P_{s0}(\frac{V}{V_0})^{N_p} \\ Q_s = Q_{s0}(\frac{V}{V_0})^{N_q} \end{cases} \tag{2}$$

where P_s is the load active power; P_{s0} is the initial load active power; N_p is the index of voltage characteristics on active power; Q_s is the load reactive power; Q_{s0} is the initial load reactive power; N_q is the index of voltage characteristics on reactive power.

From the foregoing analysis, the power equations in tie line can be described as follows:

$$\begin{cases} P_l = P_e - P_s \\ Q_l = Q_e - Q_s \end{cases} \tag{3}$$

where P_l is the active power in tie line; Q_l is the reactive power in tie line; P_e is the generator active power; Q_e is the generator reactive power.

3. Methodology of the Proposed Equivalence

3.1. Frame of the Equivalence Method

Synchronized phasor measurement technology has been widely used in the power system dynamic monitoring [21–23]. Equivalent model parameters are obtained by making use of dynamic information in the tie line during the disturbance based on the parameter estimation. In this paper, the measurements of voltage, frequency, active power and reactive power in the tie line are extracted as inputs and then the equivalent model parameters are identified, which is $\alpha = [T_j, x_d, x'_d, T'_{d0}, K_v, D, P_{s0}, Q_{s0}, N_p, N_q]$. The overall framework is shown in Figure 2.

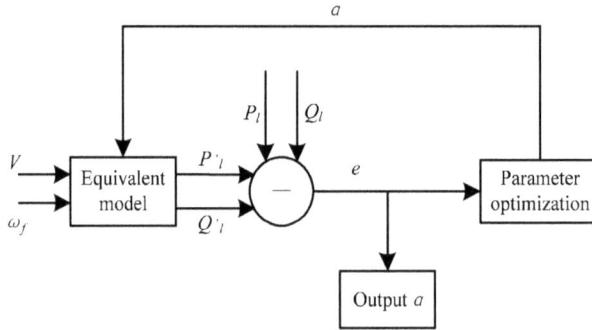

Figure 2. The equivalent schematic diagram of hydropower generator group.

The processes are as follows: firstly, initialize a suspicious parameter α; secondly, take the voltage and frequency responses in the tie line as inputs; thirdly, solve the equations of equivalent model and get active and reactive power; then compare measured and calculated values of the active and reactive power and the error e is generated; finally compare the error with the set threshold, namely if the error is greater, update parameter α by the optimization algorithm and repeat the process; otherwise, output α as the equivalent model parameters.

The electrical power outputs are calculated with the following equations:

$$\begin{cases} P_{e(k)} = \dfrac{E'_{(k)} V_{(k)}}{x'_d} \sin(\delta_{(k)}) \\ Q_{e(k)} = \dfrac{E'_{(k)} V_{(k)} \cos(\delta_{(k)}) - V^2_{(k)}}{x'_d} \end{cases} \tag{4}$$

The error is calculated with the Euclidean distance formula:

$$e = \sum_{k=1}^{N} \sqrt{\left(\frac{P_{l(k)} - P'_{l(k)}}{P_{l(1)}}\right)^2 + \left(\frac{Q_{l(k)} - Q'_{l(k)}}{Q_{l(1)}}\right)^2} \tag{5}$$

where N is the total number of sampling sites; P_1 is the measured value of active power; P'_1 is the calculated value of active power; Q_1 is the measured value of reactive power; Q'_1 is the calculated value of reactive power.

In order to obtain the electrical power outputs, the δ and E' from different times have to be calculated. The two most classic numerical methods are the improved Euler method and Runge-Kutta method [24]. The 4th order Runge-Kutta method is used widely when solving ordinary differential equations because of its briefness and accuracy [25–27]. In this paper, the 4th order Runge-Kutta method is used to solve generator dynamic equations and the state variables of time $k + 1$ moment such as ω, δ, E' can be calculated by time k moment. The initial value of state variables can be determined by the steady-state phasor diagram. The power outputs in the tie line are obtained with the certain equivalent model parameters based on Equations (1)–(4). The frame is shown in Figure 3.

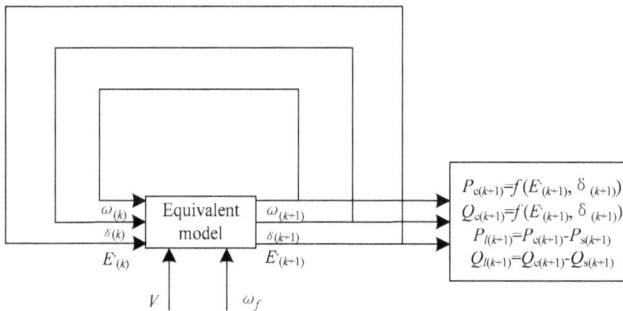

Figure 3. The schematic diagram of power in tie line.

3.2. Dynamic Multi-Swarm Particle Swarm Optimizer Algorithm

The PSO algorithm is widely used in the optimization of power system because of its good global convergence, easy operation, high efficiency and few parameter settings [28–30]. In this paper, equivalent model parameters are identified by the DMS-PSO algorithm [31].

The optimal solution is found by initializing a group of random particles, iterating them repeatedly with the basic method of PSO. In each iteration process, the particles are updated by tracking two "extremes": one is the optimal solution found by an individual (personal best, that is *pbest*), and the other is the optimal solution found by the entire group (group best, that is *gbest*). Each particle updates its velocity and position respectively according to the following equations.

$$\begin{cases} v_{n+1} = wv_n + c_1 \times rand \times (pbest_n - x_n) + c_2 \times rand \times (gbest_n - x_n) \\ x_{n+1} = x_n + v_{n+1} \end{cases} \tag{6}$$

where v_n is the velocity vector of the particle on n-th iteration; w is the weight factor; $c_{1,2}$ is the learning factor; *rand* is a random number between 0 and 1.

In order to achieve better results, the following three improvements are used based on the basic PSO algorithm.

The first strategy is DMS-PSO in order to slow down convergence speed and to increase diversity to enhance the global search capabilities. The effect is better on complex problems when PSO with small neighborhoods is used according to many reported researches of PSO. Therefore, in DMS-PSO, the population is divided into small sized swarms. Each particle seeks for better solution in the search space in each small sized swarm. After every R iterations, the population will be regrouped randomly and then the search will be started in the new swarms [32].

The second strategy is particle non-uniform mutation for the sake of improving self-help capabilities when the particles get into a local optimum. When the stable frequency of the value of *gbest* objective function reaches a certain threshold, current vector is randomly generated so that the particles have a chance to escape from the local optimum and search for the global optimum. The mutation equations are:

$$x'_{nj} = \begin{cases} x_{nj} + (U_j - x_{nj})(1 - b^{(1-\frac{n}{n_{max}})}) & \text{if} \quad l = 0 \\ x_{nj} - (x_{nj} - L_j)(1 - b^{(1-\frac{n}{n_{max}})}) & \text{if} \quad l = 1 \end{cases} \tag{7}$$

where x'_{nj} is the mutated x_{nj}; l is the random number of 0 or 1; x_{nj} is the j-dimensional component of the n-th particle; U_j is the upper limit of x_{ij}; L_j is the lower limit of x_{ij}; b is the random number of 0–1.

The third strategy is decreasing the inertia weight linearly for balancing the local search and global search capabilities. It is conducive to a global search in the earlier stage when w is larger while being conducive to local development in the later stage when w is smaller. The formula is:

$$w = w_{max} - \frac{(w_{max} - w_{min})n}{n_{max}} \tag{8}$$

where w_{max} is the upper limit of inertia weight which is set to 0.9; w_{min} is the lower limit of inertia weight which is set to 0.4; n is the current iteration number; n_{max} is the maximum number of iterations.

The process of solving the dynamic equivalent model parameters based on the above-mentioned method with DMS-PSO is shown in Figure 4.

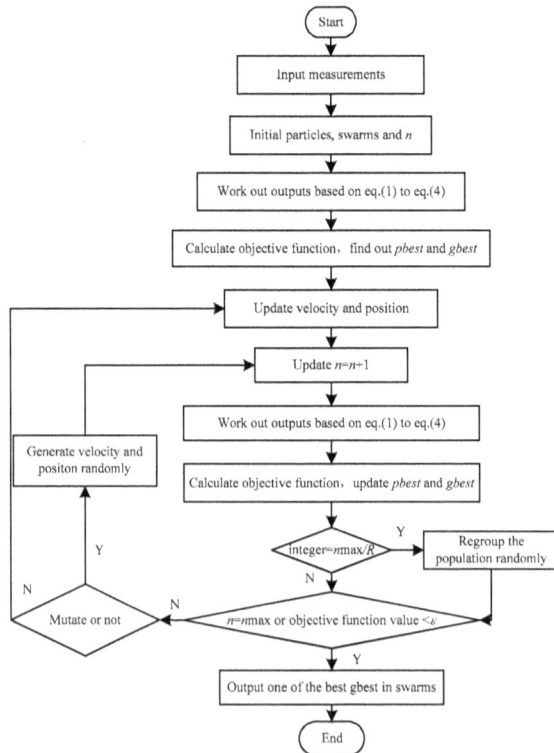

Figure 4. Flow chart of parameter identification based on dynamic multi-swarm particle swarm optimizer (DMS-PSO) algorithm.

Specific steps: (1) Input measurements, namely, V, ω_f, P_1 and Q_1 in the tie line; (2) Initialize the parameters of each particle, swarm and iteration number; (3) Work out the outputs based on Equations (1)–(4); (4) Calculate the objective function and confirm the initial value of *pbest* and *gbest*; (5) Update iterations and velocity and position of the particle; (6) Update iterations; (7) Work out the outputs based on Equations (1)–(4); (8) Calculate the objective function and then update *pbest* and *gbest*; (9) Determine whether $n_{\max}/R$ is integer or not. If the answer is Yes, regroup the population randomly, otherwise go on; (10) Determine whether the terminal condition is satisfied or not. If the answer is Yes, stop calculating and output *gbest*, otherwise determine whether mutate or not. If the answer is Yes, generate velocity and position randomly and then return to (6), otherwise return to (5).

4. Case Study

4.1. Sensitivity Analysis of Equivalent Model Parameters

Each parameter of the equivalent model has an impact on the result of the equivalent model. The sensitivity evaluation formula of each parameter is:

$$R_{\text{sensitivity}\_j} = \frac{\partial H}{\partial X_j} \tag{9}$$

where H is the residual error of power responses with the parameters; X_j is the j-th parameter; $R_{\text{sensitivity}\_j}$ is the sensitivity of the j-th parameter.

The formula of the residual error of power responses is:

$$H = \sum_{k=1}^{N} \left\{ \left[\frac{P(X_j,k) - P(X_0,k)}{P(X_0,1)} \right]^2 + \left[\frac{Q(X_j,k) - Q(X_0,k)}{Q(X_0,1)} \right]^2 \right\} \tag{10}$$

where $P(X_j, k)$ is the active power with changed X_j at k moment; $P(X_0, k)$ is the active power with basic X_0 at k moment; $P(X_0, 1)$ is the active power with basic X_0 parameter at the initial time; $Q(X_j, k)$ is the reactive power with changed X_j at k moment; $Q(X_0, k)$ is the reactive power with basic X_0 at k moment; $Q(X_0, 1)$ is the reactive power with basic X_0 at initial time.

In order to make the sensitivity of each parameter comparable, the range of the parameters change from 80% to 120% based on the basic value. The basic values which are obtained from a real group in a power system are shown in Table 1 in Section 4.2; these are actual values. The results of sensitivities are shown in Figure 5. The slope of each curve in Figure 5 is the parameter sensitivity.

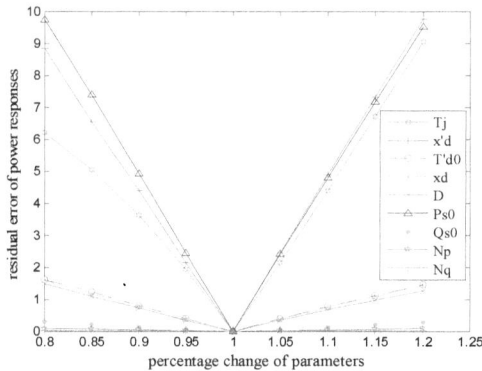

Figure 5. Residual error of power responses with different parameters.

As shown in Figure 5, active power load P_{s0}, inertia time constant T_j and d-axis transient reactance x'_d are the three most sensitive parameters. Initial load reactive power Q_{s0}, index of voltage characteristics on active power N_p, and index of voltage characteristics on reactive power N_q are the three lowest sensitivity parameters.

4.2. Identifiability Verification

The parameters of the generator and the load are identified in an infinite system when the above method is applied. The contrast of identified value and actual value are shown in Table 1 while the curves of output between identified value and actual value are shown in Figure 6. A three-phase short circuit occurred in 1 s in a transmission line and the disturbance disappeared in 1.1 s.

Table 1. Comparison of the identified parameters and the actual parameters of the model (p.u.).

Parameter	Identified Value	Actual Value	Parameter	Identified Value	Actual Value
T_j	50.96	49.50	D	0.165	0.2
x_d	0.1038	0.1460	P_s	0.987	1
x'_d	0.0512	0.0523	Q_s	0.298	0.5
T''_{d0}	8.682	10.430	N_p	1.143	1.5
K_v	0.3672	-	N_q	1.767	1.2

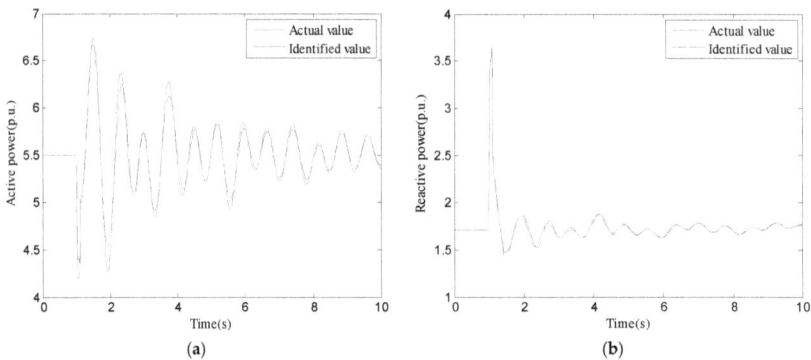

Figure 6. (a) Active power curves based on identified parameters and actual parameters; (b) reactive power curves based on identified parameters and actual parameters.

Table 1 shows that the main parameters in the identified and actual models such as T_j, x'_d, P_{s0} are very close. Certain parameters deviate from the actual values because these parameters have less influence on the dynamic characteristic and the simplified excitation model is introduced. Besides, the accuracy of the dynamic process is not affected because this method focuses on global optimization. As shown in Figure 6, the identified and the actual response curves have high fitting degree and the error is very small.

Considering noise is included in measurements in the actual power system, Gaussian noise with 50 dB signal-to-noise ratio (SNR) is added to the inputs when the same experiment is done. The comparison of values and curves are shown in Table 2 and Figure 7.

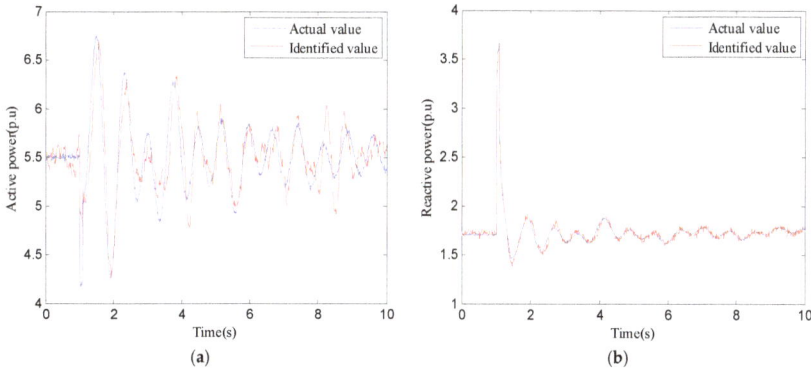

Figure 7. (a) Active power curves based on identified parameters and actual parameters; (b) reactive power curves based on identified parameters and actual parameters.

Table 2. Comparison of the identified parameters and the actual parameters of the model for 50 dB signal-to-noise ratio (SNR) inputs (p.u.).

Parameter	Identified Value	Actual Value	Parameter	Identified Value	Actual Value
T_j	52.34	49.50	D	0.1297	0.2
x_d	0.2156	0.1460	P_s	1.1681	1
x'_d	0.0586	0.0523	Q_s	0.2743	0.5
T''_{d0}	17.35	10.430	N_p	2.292	1.5
K_v	2.2693	-	N_q	1.987	1.2

It shows that the main parameters identification of T_j, x'_d, P_{s0} are still relatively accurate in Table 2, and the trends of response curves are consistent in Figure 7. Therefore, it can be said that the DMS-PSO algorithm used for parameter identification is completely feasible when the noise of input measurements is not big. It can filter out the noise at first and then identify the parameters based on the proposed method when the noise is big.

4.3. Equivalence with Simulation Data

Because it is a hydroelectric energy rich region, the hydropower capacity account for more than 60% of the total installed capacity in Sichuan. It is distributed in nearly 10 hydropower stations, a total of 23 hydroelectric generating units with a total capacity of 1578 MW in a certain area in Sichuan. The 5th order electromechanical transient generator and ZIP load models are represented in this area. Take the 500 kV hydropower channels in the area as an example. The proposed method in this paper is used to construct a dynamic equivalent model for the hydropower generator group. Parameters of equivalent model are shown in Table 3. The dynamic response characteristics curves in the tie line of equivalent model and the negative load model are compared with the actual dynamic response characteristics for the hydropower generator group in Figure 8. Negative load is expressed as a constant impedance model which is commonly used in engineering. Moreover, the three phase short circuit occurred in 1 s in a 500 kV Shimian-Ya'an transmission line and broke this line in 1.1 s. The meaning of "before equivalence" indicates measurements obtained from the original system.

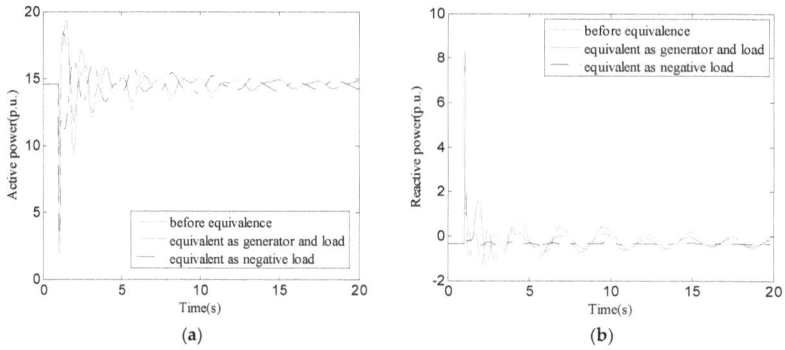

Figure 8. (**a**) Active power curves about different model; (**b**) reactive power curves about different model.

Table 3. Parameters of equivalent model (p.u).

Parameter	T_j	x_d	x'_d	T'_{d0}	K_v
Value	68.3137	0.5811	0.0455	15.3374	13.8422
Parameter	D	P_{s0}	Q_{s0}	N_p	N_q
Value	0.3093	0.7765	0.5897	0.5982	1.8078

As can be seen from Figure 8, the dynamic response characteristics cannot be described accurately when the hydropower generator group is represented as a negative load. This is determined by the essential differences between generator and load models. On the other hand, the trends of dynamic responses are consistent before and after the equivalence, and the fitting rate is higher than the equivalence as a negative load.

The simple map representing the positions which are mentioned in this paper is shown in Figure 9. The partial results of power flow in internal system before and after equivalence are listed in Table 4. It is shown that relative errors are within 0.3% and the system after equivalence can reflect the steady state operation.

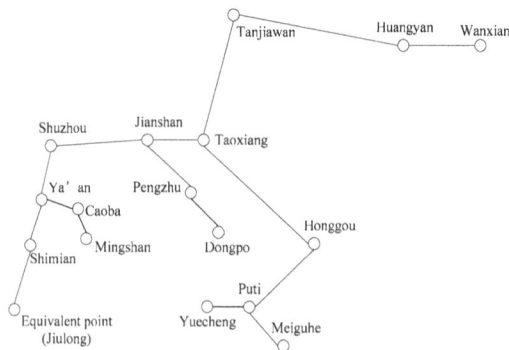

Figure 9. The simple map.

Table 4. The part of power flow results before and after equivalence (p.u.).

Position	Type	Before Equivalence	After Equivalence	Relative Errors
Jiulong500–Shimian500	Active power	14.5465	14.55	0.024%
	Reactive power	−0.34845	−0.34935	0.25%
Jiulong500	Voltage amplitude	1.00641	1.00637	−0.0039%
Jianshan500–Pengzhu500	Active power	4.41634	4.41727	0.0047%
	Reactive power	1.41196	1.41087	0.079%
Jianshan500	Voltage amplitude	0.97591	0.97589	−0.0020%
Huangyan500–Wanxian500	Active power	15.90473	15.90617	0.0091%
	Reactive power	−1.177	−1.17682	0.015%
Huangyan500	Voltage amplitude	0.98683	0.98682	−0.001%
Meiguhe220–Puti220	Active power	0.53659	0.53659	0%
	Reactive power	0.49844	0.49845	0.0020%
Puti220	Voltage amplitude	0.9913	0.9913	0%
Caoba220–Mingshan220	Active power	0.29271	0.29271	0%
	Reactive power	−0.16314	−0.16313	0.0061%
Caoba220	Voltage amplitude	0.98993	0.98991	−0.0020%
grid	Frequency	1	1	0%

The dynamic response characteristics curves on the interface between Huangyan in Sichuan and Wanxian in Chongqing before and after the equivalence are compared in Figure 10. As we can see from the figure, the transient response curves are similar and the system after equivalence can be used for transient stability analysis.

In order to verify the quality of the model reduction and parameter identification, the comparisons are done in before and after equivalence systems under other five faults, respectively. The fault descriptions and relative errors in the tie line are shown in Table 5.

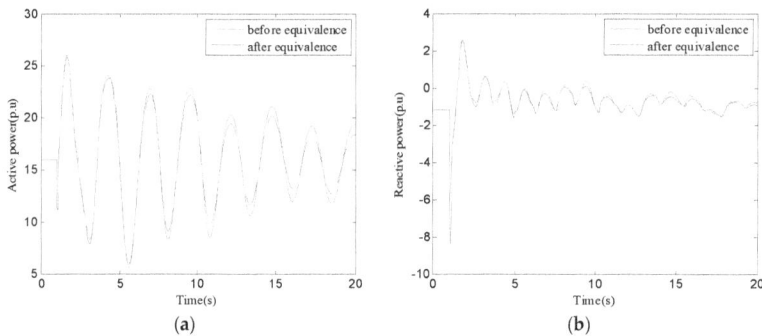

(a)

(b)

Figure 10. (**a**) Active power curves between before and after equivalence; (**b**) reactive power curves between before and after equivalence.

Table 5. Parameters of equivalent model (p.u).

Number	Fault Descriptions	Relative Errors
1	Three phase short circuit occurred in 1 s in a 500 kV Yuecheng-Puti transmission line and broke this line in 1.1 s	40.49
2	Three phase short circuit occurred in 1 s in a 500 kV Nantian-Dongpo transmission line and broke this line in 1.1 s	39.65
3	Three phase short circuit occurred in 1 s in a 500 kV Shuzhou-Danjing transmission line and broke this line in 1.1 s	34.17
4	Three phase short circuit occurred in 1 s in a 500 kV Ya'an-Jianshan transmission line and broke this line in 1.1 s	37.48
5	Three phase short circuit occurred in 1 s in a 500 kV Tanjiawan-Nanchong transmission line and broke this line in 1.1 s	28.56

As can be seen from Table 5, the maximum relative error is under No. 1 fault. Under No. 1 fault, for example, the dynamic responses characteristics curves in tie line and interface between Huangyan and Wanxian before and after equivalence are compared in Figures 11 and 12. The oscillation frequency, amplitude and the shape of the curve trend are consistent according to Figures 11 and 12. As can be seen in this section, the equivalent system keeps the dynamic characteristic of the original system and the effect of dynamic equivalence is good. The constructed model based on the proposed method can be applicable to a variety of situations and it keeps a certain robustness.

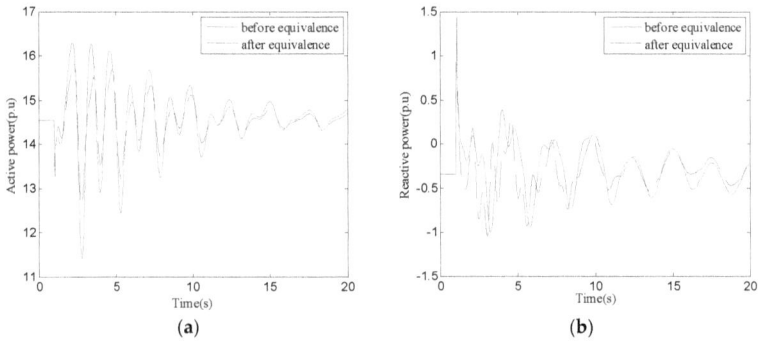

Figure 11. (**a**) Active power curves between before and after equivalence; (**b**) reactive power curves between before and after equivalence.

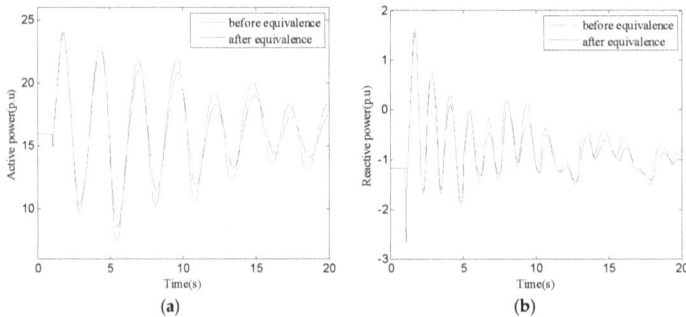

Figure 12. (**a**) Active power curves between before and after equivalence; (**b**) reactive power curves between before and after equivalence.

4.4. Equivalence with PMU Data

A small and medium hydropower generator group is connected to the grid through 220 kV and lower voltage substation, and connected to the main power grid by 500 kV substation. A disturbance occurred at 2:16:59 on October 1, 2015 in the 500 kV transmission line of Jiulong-Shimian in the Sichuan power grid. The fault waves from the 220 kV transmission line of Shaping-Jiulong were recorded by PMU. Based on the measurements and the equivalent method, the equivalent model of small and medium hydropower generator group in 220 kV Shaping area is constructed. The parameters of the model are shown in Table 6 while the transient response curves are shown in Figure 13.

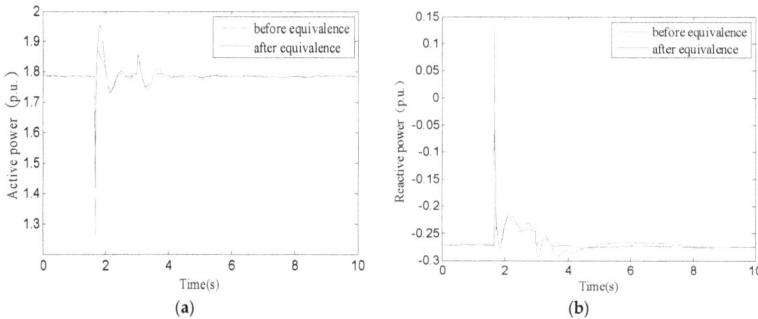

Figure 13. (**a**) Active power curves between before and after equivalence; (**b**) Reactive power curves.

Table 6. Parameters of equivalent model (p.u).

Parameter	T_j	x_d	x'_d	T'_{d0}	K_v
Value	6.0684	0.5548	0.1475	1.6488	4.2420
Parameter	D	P_{s0}	Q_{s0}	N_p	N_q
Value	0.1982	1.1065	3.1636	0.8321	0.6005

The Figure 13 shows that the transient response curves have a good fitting. The small and medium hydropower generator group can be equivalent as a 3rd order electromechanical transient generator model with damping and a static characteristic load model considering voltage in a real-world situation.

In summary, the proposed equivalent method based on DMS-PSO is a good method to identify the parameters of the dynamic equivalent model of the hydropower generator group. It is not only fast and simple, but also accurate and practical. When a disturbance occurs, the dynamic equivalent model can be obtained, and this only relies on the phasor measurements. The proposed method has practical values for power system simulation.

5. Conclusions

In this paper, a measurement-based dynamic equivalent model and parameters identification method are proposed and verified for small and medium hydropower generator group. The optimal equivalent model parameters are obtained by combining the nonlinear 4th order Runge-Kutta method and DMS-PSO algorithm. The proposed method is verified respectively with the simulation and actual data. The dynamic responses with identified equivalent models are consistent with the actual dynamic responses, which demonstrate that the equivalent model can be used in the transient stability analysis. Besides, the proposed measurement-based equivalent model is more accurate and practical by modeling the small and medium hydropower generator group with the 3rd order generator model and a static load model than modeling it as a negative load. In the future, online application of the proposed approach will be studied in the actual power system.

Acknowledgments: This study was funded by the Sichuan Electric Power Research Institute.

Author Contributions: Hu Bowei and Liu Xinyu wrote the dynamic equivalent program; Ding Lijie and Sun Jingtao provided the dynamic response data; Hu Bowei and Sun Jingtao performed the simulation and analysis; Hu Bowei and Wang Xiaoru wrote the paper.

Conflicts of Interest: The authors declare no conflict of interest.

References

1. Ni, Y.X.; Chen, S.S.; Zhang, B.L. *Theory and Analysis of Dynamic Power System*; Tsinghua University Press: Beijing, China, 2002; pp. 197–208.
2. Liu, Y.; Li, X.; Wang, Y. Research on small signal stability of power system with distributed small hydropower. In Proceedings of the Innovative Smart Grid Technologies-Asia, Tianjin, China, 21–24 May 2012; IEEE: New York, NY, USA, 2012.
3. Chang, Y.P.; Zhi, D.; Yang, D.J.; Zhao, H.S.; Zhang, Y.F.; Yang, H. An equivalent modeling for small and medium-sized hydropower generator group considering excitation and governor system. In Proceedings of the Power and Energy Engineering Conference, Hong Kong, China, 6–10 December 2014; IEEE: New York, NY, USA, 2014.
4. Xu, Z.D.; Sun, G.C.; Pan, R.R.; Xu, N.; Li, C.Q.; Ma, H.Z. An equivalent modeling for synthesis load of distributed network with small hydropower. In Proceedings of the 2014 4th International Workshop on Computer Science and Engineering, WCSE 2014, Dubai, UAE, 22–23 August 2014; Science and Engineering Institute: Bristol, UK, 2014.
5. Wang, M.; Wen, J.Y.; Hu, W.B.; Ruan, S.W.; Li, X.P.; Sun, J.B. A dynamic equivalent modeling for regional small hydropower generator group. *Power Syst. Prot. Control* **2013**, *41*, 1–9.
6. Joo, S.K.; Liu, C.C.; Choe, J.W. Enhancement of coherency identification techniques for power system dynamic equivalents. In Proceedings of the Power Engineering Society Summer Meeting, Vancouver, BC, Canada, 15–19 July 2001; IEEE: New York, NY, USA, 2001.
7. Oscar, Y.L.; Fette, M. Electromechanical identity recognition as alternative to the coherency identification. In Proceedings of the 39th International Universities Power Engineering Conference, Bristol, UK, 8 September 2004; IEEE: New York, NY, USA, 2004.
8. Nath, R.; Lamba, S.S. Development of coherency-based time-domain equivalent model using structure constraints. *IEEE Proc. C Gener. Trans. Distrib.* **1986**, *133*, 165–175. [CrossRef]
9. Ourari, M.L.; Dessaint, L.A.; Van-Que, D. Dynamic equivalent modeling of large power systems using structure preservation technique. *IEEE Trans. Power Syst.* **2006**, *21*, 1284–1295. [CrossRef]
10. Ourari, M.L.; Dessaint, L.A.; Van-Que, D. Generating units aggregation for dynamic equivalent of large power systems. In Proceedings of the IEEE Power Engineering Society General Meeting, Denver, CO, USA, 10 June 2004; IEEE: New York, NY, USA, 2004.
11. Ma, J.; Valle, R.J. Identification of dynamic equivalents preserving the internal modes. In Proceedings of the 2003 IEEE Bologna Power Tech Conference Proceedings, Bologna, Italy, 23–26 June 2004; IEEE: New York, NY, USA, 2004.
12. Price, W.W.; Ewart, D.N.; Gulachenski, E.M. Dynamic equivalents from on-line measurements. *IEEE Trans. Power Appar. Syst.* **1975**, *94*, 1349–1357. [CrossRef]
13. Azmy, A.M.; Erlich, I. Identification of dynamic equivalents for distribution power networks using recurrent ANNS. In Proceedings of the Power Systems Conference and Exposition, Bristol, UK, 10–13 October 2004; IEEE: New York, NY, USA, 2004.
14. Azmy, A.M.; Erlich, I.; Sowa, P. Artificial neural network-based dynamic equivalents for distribution systems containing active sources. *IEEE Proc. Gener. Trans. Distrib.* **2004**, *151*, 681–688. [CrossRef]
15. Rahim, A.H.M.A.; Al-Ramadhan, A.J. Dynamic equivalent of external power system and its parameter estimation through artificial neural network. *Int. J. Electr. Power Energy Syst.* **2002**, *24*, 113–120. [CrossRef]
16. Zali, S.M.; Milanovic, J.V. Dynamic equivalent model of Distribution Network Cell using Prony analysis and Nonlinear least square optimization. In Proceedings of the 2009 IEEE Bucharest Power Tech, Bucharest, Roman, 28 June 2009; IEEE: New York, NY, USA, 2009.

17. Zhou, Y.; Wang, K.; Zhang, B.H. A real-time dynamic equivalent solution for large interconnected power systems. In Proceedings of the Electric Utility Deregulation and Restructuring and Power Technologies, 2011 4th International Conference, Weihai, China, 6–9 July 2011; IEEE: New York, NY, USA, 2011.

18. Shi, H.B.; Hu, B.W.; Sun, J.T. Research on hydropower generator group equivalence and parameter identification based on PSASP calling and optimization algorithm. In Proceedings of the 2014 International Conference on Power and Energy, Shanghai, China, 29–30 November 2014; Taylor & Francis Group: London, UK, 2014.

19. Ju, P. *Theory and Method of Power System Modeling*; Science Press: Beijing, China, 2010; pp. 276–291.

20. Yang, Q.; Guan, L.; Wang, T.W. Influence on the performance of dynamic equivalence based on equivalent model. In Proceedings of the 24th Annual Conference Proceedings about Power System and Automation in China, Beijing, China, 10 October 2008.

21. Shi, D.; Tylavsky, D.J.; Koellner, M.; Logic, N.; Wheeler, D.E. Transmission line parameter identification using PMU measurements. *Eur. Trans. Electr. Power* **2011**, *21*, 1574–1588. [CrossRef]

22. Wu, S.X.; Zhang, B.M.; Wu, W.C.; Sun, H.B. Identification and validation for synchronous generator parameters based on recorded on-line disturbance data. *Power Syst. Technol.* **2012**, *36*, 87–93.

23. Chakhchoukh, Y.; Vittal, V.; Heydt, G.T. PMU based state estimation by integrating correlation. *IEEE Trans. Power Syst.* **2014**, *29*, 617–626. [CrossRef]

24. Jin, Y.Q.; Chen, Y.; Wang, D.M. *Numerical Method*; China Machine Press: Beijing, China, 2009; pp. 208–217.

25. Wang, W.H. Identification Based Dynamic Equivalents of Power Systems Interconnected with Three Areas. Master's Thesis, Hohai University, Nanjing, China, 2007.

26. Zhang, N. Research on the Identification of Synchronous Generator Parameters Based on Phasor Measurement. Master's Thesis, North China Electric Power University, Baoding, Hebei, China, 2007.

27. Shen, L.X. Parameters Identification for Power Load Models Based on Improved Particle Swarm Optimization Algorithm. Master's Thesis, Dalian Maritime University, Dalian, China, 2013.

28. Kermedy, J.; Eberhart, R. Particle swarm optimization. In Proceedings of the IEEE International Conference on Neural Networks, Piscataway, NJ, USA, 27 November 1995; IEEE: New York, NY, USA, 1995.

29. Li, Z.K.; Chen, X.Y.; Yu, K. Hybrid particle swarm optimization for distribution network reconfiguration. *Proc. CSEE* **2008**, *28*, 35–41.

30. Zhu, Y.W.; Shi, X.C.; Dan, Y.Q.; Li, P.; Liu, W.Y.; Wei, D.B.; Fu, C. Application of PSO algorithm in global MPPT for PV array. *Proc. CSEE* **2012**, *32*, 42–48.

31. Zhao, S.Z.; Liang, J.J.; Suganthan, P.N. Dynamic multi-swarm particle swarm optimizer with local search for large scale global optimization. In Proceedings of the IEEE Congress on Evolutionary Computation, Hong Kong, China, 1–6 June 2008; IEEE: New York, NY, USA, 2008.

32. Liang, J.J.; Suganthan, P.N. Dynamic multi-swarm particle swarm optimizer. In Proceedings of the 2005 IEEE Swarm Intelligence Symposium, SIS 2005, New York, NY, USA, 8–10 June 2005; IEEE: New York, NY, USA, 2005.

Article

Anti-Windup Load Frequency Controller Design for Multi-Area Power System with Generation Rate Constraint

Chongxin Huang [1,*], Dong Yue [1], Xiangpeng Xie [1] and Jun Xie [2]

[1] Institute of Advanced Technology, Nanjing University of Posts and Telecommunications,
    Nanjing 210023, China; medongy@vip.163.com (D.Y.); xiexiangpeng1953@163.com (X.X.)
[2] College of Automation, Nanjing University of Posts and Telecommunications, Nanjing 210023, China;
    jxie@njupt.edu.cn
* Correspondence: huangchongxin@foxmail.com; Tel.: +86-25-5879-7877

Academic Editor: Ying-Yi Hong
Received: 25 December 2015; Accepted: 21 April 2016; Published: 29 April 2016

Abstract: To deal with the problem of generation rate constraint (GRC) during load frequency control (LFC) design for a multi-area interconnected power system, this paper proposes an anti-windup controller design method. Firstly, an H_∞ dynamic controller is designed to obtain robust performance of the closed-loop control system in the absence of the GRC. Then, an anti-windup compensator (AWC) is formulated to restrict the magnitude and rate of the control input (namely power increment) in the prescribed ranges so that the operation of generation unit does not exceed the physical constraints. Finally, the anti-windup LFC is tested on the multi-area interconnected power systems, and the simulation results illustrate the effectiveness of the proposed LFC design method with GRC.

Keywords: load frequency control (LFC); generation rate constraint (GRC); anti-windup control; robust controller

1. Introduction

In a multi-area interconnected power system, it is important for the system's operation to keep the active power balance and regulate the tie-line power at the scheduled value. Load frequency control (LFC) plays several key roles in the active power control of the interconnected power system [1,2], such as counteracting the load fluctuation, stabilizing the system frequency, regulating the tie-line power, and narrowing the area control error (ACE). Thus, the LFC is vital for the security and stability of power system.

On the issue of LFC design for a power system, an amount of work has been done in recent years. The conventional LFC usually adopts the proportional-integral (PI)-type controller because it has simpler structure and fewer tuning parameters. However, this kind of controller has shortcomings in terms of coping with the operating point change and the load disturbance, since they are designed on nominal operating points with fixed parameters. In order to obtain better performance of the PI-type LFC, the parameter optimization methods of the PI-type controller are proposed in [3–5]. To enhance the robustness and reliability of the control system, some fuzzy-logic-based LFC methods are introduced in [6–9]. In addition, some advanced control technologies are utilized to improve LFC performance, such as sliding mode methods [10–12], optimal or suboptimal feedback control methods [13–16], and robust control methods [17–19]. Considering the delay in the open communication network, the authors in [20–22] analyze the influence of time delay on the LFC and present the relevant controller design methods. To guarantee compliance with the control performance standards (CPS) of North American Electric Reliability Council (NERC) and reduce wear and tear of generators, a decentralized model predictive control method is used to deal with the LFC problem [23]. For accommodating unexpected load change and faults, the supervisory control strategies in [24,25] are proposed to solve the load and frequency set-point problem. It is well known that due to physical limitations, generation units have inherent generation rate constraints (GRC), such as ramp rate constraints and upper-lower bound constraints. If GRC is not considered adequately in LFC design, the controller will not yield excellent performances, and even the closed-loop system stability may be destroyed under disturbances [26–28]. In the aforementioned research work, some studies make tentative consideration on the GRC problem. In [1,7,8,12], GRC is considered in the simulation, but neglected in the controller design. Therefore, the validity of these methods to deal with GRC lacks theoretical support. Towards LFC design with GRC, the extended integral control method in [26], the biased PI dual mode control method in [27], the Type-2 fuzzy approach in [28], and the anti-GRC PI-type controller in [29,30] are adopted to deal with the GRC problem. Unfortunately, the strict mathematical proof in the above methods is still absent.

Focusing on the LFC design with GRC, this paper proposes an anti-windup LFC design method for the multi-area interconnected power system. The designed LFC consists of a robust H_∞ controller and an anti-windup compensator (AWC). The former is used to guarantee the stability and robustness of the closed-loop system without constraints, and the latter takes charge of restricting the rate and magnitude of control input in the prescribed ranges to make the operation of generation unit meet the GRC requirement. For verifying the proposed method, several multi-area interconnected power systems are employed for testing. The comparative simulation results show that the performances of the LCF are improved by the design method of this paper.

2. Load Frequency Control Model

The large interconnected power system is usually partitioned into several areas for management and control. Generally, for reducing the difficulty in the LFC design, each area in the LFC model is simplified to be an equivalent generator with a turbine and a governor shown in Figure 1. The dynamics of the generator, the turbine and the governor are described by three first-order inertial processes, respectively. In addition, since the generation unit has the physical operation limitations, the LFC model includes the GRC, namely the ramp rate and the upper-lower bound constraints of the generation units. As one knows, the GRC may generate adverse impact on the LFC performances if the GRC is not considered sufficiently in LFC design.

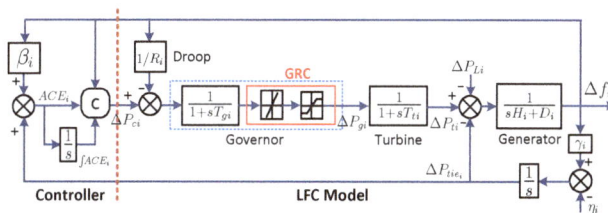

Figure 1. Load frequency control (LFC) diagram of Area *i*. ACE: area control error; GRC: generation rate constraint.

The GRC of LFC model shown in Figure 1 includes the magnitude and rate saturation of the states. This kind of state saturation nonlinearity causes much difficulty in controller design.

Remark 1. *For a real power system, we know the fact that the generation unit will operate in the linear region (without touching the saturation bounds), if the power increment is limited in the magnitude and rate ranges appropriately. In other words, the generation unit can meet its GRC when the proper rate and magnitude constraints are imposed on the control signal of the LFC. Based on the above fact, the LFC design with GRC can be solved through dealing with the problem on the controller synthesis subject to the magnitude and rate* *saturation of the control. Assuming that the deigned controller makes the generation unit operate in the linear region, the nonlinear GRC of the LFC model can be removed, and thus the original LFC model can be modified into a new one as shown in Figure 2.*

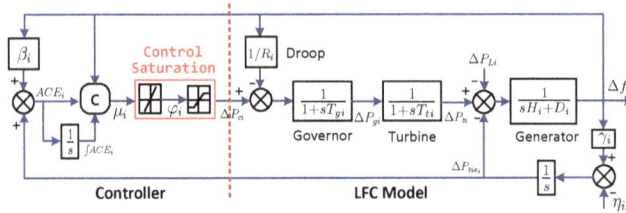

Figure 2. Modified LFC diagram of Area *i*.

According to the LFC dynamic model diagram shown in Figure 2, the state-space LFC model can be written as:

$$
\begin{cases}
\Delta \dot{f}_i = \frac{1}{H_i}\left(\Delta P_{ti} - \Delta P_{Li} - \Delta P_{tie_i} - D_i \Delta f_i\right) \\
\Delta \dot{P}_{ti} = -\frac{1}{T_{ti}}\left(\Delta P_{ti} - \Delta P_{gi}\right) \\
\Delta \dot{P}_{gi} = -\frac{1}{T_{gi}}\left(\Delta P_{gi} - \Delta P_{ci} + \frac{1}{R_i}\Delta f_i\right) \\
\Delta \dot{P}_{tie_i} = \gamma_i \Delta f_i - \eta_i \\
\dot{I}_{ACE_i} = ACE_i = \beta_i \Delta f_i + \Delta P_{tie_i} \\
y_{i1} = \Delta f_i \\
y_{i2} = ACE_i = \beta_i \Delta f_i + \Delta P_{tie_i} \\
y_{i3} = I_{ACE_i}
\end{cases}
\tag{1}
$$

where $\gamma_i = \sum\limits_{j=1, j\neq i}^{N} T_{ij}$; $\eta_i = \sum\limits_{j=1, j\neq i}^{N} T_{ij}\Delta f_j$; f_i denotes the system frequency; P_{ti} denotes the turbine power; P_{gi} denotes the governor valve; P_{ci} denotes the governor power setpoint; P_{Li} denotes the load demand; P_{tie_i} denotes the net tie-line power; ACE_i denotes the area control error; I_{ACE_i} denotes the integral of ACE_i; Δ denotes the deviation from normal value; β_i denotes the frequency bias coefficient; R_i denotes the droop coefficient; T_{gi} denotes the governor time constant; T_{ti} denotes the turbine time constant; H_i denotes the area aggregate inertia constant; D_i denotes the area load damp constant; and T_{ij} denotes the tie-line synchronizing coefficient.

Usually, we focus on the frequency deviation, the ACE and the control energy cost when evaluating the LFC performances. Thus, the controlled variables z for the H_∞ control design are selected as follows:

$$
\begin{cases}
z_{i1} = \Delta f_i \\
z_{i2} = I_{ACE_i} \\
z_{i3} = \Delta P_{ci}
\end{cases}
\tag{2}
$$

For convenience, by defining the state variables $x = [\Delta f_i, \Delta P_{ti}, \Delta P_{gi}, \Delta P_{tie_i}, I_{ACE_i}]^T \in \mathcal{R}^5$, the control variable $u = \Delta P_{ci} \in \mathcal{R}^1$, the output variables $y = [y_{i1}, y_{i2}, y_{i3}]^T \in \mathcal{R}^3$, the controlled variables $z = [z_{i1}, z_{i2}, z_{i3}]^T \in \mathcal{R}^3$, and the disturbance variables $d = [\Delta P_{Li}, \eta_i]^T \in \mathcal{R}^2$, the state-space model $\mathcal{P}$ consisting of Equations (1) and (2) can be rewritten as:

$$
\mathcal{P} : \begin{cases}
\dot{x} = Ax + B_u u + B_d d \\
y = C_y x + D_{yu} u + D_{yd} d \\
z = C_z x + D_{zu} u + D_{zd} d
\end{cases}
\tag{3}
$$

where:

$$A = \begin{bmatrix} -\frac{D_i}{H_i} & \frac{1}{H_i} & 0 & -\frac{1}{H_i} & 0 \\ 0 & -\frac{1}{T_{ti}} & -\frac{1}{T_{ti}} & 0 & 0 \\ -\frac{1}{R_i T_{gi}} & 0 & -\frac{1}{T_{gi}} & 0 & 0 \\ \gamma_i & 0 & 0 & 0 & 0 \\ \beta_i & 0 & 0 & 1 & 0 \end{bmatrix}; B_u = \begin{bmatrix} 0 \\ 0 \\ \frac{1}{T_{gi}} \\ 0 \\ 0 \end{bmatrix}; B_d = \begin{bmatrix} -\frac{1}{H_i} & 0 \\ 0 & 0 \\ 0 & 0 \\ 0 & -1 \\ 0 & 0 \end{bmatrix}$$

$$C_y = \begin{bmatrix} 1 & 0 & 0 & 0 & 0 \\ \beta_i & 0 & 0 & 1 & 0 \\ 0 & 0 & 0 & 0 & 1 \end{bmatrix}; D_{yu} = \begin{bmatrix} 0 \\ 0 \\ 0 \end{bmatrix}; D_{yd} = \begin{bmatrix} 0 & 0 \\ 0 & 0 \\ 0 & 0 \end{bmatrix}$$

$$C_z = \begin{bmatrix} 1 & 0 & 0 & 0 & 0 \\ 0 & 0 & 0 & 0 & 1 \\ 0 & 0 & 0 & 0 & 0 \end{bmatrix}; D_{zu} = \begin{bmatrix} 0 \\ 0 \\ 1 \end{bmatrix}; D_{zd} = \begin{bmatrix} 0 & 0 \\ 0 & 0 \\ 0 & 0 \end{bmatrix}$$

The control input with rate and magnitude saturations shown in Figure 2 can be defined as:

$$u = sat_m(\varphi) = \begin{cases} \overline{m}, & \varphi > \overline{m} \\ \varphi, & \underline{m} \le \varphi \le \overline{m} \\ \underline{m}, & \varphi < \underline{m} \end{cases}$$

$$\dot{\varphi} = sat_r(\mu) = \begin{cases} \overline{r}, & \mu > \overline{r} \\ \mu, & \underline{r} \le \mu \le \overline{r} \\ \underline{r}, & \mu < \underline{r} \end{cases} \tag{4}$$

where $sat(\cdot)$ denotes saturation function, $[\underline{m}, \overline{m}]$ and $[\underline{r}, \overline{r}]$ denote the magnitude bound and the rate bound, respectively.

3. Anti-Windup Load Frequency Controller Design

In this section, we design the LFC to ensure that the control input never exceeds the magnitude limit and the rate limit to meet the GRC. Based on the LFC model $\mathcal{P}$, the anti-windup schemes [31,32] are employed to synthesize the LFC in the following subsections.

3.1. Original H_∞ Controller Design

According to the anti-windup scheme, a robust H_∞ controller is designed on the basis of the LFC model $\mathcal{P}$ in absence of the control input saturation in advance. Assuming that the system (A, B_u, C_y) is controllable and observable, we can design an H_∞ dynamic controller $\tilde{\mathcal{C}}$ with the following form:

$$\tilde{\mathcal{C}} : \begin{cases} \dot{x}_{\tilde{c}} = A_{\tilde{c}} x_{\tilde{c}} + B_{\tilde{c}} u_{\tilde{c}} \\ y_{\tilde{c}} = C_{\tilde{c}} x_{\tilde{c}} + D_{\tilde{c}} u_{\tilde{c}} \end{cases} \tag{5}$$

where $x_{\tilde{c}} \in \mathcal{R}^5$ are the state variables of the controller; $u_{\tilde{c}} \in \mathcal{R}^3$ are the input variables of the controller (the measured variables of $\mathcal{P}$: $u_{\tilde{c}} = y$); $y_{\tilde{c}} \in \mathcal{R}^1$ is the output variable of the controller (the control input variable of $\mathcal{P}$: $y_{\tilde{c}} = u$); and $A_{\tilde{c}}, B_{\tilde{c}}, C_{\tilde{c}}, D_{\tilde{c}}$ are the constant matrices with appropriate dimension.

Since the robust H_∞ design method is well-known, we do not intend to repeat them. If the detailed introduction of the method is needed, one can refer to the literatures [33,34]. In this paper, we use the MATLAB/Robust Linear Matrix Inequality (LMI) Control Box [35] to solve the robust controller $\tilde{\mathcal{C}}$ directly. Here, it is assumed that the closed-loop system consisting of $\mathcal{P}$ and $\tilde{\mathcal{C}}$ is well posed and gains the prescribed H_∞ performance without consideration of the control input saturations.

3.2. Anti-Windup Compensator Design

To tackle the magnitude and rate saturations of the control input, we borrow the anti-windup control scheme [32] shown in Figure 3.

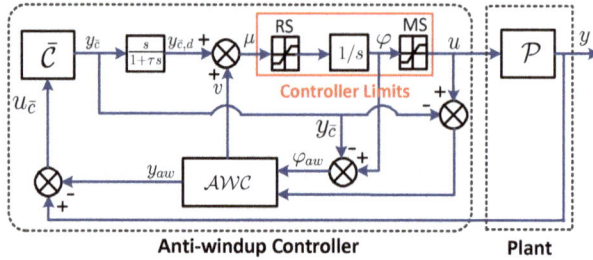

Figure 3. Structure of an anti-windup control scheme. AWC: anti-windup compensator.

In the anti-windup control approach, it is needed to compute the first-order derivative of the controller output $y_{\tilde{c}}$. Here, the differentiator s is replaced by a linear filter $\frac{s}{1+\tau s}$ with a sufficiently small constant τ, considering that the controller output may be not strictly proper. The modified controller consists of the original controller $\tilde{C}$ and the filter $\frac{s}{1+\tau s}$ can be expressed as:

$$\mathcal{C}: \begin{cases} \dot{x}_c = A_c x_c + B_c u_c \\ y_c = C_c x_c + D_c u_c \end{cases} \tag{6}$$

where $x_c = [x_{\tilde{c}}^T, x_f]^T \in \mathcal{R}^6$ are the modified controller states; $x_f \in \mathcal{R}^1$ are the filter states; $u_c = u_{\tilde{c}} \in \mathcal{R}^1$ is the modified controller input; $y_c = [y_{\tilde{c}}, y_{\tilde{c},d}]^T \in \mathcal{R}^2$ are the modified controller outputs; $y_{\tilde{c},d} \in \mathcal{R}^1$ denote approximate derivatives of $y_{\tilde{c}}$; and the parameter matrices of $\mathcal{C}$ are:

$$A_c = \begin{bmatrix} A_{\tilde{c}} & 0 \\ \frac{C_{\tilde{c}}}{\tau} & \frac{-1}{\tau} \end{bmatrix}, B_c = \begin{bmatrix} B_{\tilde{c}} \\ \frac{D_{\tilde{c}}}{\tau} \end{bmatrix}, C_c = \begin{bmatrix} C_{\tilde{c}} & 0 \\ \frac{C_{\tilde{c}}}{\tau} & \frac{-1}{\tau} \end{bmatrix}, D_c = \begin{bmatrix} D_{\tilde{c}} \\ \frac{D_{\tilde{c}}}{\tau} \end{bmatrix}$$

In Figure 3, the AWC is designed to cope with the controller limits. The AWC is formulated as follows:

$$\mathcal{AWC}: \begin{cases} \dot{x}_{aw} = A x_{aw} + B_u (u - y_{\tilde{c}}) \\ y_{aw} = C_y x_{aw} + D_{yu}(u - y_{\tilde{c}}) \\ z_{aw} = C_z x_{aw} + D_{zu}(u - y_{\tilde{c}}) \\ v = K_{aw} \begin{bmatrix} x_{aw} \\ \varphi_{aw} \end{bmatrix} \end{cases} \tag{7}$$

where $x_{aw} \in \mathcal{R}^5$ are the AWC states; $y_{aw} \in \mathcal{R}^3$ are the AWC output; $(u - y_{\tilde{c}})$ and $\varphi_{aw} = (\varphi - y_{\tilde{c}})$ serve as the AWC input; $z_{aw} \in \mathcal{R}^3$ are the AWC controlled variables; $v \in \mathcal{R}^1$ is the stabilizing signal which needs to be designed; and $K_{aw} \in \mathcal{R}^{1 \times 6}$ is the gain matrix.

The plant Equation (3), the control input limitation Equation (4), the modified Controller Equation (6), and the AWC Equation (7) are interconnected by the following relationship:

$$u_c = y - y_{aw}, \quad \mu = y_{\tilde{c},d} + v, \quad \varphi_{aw} = \varphi - y_{\tilde{c}} \tag{8}$$

From the interconnection diagram shown in Figure 3, by defining the coordinate $(x_\ell, x_c, x_{aw}, \varphi_{aw}) = (x - x_{aw}, x_c, x_{aw}, \varphi - y_{\tilde{c}})$, after some derivations, we can obtain the equivalent expression of the whole closed-loop system as follows:

$$\begin{cases} \dot{x}_\ell = Ax_\ell + B_u y_{\tilde{c}} + B_d d \\ y_\ell = C_y x_\ell + D_{yu} y_{\tilde{c}} + D_{yd} d \\ z_\ell = C_z x_\ell + + D_{zu} y_{\tilde{c}} + D_{zd} d \\ \dot{x}_c = A_c x_c + B_c y_\ell \\ y_c = C_c x_c + D_c y_\ell \end{cases} \tag{9a}$$

$$\begin{cases} \dot{x}_{aw} = Ax_{aw} + B_u[sat_m(\varphi_{aw} + y_{\tilde{c}}) - y_{\tilde{c}}] \\ \dot{\varphi}_{aw} = sat_r\left(K_{aw}\begin{bmatrix} x_{aw} \\ \varphi_{aw} \end{bmatrix} + y_{\tilde{c},d}\right) - y_{\tilde{c},d} \\ z_{aw} = C_z x_{aw} + D_{zu}[sat_m(\varphi_{aw} + y_{\tilde{c}}) - y_{\tilde{c}}] \end{cases} \tag{9b}$$

where $y_\ell = y - y_{aw}$; $z_{aw} = z - z_\ell$ denotes the mismatch between the desirable performance output z of the modified closed-loop system Equations (3), (4) and (6) and the actual performance output z_ℓ of the anti-windup closed-loop system Equations (3), (4), (6) and (7).

Theorem 1. *Given the anti-windup closed-loop system Equations (3), (4), (6) and (7), if $x_{aw}(0) = 0$ and $\varphi(0) = y_{\tilde{c}}(0)$, then the control input u of the plant never exceeds the magnitude and rate saturation bounds. Moreover, if the K_{aw} selection guarantees the asymptotic stability of the subsystem Equation (9b), then the following conclusions hold [31,32,36]:*

- *Given any response of the modified closed-loop system Equations (3), (4) and (6) such that $y_{\tilde{c}} = sat_m(y_{\tilde{c}})$ and $y_{\tilde{c},d} = sat_r(y_{\tilde{c},d})$ for all t, then $z_\ell = z$ for all t, namely, the response of the anti-windup closed-loop system coincides with the response of the modified closed-loop system;*
- *The origin of the anti-windup closed-loop system is asymptotically stable.*

Remark 2. *(1) Under the initial conditions: $x_{aw}(0) = 0$ and $\varphi(0) = y_{\tilde{c}}(0)$, obviously, the control input u meets the magnitude and the rate constraints since they are prescribed by two saturation functions; (2) If the gain matrix K_{aw} keeps the subsystem Equation (9b) stable under the foregoing initial conditions, we know that the variables $x_{aw} = 0$, $\varphi_{aw} = 0$, and $v = 0$, thus $z_{aw} = C_z x_{aw} + D_{zu}[sat_m(\varphi_{aw} + y_{\tilde{c}}) - y_{\tilde{c}}] = D_{zu}[sat_m(y_{\tilde{c}}) - y_{\tilde{c}}] = 0$ with the given assumption $y_{\tilde{c}} = sat_m(y_{\tilde{c}})$ for all t, so $z_\ell = z$ for all t is obtained based on the definition $z_{aw} = z - z_\ell$. (3) Given that K_{aw} guarantees the asymptotic stability of subsystem Equation (9b), it can be known that $x_{aw} \to 0$, $\varphi_{aw} \to 0$, $v \to 0$, $y_{\tilde{c}} \to sat_m(y_{\tilde{c}})$, and $y_{\tilde{c},d} \to sat_r(y_{\tilde{c},d})$ from Equation (9b), then the magnitude and rate saturations of the control input are ignored, and the differentiator $s/(1 + \tau s)$ offsets the integrator $1/s$ in the control loop shown Figure 3, thus the asymptotic stability of the anti-windup closed-loop system is guaranteed by the original robust H_∞ controller $\tilde{c}$.*

The proof of the above theorem is omitted in this paper, since it has been presented in the literature [31,32,36] in detail. According to the theorem, the key step for synthesizing the anti-windup controller is to design the gain matrix K_{aw} to keep the subsystem Equation (9b) stable. In terms of the recipe in [32], the gain matrix K_{aw} is selected to stabilize the following dynamic model:

$$\begin{bmatrix} \dot{x}_{aw} \\ \dot{\varphi}_{aw} \end{bmatrix} = \left(\begin{bmatrix} A & B_u \\ 0 & 0 \end{bmatrix} + \begin{bmatrix} 0 \\ I \end{bmatrix} K_{aw}\right) \begin{bmatrix} x_{aw} \\ \varphi_{aw} \end{bmatrix} \tag{10}$$

Here, the LQR method can be used to obtain K_{aw}. Obviously, K_{aw} stabilizing the dynamic model Equation (10) implies the asymptotic stability of Equation (9b), when y_c, $y_{c,d}$ and v are sufficiently small (not to cause the saturation nonlinearity).

Remark 3. *From the control diagram shown in Figure 3, we can see that the AWC is inserted between the robust H_∞ controller $\tilde{c}$ and the plant $\mathcal{P}$, thus the closed-loop system has a typical structure of the cascade* *control system. To obtain satisfied performances of the closed-loop system, it is required that the response time of the AWC in inner loop should be much shorter than that of the H_∞ controller $\bar{c}$ in outer loop. Thus, when solving the gain matrix K_{aw} based on the dynamic model Equation (10), we need to take the above requirement into account.*

4. Case Study

To test the proposed LFC design method, firstly a typical two-area interconnected power system shown in Figure 4 is selected to make simulations. In the test system, each area is represented by a equivalent generation unit with a turbine and a governor. For simplicity, it is assumed that the two areas are identical, and the corresponding parameters are as [29]: $T_{g1} = T_{g2} = 0.08$ (s), $T_{t1} = T_{t2} = 0.3$ (s), $H_1 = H_2 = 0.1667$ (pu.s), $D_1 = D_2 = 0.0083$ (pu/Hz), $T_{12} = T_{21} = \gamma_1 = \gamma_2 = 0.545$ (pu/Hz), $R_1 = R_2 = 2.4$ (Hz/pu), $\beta_1 = \beta_2 = 0.425$ (pu/Hz), rate constraint (pu/s): $[-0.0017, 0.0017]$, magnitude constraint (pu): $[-0.1, 0.1]$. Then, based on the above parameters, the two-area system shown in Figure 4 is modified into a single-area system and two three-area systems in the following simulations.

In this section, the proposed method for LFC design is compared with the methods proposed by Tan [29] and Anwar [30] for three scenarios with different load disturbances.

4.1. Scenario 1: Simulations on Single-Area System

In Scenario 1, we set the same load disturbances in the two areas of the system shown in Figure 4. By this way, each area can be treated as a single-area system, since the two areas have the same structure and parameters.

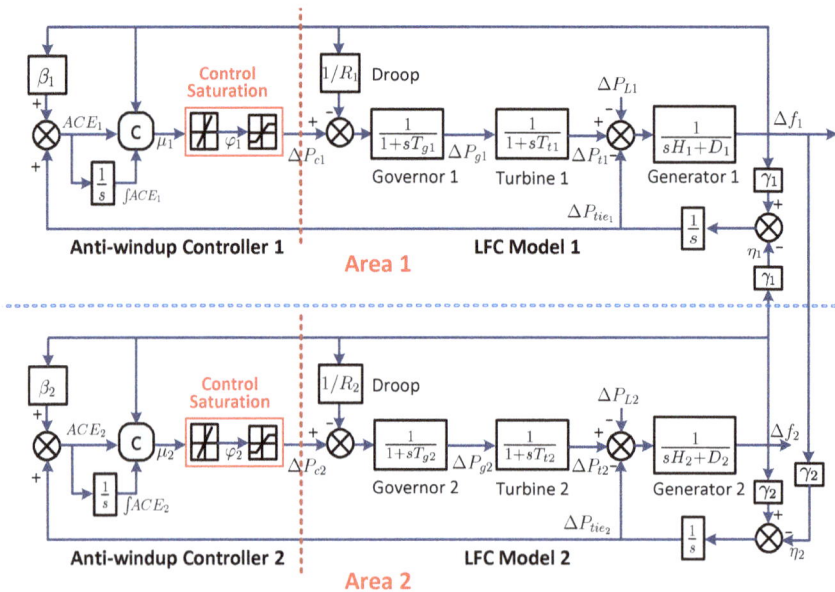

Figure 4. Diagram of a two-area interconnected power system.

Towards the LFC model of the test system, by the proposed method, we design the anti-windup LFC consisting of a robust H_∞ dynamic controller, a linear filter, a magnitude and rate saturation loop, and an AWC as follows:

The robust H_∞ dynamic controller:

$$\bar{C}' : \begin{cases} \dot{x}_{\bar{c}} = A'_{\bar{c}} x_{\bar{c}} + B'_{\bar{c}} u_{\bar{c}} \\ y_{\bar{c}} = C'_{\bar{c}} x_{\bar{c}} + D'_{\bar{c}} u_{\bar{c}} \end{cases} \tag{11}$$

where:

$$A'_{\bar{c}} = \begin{bmatrix} -9.94374 & -2.32119 & 0.620889 & 4.096447 & 312.625 \\ 21.43163 & -15.2773 & -0.43206 & 12.18535 & 1229.041 \\ 117.987 & -167.214 & -881.295 & -93.2623 & -113.066 \\ 2211.94 & -3412.12 & -1198.68 & -15702.6 & 189.4382 \\ 759871.3 & -734389 & -395045 & 84813.64 & -157822 \end{bmatrix}$$

$$B'_{\bar{c}} = \begin{bmatrix} 146.091 & 153.625 & 1.605 \\ 894.647 & 414.317 & -405.444 \\ 8964.850 & -4553.067 & -240549.496 \\ 641800.483 & -1031816.318 & 31109.807 \\ 54307506.731 & 30731255.138 & 1104996.282 \end{bmatrix}$$

$$C'_{\bar{c}} = \begin{bmatrix} -0.027 & 0.024 & -0.006 & -0.041 & -3.200 \end{bmatrix}$$
$$D'_{\bar{c}} = \begin{bmatrix} -1.134 & -1.560 & -0.006 \end{bmatrix}$$

The linear filter (approximate differentiator):

$$s/(1 + \tau s) = s/(1 + 0.01s) \tag{12}$$

The magnitude and rate bounds of the control input:

$$[\underline{m}, \overline{m}] = [-0.1, 0.1], \quad [\underline{r}, \overline{r}] = [-0.0017, 0.0017] \tag{13}$$

The AWC:

$$\begin{bmatrix} \dot{x}_{aw} \\ \dot{\varphi}_{aw} \end{bmatrix} = \left(\begin{bmatrix} A & B_u \\ 0 & 0 \end{bmatrix} + \begin{bmatrix} 0 \\ I \end{bmatrix} K'_{aw} \right) \begin{bmatrix} x_{aw} \\ \varphi_{aw} \end{bmatrix} \tag{14}$$

where $K'_{aw} = [0.908 \ 3.929 \ 1.367 \ -3.777 \ 3.162 \ 6.646]$.

Based on the designed anti-windup LFC Equations (11)–(14), we make two tests: one is for step load decrease $\Delta P_{L1} = \Delta P_{L2} = -0.015$ (pu), the other is for step load increase $\Delta P_{L1} = \Delta P_{L2} = 0.01$ (pu). The system responses to the load decrease and increase are shown in Figures 5 and 6, respectively. Considering that Area 1 and 2 have the same responses, we only illustrate the simulation results of Area 1. The concerned variables, such as frequency deviation Δf, ACE, tie-line power deviation ΔP_{tie}, control input u, and rate of control input du/dt, are shown in Figures 5 and 6.

From the results of both of the above tests, it is shown that, compared with the controllers presented by Tan and by Anwar, the proposed controller generates smaller overshoot and takes shorter settling time to force the frequency deviation and ACE to zeros. Furthermore, seeing from the control input curves, one can find that the control signal of the proposed controller in this paper meets the prescribed magnitude and rate constraint, while the control signals of Tan's and Anwar's controllers exceed the rate constraint. Here, it should be noted that ΔP_{tie} is always equal to zero since there is no tie-line power deviation between the two symmetrical areas.

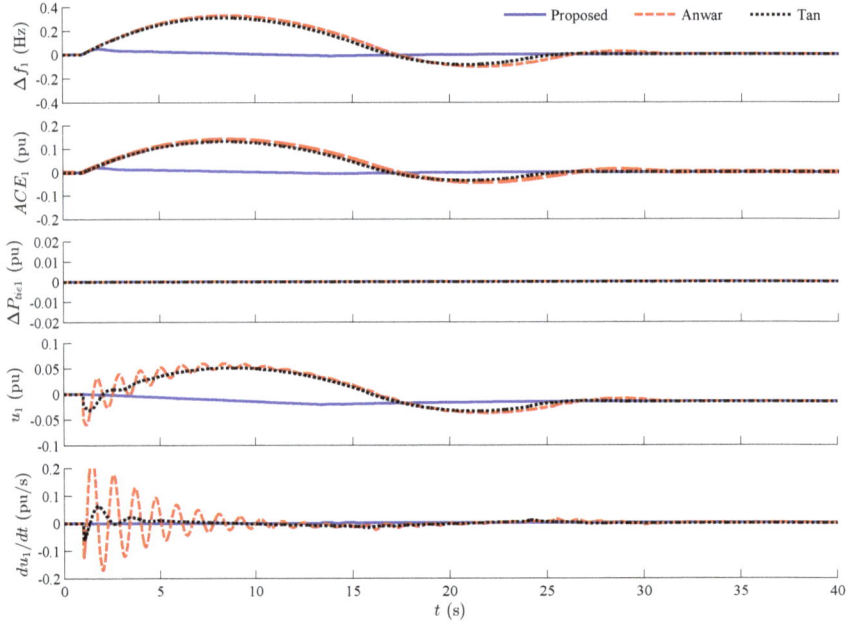

Figure 5. Results for load decrease in a single-area system.

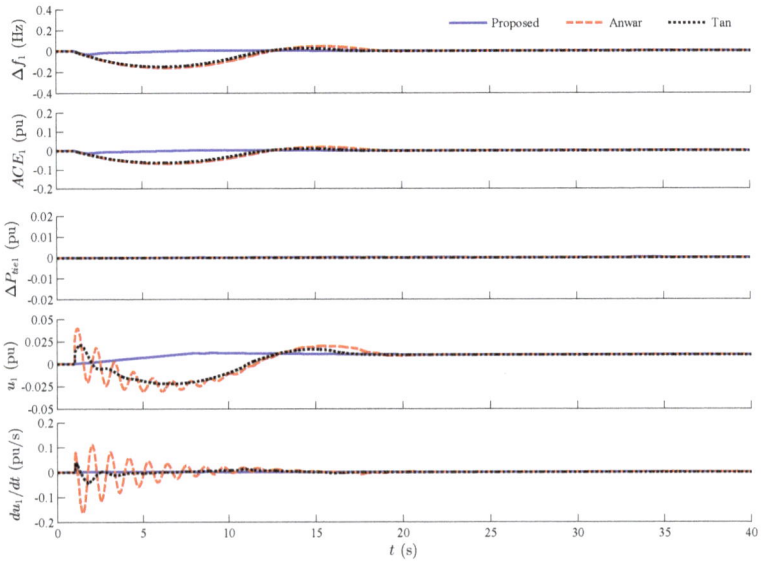

Figure 6. Results for load increase in a single-area system.

4.2. Scenario 2: Simulations on a Two-Area System

The anti-windup LFC in Scenario 2 is the same as the controller Equations (11)–(14) in Scenario 1, since the parameters of the area model are identical in both systems. In this scenario, different load disturbances are set for two tests as: $\Delta P_{L1} = 0.01$ (pu) and $\Delta P_{L2} = 0.02$ (pu) for Test 1;

$\Delta P_{L1} = 0.02$ (pu) and $\Delta P_{L2} = -0.01$ (pu) for Test 2. The simulations are performed on the two-area system directly, and the results are shown in Figures 7–10.

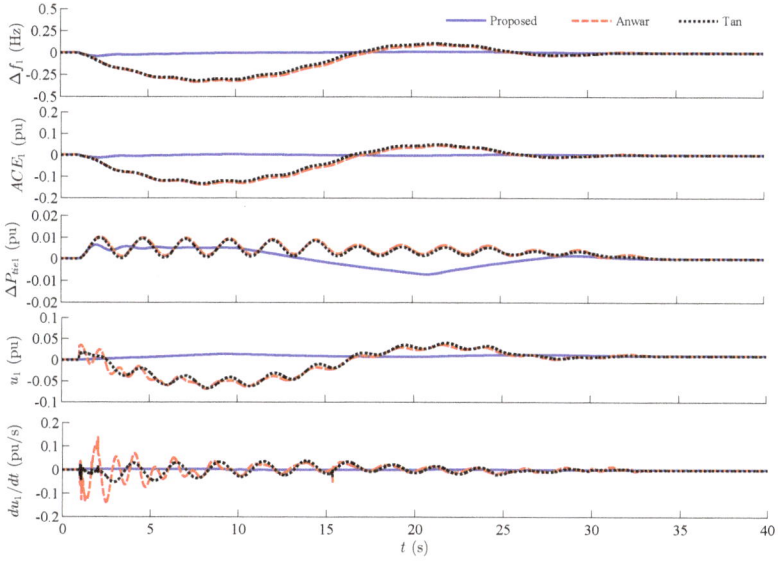

Figure 7. Results of Area 1 in a two-area system for Test 1.

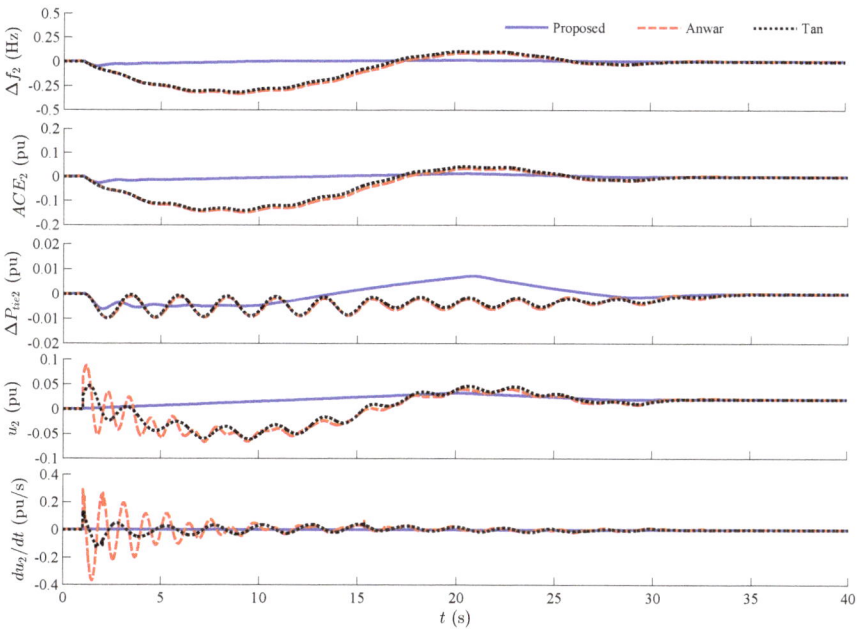

Figure 8. Results of Area 2 in a two-area system for Test 1.

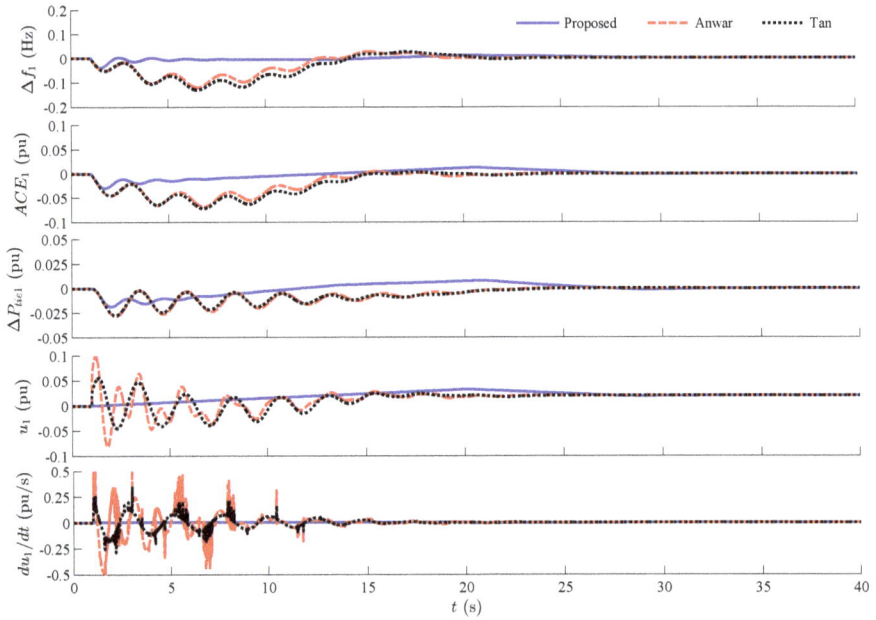

Figure 9. Results of Area 1 in a two-area system for Test 2.

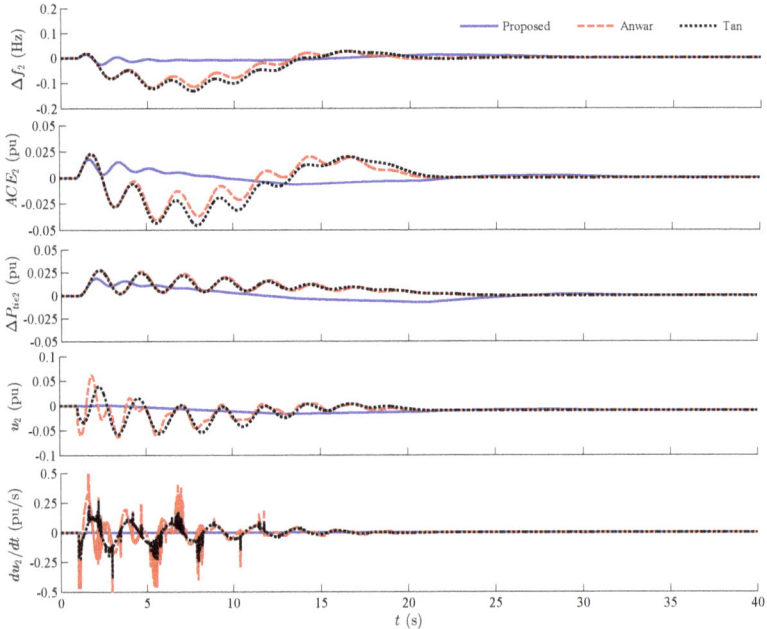

Figure 10. Results of Area 2 in a two-area system for Test 2.

From the results in Figures 7–10, it is observed that all the three controllers can drive the system frequency deviation, the ACE and the tie-line power deviation to zero, but the proposed anti-windup controller obtains more favorable performances than the other two controllers at aspect of overshoot and settling time. In addition, unlike Tan's method and Anwar's method, the proposed method avoids the undesired oscillation. Seeing the control input u, we also find that the proposed anti-windup controller can match the magnitude and rate constraint. Therefore, the system can operate in the linear region, and thus the nonlinear GRC is tackled.

By comparing the results in Scenario 1 and Scenario 2, it can be observed that the control performances of Scenario 1 are better than that of Scenario 2. The reason is that, in Scenario 1, the controller only needs to attenuate the local load disturbance, but, in Scenario 2, both the load disturbance in local area and the tie-line power disturbance from the neighboring area are needed to be restrained.

4.3. Scenario 3: Simulations on Three-Area Systems

In Scenario 3, the two-area system shown in Figure 4 is changed into two three-area interconnected systems. One is a chain-type system shown in Figure 11, the other is a delta-type system shown in Figure 12. For simplicity, we make the parameters of each area in the three-area systems be the same to those in the two-area system. In addition, the tie-line synchronizing coefficients are selected as: for the chain-type system, $T_{12} = T_{21} = T_{23} = T_{32} = 0.545$ (pu/Hz), $\gamma_1 = T_{12} = 0.545$ (pu/Hz), $\gamma_2 = T_{21} + T_{23} = 1.09$ (pu/Hz), $\gamma_3 = T_{32} = 0.545$ (pu/Hz); for the delta-type system, $T_{12} = T_{21} = T_{23} = T_{32} = T_{13} = T_{31} = 0.545$ (pu/Hz), $\gamma_1 = T_{12} + T_{13} = 1.09$ (pu/Hz), $\gamma_2 = T_{21} + T_{23} = 1.09$ (pu/Hz), $\gamma_3 = T_{31} + T_{32} = 1.09$ (pu/Hz).

Figure 11. Diagram of a three-area chain-type interconnected power system.

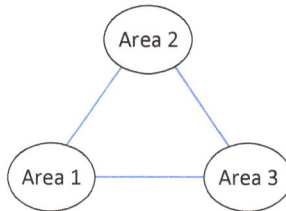

Figure 12. Diagram of a three-area delta-type interconnected power system.

For Areas 1 and 3 of the chain-type system, their anti-windup LFC are the same as the one in Scenario 1, namely Equations (11)–(14), since these area models have identical parameters in Equation (3). For Area 2, however, the anti-windup LFC needs to be redesigned, because the parameter γ_2 in the matrix A in Equation (3) is not equal to γ_1. Based on the given parameters, a robust H_∞ dynamic controller and an AWC are designed for Area 2 as follows:

The robust H_∞ dynamic controller:

$$\mathcal{C}'' : \begin{cases} \dot{x}_{\tilde{c}} = A''_{\tilde{c}} x_{\tilde{c}} + B''_{\tilde{c}} u_{\tilde{c}} \\ y_{\tilde{c}} = C''_{\tilde{c}} x_{\tilde{c}} + D''_{\tilde{c}} u_{\tilde{c}} \end{cases} \tag{15}$$

where:

$$A_{\zeta}'' = \begin{bmatrix} -9.357 & -2.267 & -2.343 & 5.018 & 535.792 \\ 18.070 & -16.938 & 0.246 & 5.616 & 1531.194 \\ 149.396 & 236.362 & -1083.630 & 183.093 & 381.424 \\ 233.018 & -2806.868 & 1859.898 & -7169.953 & -314.377 \\ 1014222.355 & -978906.090 & -830564.259 & 430071.155 & -251967.406 \end{bmatrix}$$

$$B_{\zeta}'' = \begin{bmatrix} 154.484 & 127.097 & 2.559 \\ 902.163 & -213.902 & 704.383 \\ -2720.065 & 11933.119 & -116445.141 \\ 222855.373 & -438613.914 & -30897.537 \\ 68477106.358 & 11506520.276 & 1200741.256 \end{bmatrix}$$

$$C_{\zeta}'' = \begin{bmatrix} -0.040 & 0.028 & 0.030 & -0.062 & -6.652 \end{bmatrix}$$
$$D_{\zeta}'' = \begin{bmatrix} -1.532 & -1.569 & -0.002 \end{bmatrix}$$

The AWC:

$$\begin{bmatrix} \dot{x}_{aw} \\ \dot{\varphi}_{aw} \end{bmatrix} = \left(\begin{bmatrix} A & B_u \\ 0 & 0 \end{bmatrix} + \begin{bmatrix} 0 \\ I \end{bmatrix} K_{aw}'' \right) \begin{bmatrix} x_{aw} \\ \varphi_{aw} \end{bmatrix} \tag{16}$$

where $K_{aw}'' = [0.908 \ 3.929 \ 1.367 \ -3.777 \ 3.162 \ 6.646]$.

Combining the above robust controller and AWC with the bounds of the control input and the linear filter presented in Scenario 1, we obtain the anti-windup LFC of Area 2, namely, Equations (12), (13), (15) and (16).

In summary, for the three-area chain-type system, the anti-windup LFCs are described as: Equations (11)–(14) for Area 1 and 3; Equations (12), (13), (15) and (16) for Area 2. For the three-area delta-type system, the anti-windup LFCs are expressed as Equations (12), (13), (15) and (16) for Area 1, 2, and 3, since the three areas in the delta-type system have the same parameters as Area 2 in the chain-type system.

In Scenario 3, the load disturbances are set as: in the chain-type system, $P_{L1} = 0.01$ (pu) in Area 1, $P_{L2} = -0.01$ (pu) in Area 2, and $P_{L3} = -0.01$ (pu) in Area 3; in the delta-type system, $P_{L1} = 0.02$ (pu) in Area 1, $P_{L2} = 0.01$ (pu) in Area 2, and $P_{L3} = -0.01$ (pu) in Area 3.

The results are shown in Figures 13–15 for the chain-type system and Figures 16–18 for the delta-type system. The simulations reveal that the proposed method can restrain the load disturbances, regulate the frequency of each area, and restore the tie-line power to its scheduled value. In other words, the proposed method can realize the LFC objectives of multi-area interconnected power system. Compared with Tan's method and Anwar's method, the method in this paper obtains better performances in overshoot and settling time. The magnitude and rate of the input signals are especially controlled in the predetermined ranges by the proposed anti-windup controller, which helps to reduce the wear and tear of generators and improve the stability of the closed-loop system.

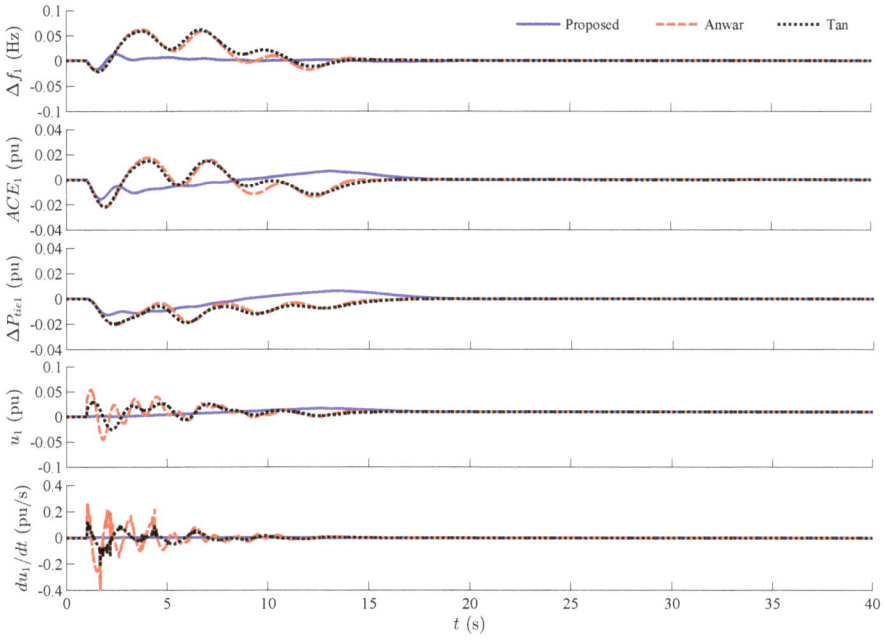

Figure 13. Results of Area 1 in a three-area chain-type system.

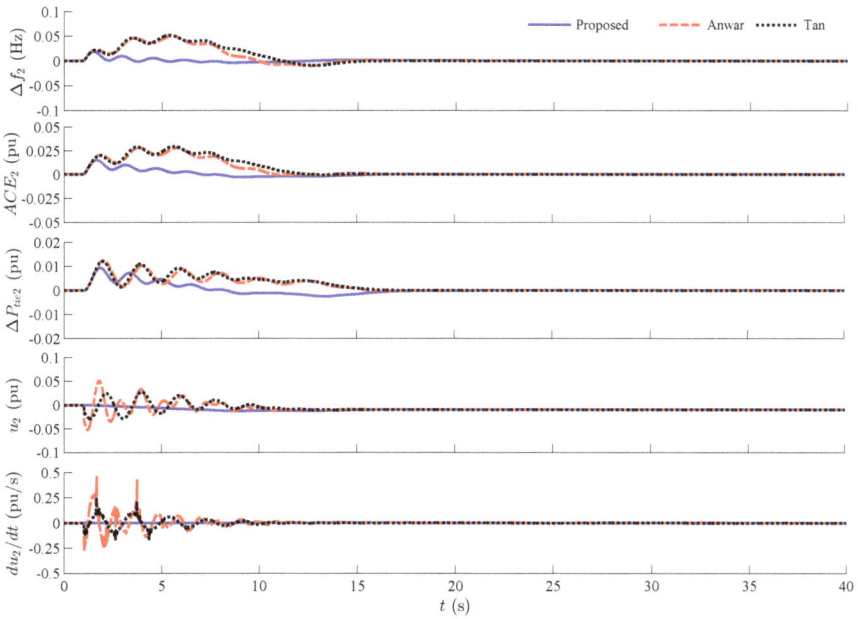

Figure 14. Results of Area 2 in a three-area chain-type system.

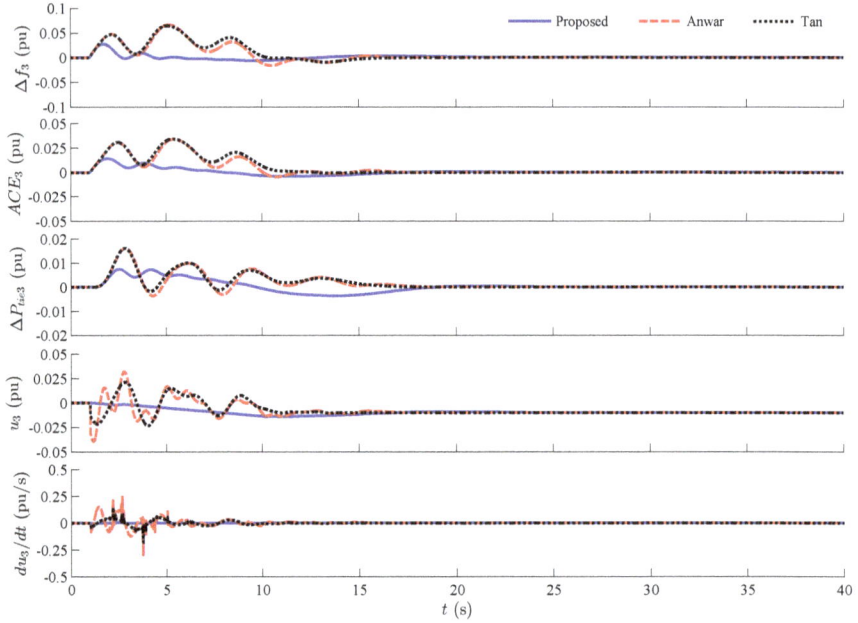

Figure 15. Results of Area 3 in a three-area chain-type system.

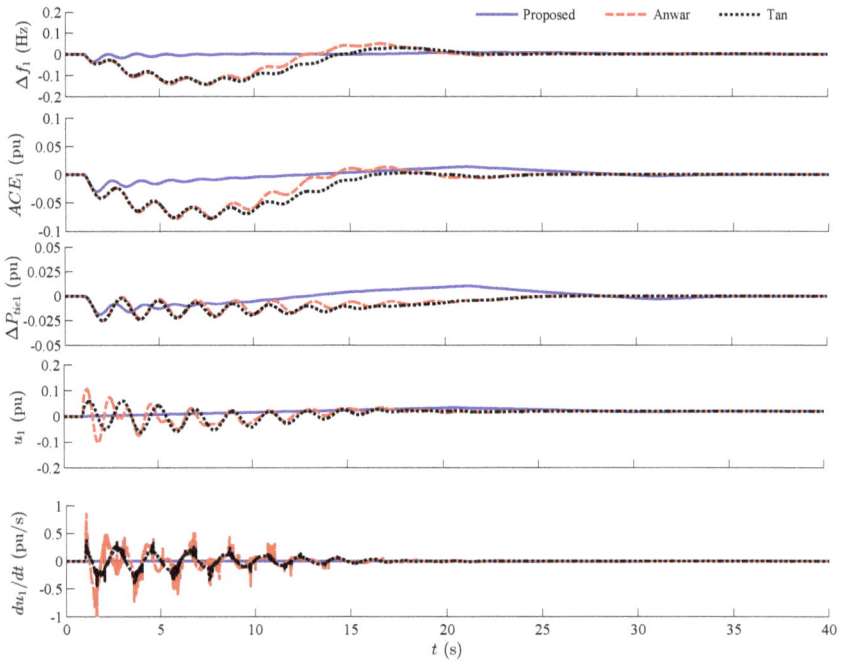

Figure 16. Results of Area 1 in a three-area delta-type system.

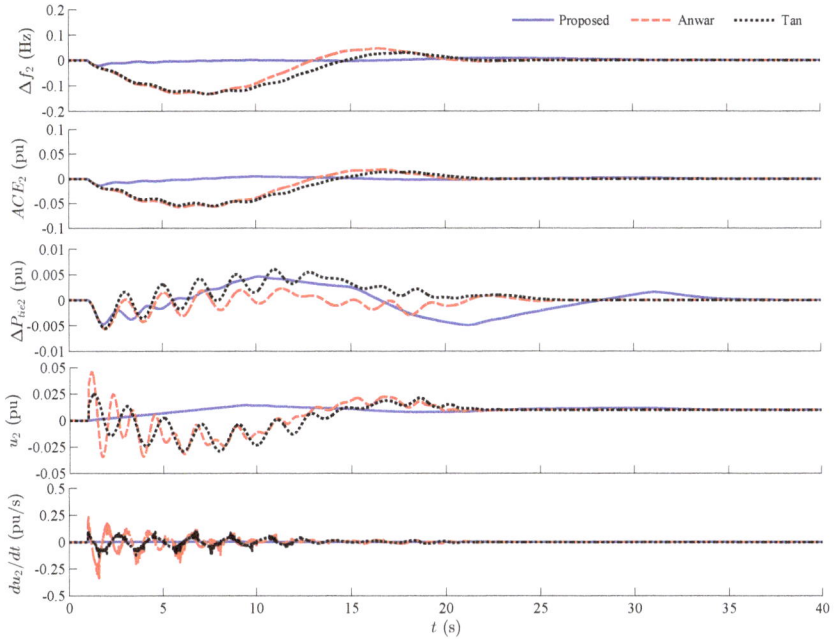

Figure 17. Results of Area 2 in a three-area delta-type system.

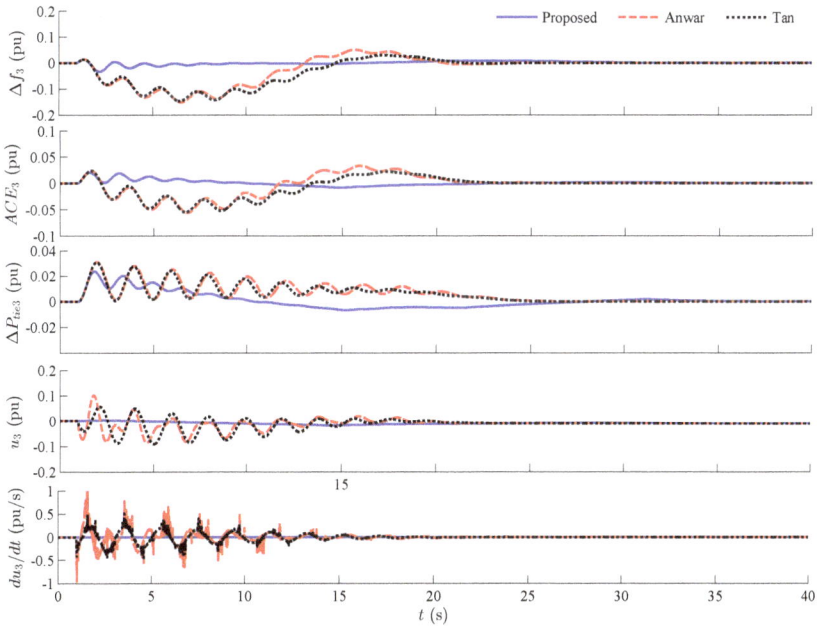

Figure 18. Results of Area 3 in a three-area delta-type system.

5. Conclusions

Towards the GRC problem in the LFC, this paper proposes an anti-windup controller design method. In the anti-windup LFC, the H_∞ dynamic controller is designed to guarantee robust performance against load disturbances and tie-line power disturbances, and the AWC is used to restrict the magnitude and rate of the control input so that the system can operate in the linear region to overcome the GRC. The simulation results show that the proposed anti-windup LFC design method effectively improves the performances against disturbances and GRC. Further work will focus on the coordination method of different LFCs to improve the overall performances of the multi-area interconnected power system.

Acknowledgments: This project is supported by the National Natural Science Foundation of China (Nos. 51507085 and 61533010) and the Scientific Fund of Nanjing University of Posts and Telecommunications (NUPTSF Grants No.NY214202 and No.XJKY14018).

Author Contributions: Chongxin Huang and Dong Yue designed the control strategy and wrote the manuscript; Xiangpeng Xie analyzed the results; Jun Xie checked the whole manuscript.

Conflicts of Interest: The authors declare no conflict of interest.

References

1. Rerkpreedapong, D.; Hasanovic, A.; Feliachi, A. Robust load frequency control using genetic algorithms and linear matrix inequalities. *IEEE Trans. Power Syst.* **2003**, *18*, 855–861.
2. Shayeghi, H.; Shayanfar, H.A.; Jalili, A. Load frequency control strategies: A state-of-the-art survey for the researcher. *Energy Convers. Manag.* **2009**, *50*, 344–353.
3. Khodabakhshian, A.; Edrisi, M. A new robust PID load frequency controller. *Control Eng. Pract.* **2008**, *16*, 1069–1080.
4. Tan, W. Unified tuning of PID load frequency controller for power systems via IMC. *IEEE Trans. Power Syst.* **2010**, *25*, 341–350.
5. Ghoshal, S.P. Application of GA/GA-SA based fuzzy automatic generation control of a multi-area thermal generating system. *Electr. Power Syst. Res.* **2004**, *70*, 115–127.
6. Cam, E.; Kocaarslan, I. Load frequency control in two area power systems using fuzzy logic controller. *Energy Convers. Manag.* **2005**, *46*, 233–243.
7. Sabahi, K.; Teshnehlab, M.; Shoorhedeli, M.A. Recurrent fuzzy neural network by using feedback error learning approaches for LFC in interconnected power system. *Energy Convers. Manag.* **2009**, *50*, 938–946.
8. Bevrani, H.; Daneshmand, P.R.; Babahajyani, P.; Mitani, Y.; Hiyama, T. Intelligent LFC concerning high penetration of wind power: Synthesis and real-time application. *IEEE Trans.Sustain.Energy* **2014**, *5*, 655–662.
9. Sahu, R.K.; Panda, S.; Pradhan, P.C. Design and analysis of hybrid firefly algorithm-pattern search based fuzzy PID controller for LFC of multi area power systems. *Int. J. Electr. Power Energy Syst.* **2015**, *69*, 200–212.
10. Al-Hamouz, Z.M.; Al-Duwaish, H.N. A new load frequency variable structure controller using genetic algorithms. *Electr. Power Syst. Res.* **2000**, *55*, 1–6.
11. Vrdoljak, K.; Perić, N.; Petrović, I. Sliding mode based load-frequency control in power systems. *Electr. Power Syst. Res.* **2010**, *80*, 514–527.
12. Al-Hamouz, Z.; Al-Duwaish, H.; Al-Musabi, N. Optimal design of a sliding mode AGC controller: Application to a nonlinear interconnected model. *Electr. Power Syst. Res.* **2011**, *81*, 1403–1409.
13. Aldeen, M.; Trinh, H. Load-frequency control of interconnected power systems via constrained feedback control schemes. *Comput. Electr. Eng.* **1994**, *20*, 71–88.
14. Alrifai, M.T.; Hassan, M.F.; Zribi, M. Decentralized load frequency controller for a multi-area interconnected power system. *Int. J. Electr. Power Energy Syst.* **2011**, *33*, 198–209.
15. Trinh, H.; Fernando, T.; Iu, H.H.C.; Wong, K.P. Quasi-decentralized functional observers for the LFC of interconnected power systems. *IEEE Trans. Power Syst.* **2013**, *28*, 3513–3514.
16. Pham, T.N.; Trinh, H.; Hien, L.V. Load frequency control of power systems with electric vehicles and diverse transmission links using distributed functional observers. *IEEE Trans. Smart Grid* **2016**, *7*, 238–252.
17. Rahmani, M.; Sadati, N. Hierarchical optimal robust load-frequency control for power systems. *IET Gener. Transm. Distrib.* **2012**, *6*, 303–312.

18. Vachirasricirikul, S.; Ngamroo, I. Robust LFC in a smart grid with wind power penetration by coordinated V2G control and frequency controller. *IEEE Trans. Smart Grid* **2014**, *5*, 371–380.

19. Dong, L.L.; Zhang, Y.; Gao, Z.Q. A robust decentralized load frequency controller for interconnected power systems. *ISA Trans.* **2012**, *51*, 410–419.

20. Jiang, L.; Yao, W.; Wu, Q.H.; Wen, J.Y.; Cheng, S.J. Delay-dependent stability for load frequency control with constant and time-varying delays. *IEEE Trans. Power Syst.* **2012**, *27*, 932–941.

21. Dey, R.; Ghosh, S.; Ray, G.; Rakshit, A. H∞ load frequency control of interconnected power systems with communication delays. *Int. J. Electr. Power Energy Syst.* **2012**, *42*, 672–684.

22. Zhang, C.K.; Jiang, L.; Wu, Q.H.; He, Y.; Wu, M. Delay-dependent robust load frequency control for time delay power systems. *IEEE Trans. Power Syst.* **2013**, *28*, 2192–2201.

23. Atić, N.; Rerkpreedapong, D.; Hasanović, A.; Feliachi, A. NERC compliant decentralized load frequency control design using model predictive control. In Proceedings of the IEEE on Power Engineering Society General Meeting, Toronto, ON, Canada, 13–17 July 2003.

24. Franze, G.; Tedesco, F. Constrained load/frequency control problems in networked multi-area power systems. *J. Frankl. Inst.* **2011**, *348*, 832–852.

25. Tedesco, F.; Casavola, A. Fault-tolerant distributed load/frequency supervisory strategies for networked multi-area microgrids. *Int. J. Robust Nonlinear Control* **2014**, *24*, 1380–1402.

26. Moon, Y.H.; Ryu, H.S.; Lee, J.G.; Song, K.B.; Shin, M.C. Extended integral control for load frequency control with the consideration of generation-rate constraints. *Int. J. Electr. Power Energy Syst.* **2002**, *24*, 263–269.

27. Velusami, S.; Chidambaram, I.A. Decentralized biased dual mode controllers for load frequency control of interconnected power systems considering GDB and GRC non-linearities. *Energy Convers. Manag.* **2007**, *48*, 1691–1702.

28. Sudha, K.R.; Santhi, R.V. Robust decentralized load frequency control of interconnected power system with generation rate constraint using Type-2 fuzzy approach. *Int. J. Electr. Power Energy Syst.* **2011**, *33*, 699–707.

29. Tan, W. Tuning of PID load frequency controller for power systems. *Energy Convers. Manag.* **2009**, *50*, 1465–1472.

30. Anwar, M.N.; Pan, S. A new PID load frequency controller design method in frequency domain through direct synthesis approach. *Int. J. Electr. Power Energy Syst.* **2015**, *67*, 560–569.

31. Forni, F.; Galeani, S.; Zaccarian, L. Model recovery anti-windup for plants with rate and magnitude saturation. In Proceedings of the European Control Conference, Budapest, Hungary, 23–26 August 2009; pp. 324–329.

32. Forni, F.; Galeani, S.; Zaccarian, L. An almost anti-windup scheme for plants with magnitude, rate and curvature saturation. In Proceedings of the American Control Conference, Baltimore, MD, USA, 30 June–2 July 2010; pp. 6769–6774.

33. Chilali, M.; Gahinet, P. H∞ design with pole placement constraints: An LMI approach. *IEEE Trans. Autom. Control* **1996**, *41*, 358–367.

34. Scherer, C.; Gahinet, P.; Chilali, M. Multiobjective output-feedback control via LMI optimization. *IEEE Trans. Autom. Control* **1997**, *42*, 896–911.

35. Gahinet, P.M.; Nemirovskii, A.; Laub, A.J.; Chilali, M. The LMI control toolbox. In Proceedings of the 33rd IEEE Conference on Decision and Control, Lake Buena Vista, FL, USA, 14–16 December 1994; pp. 2038–2038.

36. Forni, F.; Galeani, S.; Zaccarian, L. Model recovery anti-windup for continuous-time rate and magnitude saturated linear plants. *Automatica* **2012**, *48*, 1502–1513.

energies

MDPI

Article

Interaction and Coordination among Nuclear Power Plants, Power Grids and Their Protection Systems

Guoyang Wu [1,*], Ping Ju [1], Xinli Song [2], Chenglong Xie [3] and Wuzhi Zhong [2]

1   School of Energy and Electrical Engineering, Hohai University, Nanjing 210098, China; pju@hhu.edu.cn
2   China Electric Power Research Institute, Beijing 100192, China; songxl@epri.sgcc.com.cn (X.S.);
    zhongwz@epri.sgcc.com.cn (W.Z.)
3   China Nuclear Power Operation Technology Co. Ltd., Wuhan 430223, China; xiecl@cnpotech.com
*   Correspondence: guoyang_wu@sina.com; Tel.: +86-186-1383-9386; Fax: +86-10-6291-8841

Academic Editor: Ying-Yi Hong
Received: 13 January 2016; Accepted: 5 April 2016; Published: 21 April 2016

Abstract: Nuclear power plants (NPPs) have recently undergone rapid development in China. To improve the performance of both NPPs and grids during adverse conditions, a precise understanding of the coordination between NPPs and grids is required. Therefore, a new mathematical model with reasonable accuracy and reduced computational complexity is developed. This model is applicable to the short, mid, and long-term dynamic simulation of large-scale power systems. The effectiveness of the model is verified by using an actual NPP full-scope simulator as a reference. Based on this model, the interaction and coordination between NPPs and grids under the conditions of over-frequency, under-frequency and under-voltage are analyzed, with special stress applied to the effect of protection systems on the safe operation of both NPPs and power grids. Finally, the coordinated control principles and schemes, together with the recommended protection system values, are proposed for both NPPs and grids. These results show that coordination between the protection systems of NPPs and power networks is a crucial factor in ensuring the safe and stable operation of both NPPs and grids. The results can be used as a reference for coordination between NPPs and grids, as well as for parameter optimization of grid-related generator protection of NPPs.

Keywords: coordination between units and grids; dynamic simulation; modeling; nuclear power plant (NPP); pressurized water reactor (PWR); grid-related protection

1. Introduction

Nuclear power in China has been developing continually and is expected to grow substantially in the coming decades [1–4]. Owing to the high capacity and high safety requirements, large disturbances in electric systems may seriously influence grids and nuclear power plants (NPPs). Thus, ensuring the safety and stability of power networks and NPPs is a critical issue [5]. Numerous power outages at home and abroad indicate that the coordination between power plants and grids is a crucial factor in ensuring the safety and stability of power systems [6–12]. Some major achievements have been made in the coordination of control and protection system between units and the grid in recent years [13–19]. However, these studies have not addressed the particularities of NPPs. Simulation and analysis of the interactions between NPPs and grids are currently not considered in the design and operation of NPPs in China, and neither is the effect of protection systems on the safe operation of power grids.

To improve the performance of both units and grids under adverse conditions and to meet the overall reliability requirements, a precise understanding of the interactions between NPPs and grids is required to ensure coordination of their protection systems. For this purpose, it is important to develop a dynamic model with reasonable accuracy for NPPs. Early approaches for the modeling of NPP dynamics have been proposed. Among them, some are linear models [20–22] based on the assumption that power disturbances, such as system frequency deviation, are small, while others are extended models that take some response of the plant control and protection systems into account and are applicable for large disturbances [23–25]. The common problem of these models is the large computational complexity, which is not suitable for large-scale power system dynamic simulations. Hence, some more simplified models have been put forward [26–29]. However, because of their precision deficiency, these linearized models cannot be used for large disturbances. The bulk power system stability simulation of NPPs in China has to be based on models of thermal units rather than on NPP models because there are no appropriate NPP models, thus it does not address the particularities of NPPs.

On the other hand, NPPs are highly sensitive to fluctuations in voltage and frequency. Severe disturbances in voltage and frequency beyond the safety range of NPPs will actuate nuclear power unit relays and cause serious accidents. Because of their relatively large capacity, the tripping or load rejection of NPPs will cause a sudden substantial loss of active power and reactive power, which consequently affects system stability significantly. Reference [30] studies the dynamic responses of an NPP to power grid faults and the influence of tripping an NPP on the power grid. Reference [31] analyzes the NPP risk and grid instability in case of transmission system voltage excursions and discusses the interface requirements between NPPs and grids. Reference [32] studies the characteristics of NPPs, along with their interactions and compatibility with grids of limited capacity, and investigates remedial measures based on NPP operational experience. To get a true picture of the performance of NPPs, the effects of their protection systems, which have usually been simplified or eliminated in previous studies, should be considered in the dynamic simulation. Although references [23–25,31–34] do consider some protections of NPPs, the influences of NPP and grid protection systems have not been systematically taken into account. Thus, the interactions between NPPs and the grid cannot be accurately reflected.

This paper presents a new mathematical model of NPPs with high precision for the most widely used pressurized water reactors (PWRs) in China, so that the design and performance characteristics of NPPs can be well understood. Based on this detailed model, the interactions between NPPs and grids under the conditions of over-frequency, under-frequency, and under-voltage are analyzed in detail, with special stress applied to the effect of protection systems on both NPPs and power grids. The results show that incoordination between NPPs and grids could lead to serious consequences when the power system experiences adverse conditions. For instance, due to unreasonable settings of the under-frequency relay, tripping of some thermal plants will result in a faster decease in system frequency, which could trigger the under-speed relay of the main pump to shut down the reactor, and might even cause an unmanageable cascading reaction. Therefore, additional attention should be paid to the cooperation between NPPs and power systems.

Some coordinated control and protection principles for both NPPs and grids are proposed in this paper. The special protection systems of power systems, *i.e.*, under-frequency load shed (UFLS), under-voltage load shed (UVLS) and over-frequency generator trip (OFGT), should be considered the primary measures to ensure the safety of NPPs in case of severe frequency and voltage excursions. In addition, the relays of NPPs can be the last resort to ensure unit safety when the situation is out of control. A system-wide regulation of grid-related relays concerned with frequency, such as under-frequency relays, over-frequency relays and over-speed protect controllers (OPCs), will contribute to the safe operation of NPPs. Moreover, some protections of conventional plants, for example, under-frequency relays and under-voltage relays, should be less conservative than those of NPPs. Otherwise, the operating conditions of NPPs in the case of under-frequency or under-voltage will be degraded. Furthermore, limiters and protections of NPPs should also try to support the system stability as long as possible without degrading their own safety.

The rest of this paper is organized as follows. Section 2 proposes a new and accurate model of PWR NPP with reduced computational complexity. The effectiveness and accuracy of the model are verified in Section 3. Based on this model, Section 4 analyzes the interactions between NPPs and power grids and proposes the coordinated control principles and control schemes. In addition, case studies on a real 500 kV large-capacity power grid in China are performed in Section 5. Finally, conclusions are drawn in Section 6.

2. The Dynamic Model of a Pressurized Water Reactor Nuclear Power Plants

A new mathematical model of NPPs is developed in this section, which is applicable to the short, mid, and long-term dynamic simulation of large-scale power systems. It should be pointed out here that the terms of short, mid, and long-term dynamics are not strictly defined in the world. Normally, in China, the timescales of the short, mid, and long-term dynamics are approximately several seconds to 30 s, 30 s to 30 min, and 30 min to several hours, respectively.

2.1. Basic Modeling Considerations

Because of the complex mechanism, strong rigidity, severe nonlinearities and close coupling characteristic, a detailed simulation NPP model, such as a full-scope simulator, should be established to accurately reflect the dynamic performance of NPPs. However, a full-scope simulator is mainly used for NPP operator training and internal equipment failure simulations. Normally, a full-scope simulator includes more than 100 subsystems [35–42], and the order of the differential equations in its primary system model is more than 5000. In addition to the reactor kinetic model, reactor control system and thermal-hydraulic system, a full-scope simulator also simulates auxiliary systems, such as ventilation systems, feed water systems and spray systems. However, such an NPP model that is too complex would lead to an excessive amount of computation, and is definitely not suitable for the stability simulation of a large-scale power system, especially when NPPs account for a certain percentage of the grid.

In order to analyze the interaction between NPPs and large power grids, a simplified NPP model in this paper is put forward based on some existing models [25,26,36,37,39,40,42]. To reduce the computational complexity, the modeling approach in this study is based on the lumped parameter model. Special attention is paid to the heat exchange process, which is simplified according to the Laws of Conversation of Energy and Conversation of Mass. While the subsystems or devices that have little effect on the short, mid, and long-term dynamics of power systems are ignored. Thus, the simulation scope and computational complexity of the simplified model is much smaller than that of the full-scope simulator.

2.1.1. Simplification of a Point Reactor

In a full-scope NPP simulator, fission power and decay power are calculated using detailed models with six groups of delayed neutrons and eleven groups of radionuclides, respectively [42]. To achieve computational economy, an equivalent group of delayed neutrons and nuclear decay nuclides are used to calculate the reactor power without considering its three-dimensional distribution, which is also performed in a full-scope simulator. As far as the reactor core is concerned, the amount of calculation required for a simplified model is much less than that in a full-scope simulator.

To ensure adequate accuracy, the reactor response curves of the equivalent model and a detailed model in a full-scope simulator are analyzed under different operating situations, such as a generator trip, reactor shutdown, and load ramp increase and decrease. The tests demonstrate that the difference between the equivalent model and the detailed model is quite small with an error of less than 1% in most cases. In the generator tripping and net load rejection cases, the differences are relatively large, with the largest error in fission power of up to 7% and the maximum error in the decay heat of up to approximately 3%. Because the unit has been split with the grid in these cases, it does not affect the accuracy of power system dynamic simulation. Therefore, the equivalent model can simulate the system with relatively high calculation precision.

2.1.2. Ignoring the Pressurizer

The main function of the pressurizer is to maintain supercooling of the primary system to prevent departure from nucleate boiling. The pressurizer is not directly involved in the heat exchange, and, normally, it stabilizes at approximately 15.5 MPa. Even in the case of load rejection or scram, the primary circuit pressure can still be maintained within the control range, *i.e.*, between 13.2 MPa and 16.45 MPa, due to the pressure and water level control system. The main factors that affect the heat transfer of the primary system are the coolant flow, density and specific heat. Compared with the change in coolant density caused by temperature, the density fluctuations of the coolant due to pressure changes in the above range can be neglected. Therefore, the physical property changes are mainly influenced by changes in temperature other than those in pressure. Thus, the pressure of the primary system can be simplified as a constant, which will cause little deviation of the simulation results.

2.1.3. Introduction of the Flow Rate, Specific Heat and Density of the Coolant

Most existing NPP models assume that the temperature and flow rate of the primary coolant remain unchanged, therefore, the coolant density and specific heat can be processed as constants. However, the coolant flow rate is directly affected by the voltage and frequency of the primary pump, while the density and specific heat have temperature-dependent features with the corresponding design range of the coolant temperature of 291–310 °C; thus, these variables cannot be simply processed as constants. Otherwise, they will result in an error of up to 10%–20% according to the field measurements and full-scope NPP simulations. Therefore, this paper introduces these important variables into the new NPP model.

2.1.4. Simplification of the Feed Water System

The steam generator (SG) is the energy exchange center between the primary system and the secondary system, where the PWR coolant transfers heat to the secondary system to drive the steam turbine through a U-shaped metal tube. The method by which the energy is exchanged between the primary system, and the secondary system is the main difference between a conventional unit and a nuclear power unit.

Among the variables in the feed water system [41], the feed water flow rate and enthalpy have the largest influences on the heat transfer in the SG. During normal operation, the feed water flow can be assumed to be equal to the steam flow due to the SG water level control system. The feed water enthalpy is subject to the turbine steam extraction rate, which is closely linked to the unit power level. Thus, the feed water enthalpy can be set by the unit power level according to the unit heat balance diagram. The site measured data from the full scope simulator show that the error of this method is generally less than 1 percent, with the maximum error of less than 1.6 percent, which can be satisfied with the precision requirements of dynamic simulation.

2.1.5. Equivalence of the Thermodynamic Process in Turbine

Reference [25] established a turbine model for PWR NPPs, but the complex model with large computational requirements is not suitable for the dynamic simulation of large-scale power grids. Compared with the turbines of thermal units, a PWR unit turbine has lower initial steam parameters and a larger volume flow rate; however, there are no significant differences between the secondary system of an NPP and that of conventional units. Therefore, in this work, the NPP secondary system, normally including an electrical generator, exciter, stabilizer, steam turbine and governor, is modeled based on on-site measuring data with reference to the thermal unit model [43,44].

Based on the principle of conservation of energy, the thermal power of the secondary system is calculated by multiplying the steam flow and the difference between the enthalpies of steam and feed water. In addition, the thermal efficiency of the steam flow should be considered during the process. On the premise of definite enthalpy, there is a good linear relationship between the turbine power and steam flow. Therefore, accurately simulating the governor and the bypass system, and hence calculating the steam flow according to the corresponding the valve opening, are the keys to accurately determining the turbine power. The results show that the turbine power error between the new model and that of the full-scope simulator is less than 0.6%.

2.2. Model of a Pressurized Water Reactor Nuclear Power Plant

The PWR NPP model developed in this work consists of a reactor, primary system, secondary system and the corresponding protection system. This paper mainly discusses the primary system of the NPP depicted in Figure 1 together with its protection system.

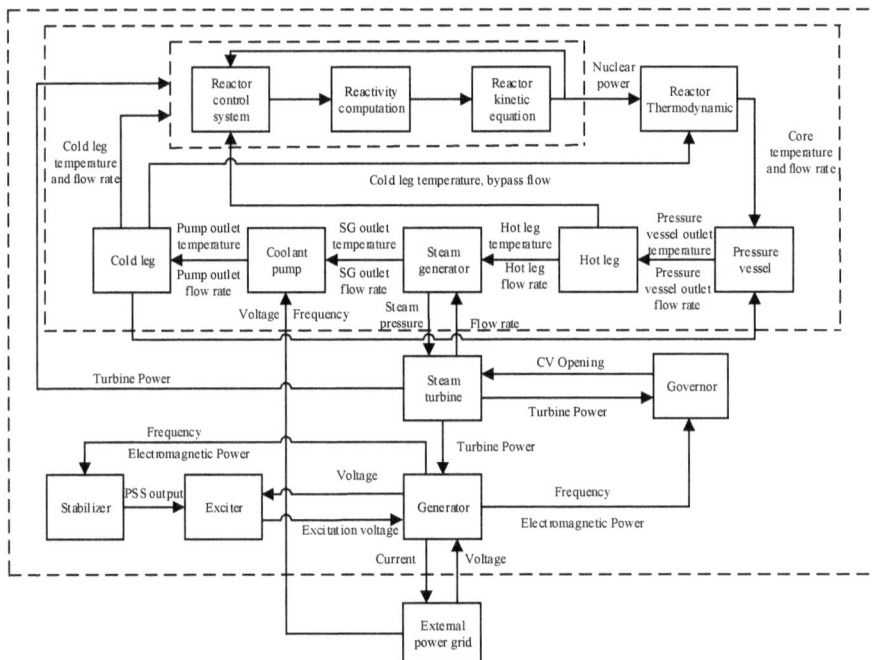

Figure 1. The principle chart of a pressurized water reactor (PWR) nuclear power plant (NPP).

2.2.1. Point Reactor Kinetic Equation

As mentioned above, the point reactor model in a full-scope simulator is quite complicated [42], which is not necessary in power system dynamic simulations. To reduce the computational complexity, the neutron kinetics should be simplified without considering the three-dimensional distribution. Nuclear power is proportionate to the neutron flux density. Thus, the dynamic characteristics of an NPP reactor can be modeled by the equivalent delayed neutrons as follows:

$$\frac{dP_f}{dt} = \frac{R - \beta}{\Lambda} P_f + \lambda C \tag{1}$$

$$\frac{dC}{dt} = \frac{\beta}{\Lambda} P_f - \lambda C \tag{2}$$

$$\frac{dF_d}{dt} = \lambda_d k P_f - \lambda_d F_d \tag{3}$$

$$P_f = P_f + F_d \tag{4}$$

where P_r, P_f, and F_d are the total reactor power, fission power and decay heat, respectively. R is the reactivity, and β is the equivalent fraction of delayed neutrons. Λ is the neutron generation time, and C is the neutron density. λ and λ_d are the equivalent decay constants of delayed neutron and decay heat, respectively.

The decay heat accounts for approximately 7.7% of the total reactor power. Because this value is relatively small, it is ignored in most power system stability research related to NPPs. However, it should be taken into account when considering large disturbances.

2.2.2. Reactor Control System

The reference NPP normally operates in the reactor-follow-turbine mode, which is realized by combining the power regulating system and temperature regulating system. The former is an open control system, which calculates the G bank position to determine the reactor power roughly according to the load set point and frequency control signal, and the latter is a closed control system, which calculates the R bank position to ensure the fine adjustment of the average temperature of the coolant according to the steady state program. On the basis of the reference [39], the reactor control system is simplified according to the operation principles of the full-scope simulator. Functions unrelated to heat transfer are ignored, such as measurement, signal transfer and pressure control. Block diagrams of these control systems are shown in Figures 2 and 3 respectively.

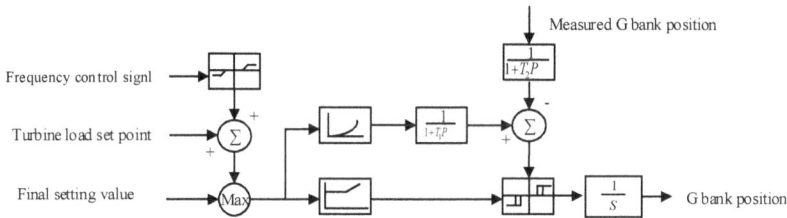

Figure 2. A schematic diagram of the power regulating system.

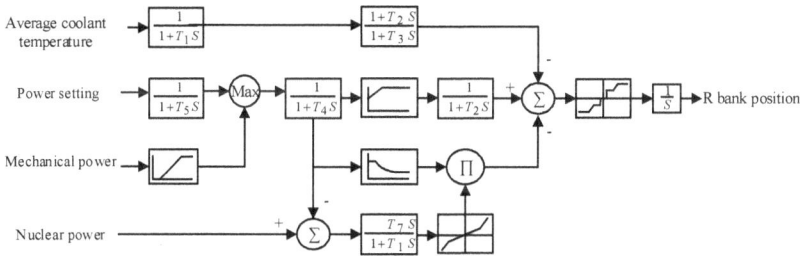

Figure 3. A schematic diagram of the temperature regulating system.

2.2.3. Reactor Thermodynamics

The reactor thermodynamics are simulated on the basis of reference [25]. However, the flow rate, specific heat and density of the primary coolant in reference [25] are processed as constants, which will result in an error of up to 10%–20%. To improve the simulation accuracy, these variables are introduced into the new model. A reactor thermodynamic model is employed to define the heat transfer in the reactor. The temperature of the fuel lump will increase when it absorbs fission energy. It will then transfer this energy to the coolant. According to the laws of conservation of energy and mass, the reactor thermodynamic model can be described as follows:

$$\frac{dT_f}{dt} = \frac{X_c P_r}{m_f C_{Pf}} + \frac{hA}{m_f C_{Pf}} \left(T_{ac} - T_f \right) \tag{5}$$

$$\frac{dT_{ac}}{dt} = \frac{(1 - X_c) P_r}{m_c C_{pc}} + \frac{hA}{m_c C_{pc}} \left(T_f - T_{ac} \right) + \frac{u_{ac}}{m_c} \left(T_{cl} - T_{aco} \right) \tag{6}$$

where T_f, T_{ac}, T_{aco} and T_{cl} are the temperatures of the fuel lump, core, outlet of the core, and cold leg, respectively. X_c is the percentage of heat released by the fuel lump. m_f and m_c are the masses of the fuel and coolant, respectively. C_{pf} and C_{pc} are the specific heats of the fuel and coolant, respectively. h and A are the heat transfer coefficient and area, respectively. u_{ac} is the mass flow rate in the core.

2.2.4. Reactor Coolant Pump

The reactor coolant pump [26], which is one of the most important pieces of equipment in a PWR NPP, is the only component in the primary system directly affected by the grid state and the main channel for the reactor transients caused by power system disturbances. The changes in coolant flow caused by changes in the main pump speed are determined by the system frequency and voltage disturbances, which cause transients in the primary system. Therefore, the main coolant pump can be modeled with Equations (7) and (8) as follows.

$$T_{jp} \frac{d\omega_p}{dt} = M_e - M_m \tag{7}$$

$$\frac{u_v}{u_n} = \frac{\omega_p}{\omega_n} \tag{8}$$

where ω_p and ω_n are speed and rated speed of the coolant pump, respectively; M_e and M_m are the electric and mechanical torque, respectively; u_v is the coolant flow rate; and u_n is the rated coolant flow rate.

2.2.5. Steam Generator Model

The SG is the hub of energy exchange between the primary system and secondary system. The high-pressure and high-temperature coolant enters the SG through a U-shaped metal tube and transfers heat to the secondary system. Then, the feed water of the secondary system absorbs the heat and evaporates into saturated steam to drive the steam turbine. According to the law of conservation of energy, the primary side, the metal of the U-shaped tube and the secondary side can be modeled by Equations (9)–(11). By assuming that the water supply rate always equals the steam rate, the steam pressure can be calculated using the temperature of the secondary side.

$$m_p C_{pp} \frac{dT_p}{dt} = h_p A_p \left(T_m - T_p \right) + u_p C_{pp} \left(T_{hl} - T_{po} \right) \tag{9}$$

$$m_p C_{pp} \frac{dT_p}{dt} = h_p A_p \left(T_m - T_p \right) + u_p C_{pp} \left(T_{hl} - T_{po} \right) \tag{10}$$

$$m_w C_{sw} \frac{dT_s}{dt} = h_s A_s \left(T_m - T_s \right) + f_{stm} \left(h_{in} - h_{out} \right) / 3 \tag{11}$$

where m, C and T are the coolant mass, specific heat and temperature, and the subscripts p, m, and S denote the primary side, secondary side, and U-shaped tube, respectively; T_{hl} is the coolant temperature of the hot leg; T_{po} is the outlet coolant temperature of the SG on the primary side; f_{stm} is the steam flow rate of the inlet of the turbine; h_{in} and h_{out} are the water enthalpy of the secondary inlet and outlet of the SG.

2.2.6. Turbine Bypass System Model

Due to the limited adjustment ability of the reactor control system, the decreasing speed of the reactor power will be restricted even in a sudden large disturbance such as load rejection. To ensure the safety of the reactor, a steam dump system should be started to vent the excess hot vapor and to maintain the power balance between the reactor and turbine. The new turbine bypass system model is derived from Reference [37]. To simplify the control logic, 15 by-pass valves are equivalent to two virtual valves. A principle diagram of the turbine bypass system is shown in Figure 4.

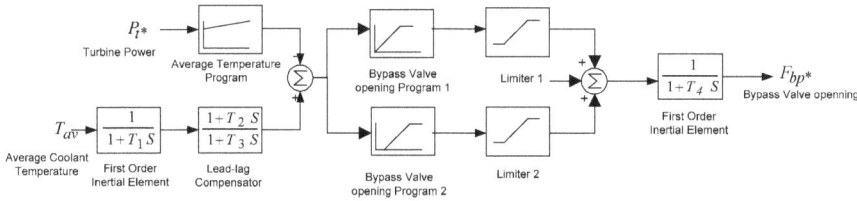

Figure 4. A principle diagram of the steam dump system.

2.2.7. Protection System of a Nuclear Power Plants

Because of their high safety and reliability requirements, NPPs are equipped with complex and high-performance protection systems, including reactor protection, turbine protection, and generator protection. This paper focuses on the protective functions that prevent abnormal behaviors caused by the grid, such as changes in voltage or frequency, and does not consider relay behaviors caused by equipment failures, e.g., failures of the condenser, steam dump system, or water level regulation system. The protection system of the NPP studied in this paper is shown in Figure 5.

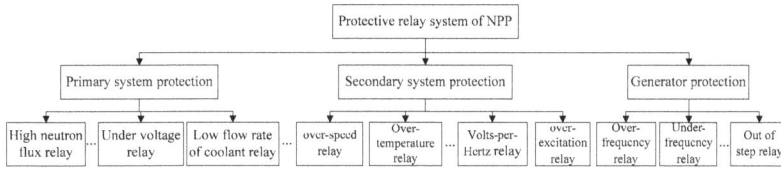

Figure 5. Hierarchical chart of NPP protection system.

3. Validation of the Pressurized Water Reactor Nuclear Power Plants Model

This model is introduced into the power system department full dynamic simulation program (PSD-FDS) by the China Electrical Power Research Institute, which is applicable to the simulation of power system short, mid, and long-term dynamics. The required data are taken from design and final safety analysis reports of the Fangjiashan (Haiyan, China) NPP. For safety reasons, it is infeasible to model NPPs based on field tests in China at present. To verify the effectiveness of the new model, the actual full-scope simulator of the Fangjiashan NPP is used as a reference in this paper, which can represent the nuclear reaction process and fluid thermal-hydraulic phenomena with high accuracy. Compared with the full-scope simulator, the change tendencies of the major variables during the transient process are consistent in both models with maximum transient errors of less than 10%, while the static errors of the main parameters of the simplified mode are less than 1%. Figure 6 shows a schematic diagram of a single unit infinite system, in which NPP G1 with a capacity of 1089 MW is connected to the infinite system G2 through a step-up transformer T1 and 500 kV transmission lines. A comparison between the simulation results of this model and the observed responses of the main process parameters in the full-scope simulator shows that they are in satisfactory agreement. Due to limited space, this paper enumerates three typical working conditions including load ramp decrease, load step decrease and net load rejection.

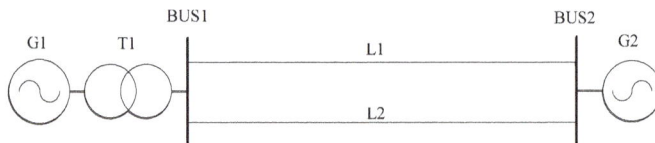

Figure 6. The single unit infinite system diagram.

3.1. Load Ramp Decrease

At 50.0 s, the generator load request begins to ramp from 100.0% full power (FP) to 50.0% FP at a rate of 50 MW/min. Owing to governor response, the control valves (CVs) begin to throttle down. Therefore, the steam flow through the turbine decreases, while the steam pressure and temperature in the SG increase. Thereby, the heat exchange between the primary side and the secondary side decreases, and the coolant temperature increases. During this process, the control rods insert into the reactor core smoothly. Finally, the reactor power stabilizes at 51.8% FP, and the turbine power steadily decreases to 50.0% FP at approximately 750 s.

Figure 7 shows that the turbine power can steadily track the load request, while the reactor power can smoothly follow the turbine power changes. There is no major oscillation and overshoot during the process. In addition, the dynamic characteristics agree well with the results of the actual unit.

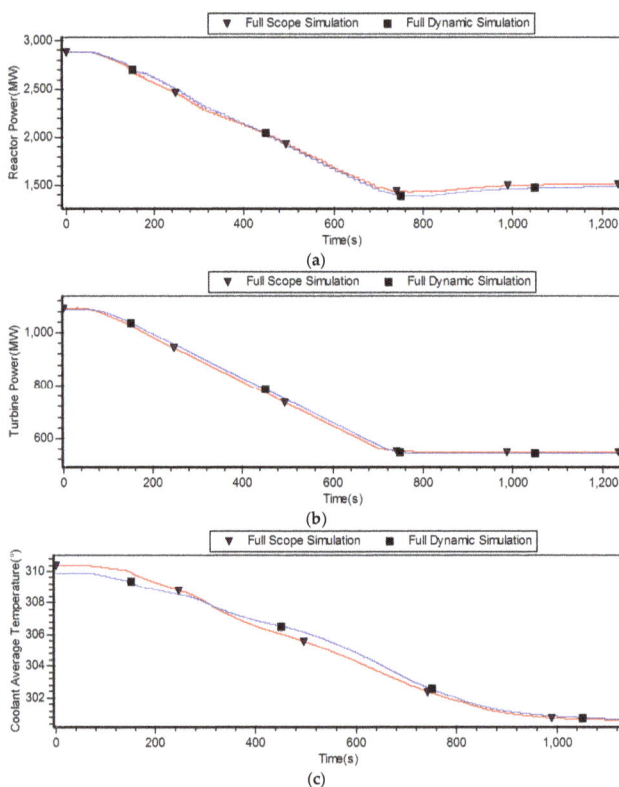

Figure 7. Dynamic responses of NPP G1 during a load ramp decrease. The load request of NPP G1 begins to ramp at 50.0 s from 100.0% full power (FP) to 50.0% FP at a rate of 50 MW/min: (**a**) Reactor power; (**b**) turbine power; and (**c**) coolant average temperature.

3.2. Load Step Decrease

At the beginning of the load step, which occurs at 85.0 s, the generator load request sharply decreases from 100% to 90% of nominal power. To follow this evolution, the CVs partly close to reduce the steam flow rate by 10%. Therefore, the steam pressure, SG temperature and coolant temperature increase, and the nuclear power begins to decrease because the reactor power set point of the reactor regulation system is determined by the load request. Meanwhile, the temperatures of the core fuel and the coolant will eventually decrease.

Compared with load ramp, load step has a similar response but at a faster rate. Moreover, the variation of the reactivity feedback introduced by the core fuel temperature, coolant temperature and reactor regulation system lags behind the change in the reactor power, which results in some overshoot. When the reactivity reverses from negative to positive, the reactor power will gradually increase and eventually remain stable at approximately 91.1% FP (Figure 8).

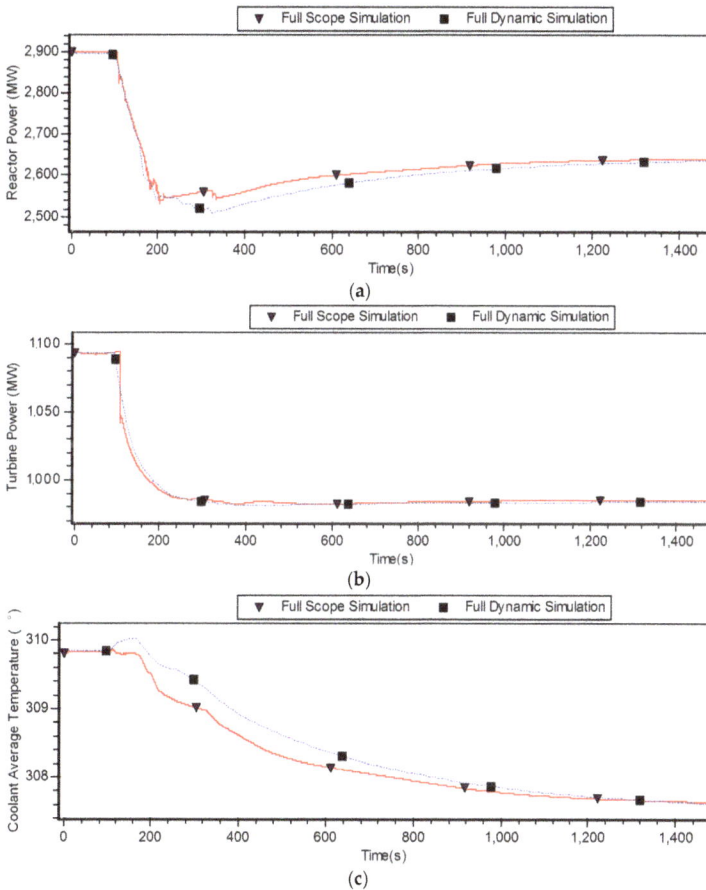

Figure 8. Dynamic responses of NPP G1 during a load step decrease. The load step occurs at 85.0 s from 100.0% to 90.0% nominal power: (**a**) Reactor power; (**b**) turbine power; and (**c**) coolant average temperature.

3.3. Net Load Rejection

NPP G1 turns to its house load from rated power when the high voltage circuit breaker trips suddenly at 15.0 s. Due to a sudden drop in the electromagnetic power, severe mismatch between the mechanical energy and electrical load will cause the turbine to speed up at the beginning. The excess input energy is cut off by closure of the CVs and intercept valves (IVs) actuated by the OPC at 16.08 s. This leads to sharp increases in steam pressure, primary coolant temperature and fuel temperature. Such an increase together with the insertion of control rods introduces negative reactivity feedback, which helps to reduce the reactor power. Figure 9 shows the reactor power perfectly stabilizes at 30% FP in approximately 1000.0 s.

The mechanical power can rapidly decrease after the operation of the OPC, and the turbine bypass system can effectively limit the secondary system steam pressure to void the opening of safety valves. Thus, the NPP is able to withstand the large transients.

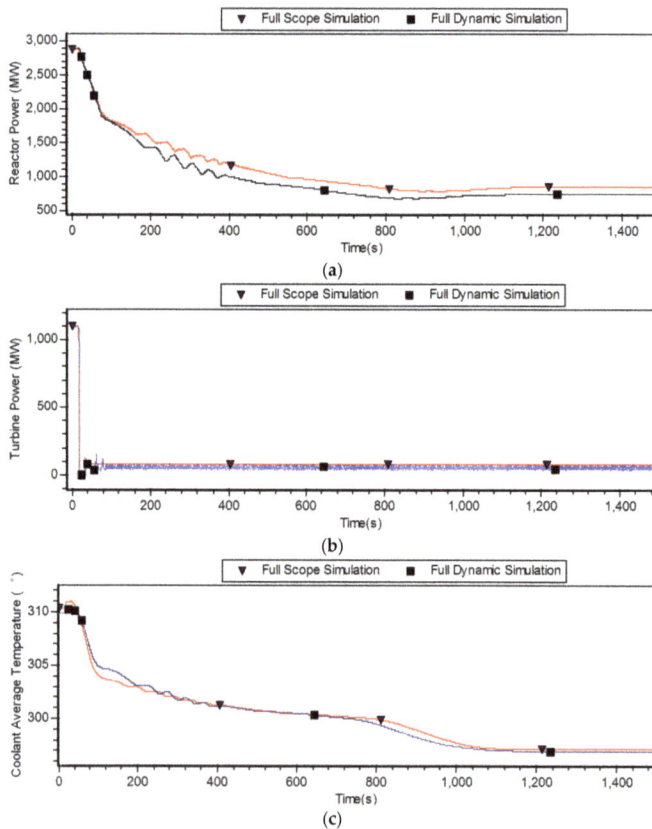

Figure 9. Dynamic responses of NPP G1 during a net load rejection. NPP G1 turns to its house load because of the tripping of the high voltage circuit breaker at 15.0 s. The disturbance responses calm down in approximately 1000.0 s: (**a**) Reactor power; (**b**) turbine power; and (**c**) coolant average temperature.

4. Interaction and Coordination among Nuclear Power Plants, Power Grids and Their Protection Systems

To investigate the coordination among NPPs, grids and their protection systems, an equivalent two-machine power system is used to analyze the dynamic performance under different types of disturbances. Actually, the mechanisms may also apply to large power systems. Although high voltages do harm transformers and large-capacity motors, compared with over-frequency, under-frequency and under-voltage, there is no substantial difference between the influence caused by over-voltage of NPPs and that of thermal power plants. Therefore, this paper will not discuss this situation in detail.

4.1. Under-Frequency

This subsection mainly analyzes the interaction between NPPs and grids under the condition of under-frequency, and proposes the corresponding coordinated control principles and schemes.

4.1.1. Dynamic Interaction with Decreasing Frequency

A decrease in frequency of the grid G2 leads to an increase in the electromagnetic power of NPP G1 and introduces a negative input to the proportional-integral-differential (PID) governing system, which turns down the CVs to reduce the mechanical power. Meanwhile, the coolant pump speed will slow down, causing the reactor coolant flow to decrease. Thus, energy exchange between the primary system and secondary system decreases and the core temperature increases, which may lead to negative reactivity feedback to decrease the reactor power.

When the governor extends beyond the dead zone, the PID input changes from negative to positive. Then, the mechanical power and nuclear power reverse to increase, and the frequency gradually recovers. During this process, there is no action taken by the reactor power control rods because they are limited by their upper bounder. Thereafter, the electromagnetic power and mechanical power decrease slowly, while the nuclear power is stabilized eventually at a point slightly higher than the rated power. Figure 10 shows the simulation results of a drop in grid frequency at a rate of 1 Hz/min.

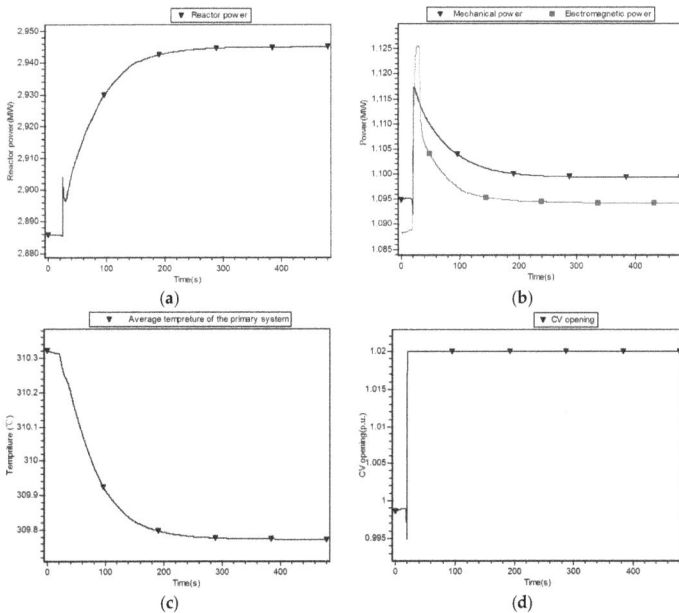

Figure 10. Simulation results of a drop in grid frequency. At 85.0 s, the system frequency begins to decrease from 50.0 Hz to 49.8 Hz at a rate of 1 Hz/min: (**a**) Reactor power; (**b**) mechanical power and electromagnetic power; (**c**) average temperature of the primary system; and (**d**) control valves (CV) opening.

In some cases, the grid may sustain an off-nominal frequency for a long period, and the turbine blades may become damaged because of resonance. Such operations have cumulative effects and are permitted only for a specific period during the life of a unit. Therefore, if the frequency decreases below the set value, the under-frequency relay will separate the unit from the grid or possibly trip the reactor when the cumulative time reaches the predetermined value.

A sharp decrease in frequency distinctly affects the auxiliary equipment of an NPP. For example, if the main pump output flow decreases quickly, which should be maintained within trip limits to avoid the action of the primary system under-flow relay and main pump under-speed relay, the cooling capacity of the fuel assembly will rapidly decrease, possibly causing an emergency shutdown because of the burning down of the fuel elements. Low-frequency conditions may also initiate Volts-per-Hertz protections.

4.1.2. Coordination across Under-Frequency Relays, Under-Frequency Load Shed Relays and Main Pump Relays

When system frequency decreases, generator units should keep running in the grid to help it recover as long as possible to maintain their own safety. For the historical reason of plant-grid separation in China, generator protections are always set according to individual situations without coordinating between units and grids. Therefore, in terms of under-frequency relay settings, manufacturers and power plant utilities tend to be conservative to ensure the safety of units.

Restarting a nuclear power unit after a reactor trip or turbine trip normally has a considerably higher price than running at low frequency for a short time. Thus, the frequency settings of the under-frequency relays of NPPs are typically lower than those of other generators in China. When the system frequency decreases sufficiently, the under-frequency relay of some thermal power units may trip first, resulting in a faster decrease in system frequency, which could trigger the under-speed relay of the main pump and cause the tripping of the nuclear unit.

In cases of prolonged under-frequency conditions, the under-frequency relay may deteriorate the active power shortage and possibly trigger UFLS relays, thus causing an unmanageable cascading reaction. Therefore, an effective graded load shedding scheme should be determined based on the system frequency to prevent the NPP from tripping or becoming isolated.

Thus, UFLS can be considered the primary protection for turbo-generators to ensure the safe operation of power plants. The duration of under-frequency, as we recommend, can be approximately 5% less than the corresponding uptimes shown in Table 1. In addition, under-frequency protection should be used as the last resort to prevent damage to the turbine. Its action time should be as long as possible on the premise of unit safety. Normally, it should be set at a value not less than 10 s with the operating value lower than 47.5 Hz and not less than 5 s with the operating value lower than 47.0 Hz. Last but not least, the under-frequency protection for thermal power plants cannot be set more conservatively than that of NPPs.

Table 1. Frequency operation capability of large turbo-generators according to the grid operation code of China.

FREQUENCY RANGE (Hz)	Cumulative Total Up-Time (min)	Up-Time (s)
51.0–51.5	>30	>30
50.5–51.0	>180	>180
48.5–50.5	Continuous operation	
48.5–48.0	>300	>300
48.0–47.5	>60	>60
47.5–47.0	>10	>20
47.0–46.5	>2	>5

4.2. Over-Frequency

In this subsection, we discuss the interaction and coordination among NPPs, grids and their protection systems in the case of over-frequency.

4.2.1. Dynamic Interaction with Increasing Frequency

Conversely, an increase in grid frequency leads to a decrease in the electromagnetic power and an increase in mechanical power of NPP G1 (Figure 6) at first. Meanwhile, such a frequency increase will also affect the primary coolant flow by accelerating the pump. After the governor overcomes the dead band, the mechanical power begins to decrease, and the control rods begin to insert into the core to reduce the reactor power.

The governor response, reactor control system, and reactivity feedback will work together to reduce the generation. The frequency will then gradually return to an acceptable level and eventually reach a new equilibrium point. The dynamic response of NPP G1 during the process of rise in grid frequency is shown in Figure 11.

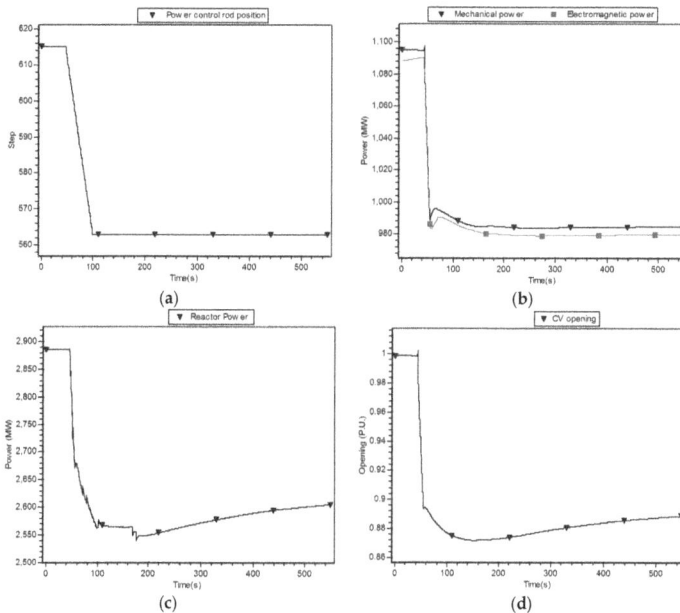

Figure 11. Simulation results of the rise in grid frequency. At 42.3 s, the system frequency begins to increase from 50.0 Hz to 52.0 Hz at a rate of 1 Hz/min: (**a**) G bank position; (**b**) electromagnetic power and mechanical power; (**c**) reactor power; and (**d**) CV opening.

When a significant oversupply of active power exists in the grid, the grid cannot ensure that frequency remains within the normal operating range through the speed governor and reactor control system alone. Without appropriate measures, such as generation reduction, the grid will enter an over-frequency condition. To avoid severe accidents, NPPs should be locked from the power system by over-frequency relays and turned to house loads.

If the frequency rises to 1.03 times the rated value, an OPC must address this situation and reduce generation to prevent the turbine from over-speeding. However, during the transient process, the reactor power level remains unchanged, thus indicating a mismatch between the steam flow generated by the SG and the steam flow passing through the CVs and IVs. The excess steam is emptied into the steam dump condenser to avoid opening the safety valves of the SG. With its CVs and IVs closed, the nuclear power unit reduces its turbine speed gradually, and the opening of the bypass valve decreases until the turbine speed returns to normal.

4.2.2. Coordination across over-Frequency Relays, Over-Speed Protect Controllers and Over-Frequency Generator Trip Relays

Because of their relatively large capacities, NPPs more easily over-speed than conventional plants, particularly in islanded systems. Increases in the system frequency jeopardize the safe operation of an NPP and its auxiliary equipment. However, an OPC alone does not help reduce excess power because the power in the turbine remains essentially unchanged after the action of the OPC. Worse, the OPC may act repeatedly, causing fluctuations in the system frequency, which is harmful to system stability. Thus, we recommend blocking the OPC of an NPP when it is running in the power system and have it work properly only when it is stepped out of the grid.

To control the frequency of the power grid, initiatives such as generation reduction along with OFGT of thermal units or hydro units must be implemented. To ensure that OFGT acts before OPC, the first round of OFGT should be set at less than 51.5 Hz, for instance, a value of 51.4 Hz would be appropriate. The amount of power reduction should be carefully calculated to maintain the system frequency under 50.5 Hz. According to our study, a three-round OFGT scheme can be suitable for different situations of excessive active power to avoid significant changes in power flow.

In case of over-frequency, the over-frequency relay of conventional generators should be triggered before that of NPPs, and the disconnection of NPPs from the grid should be the last resort under severe conditions. Therefore, particular attention should be paid to the coordination between them. The safety of the NPP itself must be ensured if the system frequency is out of control. Moreover, plant trips caused by over-frequency lead to under-frequency, which will eventually result in cascading load shedding; this situation should be avoided.

In addition to the safety of NPPs, OPC schemes of thermal plants should also consider their influence on power grids. Although most units in China are equipped with OPCs, the control schemes of different units are actually uniform at present. In case of system-wide high frequency, the output of large-capacity units drops sharply at the same time, which may result in cascading accidents of under-frequency. Thus, the control schemes of OPC should be different for each unit. To achieve greater flexibility, we tend to maintain the OPC setting at 103%, although it can be set slightly greater than 103%, and to increase the action time of OPC to 1–3 s according to the vertiginous rate and stagnant rate of the speed governor system.

4.3. Under-Voltage

The dynamic interaction between NPPs and grids under the situation of under-voltage is studied in this subsection, and then the corresponding coordinated control principles and schemes between protection systems of NPPs and grids are also proposed.

4.3.1. Dynamic Interaction with Decreasing Voltage

A depression in grid voltage has an effect on the performance of NPPs mainly through the transients of generators, transformers and motors. For instance, low voltage may slow down the main pump speed to reduce coolant flow and will therefore set up transients in the primary system. In addition, a long time low voltage situation may also trigger relative limiters and protections, leading to serious consequences, such as load shedding and generator tripping.

As an example, Figure 12 shows the dynamic response of NPP G1 (Figure 6) during the voltage dip process of BUS2 from 550.0 kV. When the voltage reaches 450.0 kV, the under-voltage relay of G1 trips the high voltage side of transformer T1, and NPP G1 turns to the house load. Although there is a sudden drop in the electromagnetic power, the mechanical power cannot be changed quickly. Thus, the turbine begins to accelerate. When the turbine speed reaches 1.03 times the rated value, the OPC acts to close the CVs, and the turbine power decreases rapidly. At the same time, the steam bypass system works as the outlet of reactor power. Therefore, reactor power can be reduced smoothly. Finally, it stabilizes at approximately 30% of the rated power.

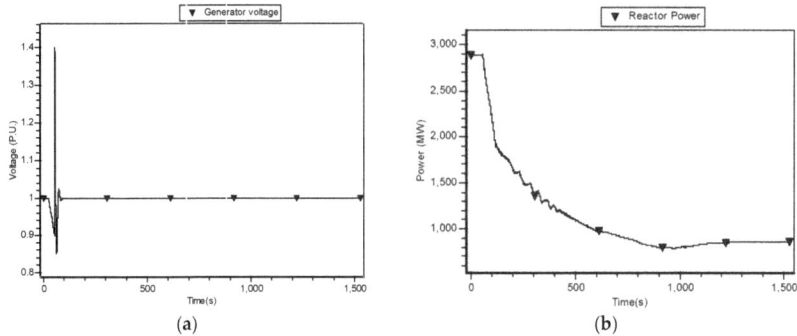

Figure 12. Simulation results of a drop in grid voltage. At 23.8 s, the grid voltage begins to decrease from 550.0 kV to 450.0 kV at a rate of 100 kV/min: (**a**) Generator voltage; and (**b**) reactor power.

In case of a voltage dip for a prolonged period, the automatic voltage regulator (AVR) equipment of the excitation system will increase the excitation current. To avoid overheating, generators are typically equipped with an over-excitation limiter (OXL) and protection.

If the set point of the field current is relatively low, e.g., slightly higher than the rated value, the OXL will decrease the field current to a preset value soon after the occurrence of an electric fault. Only when the generator field windings dissipate all of the accumulated heat, could the field current start to increase again. However, the set points of the OXL and over-excitation protection (OXP) of many NPPs in China are significantly lower than the national standard. For example, the per-unit field current setting of an OXL is often set to 1.05 with a delay time of 30 s, whereas the settings for an OXP are 1.1 and 2 s. The OXL will be triggered repeatedly if the set point is extremely low and may deny the stable operation of NPPs and grids, thus inducing mechanical fatigue in the control system and reducing the effective life of the unit.

The main pump will develop a low electromagnetic torque if the voltage decreases. Thus, the output flow of the main pump will also decrease, possibly causing a reactor trip on under-flow or under-speed. The higher the value to which the OXL is set, the longer the OXL will be required to act after a fault and to dissipate the accumulated heat of field winding. During this period, the excitation current will be limited to the predetermined value and will not respond to the decrease in voltage. Despite the system frequency remaining unchanged, a sustained decrease in voltage at this time may trigger the under-speed relay of the main pump and eventually cause significantly adverse consequences, such as generator trips or even reactor trips.

4.3.2. Coordination across Over-Excitation Limiters, Main Pump Relays and Under-Voltage Load Shed Relays

In case of low voltage, the excitation current should remain at a relative high value for an extended duration without degrading the safety of the NPPs. Suitable values for the OXL and protection are crucial for maintaining stable system voltage. Therefore, these values should be set carefully according to national standards.

The set point and delay time of the OXL and protection should be determined to ensure that the limiter acts before the protection to prevent rotor overheating. According to the "Specification for Excitation System for Large Turbine Generators" (China industry standard No. DL/T843), an OXL should be set at 2 times the rated current with a delay of 10 s. Settings that are too conservative will result in unnecessary action or an accidental tripping of the excitation system, while excessively high settings will harm the rotor winding.

In addition, if the voltage of the NPP cannot recover to its normal value after the operation of an OXL, the UVLS must be implemented to cut a certain amount of load several times to restore the voltage. Otherwise, a low voltage of 0.7 will definitely trigger the under-speed relay of the main coolant pump and trip the reactor. Thus, the characteristics of these relays should be taken into account in the setting of UVLS relays.

5. Case Studies

This research is performed on a real 500 kV bulk power grid in southeast China. Based on the above-mentioned new model, relatively accurate dynamic performances of the NPPs and power systems used in the case study can be calculated under the conditions of over-frequency, under-frequency and under-voltage, with special stress applied to the effects of protection systems on both NPPs and power grids. To ensure the safe operation of both NPPs and power grids and to improve their performances during adverse conditions, different control and protection schemes are discussed in this paper. The results show that the lack of coordination between the control and protection systems of NPPs and grids could lead to serious consequences in case of severe frequency and voltage excursions. Thus, additional attention should be paid to the cooperation between them.

The local power grid is incorporated into a 500 kV main grid by 500 kV transmission lines Lin11 and Lin12 between substations 4 and 8 (Figure 13). Generators G1 and G2 are NPPs with an installed capacity of 1089 MW and are connected to the local grid through 500 kV double-circuit lines. In addition to the 500 kV tie lines for NPPs, the main substations around the NPPs all have more than four 500 kV outlets and are closely interrelated with the local grid.

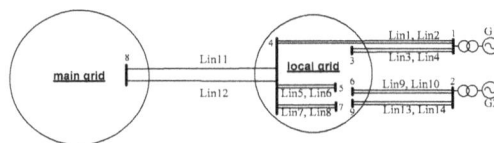

Figure 13. Network diagram of the power systems case.

5.1. Under-Frequency Case

The local grid generates electrical energy of 16,430 MW, which cannot satisfy the load demand of 18,150 MW. The shortage is balanced by borrowing active power from the major system through transmission lines Lin11 and Lin12. If an N-2 fault occurs in Lin11 and Lin12, the parallel transmission lines are tripped, causing the local grid to become isolated. An active power shortage of 1700 MW would occur, and the frequency deviation would rise to approximately 3 Hz. The frequency of the islanded system eventually recovers to approximately 48.7 Hz, but it is nevertheless too low to maintain the long-term stable operation of the system.

If the under-frequency relays of thermal power plants are set too conservatively, they will be triggered to reduce the power generation by 3200 MW during the decrease in system frequency. This process speeds up the rate of frequency decrease. Because the rotation speed of the coolant pump is determined by the difference between the electromagnetic torque and the mechanical torque, which depend on the frequency and voltage of the auxiliary power bus, it declines rapidly as the frequency decreases. When the pump speed reaches the trip limit, it will trigger the under-speed relay of the main coolant pump of the NPP and cause serious accidents, such as reactor trips or system frequency collapses. To maintain the stability of the system frequency, such schemes as automatic load shedding based on frequency decreases must be used to eliminate the power shortage.

After a load shedding of 1420 MW, combined with frequency control and the load characteristics of the frequency, the frequency is gradually restored to approximately 49.8 Hz; thus, the pump speed also returns to normal, which satisfies the system requirements for long-term stable operation without initiating the under-frequency relay and other NPP relays. The dynamic response curves are shown in Figure 14.

5.2. Over-Frequency Case

The generating capacity of the local grid is 21,300 MW with an estimated load of 18,700 MW. The local grid transports 2500 MW to the main power grid through 500 kV transmission lines Lin11 and Lin12. The local grid is separated from the major network by tripping 500 kV transmission lines Lin11 and Lin12 caused by an N-2 fault. If the generation is not reduced, the power grid would enter a condition of over-frequency, generating a surplus power of approximately 2500 MW.

The frequency curves are plotted in Figure 15, corresponding to cases without protective relay models, with relay models for units only, with uniform OPC control schemes with a generation reduction of 957 MW, and with varied control OPC schemes with a generation reduction of 957 MW. The increase of frequency means there is surplus active power. Because NPPs normally operate in the reactor-follow-turbine mode, frequency changes will directly cause the reactor control system to change the position of the control rod bank to meet the power request. However, due to the limited adjustment ability of the reactor control system, once the frequency increase triggers the OPC, the steam dump system should be started to ensure the safety of the reactor, which is quite different from thermal units.

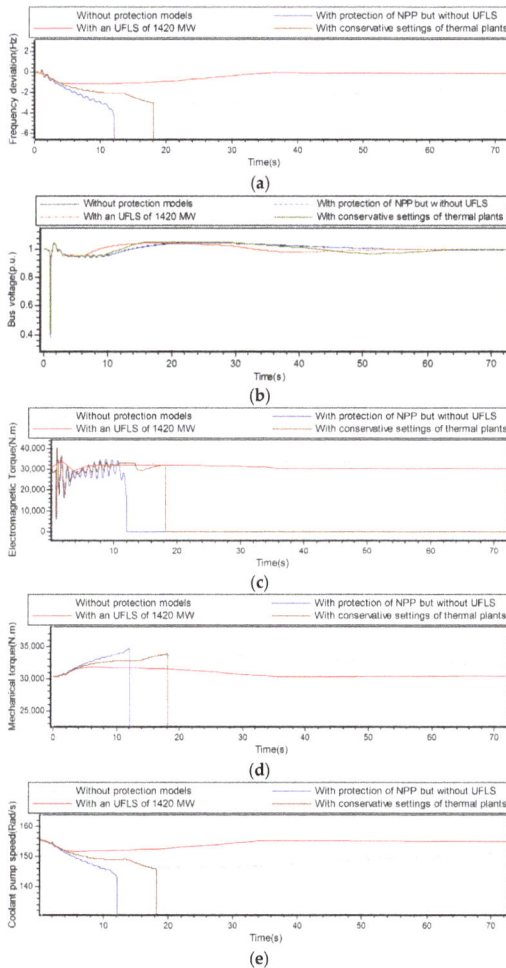

Figure 14. *Cont.*

(f)

(g)

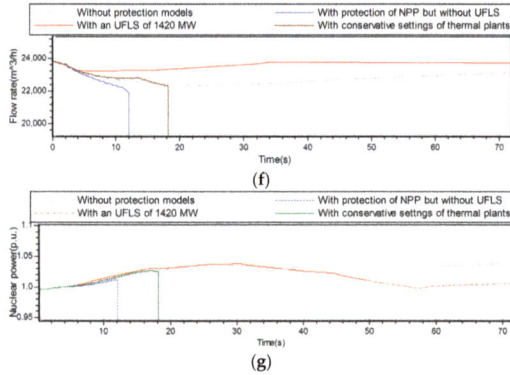

Figure 14. Dynamic response curves of NPP G1 in the case of under-frequency. When transmission lines Lin11 and Lin12 are tripped, the local grid become isolated and an active power shortage of 1700 MW occurs. This figure presents the dynamic responses of NPP G1 with different control schemes: (**a**) Frequency deviation; (**b**) bus voltage; (**c**) electromagnetic torque; (**d**) mechanical torque; (**e**) coolant pump rotation speed; (**f**) flow rate; and (**g**) nuclear power.

Apparently, the real dynamic characteristics of NPPs and power systems cannot be indicated without considering the protection models. When the frequency of the isolated grid increases to 51.5 Hz, the apparent excess of active power repeatedly triggers OPCs. To avoid any adverse effect on the stability of the system frequency, measures should be implemented. After a generation reduction of 957 MW, along with the governor systems of other generators, excessive power begins to decrease, and the frequency is restored to approximately 50.3 Hz, which can satisfy system requirements without trigging the NPPs. Figure 15 shows that varied OPC control schemes other than uniform schemes can greatly help the frequency return smoothly to normal.

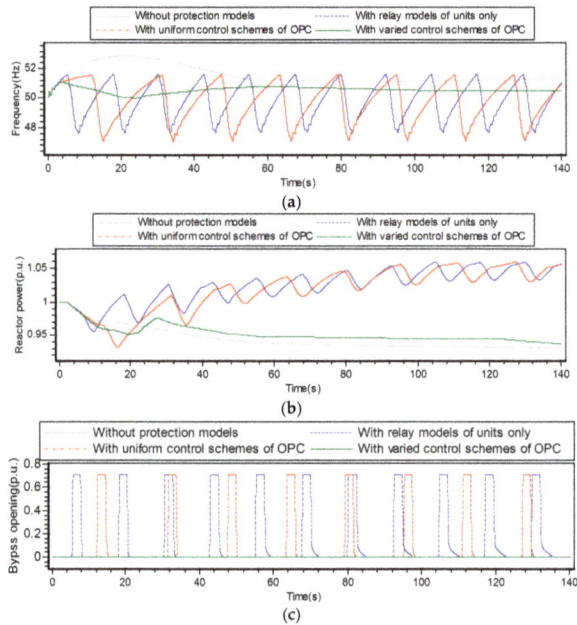

(a)

(b)

(c)

Figure 15. *Cont.*

(d)

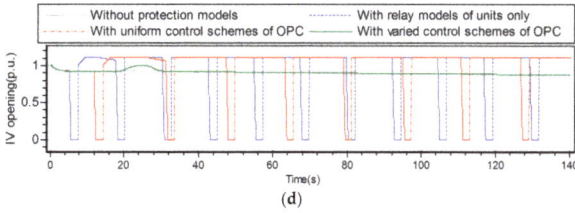

Figure 15. Dynamic response curves of NPP G1 in the case of over-frequency. When transmission lines Lin11 and Lin12 are tripped, the local grid is separated from the major grid with a surplus power of approximately 2500 MW. This figure presents the dynamic responses of NPP G1 with different control schemes: (**a**) Frequency of NPP G1; (**b**) reactor power of NPP G1; (**c**) bypass opening; and (**d**) IV opening.

5.3. Under-Voltage Case

During the maintenance of parallel transmission lines Lin5 and Lin6, the total load demand of the grid is 26,430 MW, and its generating capacity is 25,650 MW. The local grid must extract active power from the major system to address the mismatch. The power angle, frequency, and voltage of the NPP remain stable after an N-2 fault occurs in Lin7 and Lin8. However, approximately 3300 MW of active power is transferred to Lin1, Lin2, Lin3, and Lin4, and the reactive power loss increases sharply. The voltages of other units, such as NPP G1, significantly decrease, which causes an obvious increase in the field current for a prolonged period and eventually activation of the OXL.

As shown in Figure 16, the OXL will act repeatedly if its set point is too low, e.g., 1.05, which is likely to cause mechanical fatigue of the control system and instability of the NPPs. Meanwhile, the OXL will begin to operate until 48.5 s after the fault when the set point is increased from 1.05 to 1.4. The 500 kV bus voltage and the terminal voltage of NPP G1 decrease as the excitation current decreases. As mentioned above, the rotation speed of the main pump is mainly affected by the system frequency and bus voltage. Although the system frequency only fluctuates slightly, there is a relatively large drop of voltage, which increases the difference between the mechanical torque and the electromagnetic torque, and slows down the pump speed. These processes eventually initiate an under-speed relay of the main coolant pump at 67.7 s, consequently causing an emergency reactor trip. To avoid significant accidents, a load shedding of 1290 MW can be implemented by the receiving-end power grid to ensure that the voltage of NPP G1 stays within a reasonable range without initiating the OXL.

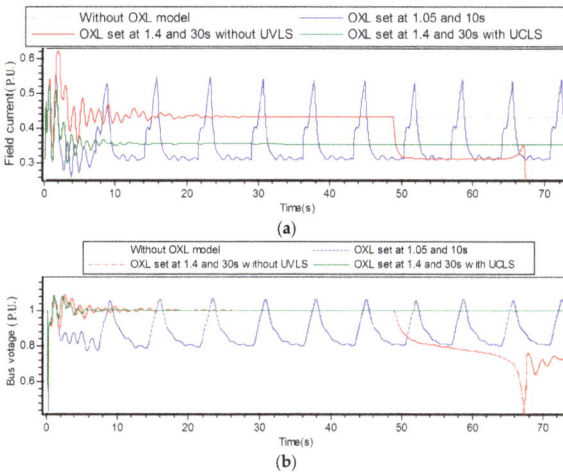

(a)

(b)

Figure 16. *Cont.*

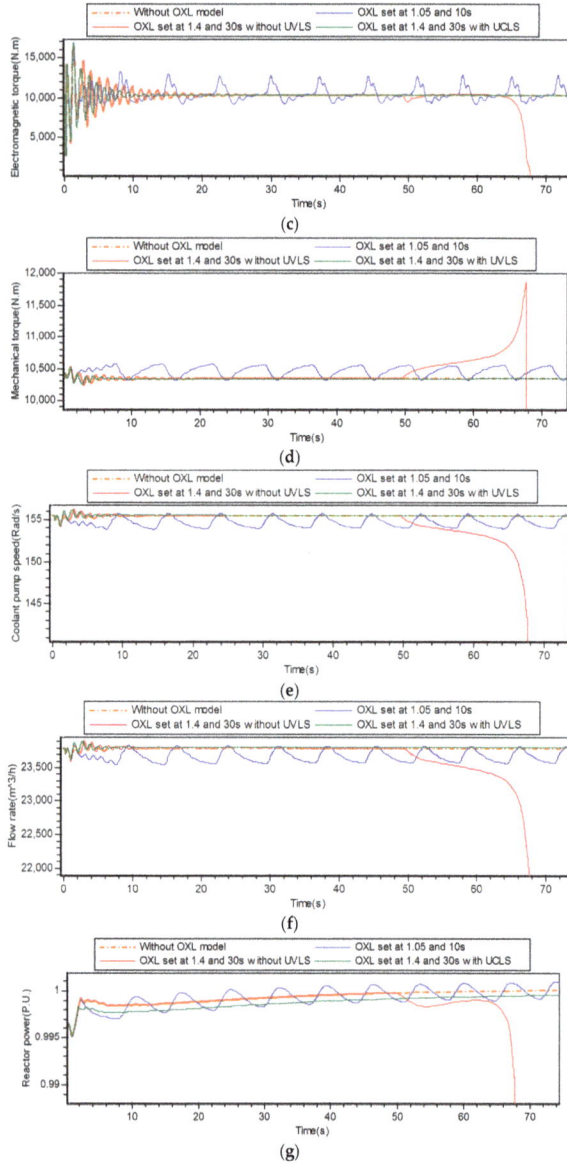

Figure 16. Dynamic response curves of NPP G1 in the case of under-voltage. After an N-2 fault occurs in Lin7 and Lin8, approximately 3300 MW of active power in the local grid is transferred to Lin1, Lin2, Lin3, and Lin4. The reactive power loss increases sharply, and the voltages of some units, such as NPP G1, significantly decrease. This figure presents the dynamic responses of NPP G1 with different control schemes: (**a**) Field current; (**b**) bus voltage; (**c**) electromagnetic torque; (**d**) mechanical torque; (**e**) coolant pump rotation speed; (**f**) coolant flow rate; and (**g**) nuclear power.

6. Conclusions

This paper proposes a new model with relatively high precision for the most widely used PWR NPP in China, which is relatively simple and can be applied to the simulation of the short, mid-term and long-term dynamics of large-scale power systems. The effectiveness is verified by an actual NPP full-scope simulator. Based on this model, this paper investigates the interactions between NPPs and power grids under abnormal conditions of over-frequency, under-frequency and under-voltage, with a focus on the coordination of grid-related NPP protections and grid protection systems. Deficiencies in the coordination of the protection systems between NPPs and grids may cause accidents with potentially disastrous consequences. To achieve overall reliability, a high priority should be assigned to the system-wide coordination of protection systems between NPPs and grids. Therefore, the control principles and schemes of the coordination between the protection systems of NPPs and grids, together with the recommended protection values for both NPPs and grids, are put forward in this paper. The results can be used as a reference for the study of coordination between NPPs and grids and for the parameter optimization of grid-related generator protection of NPPs.

Acknowledgments: This work is supported by National Key Basic Research Program of China (973 Program) (No. 2013CB228204) and Science and Technology Project of the State Grid Corporation of China (SGCC) (No. XTB17201500050). The authors are also grateful to the SGCC for kindly providing data from the power system to perform the study.

Author Contributions: Guoyang Wu conceived and designed the research, developed the NPP model and prepared the manuscript. Ping Ju conceived the research and revised the manuscript. Xinli Song carried out data analysis and revised the manuscript. Chenglong Xie contributed to the modeling of NPPs, Wuzhi Zhong performed the data analysis. Guoyang Wu is the first author in this manuscript. All authors approved the publication.

Conflicts of Interest: The authors declare no conflict of interest.

References

1. *Medium- and Long-Term Nuclear Power Development Plan*; National Energy Administration of China: Beijing, China, 2012.
2. Energy Information Administration. *International Energy Outlook 2009*; U.S. Department of Energy: Washington, DC, USA, 2009.
3. Zhou, Y. Why is China going nuclear? *Energy Policy* **2010**, *38*, 3755–3762. [CrossRef]
4. Adamantiades, A.; Kessides, I. Nuclear power for sustainable development: Current status and future prospects. *Energy Policy* **2009**, *37*, 5149–5166. [CrossRef]
5. Trehan, N.K.; Saran, R. Nuclear Power Revival. In Proceedings of the IEEE Nuclear Science Symposium Conference, Portland, OR, USA, 19–25 October 2003.
6. Maldonado, G. The Performance of North American NPPs during the Electric Power Blackout of August 14, 2003. In Proceedings of the IEEE Nuclear Science Symposium Conference, Rome, Italy, 16–22 October 2004; pp. 4603–4606.
7. Lee, C.H.; Chen, B.K.; Chen, N.M.; Liu, C.W. Lessons learned from the blackout accident at a nuclear power plant in Taiwan. *IEEE Trans. Power Deliv.* **2010**, *25*, 2726–2733. [CrossRef]
8. *Analysis Disruption of the Day 10/11/2009 at the 22 h 13 m Involving the Shutdown of Three Circuits LT 765 kV Itaberá-Ivaiporã*; Report ONS RE-3-252/2009; Brazil ONS: Rio de Janeiro, Brazil, 2009.
9. *Final Report on the August 14, 2003 Blackout in the United States and Canada: Causes and Recommendations*; US-Canada Power System Outage Task Force: Ottawa, ON, Canada, 2004.
10. Netz, E.O.N. Report on the Status of the Investigations of the Sequence of Event s and Causes of the Failure in the Continental European Electricity Grid on Saturday after 22:10 hours, Bayreuth, Germany, 4 November 2006.
11. *Report of the Enquiry Committee on Grid Disturbance in Northern Region on 30th July 2012 and in Northern, Eastern & North-Eastern Region on 31st July 2012*; India Ministry of Power: New Delhi, India, 2012.
12. Meng, D. China's Protection Technique in Preventing Power System Blackout to World. In Proceedings of the International Conference on Advanced Power System Automation and Protection, Beijing, China, 16–20 October 2011.
13. Mu, G.; Cui, Y.; Yan, G.; Zheng, T.; Xu, G. Source-grid coordinated dispatch method for transmission constrained grid with surplus wind generators. *Autom. Electr. Power Syst.* **2013**, *37*, 24–29.

14. Mozina, C.J. Implementing NERC Guidelines for Coordinating Generator and Transmission Protection. In Proceedings of the 65th Annual Conference on Protective Relay Engineers, College Station, TX, USA, 2–5 April 2012.

15. Kharel, A.P.; Rusch, R.J.; Thornton-Jones, R. Review of Generation System Overflux Limiters and Protection and Consequences of Incorrect Settings. In Proceedings of the IEEE Power and Energy Society General Meeting, Minneapolis, MN, USA, 25–29 July 2010.

16. Schaefer, R.; Jansen, D.; McMullen, S.; Rao, P. Coordination of Digital Excitation System Settings for Reliable Operation. In Proceedings of the IEEE Power and Energy Society General Meeting, Calgary, AB, Canada, 26–30 July 2009.

17. *Research on Coordination between Unit's Grid-Related Protections and Grid*; China Electric Power Research Institute: Beijing, China, 2011.

18. *Research on Coordination between Excitation System and the Grid*; China Electric Power Research Institute: Beijing, China, 2010.

19. *Coordinated Control Technology of Unit's Grid-Related Protection and Power System Stability*; China Electric Power Research Institute: Beijing, China, 2014.

20. Schulz, R.P.; Turner, A.E. *Long-Term Power System Dynamics—Phase II*; Final Report for EPRI Research Project 764-1, Reprot No. EL-367; General Electric Company: Schenectdy, NY, USA, 1977.

21. Di Lascio, M.A.; Moret, R.; Poloujadoff, M. Reduction of program size for long-term power system simulation with pressurized water reactor. *IEEE Trans. Power Appar. Syst.* **1983**, *PAS-102*, 745–751. [CrossRef]

22. Kerlin, T.W.; Katz, E.M. *Pressurized Water Reactor Modeling for Long-Term Power System Dynamic Simulation*; Final Report for EPRI Research Project 764-4, Report No. EL-3087; Tennessee University: Knoxville, TN, USA, 1983.

23. Ichikawa, T.; Inoue, T. Light water reactor plant modeling for power system dynamic simulation. *IEEE Trans. Power Syst.* **1988**, *3*, 463–471. [CrossRef]

24. Inoue, T.; Ichikawa, T.; Kundur, P.; Hirsch, P. Nuclear plant models for medium- to long-term power system stability studies. *IEEE Trans. Power Syst.* **1995**, *10*, 141–148. [CrossRef]

25. Hu, X.; Zhang, X.; Zhou, X.; Gan, F.; Zong, W. Pressurized Water Reactor Nuclear Power Plant (NPP) Modelling and the Midterm Dynamic Simulation after NPP has been Introduced into Power System. In Proceedings of the IEEE Region 10 Conference on Computer, Communication, Control and Power Engineering (TENCON'93), Beijing, China, 19–21 October 1993; Volume 5, pp. 367–370.

26. Gao, H.; Wang, C.; Pan, W. A Detailed Nuclear Power Plant Model for Power System Analysis Based on PSS/E. In Proceedings of the Power Systems Conference and Exposition, IEEE PSCE, Atlanta, GA, USA, 29 October–1 November 2006.

27. Shi, X.; Wu, P.; Liu, D.; Xiong, L.; Zhao, J.; Zhang, Y. Modeling and Dynamic Analysis of Nuclear Power Plant Reactor Based on PSASP. In Proceedings of the Asia-Pacific Power and Energy Engineering Conference, APPEEC 2009, Wuhan, China, 27–31 March 2009.

28. Zhao, J.; Wu, P.; Liu, D. User-Defined Modeling of Pressurized Water Reactor Nuclear Power Plant Based on PSASP and Analysis of its Characteristics. In Proceedings of the Asia-Pacific Power and Energy Engineering Conference, Wuhan, China, 27–31 March 2009.

29. Arda, S.E.; Holbert, K.E. Implementing a Pressurized Water Reactor Nuclear Power Plant Model into Grid Simulations. In Proceedings of the IEEE PES General Meeting, National Harbor, MD, USA, 27–31 July 2014; pp. 1–5.

30. Zhao, J.; Liu, D.; Ouyang, L.; Sun, W.; Wang, Q.; Yang, N. Analysis of the mutual interaction between large-scale pressurized water reactor nuclear power plants and power systems. *Proc. CSEE* **2012**, *32*, 64–70.

31. Kirby, B.; Kueck, J.; Leake, H.; Muhlheim, M. Nuclear Generating Stations and Transmission Grid Reliability. In Proceedings of the 39th Noah American Power Symposium, Las Cruces, NM, USA, 30 September–2 October 2007; pp. 279–287.

32. *Introducing Nuclear Power Plants into Electrical Power Systems of Limited Capacity: Problems and Remedial Measures*; Technical Reports Series No. 271; International Atomic Energy Agency: Vienna, Austria, 1987.

33. Peters, S.; Machina, G.T.; Nthontho, M.; Chowdhury, S.; Chowdhury, S.P.; Mbuli, N. Modeling and Simulation of Degraded and Loss of Voltage Protection Scheme for Class 1E Bus of a Nuclear Power Plant. In Proceedings of the Universities Power Engineering Conference, London, UK, 4–7 September 2012.

34. Koepfinger, J.L.; Khunkhun, K.J.S. Protection of auxiliary power systems in a nuclear power plant. *IEEE Trans. Power Appar. Syst.* **1979**, *PAS-98*, 290–299. [CrossRef]
35. *Reactor Coolant System Manual for the Fujian Fuqing NPP First-Stage Project*; Nuclear Power Institute of China: Chengdu, China, 2009.
36. *Main Steam System Manual for the Fujian Fuqing NPP First-Stage Project*; China Nuclear Power Engineering Co.: Beijing, China, 2009.
37. *Turbine Bypass System Manual for the Fujian Fuqing NPP First-Stage Project*; East China Electric Power Design Institute: Shanghai, China: Shanghai, China, 2010.
38. *Turbine Governing System Manual for the Fujian Fuqing NPP First-Stage Project*; China Nuclear Power Engineering Co.: Beijing, China, 2010.
39. *Rod Control System Manual for the Fujian Fuqing NPP First-Stage Project*; Nuclear Power Institute of China: Chengdu, China, 2009.
40. *Reactor Protection System Manual for the Fujian Fuqing NPP First-Stage Project*; Nuclear Power Institute of China: Chengdu, China, 2009.
41. *Main Feedwater Flow Control System Manual for the Fujian Fuqing NPP First-Stage Project*; Nuclear Power Institute of China: Chengdu, China, 2009.
42. *Nuclear Design Report for the Fujian Fuqing NPP First-Stage Project*; Nuclear Power Institute of China: Chengdu, China, 2008.
43. *Modeling and Measurement Technology for High-Capacity Nuclear Power Unit*; China Electric Power Research Institute: Beijing, China, 2013.
44. Stott, B. Power system dynamic response calculations. *Proc. IEEE* **1979**, *67*, 219–241. [CrossRef]

energies

MDPI

Review

Smart Distribution Systems

Yazhou Jiang [1], Chen-Ching Liu [1,2] and Yin Xu [1,*]

[1] School of Electrical Engineering and Computer Science, Washington State University,
    Pullman, WA 99164, USA; yjiang1@eecs.wsu.edu (Y.J.); liu@eecs.wsu.edu (C.-C.L.)
[2] School of Mechanical and Materials Engineering, University College Dublin, Belfield, Dublin 4, Ireland
* Correspondence: yxu2@eecs.wsu.edu; Tel.: +1-409-594-7210

Academic Editor: Ying-Yi Hong
Received: 6 March 2016; Accepted: 11 April 2016; Published: 19 April 2016

Abstract: The increasing importance of system reliability and resilience is changing the way distribution systems are planned and operated. To achieve a distribution system self-healing against power outages, emerging technologies and devices, such as remote-controlled switches (RCSs) and smart meters, are being deployed. The higher level of automation is transforming traditional distribution systems into the smart distribution systems (SDSs) of the future. The availability of data and remote control capability in SDSs provides distribution operators with an opportunity to optimize system operation and control. In this paper, the development of SDSs and resulting benefits of enhanced system capabilities are discussed. A comprehensive survey is conducted on the state-of-the-art applications of RCSs and smart meters in SDSs. Specifically, a new method, called Temporal Causal Diagram (TCD), is used to incorporate outage notifications from smart meters for enhanced outage management. To fully utilize the fast operation of RCSs, the spanning tree search algorithm is used to develop service restoration strategies. Optimal placement of RCSs and the resulting enhancement of system reliability are discussed. Distribution system resilience with respect to extreme events is presented. Test cases are used to demonstrate the benefit of SDSs. Active management of distributed generators (DGs) is introduced. Future research in a smart distribution environment is proposed.

Keywords: feeder restoration; outage management; remote control capability; smart distribution system (SDSs); smart meter

1. Introduction

Electric power distribution systems are designed to deliver power from substations to customers. Efficient delivery and reliability of service are crucial measures for distribution systems. However, extreme events, such as Superstorm Sandy [1] and derecho [2], threaten the reliable operation of distribution systems, and cost the economy billions of dollars [3], e.g., an estimated loss of $52 billion due to Superstorm Sandy [2]. Destructive hurricanes, winter storms, and other extreme weather events, resulting from global climate change, may further challenge the reliable operation of distribution systems. To minimize the impact of these events on reliable power supply, governments and utilities are making an effort toward smart distribution systems (SDSs) through grid modernization [4].

Traditionally, limited information is acquired along distribution feeders with few deployed sensors. Crews are sent to gather field data and operate devices on site [5]. The lack of remote monitoring and control capability limits distribution operators' ability to monitor system operations and take control actions promptly in response to extreme events. It may take them hours to determine fault locations through field crews and trouble calls from affected customers.

The observability and controllability of distribution systems can be enhanced by adopting emerging intelligent devices and smart grid applications. With the ongoing smart grid development, smart meters and remote control switches (RCSs) are deployed nationwide in the U.S. According to [6], a total of 7661 RCSs have been installed by 2013. It is estimated that 65 million smart meters have been installed by 2015 [7]. These devices with bidirectional communication are accelerating the transformation to SDSs of the future [8]. In a SDS, a large amount of data, gathered by smart meters and intelligent electronics devices (IEDs), provides sufficient information to monitor system operations in nearly real time. Smart grid applications, such as advanced outage management and fault location, isolation, and service restoration (FLISR), are developed and integrated into Outage Management Systems (OMSs) and Distribution Management Systems (DMSs) [9]. These tools assist distribution operators in determining optimal system operation strategies and taking control actions promptly in response to a disturbance.

This paper is focused on the state-of-the-art applications of smart meters and RCSs in SDSs. The impact of extreme events on distribution systems is analyzed to highlight the importance of developing SDSs. With the availability of numerous smart meters, applications based on data gathered from these meters are discussed. A new method, called Temporal Causal Diagram (TCD), is presented for enhanced outage management by incorporating outage reports from smart meters. Fault isolation and service restoration are implemented to restore service to the interrupted customers. A state-of-the-art survey is conducted on service restoration techniques. Specifically, the spanning tree search algorithm, which determines service restoration strategies that restore the maximum amount of interrupted load with a minimum number of switching operations, is illustrated with examples. Methodologies for placement of RCSs to enhance service restoration capability are discussed. The enhancement of system reliability through added remote control capability is demonstrated. Distribution system resilience with respect to extreme events is discussed and demonstrated using numerical simulation results of the Pullman-Washington State University (WSU) distribution system. In addition, worldwide development of SDSs, especially in Europe, and active management of DGs are summarized.

2. Smart Distribution System Development in the U.S.

2.1. Enhanced System Reliability Utilizing Smart Grid Technologies

To quantify the reliability of a distribution system, the System Average Interruption Duration Index (SAIDI) is widely used by utilities [10]. Reliability of the U.S. electric energy distribution system during 2000–2012 is summarized in [11]. Take 2012 for instance. The average SAIDI in the U.S. is above 100 min without consideration of major events [11]. According to [12], SAIDI in South Korea in the same year is about 11 min. The enhanced reliability of South Korea distribution systems results from a high level of distribution automation through deploying smart grid technologies [13]. By 2012, South Korea has upgraded 48% of manual switches to RCSs and deployed DMSs widely in the distribution operating center [12].

As a result of the American Recovery and Reinvestment Act of 2009, the U.S. Department of Energy (DOE) in collaboration with power industries launched the Smart Grid Investment Grant (SGIG) program [6,14], which is aimed at modernizing electric power grids with advanced tools and smart grid techniques. Billions of dollars have been invested in modernization of power grids across the country. As of March 2013, 7661 RCSs through SGIG have been deployed on 6500 distribution feeders, which account for around 4% of the estimated 160,000 distribution feeders in the U.S. [6]. The projected SGIG expenditures at completion are summarized in Figure 1.

Ninety-nine smart grid demonstration projects have been funded from transmission systems to end-use customers. An estimated total of $4.05 billion and $1.96 billion have been invested in deploying Advanced Metering Infrastructures (AMIs) and automating electric distribution systems, respectively. These advanced tools and emerging smart grid technologies play an important role in improving system reliability. It is reported by the U.S. DOE that SAIDI of the smart grid demonstration systems is decreased by up to 56% [6].

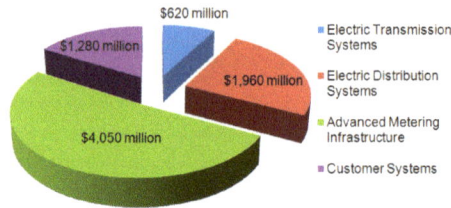

Figure 1. Estimated expenditures of the SGIG programs [14].

2.2. Smart Distribution Systems under Extreme Events

Extreme events are a threat to reliable operation of distribution systems. According to [1], power outages caused by severe weather events account for 58% of total outages since 2002. Electric power distribution infrastructures, including substations, power lines, and utility poles, were severely damaged as a result of these extreme events. During Superstorm Sandy, 50 substations, 2100 transformers, and 4500 utility poles in the Long Island Power Authority (LIPA)'s service territory experienced damage [15]. Power recovery to 84 percent of the total interrupted customers was not achieved until six days after the occurrence of Sandy [15]. The devastating effect of extreme events was demonstrated by Hurricane Irene as well. When Irene moved up along the East Coast from South Carolina to Maine, 6.69 million customer outages were reported with an estimated damage of nearly $15.8 billion [15,16].

Distribution automation and smart grid technologies enhance the ability of a distribution system to withstand extreme events and restore power supply to interrupted customers efficiently after major outages. It is reported in [14] that Electric Power Board of Chattanooga (EPB), serving about 170,000 customers in Tennessee and Georgia, successfully improved outage responses by using RCSs. When nine tornados ripped through EPB territories in 2011, with 122 RCSs in service, damaged distribution lines were isolated remotely and power was restored to interrupted customers promptly. A total of 250 truck rolls were avoided and thousands of hours of outage time were saved. It is expected by EPB that upon the completion of the SGIG program, the power outage duration can be further lowered by more than 40 percent and system reliability can be increased, which is worth more than $35 million per year [14]. Besides deploying RCSs, utilities across the U.S. installed numerous smart meters at customers' premises. The potential application of outage reports from smart meters was demonstrated by Philadelphia Electric Company (PECO) [6]. By leveraging smart meter outage reports for advanced outage management, PECO was able to restore electrical service to interrupted customers 2–3 days sooner than would have been in the event of Superstorm Sandy, and more than 6000 truck rolls were avoided [7].

3. Smart Metering Technology

In a SDS, smart meters play an important role in acquiring data and enhancing the situational awareness of system operators. A smart meter is an electronic device with two-way communications that can automatically transmit customers' energy consumption data as well as system operation information to the distribution operating center [17]. The available information includes the values of voltage and real and apparent power [18]. A large amount of data from smart meters provides distribution operators with near real-time monitoring of system operations. The architecture for integration of smart meters in the distribution operating center is illustrated in Figure 2. Smart meters are set up to collect data every 5 min and transmit the information via a Local Area Network (LAN) to a Meter Data Management System (MDMS) as often as 15 min or more infrequently according to the data usage [17]. Peer-to-Peer (P2P) and Zigbee technologies are widely used in a LAN [17]. To pass on the data to the central collection point in the distribution operating center, a Wide Area Network (WAN), e.g., Power Line Carrier (PLC), is adopted [17,19,20]. The data is then processed at the distribution operating center for customer billing, outage management, and other operational purposes.

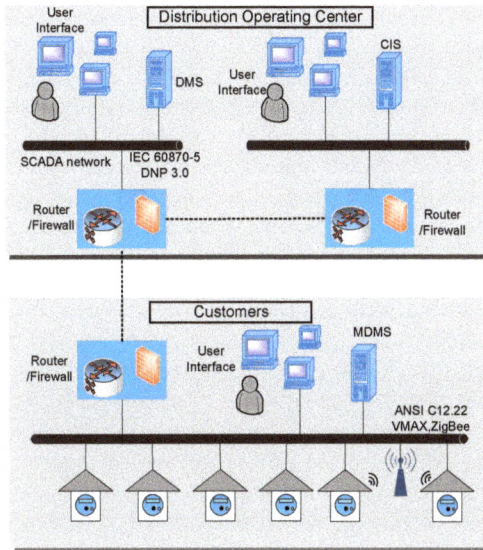

Figure 2. The architecture for integration of smart meters in the distribution operating center.

By 2015, an estimated total of 65 million smart meters have been installed in the U.S. [7] and this number keeps increasing. As more smart meters with communications are deployed, concerns regarding data security and integrity arise [21,22]. To address these issues, a comprehensive set of cyber security guidelines has been published by the National Institute of Standards and Technology (NIST) [23,24]. It is required that system vendors comply with new requirements to address remote access, authentication, encryption, and privacy of metered data and customer information. Besides these guidelines, research related to cyber security of smart meters is being conducted to prevent metering infrastructures from cyber-attacks [25–28]. Issues concerning health effects of radio frequency exposure from smart meters are also raised [29]. It is reported by California Council on Science and Technology that radio frequency exposure resulting from smart meters, when installed and maintained properly, is lower than common household electronic devices, such as cell phones and microwave ovens [30]. To standardize the radio exposure from transmitters, the Federal Communications Commission (FCC) has issued Maximum Permissible Exposure (MPE) limits for field strength and power density of radio frequency electromagnetic fields [31]. All smart meters are mandated to comply with the FCC rules.

Smart meters enable near real-time data collection and remote control capability, which provide great opportunities for utilities to improve distribution system management and operations. Applications of smart meters include:

(1) Automatic customer billing: compared with kilo-watt meters, which need tedious on-site meter reading work, smart meters automatically send energy consumption data to the utility. It is reported by Avista Utilities, based in Spokane, WA, that developed customer billing portals using smart meter data reveal significant benefit to utilities as well as customers. The benefit includes recording the billing history, analyzing the bill to identify ways to increase energy efficiency, and acting as online home energy advisor to outline ways to save energy [32].

(2)   State estimation of distribution systems: numerous data from smart meters can be used for state estimation of distribution systems [33]. Different methods, such as weighted least square (WLS) [34], Bayesian network [35], graph theory [36], and machine learning [37], are proposed for state estimation.

(3)   Volt/VAR management: voltage and reactive power management is essential for utilities to minimize power losses while maintaining an acceptable voltage profile along the distribution feeder under various loading conditions [38,39]. The near real-time voltage measurements from smart meters can be used as inputs for Volt/VAR controls to support decision-making, such as switch-on/off of capacitor banks and adjustment of voltage regulator tap positions.

(4)   Remote connect/disconnect: two-way communications of smart meters enable distribution operators to remotely connect and disconnect meters. If a customer defaults on electricity payment, a command to the smart meter can quickly cut the customer's power supply. Connect/disconnect functions of smart meters provide distribution operators with more remote control capability to reduce dispatching field crews [14].

(5)   Demand response: demand response is aimed to reduce the peak load [40], which avoids utilities from purchasing electricity at a high cost and delays the construction of new power substations. According to the U.S. Federal Energy Regulatory Commission (FERC), an estimation of 41,000 MW power is reduced through existing demand response programs in 2008 [41]. Different methods for demand response are proposed based on varying electricity prices [42–44] or incentives [45]. These demand response programs can be implemented through smart meters to control appliances so as to change customers' energy consumption patterns.

(6)   Load modeling and forecasting: accurate load modeling and forecasting is crucial for system operations and resource planning [46]. Using data from smart meters, the daily energy consumption pattern of each customer can be identified. The loading profile of each distribution transformer is determined through aggregating energy consumption from customers downstream. The temporal relationship among different load patterns can be used for load forecasting [47].

4. Enhanced Outage Management System

In a distribution system, a fault and the resulting activation of protective devices can lead to a power outage. Detection of a fault and identifying the activated protective devices to isolate the fault are key functions of an OMS. Once the faulted line section and activated protective devices are determined, field crews are dispatched to locate the fault. Fault isolation and service restoration can then be performed with remote and/or manual switches. An advanced OMS using outage reports from smart meters can help distribution operators determine the faulted line section promptly, facilitating the service restoration process.

4.1. OMS Based on Trouble Call Handling and Meter Polling

Traditionally, distribution systems have few data acquisition points beyond the substation. When a power outage occurs, distribution operators rely on trouble calls from interrupted customers to determine the outage area. Once an outage report from a customer is received, the OMS will match the phone number from the originating call to the specific customer location, and determine the distribution transformer and protective devices on the corresponding feeder. When sufficient trouble calls are collected, the OMS predicts the actuated protective device(s) and faulted line section(s). However, it may take hours to collect a sufficient number of trouble calls. This trouble call handling prolongs the identification of outage scenarios, leading to long outage durations for affected customers.

Automatic Meter Reading (AMR) technology gives another way to access customer information [48,49]. When an outage occurs, AMR meters are polled by distribution operators. Their responses are used to determine if the corresponding customers experience a power outage. An algorithm to poll AMR meters strategically is proposed in [48]. Data from AMR meters, trouble calls, and SCADA systems are put together for accurate outage management in [5]. Regarding the low quality of AMR meter data, an intelligent data filter is proposed in [49]. Generally, AMR is considered a supplement to trouble calls. Meter polling helps reduce the time to collect evidence for identification of the outage scenario, but does not significantly enhance the intelligence of an OMS. Moreover, outage management methods based on trouble call handling and AMR meter polling have limited capabilities in handling complex outage scenarios with multiple faults and protection miscoordination.

4.2. Enhanced Outage Management Based on TCD

Outage reports from smart meters provide an efficient way to identify outage areas [50]. When a smart meter experiences the loss of power, an outage report will be sent automatically to the distribution operating center. Since service can be restored to customers by automatic reclosing of protective devices after a temporary fault, distribution operators are more interested in using smart meter outage reports for permanent fault scenarios. Thus, smart meters are set to wait for a time period, in seconds, to send outage reports after experiencing the loss of power [51]. The clock embedded within smart meters records the time when outage reports are sent. The time-stamped reports from smart meters can be used for efficient and accurate outage management. In practice, a distribution system may experience multiple faults. Additionally, protection miscoordination and missing outage reports from smart meters may occur. Therefore, accurate identification of outage scenarios is a challenging task.

In the authors' previous work [52], a multiple-hypothesis method incorporating smart meter data, is proposed to handle complex outage scenarios with multiple faults, failures of fault indicators, protection miscoordination, and missing outage reports from smart meters. Hypotheses are generated based on the hypothetical problems in an outage scenario, *i.e.*, the number of faults, fault indicator failures, and protection miscoordination pairs. For each hypothesis, an optimization model based on linear integer programming (IP) is used to determine the most credible outage scenario. Theoretically, infinite hypotheses are supported by the same evidence. By ignoring unlikely hypotheses, such as a scenario with a large number of faults simultaneously, a finite-size set of candidate hypotheses can be obtained. However, the number of candidate hypotheses can still be large due to various combinations of multiple factors mentioned above, leading to high computational complexity in determining the most credible hypothesis.

In this section, a new method for outage management, called TCD, is introduced. The distribution feeder shown in Figure 3 is used for illustration. Two automatic reclosers sectionalize the entire feeder into three segments. Four laterals supply a total of 14 customers. For each customer, a smart meter is installed. The smart meters and automatic reclosers are equipped with two-way communications. Suppose a permanent fault occurs at 9:10:13:000 a.m., as indicated by the bolt in Figure 3. Recloser R2 opens to isolate the fault. Customers served by laterals L2, L3, and L4 experience a power outage. Consequently, outage reports from smart meters SM4-SM14, and flags from reclosers R1 and R2 are transmitted to the distribution operating center. Due to failures of smart meter hardware, software, and/or firmware as well as communication problems, OMS may not receive outage reports from some smart meters that experience the loss of power.

A TCD can be constructed based on the topology of the feeder as well as locations of protective devices and smart meters. The TCD for the distribution feeder in Figure 3 is shown in Figure 4. White vertices represent protective devices while smart meters are indicated by grey nodes. It can be seen that a TCD is actually a graph-theoretic tree. The vertex representing the circuit breaker is the root. The vertices representing smart meters are the leaves. Other vertices represent protective devices on the feeder. The spatial connections (upstream and downstream relationship) of these devices are represented by edges in the TCD. For example, the vertex, representing recloser R2, is directly connected with fuses F2 and F3 by an edge as recloser R2 is right upstream fuses F2 and F3. A status, *i.e.*, 1 or 0, is assigned to each leaf. If the distribution operating center receives an outage report from a smart meter, the status of the corresponding vertex is 1. In contrast, a leaf with the status 0 indicates no outage report is received from the corresponding smart meter. Note that multiple smart meters downstream the same protective devices can be lumped into one vertex in the TCD.

Therefore, one leaf in the TCD may be mapped to two or more smart meters in the distribution system. For example, the leaf connected to vertex F1 represents three smart meters that do not report an outage. The attribute *t* of an edge entering a leaf indicates the outage reporting time of a smart meter. Using the Escalation method [5], the number of smart meter outage notifications correlated with each protective device is determined. Three indices, *i.e.*, N_{outage}, N_{meter}, and *Cred.* are associated with each vertex that represents a protective device. Another two indices, *i.e.*, *T* and ΔT, are defined for each edge connecting protective devices. These indices are defined as follows:

N_{outage}: number of smart meters downstream the protective device reporting a power outage;
N_{meter}: total number of smart meters downstream the protective device;
Cred.: percentage of downstream smart meters reporting an outage, $Cred. = N_{outage}/N_{meter}$;
T: time of occurrence of a fault;
ΔT: time window to select the smart meter notifications corresponding to an outage.

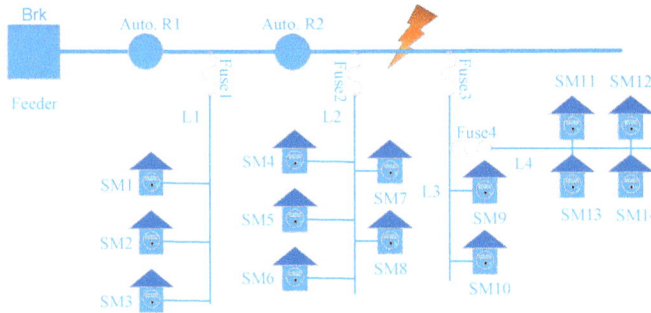

Figure 3. Configuration of a simple distribution system.

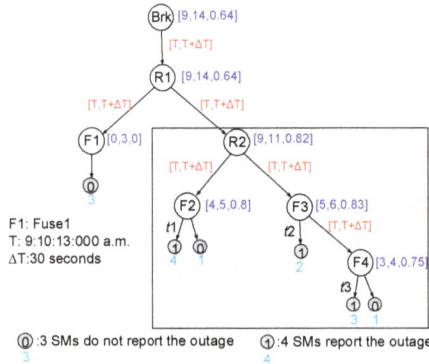

Figure 4. Temporal Causal Diagram (TCD) for outage management.

The timing attribute of the edges in the TCD establishes the temporal relationship of the outage scenario and evidence for outage management. The fault occurrence time can be obtained from the digital event recorder at substations with a resolution of half a cycle. For the given outage scenario, *T* is equal to 9:10:13:000 a.m. The time window, ΔT, defines the length of a time period used to filter outage reports from smart meters. Only outage reports with *t* falling in the range of *T* to *T* + ΔT are used in the TCD for fault scenario identification. In the example, ΔT is chosen to be 30 s. Assuming there is an outage report with time stamp *t* at 9:01:13:000 a.m., the outage report is abandoned as it is not in the range of 9:10:13:000 a.m. to 9:10:43:000 a.m.

The attributes N_{outage}, N_{meter}, and *Cred.* can be used to identify the activated protective device. Take recloser R2 as an example. Recloser R2 is correlated with nine outage reports ($N_{outage} = 9$) with a total of 11 smart meters downstream ($N_{meter} = 11$). The credibility of R2 is *Cred.* $= 9/11 = 0.82$. The attributes of other nodes are shown in Figure 4. It can be seen that breaker Brk and reclosers R1 and R2 are all correlated with the maximum number of outage reports. However, recloser R2 has the highest credibility of 0.82. Therefore, recloser R2 is identified as the activated protective device.

By assuming that only one fault occurs and all protective devices operate correctly, it can be concluded that recloser R2 opens to isolate the fault, and the fault location is downstream recloser R2 and upstream fuses F2 and F3. Meanwhile, with communication capabilities, the status of automatic reclosers as well as the current flowing through is transmitted to the distribution operating center. The data can help to validate the activated protective device identified from smart meter outage reports. In the example, recloser R2 reports an open status to the distribution operating center, which is consistent with the inference based on smart meter data. In addition, two missing outage reports are identified based on the fact that only nine out of 11 smart meters report an outage correctly.

Note that the TCD alone cannot identify the fault scenarios if protection miscoordination is considered. For example, if fuse F3 and recloser R2 are not coordinated, a fault downstream fuse F3 may cause recloser R2 to open. In this case, information from fault indicators will be useful. Study on integrating fault indicators with communication capabilities in the TCD is needed.

5. Distribution System Restoration

Distribution system restoration is intended to promptly restore as much load as possible in areas where electricity service is interrupted following an outage. It plays a critical role in SDSs. By operating normally open tie switches and normally closed sectionalizing switches, system topologies are altered in order to restore power supply to interrupted customers. A well-designed service restoration strategy can restore the maximum amount of load with a minimum number of switching operations [53]. Thus, the outage duration is shortened and system reliability is enhanced.

5.1. Service Restoration Procedure

Following an outage, the fault is located using the methods described in Section 4. Faulted zones are isolated by opening adjacent switches. Then the actuated breaker or recloser is reclosed to bring service back to the customers upstream the fault. The loads downstream the fault are picked up by other feeders, distributed generators, or microgrids through feeder reconfiguration. Constraints, such as limits on bus voltages and capacity of transformers, need to be evaluated. After the faulted component is repaired, switching actions will be taken to bring the distribution system back to its normal topology.

5.2. Service Restoration Algorithms

Distribution system service restoration is a multi-objective, mix-integer non-linear optimization problem with a number of constraints, including topological and operational constraints [54]. Due to its combinatorial nature, the service restoration problem is NP-hard. It is a challenge to develop an efficient algorithm to develop service restoration plans, especially for large-scale distribution systems with numerous components. Various methods have been proposed, including mathematical programming [55], heuristic search [56], expert systems [57], fuzzy logic [58], and multi-agent systems [59], to determine a final system configuration that restores the maximum amount of load.

DGs and microgrids in SDSs can serve as power resources to restore more loads when power from utilities is insufficient or unavailable. Optimal islanding strategies using DGs and micro-resources for service restoration are proposed in [60–64]. Methods, such as branch and bound [61], weighted graph [62], and layered directed tree model [63], are investigated to determine the islanding range.

In this section, the spanning tree search algorithm proposed in [53] is introduced. Spanning tree is a graph-theoretic term that refers to a connected graph containing all nodes in the distribution system without loops. A searching procedure is designed to identify the post-outage distribution system topology that will restore the maximum amount of load with a minimum number of switching operations while satisfying all constraints. The flow chart of the spanning tree search algorithm is shown in Figure 5.

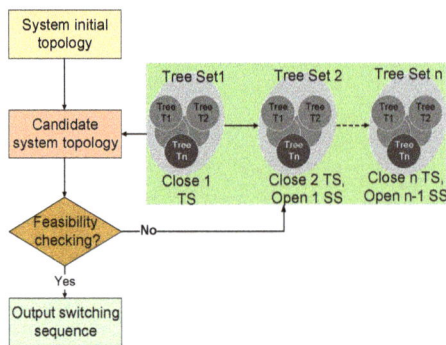

Figure 5. Flowchart of the spanning tree search algorithm.

A distribution network is formed by interconnected distribution feeders. The feeders are connected with each other through normally open tie switches. If a zone is modeled as a vertex and a switch is viewed as an edge, the distribution network can be represented as a connected graph. A spanning tree in the graph can represent the radial structure of a distribution network if all root nodes are lumped into one source node.

By operating a switch pair, *i.e.*, opening a normally closed sectionalizing switch and closing a normally open tie switch, the original spanning tree is transformed to another one that serves as a candidate topology for the post-outage network. The spanning tree search algorithm looks for all candidate network topologies iteratively. The search process starts from the original network topology. In the first iteration, it searches for candidate network topologies that can be obtained by operating one pair of switches. Unbalanced three-phase power flow calculations are performed to check the feasibility of candidate network topologies. If there is a radial network topology that satisfies all the operational constraints, output the result; otherwise, move to the next iteration. Candidate network topologies that can be obtained by operating two, three, or more switch pairs are identified and evaluated in the second, third, and subsequent iterations, respectively. If all potential network topologies are explored and there is no configuration that satisfies all operational constraints, the partial load restoration in the out-of-service area is selected. The minimum amount of load to relieve the overload is identified for the network topology with the least severe overloading condition. Based on the previous power flow calculation results, if the lowest node voltage in the system is defined as the minimum node voltage for this network topology, the network topology with the highest minimum node voltage is the least severe overload topology.

5.3. Test Case

A 4-feeder, 1069-bus system, as shown in Figure 6, is used as the test case. The test system is based on the Taxonomy "R3-12.47-2" developed by Pacific Northwest National Laboratory (PNNL) [53,65]. The voltage level of the unbalanced system is 12.47 kV. The real and reactive load on each feeder is 17.467 MW and 2.362 MVar, respectively. The four feeders are interconnected through seven normally open tie switches. Four microgrids are connected to the test system, as shown in Figure 6. The real power limits for microgrids 1–4 are 5.15, 1.65, 2.5, and 1.0 MW, respectively, and the reactive power limits are 2.25, 0.95, 1.75, 0.55 MVar, respectively.

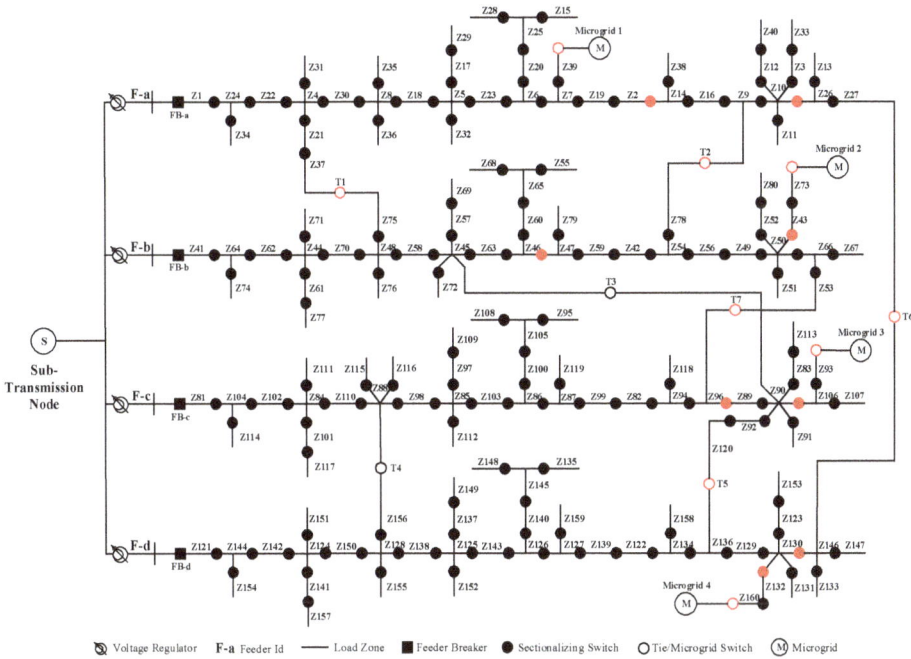

Figure 6. One-line diagram of the 4-feeder 1069-node test system.

Two scenarios are simulated. The spanning tree search algorithm is used to generate service restoration strategies for both scenarios. Restoration strategies with and without microgrids are compared. The results are summarized in Table 1. In scenario 1, a fault occurs at zone Z139. Without microgrids, a partial restoration strategy will be applied, with 315.04 kVA load at feeder F-b remaining interrupted.

Table 1. Service restoration strategies proposed by spanning tree search algorithm.

Index	Fault Location	Switching Operations without Microgrids	Switching Operations with Microgrids
1	Z139	Open: Z46–Z47, Z96–Z89 Close: Z136–Z120, Z53–Z96, Z45–Z90 Partial Restoration, 315.04 kVA load should be shed at F-b	Open: Z50–Z43, Z90–Z92 Close: Z45–Z90, Z73–Microgrid 2, Z136–Z120
2	Z23	Open: Z49–Z50, Z90–Z92 Close: Z78–Z9, Z53–Z96, Z136–Z120	Close: Z39–Microgrid 1

With the help of Microgrid 2, full restoration is achieved. In scenario 2, a fault is assumed to be located at zone Z23. With or without microgrids, all load in the outage area is successfully restored. However, by using Microgrid 1 for service restoration, only one switching operation is needed after fault isolation, while five switching operations are needed for the case without microgrids. In summary, microgrids enhance service restoration capability of a distribution system in two ways, restoring service to more load and reducing the number of switching operations.

6. Remote Control Capability in Smart Distribution Systems

With the ongoing development of SDSs, field devices with communication capabilities, such as RCSs, are installed in distribution systems. Two-way communications enable distribution operators to operate these devices remotely. As a result, system topologies can be altered quickly to minimize power losses or restore interrupted load.

6.1. Placement of Remote-Controlled Switches

Remote control capability is crucial for SDSs [66]. With more RCSs, distribution operators can perform service restoration faster. Since installation of RCSs is costly, the number of RCSs installed will be limited. On the other hand, critical switches that are most likely to be used for service restoration must be identified and upgraded. Therefore, the placement of RCSs must take into account the functional requirements and cost benefit [67]. Usually, switch placement is formulated as a constrained nonlinear mix-integer optimization problem [67–75]. Heuristics, such as fuzzy logic approach [68], genetic algorithm [69], and immune algorithm [70], are used to obtain a near-optimal solution with acceptable computational performance.

The RCS placement problem can be formulated for two scenarios. Most research is focused on optimal installation of RCSs in a distribution network without any switch deployed. In existing distribution systems, manual switches have already been installed. Utilities are interested in upgrading manual switches to RCSs to enhance service restoration capability. Given a limited budget, manual switches should be upgraded in such a way that the service restoration capability is maximized. A method based on weighted set cover is proposed in [76], which is intended to maximize restoration capability of an existing distribution system by upgrading a minimum number of manual switches to RCSs. The universal set of single-fault scenarios is considered. The concepts of load group and switch group are proposed to describe potential restoration schemes. With these concepts, the RCS placement is transformed into a weighted set cover problem. A greedy algorithm is then used to determine a near-optimal solution for the weighted set cover problem. Finally, the RCS placement plan is obtained by mapping the selected switch groups back to actual switches in the target distribution system.

6.2. Improve System Reliability with Remote Control Capability

System Average Interruption Frequency Index (SAIFI) and System Average Interruption Duration Index (SAIDI) are widely used by utilities for reliability assessment. SAIFI and SAIDI are defined by:

$$\text{SAIFI} = \frac{\text{total number of customer interruptions}}{\text{total number of customers served}} \tag{1}$$

$$\text{SAIDI} = \frac{\text{sum of all customers interruotion durations}}{\text{total number of customers served}} \tag{2}$$

According to the 2014 Electric System Reliability Annual Reports [77], SAIFI and SAIDI only take into account sustained power outages, which last for five minutes or longer. Without remote control capability, switches need to be operated manually by field crews, which may take minutes to hours for truck rolls and on-site switching operations. Compared with manual switches, RCSs are remotely operated by distribution operators within seconds. As a result, fast fault isolation and service restoration can be achieved. System reliability is improved with reduced outage duration.

The 4-feeder distribution system in Figure 7 is used to show the benefit from adding remote control capability into distribution systems. Some assumptions are given as follows.

(1)   the mean time to operate a manual switch is 90 min;
(2)   the mean time to operate a remote-controlled switch is 1 min;
(3)   the permanence failure rate of all zones is 0.02 per year;

(4)   the mean time to repair the damaged component is 4 h.

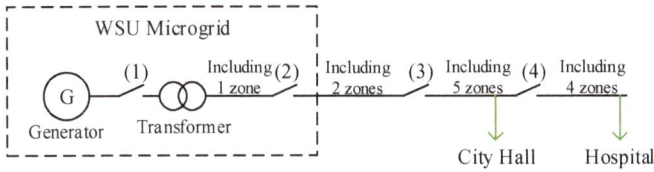

Figure 7. Restoration path from WSU generator to critical load in Pullman distribution system.

Suppose that the four feeder breakers are remote-controlled while other switches are manual operated. Using the method in [76], 17 switches are selected for upgrade, including eight sectionalizing switches, five tie switches, and four microgrid switches. The selected switches are indicated by red color in Figure 7. The results are summarized in Table 2.

Table 2. Switches selected for upgrade.

Tie Switches	T1, T2, T5, T6 and T7
Microgrid switches	Z39–Microgrid 1, Z73–Microgrid 2, Z93–Microgrid 3, and Z160–Microgrid 4
Sectionalizing switches	Z2–Z14, Z10–Z26, Z46–Z47, Z50–Z43, Z90–Z106, Z96–Z89, Z130–Z146 and Z130–Z132

Consider the universal set of single-fault scenarios. Service restoration schemes are obtained using the spanning tree search algorithm discussed in Section 5.2. For each scenario, the outage duration of each customer is recorded. SAIDI and SAIFI are calculated and shown in Table 3. Note that a power outage lasting shorter than 5 min is considered momentary and not counted in the calculation of SAIDI and SAIFI. It can be seen that by upgrading the selected switches to RCSs, SAIDI is improved by 75.70% and SAIFI 16.05%.

Table 3. Reliability improvement from installing RCSs.

Index	Without RCSs	With RCSs	Improvement
SAIDI (minute/year)	181.72	44.17	75.70%
SAIFI (/year)	0.7800	0.6548	16.05%

7. Distribution System Resilience with Respect to Extreme Events

7.1. Approaches to Resilient Distribution Systems

Extreme events, such as major hurricanes and earthquakes, have a great impact on distribution systems, resulting in extended outages to customers. Distribution system resilience with respect to extreme events is considered by the U.S. Department of Energy (DOE) as an essential characteristic of the future SDSs [78]. According to U.S. Presidential Policy Directive 21 [79], resiliency is "the ability to prepare for and adapt to changing conditions and withstand and recover rapidly from disruptions". For a distribution system, resiliency means the ability to avoid severe damages to the distribution infrastructure caused by extreme events and to restore interrupted loads efficiently after major outages [80]. Approaches to a resilient distribution system has been summarized into four categories in [80], *i.e.*, construction programs, maintenance measures, smart grid techniques, and other approaches. Smart grid techniques play an essential role in resilient distribution systems. In this section, a smart grid application, that is, distribution service restoration, and its contributions to the enhancement of resilience will be discussed.

After an extreme event, electric power from utilities may not be available due to damages on transmission lines or substations serving a distribution system. DGs and microgrids connected to the distribution system become valuable resources for service restoration to critical loads. In this paper, critical loads refer to those necessary for maintaining basic societal functions, such as hospitals and street lighting.

Utilization of DGs, energy storages, electric vehicles (EVs), and microgrids for service restoration has been studied [60,61,81–85]. In [61], a multi-stage restoration procedure is proposed to maximize the amount of load restored by DGs. In [60], a decentralized multi-agent system (MAS) is proposed to restore interrupted loads using DGs and EVs. The potential to use microgrids to serve critical load on neighboring distribution feeders when utility power is unavailable has been shown in [81,82]. Procedures for microgrid-assisted service restorations are proposed in [83,84], along with practical considerations. Microgrids are used as fast-ramping sources to restore a regional grid after natural disasters in [85].

7.2. Test Case

To evaluate the feasibility of using microgrids to serve critical loads in the distribution system after an extreme event, the Pullman-WSU distribution system [82,86] is used as a study case. South Pullman substation, where five Avista Utilities feeders start, and the portion of WSU microgrid served by these feeders are considered. Two critical loads are connected on the Avista Utilities feeders, *i.e.*, the Pullman Hospital and City Hall. Three DGs, one diesel generator and two natural gas generators, with a total capacity of 3.75 MW, are installed in the WSU microgrid. Under normal operating conditions, Avista supplies power to the WSU campus and Pullman Hospital and City Hall, while the WSU generators are operating as a backup.

Suppose that an extreme event occurs at the South Pullman substation. As a result, the 5 feeders served by the substation are out of electricity service. Moreover, no power source in the Avista system is available. Therefore, the three DGs in the WSU microgrid will be used to restore service to the Hospital and City Hall. A restoration path, as shown in Figure 7, is identified [86]. All non-critical load that is connected to the restoration path through a switching device is disconnected from the restoration path by opening the corresponding switching devices. It can be seen that there are 12 zones and four normally open tie switches on the restoration path. During the restoration process, these zones will be re-energized by closing tie switches in sequence. The City Hall and Hospital will be restored in the third and fourth steps, respectively.

Unbalanced three-phase power flow calculations are performed using the GridLAB-D software tool [87] after each switching action. The output power of the WSU generators and voltages at the critical loads are given in Tables 4 and 5 respectively. It can be seen that the generator output power does not exceed the maximum capability and the voltages at the critical loads are close to the nominal voltage during the restoration process.

Table 4. Output Power of WSU Generators.

Step	Active Power (kW)	Reactive Power (kVar)	Apparent Power (kVA)
1	137.8	45.44	145.1
2	429.6	142.7	452.7
3	621.1	436.6	795.1
4	1325	455.5	1401

Table 5. Voltages at the Critical Loads *.

Step	1		2		3		4	
Critical Load	Voltage (kV)	Voltage (p.u.)	Voltage (kV)	Voltage (p.u.)	Voltage (kV)	Voltage (p.u.)	Voltage (kV)	Voltage (p.u.)
City Hall	0	0	0	0	7.914	0.993	7.889	0.99
Hospital	0	0	0	0	0	0	7.873	0.988

* The nominal voltage is 7.967 kV.

8. Smart Distribution System Development around the World

To reap the benefit of SDSs, countries over the world are enhancing automation of distribution systems through deploying intelligent devices and smart grid technologies. Worldwide deployment of smart meters is expected to be around 830 million by 2020 [88]. An estimated total of 154.7 million smart meters will be deployed in Europe by 2017 [89]. In Asia, governments are making efforts to deploy smart meters for demand side management. Take South Korea as an example. South Korean government plans to install smart meters in half of the households across the country by 2016 and replace the remaining analog meters by 2020 [90]. The increased level of distribution automation helps to accommodate integration of DGs, e.g., PV modules and wind turbines, into distribution networks. In Japan, a smart grid program, named Eco-Model Cities, has been launched to develop next-generation energy and social systems using low carbon technology [90]. Real-time energy management systems for homes and buildings have been developed to integrate PV and enable demand response [90]. The potential benefits of small-scale DGs are recognized by the European energy industry [91,92]. The remaining of this section will be focused on the SDS development in Europe and active management of DGs in SDSs.

8.1. Smart Distribution System Development in Europe

In 2007, the Europe Council set up the 20:20:20 objective of reducing greenhouse gas (GHG) emissions by 20%, increasing the share of renewable energy to 20%, and making 20% improvements in energy efficiency by 2020 [90]. To meet this goal, two primary efforts are made by European countries, *i.e.*, the deployment of smart meters and integration of DGs into distribution networks. The European Union (EU) aims to deploy around 200 million smart meters for electricity and 45 million for gas by 2020 [93]. These smart meters will help save energy consumption. For example, Greece expects 5% energy saving by deploying and making proper use of smart meters [94]. Supported by the EU Research Framework Programmes (FPs) 5–7, contributions have been made in the corresponding projects to design and validate new system architectures and advanced components towards the future European electricity networks with a high penetration level of DGs [95]. For instance, a demonstration project, *Active Distribution Networks with Full Integration of Demand and Distributed Energy Resources*, is aimed to explore active demand management and integration of DGs into distribution networks [90].

8.2. Active Management of DGs in Smart Distribution Systems

Renewable-based DGs, such as PV modules, and energy storage systems (ESSs) are connected to distribution systems through power electronic converters and interface protection systems (IPSs). The power electronic converters can be used to control the real and reactive power of a DG or ESS, e.g., the maximum power point tracking (MPPT) for PV [96]. The IPS protects the DG from abnormal conditions, such as faults on the distribution feeders and unintentional islanding [97]. Traditionally, the IPS monitors voltage and current at the point of common coupling (PCC) of the DG and makes decisions with local information. Such a passive scheme can cause problems. For example, the DG may continue injecting power into the distribution feeder when an unintentional islanding or external fault condition occurs [98].

By introducing communication between DGs and the distribution system control center, active management of DGs can be achieved [99]. Distribution system operators can be aware of status of DGs in the distribution system by collecting information from the DG IPSs and send control commands to DG IPSs via the communication infrastructure. Consequently, a distribution system operator can remotely disconnect DGs in the network when an abnormal condition occurs. In addition, with the communication capability, the power electronic converters of DGs can also be remotely controlled to regulate voltage or reduce the net load seen by the utilities [98].

An economic and effective solution for active IPS of DGs is proposed in [98,100]. An IPS and a communication bridge are designed. The narrow-band (NB) power line communication (PLC) technology is used to build the communication links between DGs and distribution system control center at a low cost for installation and no cost to communication providers. The effectiveness NB-PLC-based solution has been evaluated by on-site experiments [98].

9. Conclusions

The worldwide development of SDSs are introduced, particularly the progress in the U.S. and Europe. Several state-of-the-art smart grid technologies are reviewed. Smart meters and their potential applications are summarized. For enhanced outage management, TCD incorporates outage reports from smart meters to accurately identify the fault location. Service restoration strategies are determined by the spanning tree search algorithm to restore as much load as possible after an outage. A novel method for placement of RCSs to enhance distribution system restoration capability is introduced. Improved system reliability from the installation of RCSs is reported. By using microgrids to serve critical loads, the resilience of a SDS with respect to extreme events is enhanced, which is illustrated by numerical simulations of a real distribution system. Technologies for active management of DGs are presented.

It is envisioned that more smart grid technologies will be adopted by utilities in the future, enabling an efficient, economic, reliable, and resilient distribution system. As the penetration of renewable energy sources (RES), such as PV modules, continues to increase and reaches a significant level, new technologies will be needed to deal with uncertainties induced by them. Battery energy storage systems and electric vehicles will play an important role in energy management of the future SDSs. Smart buildings and smart homes will enable flexible loads to participate in demand response programs. Market mechanisms should be designed to accommodate these new participants. Interdependency between electrical systems and other critical infrastructures, such as communication, transportation, and natural gas systems, will become more important in the context of resiliency. Research on the cooperation of multiple infrastructures needs to be conducted. Finally, as various kinds of communication networks are deployed in SDSs, cyber security and privacy of customers become an important issue that needs to be addressed. Technologies for cyber-attack detection and defense should be developed and implemented to protect utilities and customers from malicious intrusions.

Acknowledgments: The research is supported by U.S. National Science Foundation Grant CNS-1329666, "CPS: Synergy: Collaborative Research: Diagnostics and Prognostics Using Temporal Causal Models for Cyber Physical Systems-A case of Smart Electric Grid" and the Office of Electricity, U.S. Department of Energy (DoE) microgrid research and development (R&D) program through Pacific Northwest National Laboratory (PNNL).

Author Contributions: The work is done by Yazhou Jiang and Yin Xu under the supervision of Chen-Ching Liu.

Conflicts of Interest: The authors declare no conflict of interest.

References

1. U.S. Department of Energy. Economic Benefits of Increasing Electric Grid Resilience to Weather Outages. Available online: http://energy.gov/sites/prod/files/2013/08/f2/Grid%20Resiliency%20Report _FINAL.pdf (accessed on 30 August 2013).

2. The GridWise Alliance. Improving Electric Grid Reliability and Resilience: Lessons Learned from Superstorm Sandy and Other Extreme Events. Available online: http://www.gridwise.org/documents/ ImprovingElectricGridReliabilityandResilience_6_6_13webFINAL.pdf (accessed on 30 June 2013).

3. U.S. Department of Energy. Smart Grid Investments Improve Grid Reliability, Resilience, and Storm Responses. Available online: http://energy.gov/oe/downloads/smart-grid-investments-improve-grid-reliability-resilience-and-storm-responses-november (accessed on 31 November 2014).

4. Southern California Edison. Southern California Edison Smart Grid Strategy & Roadmap. Available online: https://www.smartgrid.gov/document/southern_california_edison_smart_grid_strategy_roadmap (accessed on 30 November 2010).

5. Liu, Y.; Schulz, N.N. Knowledge-Based System for Distribution System Outage Locating Using Comprehensive Information. *IEEE Trans. Power Syst.* **2002**, *17*, 451–456.

6. U.S. Department of Energy. Smart Grid Investment Grant Program. Progress Report II. Available online: https://www.smartgrid.gov/files/SGIG_progress_report_2013.pdf (accessed on 5 October 2013).

7. U.S. Department of Energy. Smart Grid System Report. Available online: http://energy.gov/sites/ prod/files/2014/08/f18/SmartGrid-SystemReport2014.pdf (accessed on 18 August 2014).

8. Simard, G.; Uluski, R.; Larry, G. From Today's Distribution System to Tomorrow's Smart Distribution. Available online: http://smartgrid.ieee.org/newsletters/january-2012/from-today-s-distribution-system-to-tomorrow-s-smart-distribution (accessed on 5 January 2012).

9. Teng, J.-H.; Huang, W.-H.; Luan, S.-W. Automatic and Fast Faulted Line-Section Location Method for Distribution Systems Based on Fault Indicators. *IEEE Trans. Power Syst.* **2014**, *29*, 1653–1662. [CrossRef]

10. Chowdhury, A.A.; Koval, O.D. *Power Distribution System Reliability: Practical Methods and Applications*; John Wiley & Sons Inc.: Hoboken, NJ, USA, 2009; pp. 100–156.

11. Larsen, P.; LaCommare, K.; Eto, J.; Sweeney, J. Assessing Changes in the Reliability of the U.S. Electric Power System. Available online: http://eetd.lbl.gov/sites/all/files/lbnl-188741.pdf (accessed on 3 August 2015).

12. Na, B.-N. KPECO & Distribution Automation. In Presented at Washington State University, 22 July 2014. Unpublished.

13. Yun, S.-Y.; Chu, C.-M.; Kwon, S.-C.; Song, I.-K.; Choi, J.-H. The Development and Empirical Evaluation of the Korean Smart Distribution Management System. *Energies* **2014**, *7*, 1332–1362. [CrossRef]

14. U.S. Department of Energy. Smart Grid Investment Grant Program. Progress Report. Available online: http://energy.gov/sites/prod/files/Smart%20Grid%20Investment%20Grant%20Program%20-%20Progress %20Report%20July%202012.pdf (accessed on 5 July 2012).

15. U.S. Department of Energy. Comparing the Impacts of Northeast Hurricanes on Energy Infrastructure. Available online: http://www.oe.netl.doe.gov/docs/Northeast%20Storm%20Comparison_FINAL_041513c. pdf (accessed on 30 April 2013).

16. U.S. National Hurricane Center. Tropical Cyclone Report Hurricane Irene. Available online: http://www. nhc.noaa.gov/data/tcr/AL092011_Irene.pdf (accessed on 14 December 2011).

17. Depuru, S.S.S.R.; Wang, L.; Devabhaktuni, V. Smart Meters for Power Grid: Challenges, Issues, Advantages and Status. *Renew. Sustain. Energy Rev.* **2011**, *15*, 2736–2742. [CrossRef]

18. Itron Inc. OpenWay CENTRON Meter. Available online: https://www.itron.com/na/PublishedContent/ OpenWay%20Centron%20Meter.pdf (accessed on 3 March 2016).

19. Sendin, A.; Arzuaga, T.; Urrutia, I.; Berganza, I.; Fernandez, A.; Marron, L.; Llano, A.; Arzuaga, A. Adaption of Powerline Communications-Based Smart Metering Deployments to the Requirements of Smart Grids. *Energies* **2015**, *8*, 13481–13507. [CrossRef]

20. Sendin, A.; Pena, I.; Angueira, P. Strategies for Power Line Communications Smart Metering Network Deployment. *Energies* **2014**, *7*, 2377–2420. [CrossRef]

21. Ye, X.; Zhao, J.; Zhang, Y.; Wen, F. Quantitative Vulnerability Assessment of Cyber Security for Distribution Automation Systems. *Energies* **2015**, *8*, 5266–5286. [CrossRef]

22. Cleveland, F.M. Cyber Security Issues for Advanced Metering Infrastructure. In Proceedings of the IEEE Power and Energy Society General Meeting-Conversion and Delivery of Electrical Energy in the 21st Century, Pittsburgh, PA, USA, 20–24 July 2008.

23. The National Institute of Technology and Standards (NIST). Framework for Improving Critical Infrastructure Cybersecurity version 1.0. Available online: http://www.nist.gov/cyberframework/upload/ cybersecurity-framework-021214.pdf (accessed on 12 February 2014).

24. The National Institute of Technology and Standards (NIST). Cybersecurity for Smart Grid Systems. Available online: http://www.nist.gov/el/smartgrid/cybersg.cfm (accessed on 12 February 2014).

25. Cardenas, A.A.; Berthier, R.; Bobba, R.B.; Huh, J.H.; Jetcheva, J.G.; Grochochi, D.; Sanders, W.H. A Framework for Evaluating Intrusion Detection Architectures in Advanced Metering Infrastructures. *IEEE Trans. Smart Gird* **2014**, *5*, 906–915. [CrossRef]

26. Liu, N.; Chen, J.; Zhu, L.; Zhang, J.; He, Y. A Key Management Scheme for Secure Communications of Advanced Metering Infrastructure in Smart Grid. *IEEE Trans. Ind. Electron.* **2013**, *60*, 4746–4756. [CrossRef]

27. Sankar, L.; Rajagopalan, S.R.; Mohajer, S.; Poor, H.V. Smart Meter Privacy: A Theoretical Framework. *IEEE Trans. Smart Grid* **2013**, *4*, 837–846. [CrossRef]

28. Berthier, R.; Sanders, W.H. Specification-based Intrusion Detection for Advanced Metering Infrastructures. In Proceedings of the 17th IEEE Pacific Rim International Symposium on Dependable Computing, Pasadena, CA, USA, 12–14 December 2011.

29. Electric Power Research Institute (EPRI). Radio-Frequency Exposure Levels from Smart Meters: A Case Study of One Model. Available online: http://www.epri.com/abstracts/Pages/ProductAbstract.aspx? ProductId=000000000001022270 (accessed on 2 February 2011).

30. California Council on Science and Technology. Health Impacts of Radio Frequency Exposure from Smart Meters. Available online: https://ccst.us/publications/2011/2011smart-final.pdf (accessed on 5 April 2011).

31. Federal Communications Commission. Radio Frequency Safety. Available online: https://www.fcc.gov/general/radio-frequency-safety-0 (accessed on 7 March 2013).

32. Avista Utilities. Avista Utilities's Pullman Smart Grid Project. Available online: https://www.smartgrid.gov/document/utility_scale_smart_meter_deployments_plans_proposals.html (accessed on 5 August 2011).

33. Wang, D.; Guan, X.; Liu, T.; Gu, Y.; Shen, C.; Xu, Z. Extended Distributed State Estimation: A Detection Method against Tolerable False Data Injection Attacks in Smart Grids. *Energies* **2014**, *7*, 1517–1538. [CrossRef]

34. Haughton, D.; Heydt, G. A Linear State Estimation Formulation for Smart Distribution Systems. *IEEE Trans. Power Syst.* **2013**, *28*, 1187–1195. [CrossRef]

35. Hu, Y.; Kuh, A.; Yang, T.; Kavcic, A. A Belief Propagation based Power Distribution System State Estimator. *IEEE Comput. Intell. Mag.* **2011**, *6*, 36–46. [CrossRef]

36. Boas Leite, J.; Sanches Mantovani, J.R. State Estimation of Distribution Networks through the Real-Time Measurements of the Smart Meters. In Proceedings of the 2013 IEEE Grenoble in PowerTech (POWERTECH), Grenoble, France, 16–20 June 2013.

37. Wu, J.; He, Y.; Jenkins, N. A Robust State Estimator for Medium Voltage Distribution Networks. *IEEE Trans. Power Syst.* **2013**, *28*, 1008–1016. [CrossRef]

38. Yun, S.-Y.; Hwang, P.-I.; Moon, S.-I.; Kwon, S.-C.; Song, I.-K.; Choi, J.-H. Development and Field Test of Voltage VAR Optimization in the Korean Smart Distribution Management System. *Energies* **2014**, *7*, 643–669. [CrossRef]

39. Kolenc, M.; Papic, I.; Blazic, B. Minimization of Losses in Smart Grids Using Coordinated Voltage Control. *Energies* **2012**, *5*, 3768–3787. [CrossRef]

40. Yang, S.; Zeng, D.; Ding, H.; Yao, J.; Wang, K.; Li, Y. Multi-Objective Demand Response Model Considering the Probabilistic Characteristic of Price Elastic Load. *Energies* **2014**, *9*. [CrossRef]

41. National Action Plan for Energy Efficiency. Coordination of Energy Efficiency and Demand Response. Prepared by Charles Goldman, 2010. Available online: https://emp.lbl.gov/sites/all/files/report-lbnl-3044e.pdf (accessed on 30 January 2010).

42. Federal Energy Regulatory Commission. Assessment of Demand Response & Advanced Metering Staff Report. Available online: http://www.ferc.gov/legal/staff-reports/demand-response.pdf (accessed on 17 August 2006).

43. Severin, B.; Michael, J.; Arthur, R. Dynamic Pricing, Advanced Metering and Demand Response in Electricity Markets. Available online: http://sites.energetics.com/MADRI/toolbox/pdfs/vision/dynamic_pricing.pdf (accessed on 25 October 2002).

44. Kang, C.; Jia, W. Transition of Tariff Structure and Distribution Pricing in China. In Proceedings of the IEEE Power and Energy Society General Meeting, San Diego, CA, USA, 24–29 July 2011.

45. Cappers, P.; Goldman, C.; Kathan, D. Demand Response in U.S. Electricity Markets: Empirical Evidence. *Energy* **2010**, *35*, 1526–1535. [CrossRef]

46. Gajowniczek, K.; Zabkowski, T. Data Mining Techniques for Detecting Household Characteristics Based on Smart Meter Data. *Energies* **2015**, *8*, 7407–7427. [CrossRef]
47. Quilumba, F.; Lee, W.-J.; Huang, H.; Yang, D.; Szabados, R. Using Smart Meter Data to Improve the Accuracy of Intraday Load Forecasting Considering Customer Behavior Similarities. *IEEE Trans. Smart Grid* **2015**, *6*, 911–918. [CrossRef]
48. Fischer, R.A.; Laakonen, A.S.; Schulz, N.N. A General Polling Algorithm Using a Wireless AMR System for Restoration Confirmation. *IEEE Trans. Power Syst.* **2001**, *16*, 312–316. [CrossRef]
49. Sridharan, K.; Schulz, N.N. Outage Management through AMR Systems Using an Intelligent Data Filter. *IEEE Trans. Power Deliv.* **2001**, *16*, 669–675. [CrossRef]
50. Tram, H. Technical and Operation Considerations in Using Smart Metering for Outage Management. In Proceedings of the IEEE/PES Transmission and Distribution Conference and Exposition, Chicago, IL, USA, 21–24 April 2008.
51. Itron Inc. Managing Power Outage with OpenWay. Available online: https://www.itron.com/Published Content/Managing%20Power%20Outage%20with%20OpenWay.pdf (accessed on 5 December 2010).
52. Jiang, Y.; Liu, C.-C.; Diedesch, M.; Lee, E.; Srivastava, A. Outage Management for Distribution Systems Incorporating Information from Smart Meters. *IEEE Trans. Power Syst.* **2015**. [CrossRef]
53. Li, J.; Ma, X.-Y.; Liu, C.C.; Schneider, K. Distribution System Restoration with Microgrids Using Spanning Tree Search. *IEEE Trans. Power Syst.* **2014**, *29*, 3021–3029. [CrossRef]
54. Lim, S.-I.; Lee, S.-J.; Choi, M.-S.; Lim, D.-J.; Ha, B.-N. Service Restoration Methodology for Multiple Fault Case in Distribution Systems. *IEEE Trans. Power Syst.* **2006**, *21*, 1638–1644. [CrossRef]
55. Khushalani, S.; Solanki, J.M.; Schulz, N.N. Optimized Restoration of Unbalanced Distribution Systems. *IEEE Trans. Power Syst.* **2007**, *22*, 624–630. [CrossRef]
56. Morelato, A.L.; Monticelli, A. Heuristic Search Approach to Distribution System Restoration. *IEEE Trans. Power Deliv.* **1989**, *4*, 2235–2241. [CrossRef]
57. Chen, C.-S.; Lin, C.-H.; Tsai, H.-Y. A Rule-Based Expert System with Colored Petri Net Models for Distribution System Service Restoration. *IEEE Trans. Power Syst.* **2002**, *17*, 1073–1080. [CrossRef]
58. Lee, S.-J.; Lim, S.-I.; Ahn, B.-S. Service Restoration of Primary Distribution Systems Based on Fuzzy Evaluation of Multi-criteria. *IEEE Trans. Power Syst.* **1998**, *13*, 1156–1163.
59. Solanki, J.M.; Khushalani, S.; Schulz, N.N. A Multi-agent Solution to Distribution Systems Restoration. *IEEE Trans. Power Syst.* **2007**, *22*, 1026–1034. [CrossRef]
60. Sharma, A.; Srinivasan, D.; Trivedi, A. A Decentralized Multiagent System Approach for Service Restoration Using DG Islanding. *IEEE Trans. Smart Grid* **2015**, *6*, 2784–2793. [CrossRef]
61. Pham, T.T.H.; Besanger, Y.; Hadjsaid, N. New Challenges in Power System Restoration with Large Scale of Dispersed Generation Insertion. *IEEE Trans. Power Syst.* **2009**, *24*, 398–406. [CrossRef]
62. Feng, X.; Liang, Y.; Guo, B. A New Islanding Method for Distributed Generation and Its Application in Power System Restoration. In Proceedings of the International Conference on Advanced Power System Automation and Protection, Beijing, China, 16–20 October 2011.
63. Su, J.; Bai, H.; Zhang, P.; Liu, H.; Miao, S. Intentional Islanding Algorithm for Distribution Network Based on Layered Directed Tree Model. *Energies* **2016**, *9*. [CrossRef]
64. Colmenar-Santos, A.; Palacio, C.D.; Enriquez-Garcia, L.A.; Lopez-Rey, A. A Methodology for Assessing Islanding of Microgrids: Between Utility Dependence and Off-Grid Systems. *Energies* **2015**, *8*, 4436–4454.
65. Schneider, K.P.; Chen, Y.; Engle, D.; Chassin, D. A Taxonomy of North American Radial Distribution Feeders. In Proceedings of the IEEE & Energy Society General Meeting, Calgary, AB, Canada, 26–30 July 2009.
66. U.S. Department of Energy. Reliability Improvements from the Application of Distribution Automation Technologies—Initial Results. Available online: http://energy.gov/sites/prod/files/DistributionReliability Report_Dec2012Final.pdf/ (accessed on 30 December 2012).
67. Moradi, A.; Fotuhi-Firuzabad, M.; Rashidi-Nejad, M. A Reliability Cost/Worth Approach to Determine Optimum Switching Placement in Distribution Systems. In Proceedings of the IEEE/PES Transmission and Distribution Conference & Exhibition: Asia and Pacific, Dalian, China, 15–18 August 2005.
68. Miranda, V. Using Fuzzy Reliability in a Decision Aid Environment for Establishing Interconnection and Switching Location Policies. In Proceedings of the International Conference on Electricity Distribution, Liège, Belgium, 25–28 April 1991.

69. Assis, L.S.D.; Gonzalez, J.F.V.; Usberti, F.L.; Lyra, C.; Zuben, F.V. Optimal Allocation of Remote Controlled Switches in Radial Distribution Systems. In Proceedings of the IEEE Power and Energy Society General Meeting, San Diego, CA, USA, 22–26 July 2012.

70. Chen, C.-S.; Lin, C.-H.; Chuang, H.-J.; Li, C.-S.; Huang, M.-Y.; Huang, C.-W. Optimal Placement of Line Switches for Distribution Automation Systems Using Immune Algorithm. *IEEE Trans. Power Syst.* **2006**, *21*, 1209–1217. [CrossRef]

71. Carvalho, P.M.S.; Ferreira, L.A.F.M.; Silva, A.J.C.D. A Decomposition Approach to Optimal Remote Controlled Switch Allocation in Distribution Systems. *IEEE Trans. Power Deliv.* **2005**, *20*, 1031–1036. [CrossRef]

72. Bernardon, D.P.; Sperandio, M.; Garcia, V.J.; Canda, L.N.; Da R. Abaide, A.; Daza, E.F.B. AHP Decision-Making Algorithm to Allocate Remotely Controlled Switches in Distribution Networks. *IEEE Trans. Power Deliv.* **2011**, *26*, 1884–1892. [CrossRef]

73. Bezerra, J.R.; Barroso, G.C.; Leao, R.P.S. Switch Placement Algorithm for Reducing Customers Outage Impacts on Radial Distribution Networks. In Proceedings of the IEEE Region 10 Conference, Cebu, Philippines, 19–22 November 2012.

74. Moradi, A.; Fotuhi-firuzabad, M. Optimal Switch Placement in Distribution Systems Using Trinary Particle Swarm Optimization Algorithm. *IEEE Trans. Power Deliv.* **2008**, *23*, 271–279. [CrossRef]

75. Lim, I.; Sidhu, T.S.; Choi, M.S.; Lee, S.J.; Ha, B.N. An Optimal Composition and Placement of Automatic Switches in DAS. *IEEE Trans. Power Deliv.* **2013**, *28*, 1474–1482. [CrossRef]

76. Xu, Y.; Liu, C.-C.; Schneider, K.; Ton, D. Placement of Remote-Controlled Switches to Enhance Distribution System Restoration Capability. *IEEE Trans. Power Syst.* **2016**, *31*, 1139–1149. [CrossRef]

77. California Public Utilities Commission. Electric System Reliability Annual Report. Available online: http://www.cpuc.ca.gov/General.aspx?id=4529 (accessed on 30 December 2014).

78. U.S. Department of Energy and National Energy Technology Laboratory. Operates Resiliently Against Attack and Natural Disaster. Available online: http://www.smartgridinformation.info/pdf/1438_doc_1.pdf (accessed on 29 September 2009).

79. Office of the Press Secretary of the White House. Presidential Policy Directive—Critical Infrastructure Security and Resilience. Available online: http://www.whitehouse.gov/the-press-office/2013/02/12/presidential-policy-directive-critical-infrastructure-security-and-resil (accessed on 12 February 2013).

80. Xu, Y.; Liu, C.-C.; Schneider, K.; Ton, D. Toward a Resilient Distribution System. In Proceedings of the IEEE Power and Energy Society General Meeting, Denver, CO, USA, 26–30 July 2015.

81. Wang, Y.; Chen, C.; Wang, J.; Baldick, R. Research on Resilience of Power Systems under Natural Disasters—A Review. *IEEE Trans. Power Syst.* **2015**, *31*, 1604–1613. [CrossRef]

82. Schneider, K.; Tuffner, F.K.; Elizondo, M.A.; Liu, C.C.; Xu, Y.; Ton, D. Evaluating the Feasibility to Use Microgrids as A Resilience Resource. *IEEE Trans. Smart Grid* **2016**. [CrossRef]

83. Mohagheghi, S.; Yang, F. Applications of Microgrids in Distribution System Service Restoration. In Proceedings of the IEEE PES Innovative Smart Grid Technologies, Anaheim, CA, USA, 17–19 January 2011.

84. Ansari, B.; Mohagheghi, S. Electric Service Restoration Using Microgrids. In Proceedings of the IEEE Power and Energy Society General Meeting, National Harbor, MD, USA, 27–31 July 2014.

85. Castillo, A. Microgrid Provision of Blackstart in Disaster Recovery for Power System Restoration. In Proceedings of the IEEE International Conference on Smart Grid Communications, Vancouver, BC, Canada, 21–24 October 2013.

86. Schneider, K. Microgrids as a Resiliency Resource. In Proceedings of the 2014 IEEE International Test Conference (ITC), Seattle, WA, USA, 20–23 October 2014.

87. U.S. Department of Energy at Pacific Northwest National Laboratory. GridLAB-D, Power Distribution Simulation Software. Available online: http://www.gridlabd.org/ (accessed on 21 December 2012).

88. Telefonica. The Smart Meter Revolution: Towards a Smart Future. Available online: https://m2m.telefonica.com/multimedia-resources/the-smart-meter-revolution-towards-a-smarter-future (accessed on 31 January 2014).

89. M2M Research Series, Smart Metering in Europe. Available online: http://www.berginsight.com/reportpdf/productsheet/bi-sm9-ps.pdf (accessed on 27 March 2016).

90. Global Smart Grid Federation. Global Smart Grid Federation Report. Available online: https://www.smartgrid.gov/files/Global_Smart_Grid_Federation_Report.pdf (accessed on 31 December 2012).

91. European Commission. European Distributed Energy Resources Projects. Available online: https://ec. europa.eu/research/energy/pdf/dis_energy_en.pdf (accessed on 31 December 2004).

92. VTT Technical Research Centre of Finland. Distributed Energy Systems. Available online: http://www. vtt.fi/inf/pdf/technology/2015/T224.pdf (accessed on 12 June 2015).

93. European Commission. Smart Grids and Meters. Available online: https://ec.europa.eu/energy/en/topics/ markets-and-consumers/smart-grids-and-meters (accessed on 27 March 2016).

94. Zgajewski, T. Smart Electricity Grids: A Very Slow Deployment in the EU. Available online: http://aei.pitt.edu/63582/ (accessed on 28 February 2015).

95. European Commission. European Electricity Projects. Available online: https://ec.europa.eu/research/ energy/pdf/synopses_electricity_en.pdf (accessed on 27 March 2007).

96. Esram, T.; Chapman, P.L. Comparison of Photovoltaic Array Maximum Power Point Tracking Techniques. *IEEE Trans. Energy Convers.* **2007**, *22*, 439–449. [CrossRef]

97. Delfanti, M.; Merlo, M.; Monfredini, G.; Olivieri, V. Coordination of interface protection systems for DG applications in MV distribution networks. In Proceedings of the International Conference and Exhibition on Electricity Distribution, Stockholm, Sweden, 10–13 June 2013.

98. Cataliotti, A.; Cosentino, V.; Di Cara, D.; Guaiana, S.; Panzavecchia, N.; Tinè, G. A New Solution for Low-Voltage Distributed Generation Interface Protection System. *IEEE Trans. Instrum. Meas.* **2015**, *64*, 2086–2095. [CrossRef]

99. Colmenar-Santos, A.; Reino-Rio, C.; Borge-Diez, D.; Collado-Fernández, E. Distributed Generation: A Review of Factors that can Contribute Most to Achieve a Scenario of DG Units Embedded in the New Distribution Networks. *Renew. Sustain. Energy Rev.* **2016**, *59*, 1130–1148. [CrossRef]

100. Cataliotti, A.; Cosentino, V.; Guaiana, S.; Di Cara, D.; Panzavecchia, N.; Tinè, G. An Interface Protection System with Power Line Communication for Distributed Generators Remote Control. In Proceedings of the IEEE International Workshop on Applied Measurements for Power Systems (AMPS), Aachen, Germany, 24–26 September 2014.

energies

MDPI

Article

A Time-Frequency Analysis Method for Low Frequency Oscillation Signals Using Resonance-Based Sparse Signal Decomposition and a Frequency Slice Wavelet Transform

Yan Zhao [1,2,*], Zhimin Li [1] and Yonghui Nie [3]

[1] Department of Electrical Engineering, Harbin Institute of Technology, Harbin 150001, China; lizhimin@hit.edu.cn

[2] Department of Power Transmission and Transformation Technology College, Northeast Dianli University, Jilin 132012, China

[3] Academic Administration Office, Northeast Dianli University, Jilin 132012, China; Yonghui_n@aliyun.com

* Correspondence: zjb_112006@163.com; Tel.: +86-451-8623-0549; Fax: +86-451-8641-3641

Academic Editor: Ying-Yi Hong
Received: 5 December 2015; Accepted: 24 February 2016; Published: 3 March 2016

Abstract: To more completely extract useful features from low frequency oscillation (LFO) signals, a time-frequency analysis method using resonance-based sparse signal decomposition (RSSD) and a frequency slice wavelet transform (FSWT) is proposed. FSWT can cut time-frequency areas freely, so that any band component feature can be extracted. It can analyze multiple aspects of the LFO signal, including determination of dominant mode, mode seperation and extraction, and 3D map expression. Combined with the Hilbert transform,the parameters of the LFO mode components can be identified. Furthermore, the noise in the LFO signal could reduce the frequency resolution of FSWT analysis, which may impact the accuracy of oscillation mode identification. Complex signals can be separated by predictable Q-factors using RSSD. The RSSD method can do well in LFO signal denoising. Firstly, the LFO signal is decomposed into a high-resonance component, a low-resonance component and a residual by RSSD. The LFO signal is the output of an underdamped system with high quality factor and high-resonance property at a specific frequency. The high-resonance component is the denoised LFO signal, and the residual contains most of the noise. Secondly, the high-resonance component is decomposed by FSWT and the full band of its time-frequency distribution are obtained. The 3D map expression and dominant mode of the LFO can be obtained. After that, due to its energy distribution, frequency slices are chosen to get accurate analysis of time-frequency features. Through reconstructing signals in characteristic frequency slices, separation and extraction of the LFO mode components is realized. Thirdly, high-accuracy detection for modal parameter identification is achieved by the Hilbert transform. Simulation and application examples prove the effectiveness of the proposed method.

Keywords: low-frequency oscillation; time-frequency analysis; resonance-based sparse signal decomposition; frequency slice wavelet transform; Hilbert transform

---

1. Introduction

Low frequency oscillations (LFOs) arising from the interconnection of local power systems has caused more and more concerns over the years. They may limit the power output of generators and transmission capability of tie lines, and in some serious circumstances may cause the system to collapse [1–6]. The LFO signal is a typical non-stationary signal, which demonstrates non-stationary signal characteristics such as an amplitude that changes over time and the random emergence of excitation modes, so it is not appropriate that its oscillation character is presented in the time domain or in the frequency domain alone.

Time-frequency analysis has been a hot spot in the area of signal processing research in recent years. It is a powerful tool for analyzing non-stationary signals. Yan [7–9] proposed a new time-frequency analysis method which he named frequency slice wavelet transform (FSWT). FSWT retains the advantages of both the short-time Fourier transform (STFT) and wavelet transform (WT). FSWT can very clearly represent the damping characteristics of multi-modal signals simultaneously in the time and frequency domains [7–9], which is very suitable for dealing with LFO signals. It can analyze the oscillation character from multiple angles, which includes the determination of dominant mode, mode separation and extraction, and the 3D map expression of the LFO signal. The noise in the LFO signal could reduce the frequency resolution of FSWT analysis, which may impact the accuracy of oscillation mode identification, so it is necessary to denoise the LFO signal. RSSD can do well in the LFO signal denoising. RSSD is a sparsity-enabled signal analysis method proposed by Selesnick [10]. It is a new nonlinear signal analysis method based on signal resonance rather than on frequency or scale, as provided by the Fourier and wavelet transforms. This method expresses a signal as the sum of a "high-resonance" and a "low-resonance" component. The high-resonance component is a signal consisting of multiple simultaneous sustained oscillations with high Q-factor. In contrast, the low-resonance component is a signal consisting of non-oscillatory transients of unspecified shape and duration with low Q-factor. The insufficient damping mechanism is one of many reasons for LFO. As a result, the LFO signal is an output of underdamped systems at a specific frequency. A system with high Q-factor is said to be underdamped. The LFO signal is a high-resonance component which consists of multiple simultaneous sustained oscillations. Thus, the LFO signal is decomposed into a high-resonance component, a low-resonance component and a residual by RSSD. The high-resonance component is the denoised signal. Then FSWT is used to decompose the high-resonance component and the full band of its time-frequency distribution can thus be obtained, along with the 3D map expression and dominant mode of the LFO. After that, due to the energy distribution, the frequency intervals are chosen to accurately analyze the time-frequency features. Through the reconstructed signals in the characteristic frequency slices, separation and extraction of LFO mode components are realized. Finally, high-accuracy detection for modal parameter identification is achieved by the Hilbert transform.

2. Resonance-Based Sparse Signal Decomposition

RSSD consists of two parts: a tunable Q-factor wavelet transform (TQWT) and morphological component analysis (MCA). TQWT is a discrete-time wavelet transform for which the Q-factor is easily specified. TQWT adaptively generates the basis functions of signal decomposition. MCA is applied to separate signals into a high-resonance component and a low-resonance component according to the nonlinear characteristics so that their best sparse signal representation is established.

2.1. Q factor, Damping and LFO's Resonance

2.1.1. Signal Resonance and Q-Factor

The quality factor or Q-factor is a dimensionless parameter that describes how underdamped an oscillator or resonator is, or equivalently, it characterizes a resonator's bandwidth relative to its center frequency. For high values of Q, the following definition is also mathematically accurate:

$$Q = \frac{f_c}{B_W} \tag{1}$$

where f_c is the resonant frequency, B_w is the half-power bandwidth. Figure 1 illustrates the concept of signal resonance. Pulses 1 and 3, essentially a single cycle in duration, are low-resonance pulses because they do not exhibit sustained oscillatory behavior. Pulses 2 and 4, whose oscillations are more sustained, are high-resonance pulses. A low Q-factor wavelet transform is suitable for the efficient representation of pulses 1 and 3. The efficient representation of Pulses 2 and 4 calls for a wavelet transform with higher Q-factor.

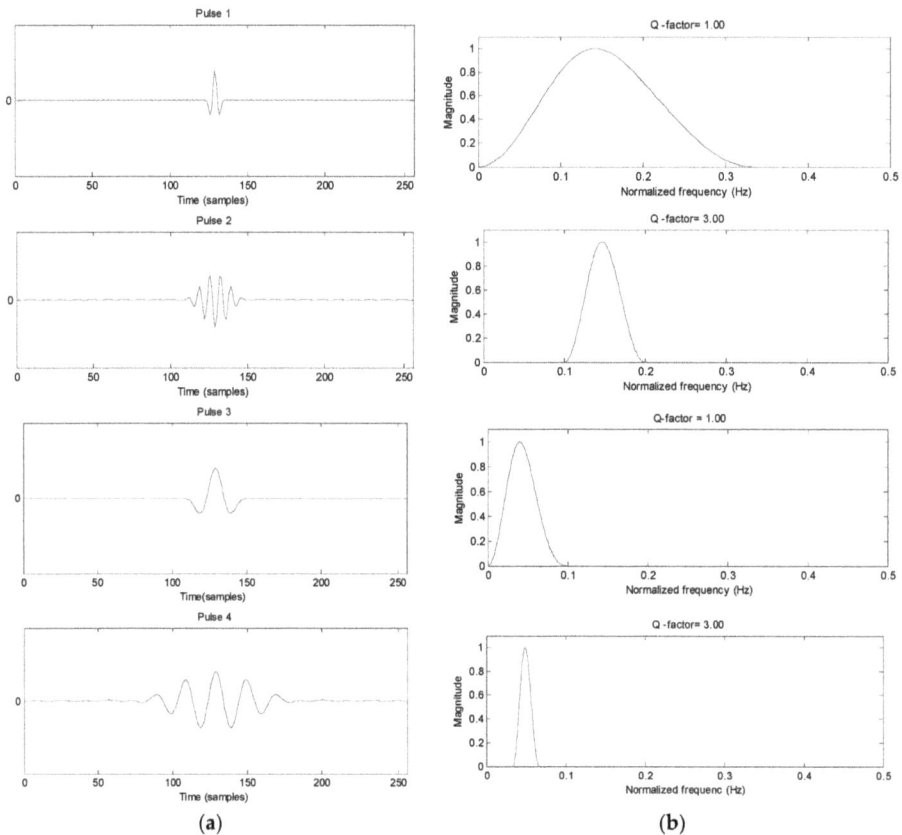

Figure 1. The resonance property of a signal. (**a**) Signals; (**b**) Frequency spectra.

2.1.2. LFO's Resonance

Higher Q indicates a lower rate of energy loss relative to the stored energy of the resonator. The oscillations die out more slowly. Resonators with high quality factors have low damping. A system with high quality factor is said to be underdamped. Underdamped systems combine oscillations at a specific frequency with a decay of the signal amplitude.

A system with low quality factor is said to be overdamped. Such a system doesn't oscillate at all, but when displaced from its equilibrium steady-state output it returns to it by exponential decay, approaching the steady state value asymptotically.

Figure 2 illustrates the resonance of the LFO signal, using Kunder's four machine two area system as an example [11]. The signal is the relative power angle swing curve of generator 4 to generator 1. Thus, the LFO signal is the output of underdamped systems at a specific frequency with high Q-factor and high-resonance properties.

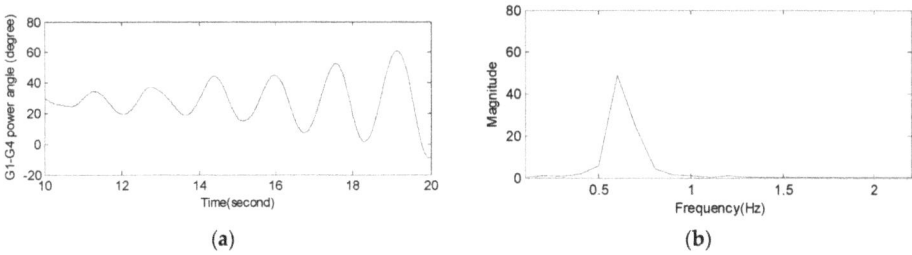

Figure 2. The resonance property of the LFO signal. (**a**) Power angle oscillation waveform of G4 relative to G1; (**b**) FFT frequency spectrum.

2.2. Tunable Q-Factor Wavelet Transform

TQWT is a flexible fully-discrete wavelet transform [12]. Hence, the transformation can be tuned according to the oscillatory behavior of the signal to which it is applied. The transformation is based on a real-valued scaling factor (dilation-factor) and implemented by a perfect reconstruction over-sampled filter bank with real-valued sampling factors [12].

In Figure 3, the sub-band signal $v_0(n)$ has a sampling rate of αf_s where f_s is the sampling rate of the input signal $x(n)$. Likewise, the sub-band signal $v_1(n)$ has a sampling rate of βf_s. LPS and HPS represent low-pass scaling and high-pass scaling, respectively.

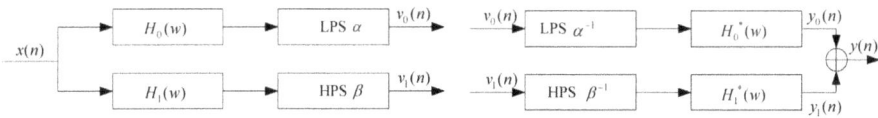

Figure 3. Analysis and synthesis filter banks for the TQWT.

The transfer functions of the blocks $H_0(\omega)$ and $H_1(\omega)$ are given by [12]:

$$H_0(\omega) = \begin{cases} 1 & |\omega| \leqslant (1-\beta)\pi \\ \theta(\dfrac{\omega + (\beta-1)\pi}{\alpha + \beta - 1}) & (1-\beta)\pi \leqslant |\omega| \leqslant \alpha\pi \\ 0 & \alpha\pi \leqslant |\omega| \leqslant \pi \end{cases} \tag{2}$$

$$H_1(\omega) = \begin{cases} 0 & |\omega| \leqslant (1-\beta)\pi \\ \theta(\dfrac{\alpha\pi - \omega}{\alpha + \beta - 1}) & (1-\beta)\pi \leqslant |\omega| \leqslant \alpha\pi \\ 1 & \alpha\pi \leqslant |\omega| \leqslant \pi \end{cases} \tag{3}$$

where $0 < \alpha < 1, 0 < \beta \leqslant 1, \alpha + \beta > 1, \theta(\omega) = 1/2(1 + \cos\omega)\sqrt{2 - \cos\omega}, |\omega| \leqslant \pi$. Scaling factors α and β can be expressed in terms of the Q-factor and redundancy r [12]: $\beta = 2/(Q+1), \alpha = 1 - \beta/r$.

The main parameters for the TQWT are the Q-factor Q, the redundancy r, and the number of stages (or levels) J. Generally, Q is a measure of the number of oscillations the wavelet exhibits. A high Q-factor wavelet transformation requires more levels to cover the same frequency range compared with a low Q-factor transformation. We use $Q = 1$ for non-oscillatory signals. A higher value of Q is appropriate for oscillatory signals. The parameter r is the redundancy of the TQWT when it is computed by infinitely many levels. Generally, a value of $r = 3$ is recommended. Meanwhile, it can be interpreted as a measure of how much spectral overlap exists between adjacent band-pass filters. Increasing r has the effect on increasing overlapping in the frequency domain of the band-pass filters constituting the TQWT. The parameter J denotes the number of filter banks. There will be $J + 1$ sub-bands: the high-pass filter output signal of each filter bank, and the low-pass filter output signal of the final filter bank.

2.3. Morphological Component Analysis

Sparse signal processing exploits sparse representations of signals for applications such as denoising, deconvolution, signal separation, classification, *etc.* Therefore, the transformations for the sparse representation of signals is the key ingredient for sparse signal processing. In the noisy case, we should not ask for exact equality. Given an observed signal:

$$x = x_1 + x_2 + n, \text{ with } x, x_1, x_2 \in R^N \tag{4}$$

where n is noise. The goal of Morphological Component Analysis (MCA) is to estimate/determine x_1 and x_2 individually. Assuming that x_1 and x_2 can could be sparsely represented in bases (or frames) Φ_1 and Φ_2 respectively, they can be estimated by minimizing the objective function:

$$J(w_1, w_2) = \lambda_1 \|w_1\|_1 + \lambda_2 \|w_2\|_1 \tag{5}$$

respect to w_1 and w_2, subject to the constraint:

$$\Phi_1 w_1 + \Phi_2 w_2 = x \tag{6}$$

Then MCA provides the estimates:

$$\hat{x}_1 = \Phi_1 w_1 \tag{7}$$

and:

$$\hat{x}_2 = \Phi_2 w_2 \tag{8}$$

We apply a variant of the split augmented Lagrangian shrinkage algorithm (SALSA) [13] to solve the MCA problem with iterative algorithm.

3. Frequency Slice Wavelet Transform

3.1. The Definition of Frequency Slice Wavelet Transform

Suppose $\hat{p}(w)$ is the Fourier transformation of the function $p(t)$. For $f(t) \in L^2(R)$, the FSWT is defined directly in the frequency domain as:

$$W(t, w, \lambda, \sigma) = \frac{1}{2\pi} \lambda \int_{-\infty}^{+\infty} \hat{f}(u) \hat{p}(\frac{u - w}{\sigma}) e^{iut} du \tag{9}$$

where the scale $\sigma \neq 0$ and energy coefficient $\lambda \neq 0$ are constants or are functions of w and t. The star "*" means the conjugate of a function. The general wavelet is a "microscope" in the time domain, but here FSWT is a "microscope" in the frequency domain, and this transformation is also named the wavelet transform in frequency domain [7,8].

$\hat{p}(w)$ also called a frequency slice function (FSF) [7]. By using the Parseval equation, if σ is not the function of the assessed frequency u then Equation (9) can be translated into its time domain [8]:

$$W(t, w, \lambda, \sigma) = \sigma \lambda e^{iwt} \int_{-\infty}^{+\infty} f(\tau) e^{-iw\tau} p^*(\sigma(\tau - t)) d\tau \tag{10}$$

We are not concerned with the definition of Equation (10) in the time domain because it is not easy to analyze in the frequency domain, even if we know the function type of $p(t)$ or its $\hat{p}(w)$. Hence, we pay more attention to the function $\hat{p}(w)$. Here, $\hat{p}(w)$ is a frequency slice function (FSF), also called a frequency modulated signal or a filter. According to the Morlet transformation idea, taking $\sigma = (\omega/\kappa)$ $(\kappa > 0)$, Equation (8) can be changed into:

$$W(t, \omega, \lambda, \kappa) = \frac{1}{2\pi}\lambda\int_{-\infty}^{+\infty}\hat{f}(u)\hat{p}^*(\kappa\frac{u-\omega}{\sigma})e^{iut}du \tag{11}$$

Note that the parameter κ in Equation (11) is a unique parameter that should be chosen in the application:

$$\kappa = \frac{\Delta\omega_p}{\eta_s} \tag{12}$$

where η_s is frequency resolution ratio. $\Delta\omega_p$ is the width of frequency window of FSF. As stated above in Equation (11), FSWT has another important property: FSWT can be controlled by the frequency resolution ratio η_s of the measured signal.

3.2. Frequency Slice Wavelet Inverse Transform

If $\hat{p}(\omega)$ satisfies $\hat{p}(0) = 1$, then the original signal $f(t)$ can be reconstructed by:

$$f(t) = \frac{1}{2\pi\lambda}\int_{-\infty}^{+\infty}\int_{-\infty}^{+\infty}W(\tau, \omega, \lambda, \sigma)e^{-i\omega(t-\tau)}d\tau d\sigma \tag{13}$$

As an important result of Equation (10), the reconstruction procedure is independent from the selected FSF. Reconstruction independency is an important feature in FSWT, but this characteristic is not allowable in traditional wavelets.

Particularly, the signal component in the time-frequency area $(t_1, t_2, \omega_1, \omega_2)$ is as below[14]:

$$f_x(t) = \frac{1}{2\pi\lambda}\int_{\omega_1}^{\omega_2}\int_{t_1}^{t_2}W(\tau, \omega, \lambda, \sigma)e^{-i\omega(t-\tau)}d\tau d\sigma \tag{14}$$

FSWT provides a new approach whereby both filtering and segmenting can be processed simultaneously in the time and frequency domain.

4. Procedure of the LFO Time-Frequency Analysis Method Using RSSD and FSWT

Figure 4 is the flow chart of the proposed method.

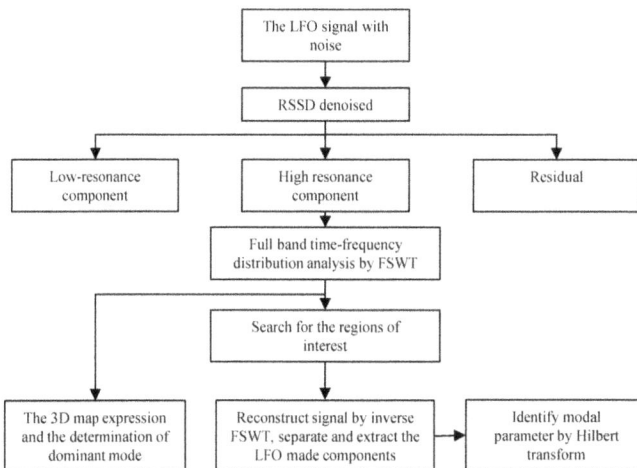

Figure 4. The flow chart of the LFO time-frequency analysis method using RSSD and FSWT.

In summary, the method procedure is divided into the following steps:

(1) Decompose the LFO signal with noise by RSSD and get a high-resonance component, a low-resonance component and a residual. The high-resonance component is de-noised LFO signal.

(2) Transform the high-resonance component by FSWT, and compute the FSWT coefficients using Equation (9), and obtain the full band of its time-frequency distribution. The 3D map expression and dominant mode of the LFO signal can be obtained. After that, search for the regions of interest, called modal domains, in which the main energy is concentrated. Chose a frequency slice $[w_i, w_i+1]$ to get an accurate analysis of the time-frequency feature. Separate and extract the LFO mode components through reconstructing signals by inverse FSWT.

(3) Identify the parameters of the LFO mode components by Hilbert transform (HT). With the HT method, the LFO mode component's oscillation frequency and damping ratio are obtained. Readers may refer to [15,16] for the details of the HT method.

5. Simulation

5.1. Denoised RSSD

5.1.1. Example 1

The LFO signal can be approximated by:

$$x(t) = \sum_{i=1}^{n} A_i e^{\sigma_i t} \cos(2\pi f_i t + \phi_i) + \lambda(t) \tag{15}$$

where A_i is the amplitude, f_i is the oscillation frequency, σ_i is damping factor, ϕ_i is the phase shift, $\lambda(t)$ is white noise and σ its standard deviation. We set a $f_s = 1000$ Hz, $T_s = 10$ s and $\sigma = 0.1$. A signal simulated with Equation (15) is described in Table 1. There are two oscillation modes of the simulated signal. One is an inter-area oscillation mode in low frequency and the other is an intra-area oscillation mode in high frequency.

Table 1. The parameters of the simulated signal in Equation (15).

LFO Mode	Amplitude (A_i)	Oscillation Frequency (f_i)	Damping Ratio (ξ_i)	Damping Factor (σ_i)	Phase Shift (Φ_i)
1	1	0.5	0.02	0.0628	60°
2	0.5	1.1	0.07	0.4837	45°

When the RSSD is applied to the simulated signal with Gaussian white noise, we get decomposition of the signal into high and a low-resonance components as seen in Figure 5. We use the parameters $Q_1 =1$, $Q_2 = 4$ and $\gamma_1 = \gamma_2 = 3$.

It can be seen from Figure 5 that this procedure separates the given signal into three signals that have quite different behavior. One signal (high-resonance component) is sparsely represented by a high Q-factor wavelet transform (Q = 4). The second signal (low-resonance component) is sparsely represented by a low Q-factor wavelet transform (Q = 1). The third signal (residual) is the signal component which is not sparsely represented in either the high Q-factor wavelet transformation or the low Q-factor wavelet transformation.

Note that the high-resonance component consists largely of sustained oscillatory behavior, and that is the denoised LFO signal, while the low-resonance component (in Figure 5, part (b)) looks like nearly zero, and almost no low-resonance component is there. The residual appears noise-like. The residual is mostly white Gaussian noise.

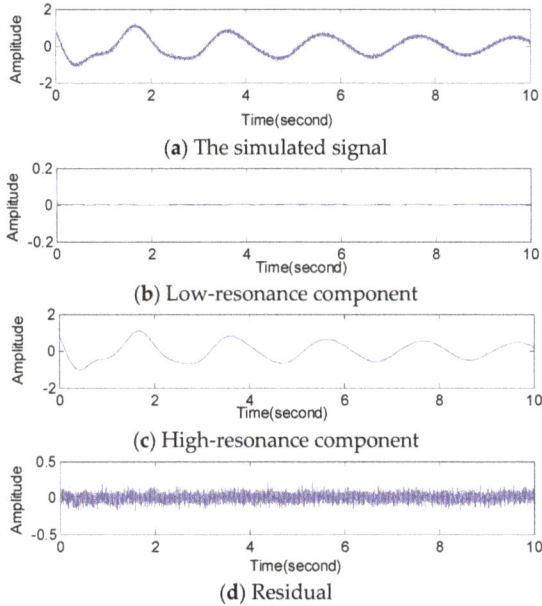

(a) The simulated signal

(b) Low-resonance component

(c) High-resonance component

(d) Residual

Figure 5. RSSD of the simulated signal with white Gaussian noise.

For quantifying the performance of RSSD de-noising, we calculated the root mean squared errors (RMSE) of the high-resonance signal and the ideal signal without noise .RMSE equals 0.6198. This proves that RSSD can remove white Gaussian noise well.

5.1.2. Example 2

Add $s(t)$ to the simulated signal of Example 1, and the LFO signal with transient impulse and noise is obtained as shown in Figure 6a. It's worth noting that the components overlap in frequency:

$$s(t) = \begin{cases} 0.1\sin 2\pi \times 0.2t & 2\,\text{s} \leqslant t \leqslant 2.04\,\text{s}, 6\,\text{s} \leqslant t \leqslant 6.04\,\text{s} \\ 0.19\sin 2\pi \times 0.5t & 4\,\text{s} \leqslant t \leqslant 4.08\,\text{s} \\ 0.08\sin 2\pi \times 0.4t & 8\,\text{s} \leqslant t \leqslant 8.06\,\text{s} \\ 0 & \text{other} \end{cases} \tag{16}$$

It can be seen from Figure 6, part (b), that the low-resonance component is separated by RSSD. The low-resonance component consists largely of transients and oscillations that are not sustained, and that is exactly the transient impulse signal. The residual is mostly white Gaussian noise. Therefore, RSSD can successfully separate complex signals whose frequency bands are overlapped but differ in resonance properties.

5.2. FSWT Analysis

5.2.1. Full Band Time-Frequency Distribution Analysis by FSWT

Let the FSF be the Gaussian function $\hat{p}(\omega) = e^{-\frac{1}{2}\omega^2}$, hence $\Delta\omega_p = \frac{\sqrt{2}}{2}$. Let $\kappa = \frac{0.707}{\eta}$ and set $\eta = 0.025$, hence $\kappa = 28.28$. The chosen parameters in FSWT are summarized as follows:

$$\hat{p}(\omega) = e^{-\frac{1}{2}\omega^2}, \ \sigma = \frac{\omega}{\kappa}, \ \eta = 0.025, \ \kappa = 28.28 \tag{17}$$

Considering the LFO's frequency range, the frequency interval is from 0 Hz to 5 Hz. The full band time-frequency distribution analysis by FSWT is shown in Figure 7. The Fourier spectrum of the high-resonance component in Figure 7a shows that the first modal signal is a clear indication of the greatest energy at 0.5 Hz and other peaks at 1.1 Hz are smaller. Figure 7b,c shows the results of the 2D or 3D map of FSWT coefficients, where all of FSWT parameters are assumed in Equation (17). The damping characteristics of the signal in time-frequency domain are more clearly revealed in Figure 7c. Figure 7 reveals that FSWT has two clearly separated peaks that indicate two main modes. The first mode has the highest amplitude at about 0.5 Hz, while the second mode and not the first has the highest amplitude at about 1.1 Hz, so the LFO dominant mode is an oscillation mode of frequency 0.5 Hz.

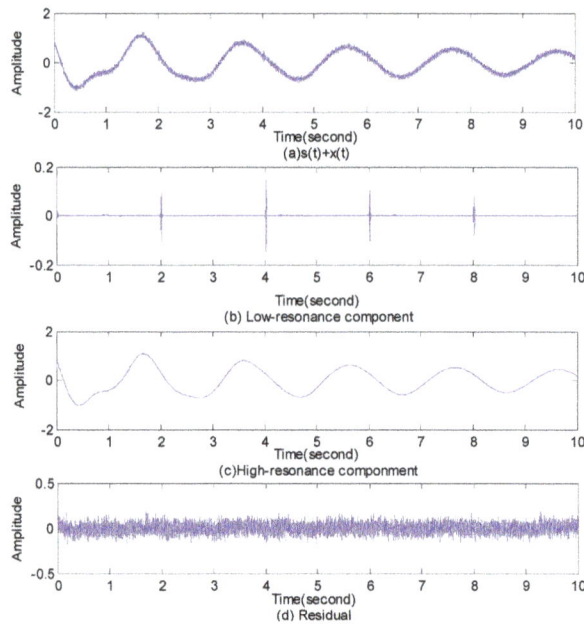

Figure 6. RSSD of the simulated signal with transient impulse and white noise.

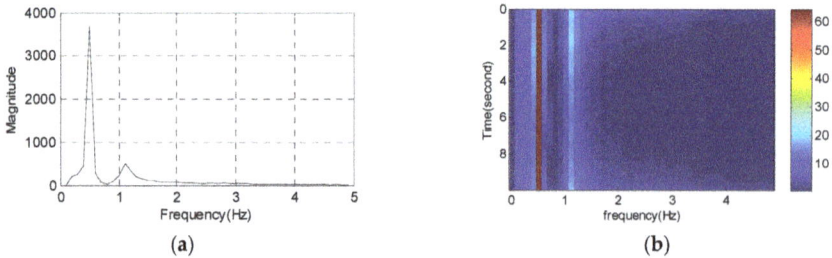

Figure 7. *Cont.*

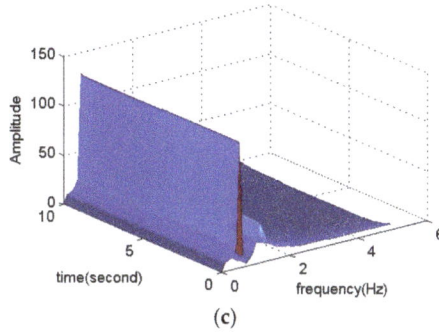

(c)

Figure 7. Full band time-frequency distribution analysis by FSWT. (**a**) Fourier spectrum; (**b**) The 2D maps of FSWT coefficients; (**c**) The 3D maps of FSWT coefficients.

5.2.2. Fine Analysis of Frequency Slices Based on FSWT

Accurate analysis of the time-frequency features is shown in Figures 8 and 9. According to the analysis results of Figure 7, the chosen frequency slices are [0, 0.8] Hz and [0.8, 2] Hz, respectively. Note that each modal signal on the 2D map of FSWT coefficients is a connected area. This is an important feature for modal signal separation. Here the segments are not strict, but it is necessary that each slice include the main energy of one modal signal. We can reconstruct the objective signal through Equation (14). The reconstructed signals shown in Figures 8c and 9c are good. There are only some differences in FSWT amplitudes. The end is not accurate reconstruction, because the reconstructed signal and the ideal signal do not perfectly match in the end, but this does not affect damping estimation, because the damping is only a ratio of the FSWT amplitudes.

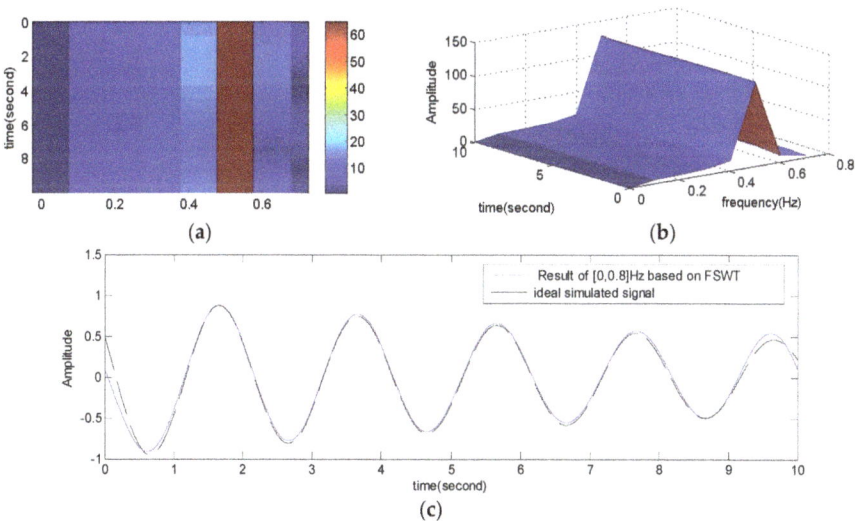

Figure 8. Fine analysis of frequency slice [0, 0.8] Hz based on FSWT. (**a**) FSWT 2D map; (**b**) FSWT 3D map; (**c**) Comparison of the reconstructed signal with the ideal signal.

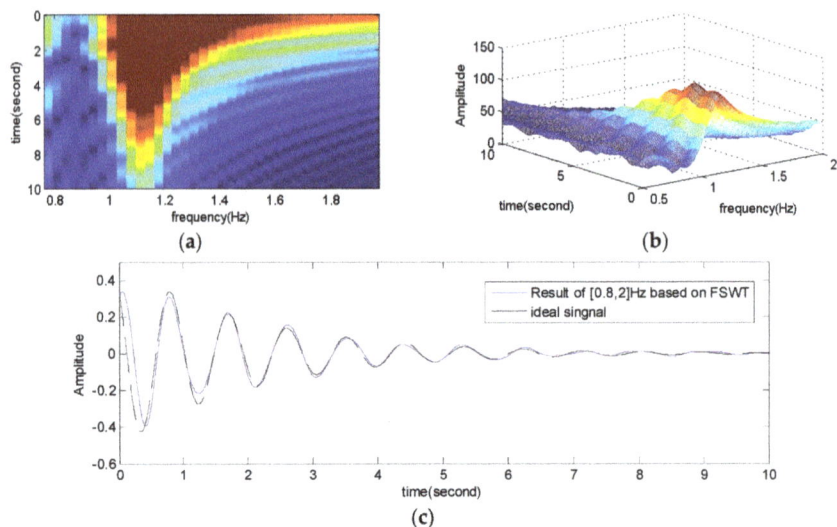

Figure 9. Fine analysis of frequency slice [0.8, 2] based on FSWT. (a) FSWT 2D map; (b) FSWT 3D map; (c) Comparison of the reconstructed signal with the ideal signal.

5.3. Identification of the Parameters of LFO Mode Components by HT

The reconstructed signals based on feature of frequency slice are the separated and extracted LFO mode components. We identify the parameters of mode components by the Hilbert transform. The results is shown in Table 2. The results shows that the identified results are relatively accurate and the error of the method is less than 5%.

Table 2. LFO mode components parameters after RSSD denoising ($\sigma = 0.1$).

Mode		True Value	FSWT-HT	
			Identification Result	Error (%)
1	Oscillation frequency (Hz)	0.5000	0.4980	0.40
	Damping ratio	0.0200	0.0196	2.00
2	Oscillation frequency (Hz)	1.1000	1.1101	0.92
	Damping ratio	0.0700	0.0666	4.86

5.4. Impact of Noise on the Accuracy of FSWT-HT

We change the noise level of Equation (15), and let the standard deviation (σ) equal 0.5, 1, 1.5 or 2, respectively. The full band time-frequency distribution analysis and fine analysis of the frequency-slice interval are shown in Figure 10.

As shown in Figure 10, noises do not affect the results of full band time-frequency analysis seriously. The time-frequency characteristics are not clear when the noise standard deviation $\sigma = 2$. This proves the FSWT method presents good performance against noise to a certain degree, but noise affects the amplitude of reconstructed signals, especially for the reconstructed signals of frequency from 0 to 0.8 Hz. Frequency slice equals a bandpass filter which removes other frequency interference, but the white noise of the frequency slice has not been completely removed, which may impact the accuracy of oscillation mode identification, especially for damping ratio. Results are shown in Table 3. Compared with Table 2, it is necessary to denoise the LFO signal by RSSD.

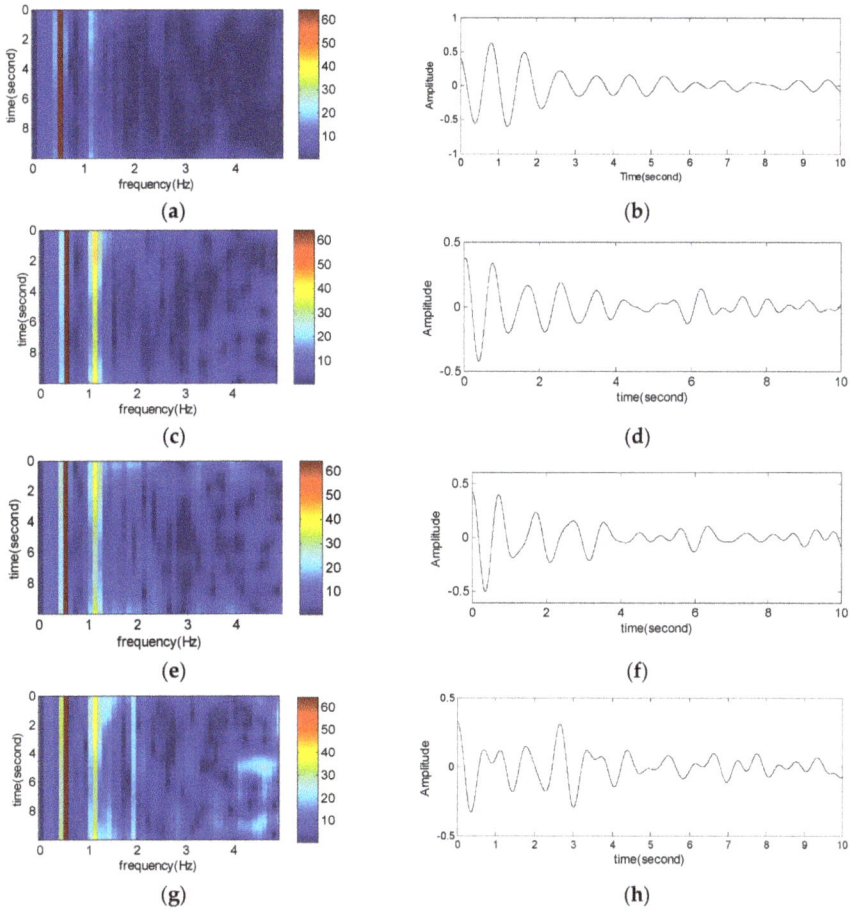

Figure 10. Full band time-frequency distribution and Fine analysis of frequency slice based on FSWT. (**a**) FSWT 2D map (σ = 0.5); (**b**) Reconstructed signal of [0, 0.8] Hz; (**c**) FSWT 2D map (σ = 1); (**d**) Reconstructed signal of [0, 0.8] Hz; (**e**) FSWT 2D map (σ = 1.5); (**f**) Reconstructed signal of [0, 0.8] Hz; (**g**) FSWT 2D map (σ = 2); (**h**) Reconstructed signal of [0, 0.8] Hz.

Table 3. Result of identification with white noise.

Mode	Standard Deviation (σ)	Oscillation Frequency (Hz)	Error (%)	Damping Ratio	Error (%)
1	0.1	0.4980	0.400	0.0188	6.000
2		1.1101	0.92	0.0657	6.143
1	0.5	0.4978	0.440	0.0190	5.000
2		1.1031	0.282	0.0637	9.000
1	1	0.4968	0.640	0.0181	9.500
2		1.1131	1.20	0.0613	12.429
1	1.5	0.4975	0.500	0.0171	14.500
2		1.1156	1.418	0.0571	18.428
1	2	0.4977	0.460	0.0170	15.000
2		1.1221	2.009	0.0470	32.857

5.5. Comparison with Low Pass Filtering

We design a Butterworth low-pass filter with less than 3 dB of ripple in the passband, defined frequency from 0 to 2.5 Hz, and at least 20 dB of attenuation in the stopband, defined from 150 Hz to the Nyquist frequency (500 Hz). Transfer function is:

$$H(z) = \frac{0.0487(1 + Z^{-1})}{1 - 0.9026Z^{-1}} \tag{18}$$

The frequency response of the filter is plotted that illustrates in Figure 11.

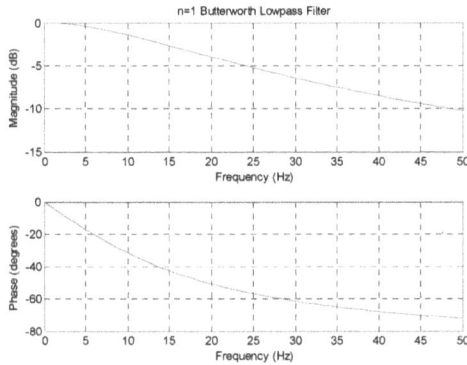

Figure 11. Frequency characteristics of a Butterworth low-pass filter.

After the simulated signal with white Gaussian noise of Equation (15) is processed through a low-pass filter, we obtain a signal with colored Gaussian noise, which can be seen in Figure 12. Choose frequency slices are [0, 0.8] Hz and [0.8, 2] Hz, respectively. The reconstructed signals are shown in Figure 13.

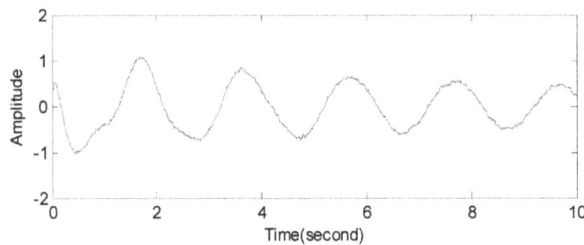

Figure 12. The simulated signal with colored noise after low-pass filtering.

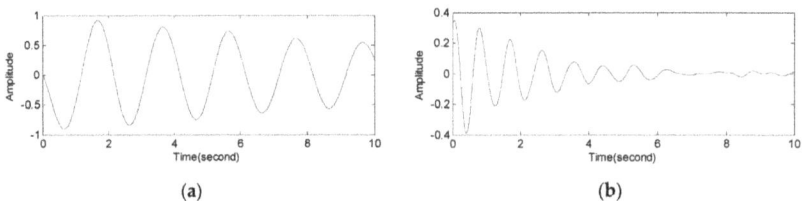

(a)   (b)

Figure 13. Fine analysis of the simulated signal with colored noise based on FSWT. (**a**) Frequency slice [0, 0.8] Hz; (**b**) Frequency slice [0, 0.8] Hz.

Comparing Figures 5c and 12, it is clear that the RSSD method denoises the LFO signal better. It can be seen that although the white Gaussian noise changes to colored one, low frequency noise interference still exists. The reconstructed results are similar to Example 1 with white Gaussian noise.

It is thus indicated that low-pass filtering is not effective as a denoising pretreatment of FSWT. On the one hand, it is also shows that both colored and white noise may impact the accuracy of the oscillation mode identification of FSWT-HT and denoising the LFO signal by RSSD is necessary.

5.6. Comparison with Other Parameter Identification Methods

In order to validate the effectiveness of the proposed method, it is compared with the Prony [17], stochastic subspace identification (SSI) [18], empirical mode decomposition (EMD) and SSI (EMD-SSI) [19], and RSSD-SSI methods. The simulated signal is the one of Equation (15), which the standard deviation (σ) equals 0.5. The number of simulation trials is 100. The results are shown in Table 4.

Table 4. Result of identification with Gaussian noise ($\sigma = 0.5$).

Method	Mode	Oscillation Frequency (Hz)	Error (%)	Damping Ratio	Error (%)
Prony	1	0.4502	9.960	0.0240	20.00
	2	1.1616	5.609	0.0580	17.14
SSI	1	0.4766	4.680	0.0222	11.00
	2	1.1573	5.209	0.0753	7.571
EMD-SSI	1	0.4860	2.800	0.0214	7.000
	2	1.1502	4.727	0.0743	6.142
RSSD-SSI	1	0.5060	1.200	0.0206	3.000
	2	1.1302	2.745	0.0730	2.857
RSSD-FSWT	1	0.4978	0.440	0.0192	4.000
	2	1.1030	0.272	0.0667	4.714

As can be seen from Table 4 below, the identified results of the RSSD-FSWT and RSSD-SSI methods are ranked most accurate, followed by EMD-SSI, SSI and Prony. The Prony method is the worst. Since traditional Prony method is sensitive to the noise of data [20,21]. The existence of noise in the signal will leads to a larger error in the results of Prony algorithm.

SSI is a novel approach that developed in recent years. It can identify modal parameters of linear structures from ambient structure vibrations [22]. The SSI method is recently applied to the identification of LFO modes. For singular value decomposition (SVD) is the main step of SSI, the method is more satisfactory than traditional methods, as its higher accuracy and better noise immunity. There are two stable oscillation modes, and considering the noise, the order of SSI can be set to 6, the frequency tole is 1%. But if the noise is strong (the signal-to-noise ratio,SNR is lower than 10 dB), denoising pretreatment is necessary [23].The SNR of this example is 0.2210 dB. So in this case, the result of SSI is better than Prony, but it is not as good as the EMD-SSI and RSSD-SSI methods.

The EMD-SSI method means that the signal denoised by EMD first,and then the SSI method is employed to identify modal parameters of the LFO signal. The similar RSSD-SSI method means SSI after RSSD. Compared the results between the EMD-SSI and RSSD-SSI methods, the latter is better than the former. The EMD method is a relatively new tool for the nonlinear and non-stationary signals analysis, and it can adaptively decompose signals into several intrinsic mode functions (IMFs) according to its characteristic time scale. It starts from the data itself to decompose the signals in spatial domain, so it can discriminate the signals from the noise. But the EMD method contains some theoretical problems, such as envelop and mean curve fit, boundary effect, modes mix and so on, which affect its application. The result of EMD threshold de-noising[24] is showed in Figure 14. The LFO signal is decomposed into 16 IMFs and a residue.

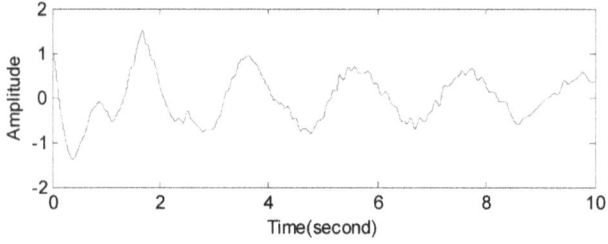

Figure 14. The simulated signal after EMD threshold de-noising.

From Table 4, both of the RSSD-SSI and RSSD-FSWT methods have high accuracy in the identification of the oscillation frequency and damping ratio. RSSD-SSI is the best at estimating damping ratio while RSSD-FWST gives a better estimate of oscillation frequency.

6. Engineering Application

In this section, the proposed method is applied to an engineering application. In order to validate the effectiveness of the proposed method, it is compared with the RSSD-SSI and EMD-SSI methods [19]. The measured power oscillation waveform in Figure 15 is the active power data of the Jiangjiaying-Gaojiang 1st line from a 3.29 large-disturbance testing scheme for the Northeast China power grid. The data have been filtered by low-pass filtering.

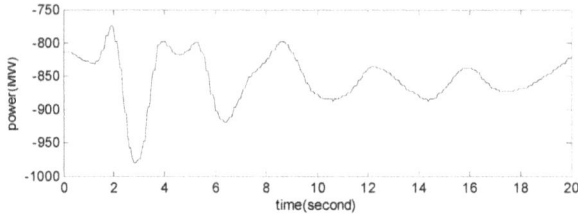

Figure 15. Active power of the Jiangjiaying-Gaojiang 1st line.

Firstly, RSSD is applied to active power data; the high-resonance component in Figure 16b is the denoised data. We used the parameters $Q_1 = 1$, $Q_2 = 4$, $\gamma_1 = \gamma_2 = 3$.

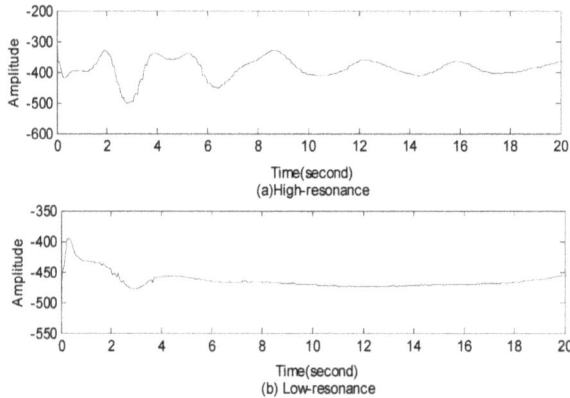

Figure 16. *Cont.*

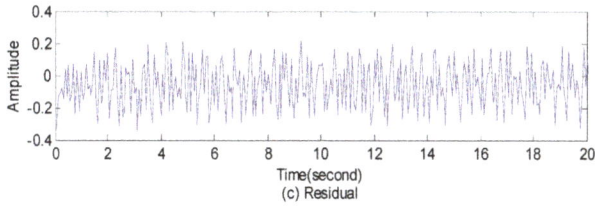

Figure 16. RSSD of power oscillation waveform of the Jiangjiaying-Gaojiang line 1st.

Full band time-frequency distribution analysis by FSWT is shown in Figure 17. Three frequency component amplitudes are prominent. They are 0.54 Hz, 0.25 Hz, and 0.12 Hz, respectively. There are three main modes. The highest amplitude is at about 0.25 Hz, so the LFO dominant mode is the 0.25 Hz oscillation mode.

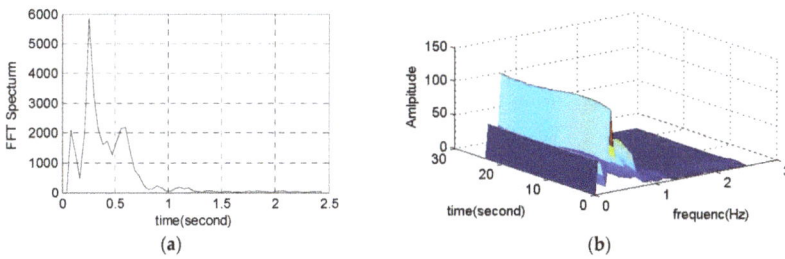

Figure 17. Full band time-frequency distribution analysis by FSWT. (**a**) FFT spectrum; (**b**) FSWT 3D map.

Frequency slice intervals are chosen [0, 0.2] Hz, [0.2, 0.5] Hz, and [0.8, 2] Hz, respectively. The multi-modals LFO signal can be separated into three single modes. The reconstructed signals are shown in Figure 18.

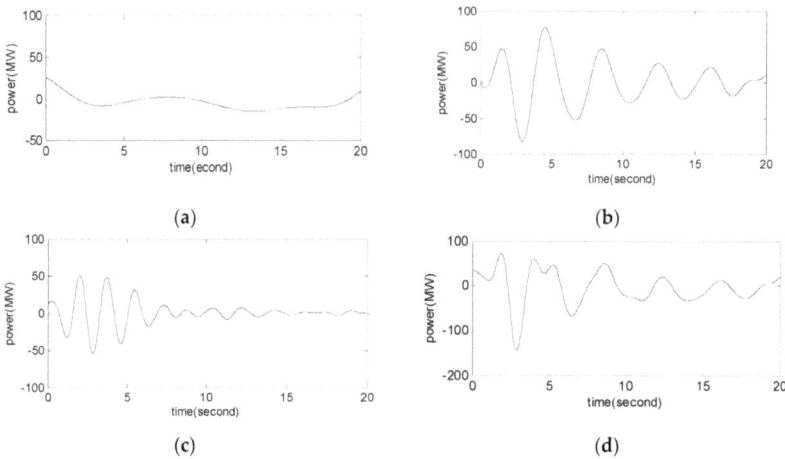

Figure 18. Fine analysis of frequency slices based on FSWT. (**a**) Reconstructed signal of [0, 0.2] Hz; (**b**) Reconstructed signal of [0.2, 0.5] Hz; (**c**) Reconstructed signal of [0.8, 2] Hz; (**d**) Reconstructed signal = (**a**) + (**b**) + (**c**).

We identify the parameters of three mode components by the Hilbert transform. The results are shown in Table 5. The results of parameters identified by the RSSD-SSI and EMD-SSI methods are also shown in Table 5. The fourth modal component is identified by EMD-SSI, but not detected by RSSD-FSWT and RSSD-SSI. The fourth modal component may be the false component which produced during the process of cubic spline curve fitting using EMD. Even if the fourth modal component is existent, the oscillation frequency (0.0411 Hz) is rather low, and the damping ratio is relatively high. Thus, the energy of the fourth modal component is very small and has negligible impact on the system. In other words, it does not play an important role in engineering application. Furthermore the RSSD-FSWT method has as high accuracy in the identification of the oscillation frequency and damping ratio as the other two methods.

Table 5. Parameters of low frequency oscillation mode.

	Mode	Oscillation Frequency (Hz)	Damping Ratio
1	RSSD-SSI	0.5356	0.0556
	RSSD-FSWT	0.5244	0.0542
	EMD-SSI	05354	0.0571
2	RSSD-SSI	0.2510	0.0541
	RSSD-FSWT	0.2508	0.0538
	EMD-SSI	0.2658	0.0562
3	RSSD-SSI	0.1220	0.0149
	RSSD-FSWT	0.1215	0.0145
	EMD-SSI	0.1199	0.0123
4	RSSD-SSI	-	-
	RSSD-FSWT	-	-
	EMD-SSI	0.0411	0.0993

7. Concluding Remarks

In this paper, a time-frequency analysis method using RSSD and FSWT is proposed. The main conclusions are as follows:

(1) RSSD can remove most noise of LFO signals which improved accuracy of LFO modal parameter identification.

(2) FSWT enables the shape of characteristics of LFO modal signal in time-frequency domain to be clearly visible.

(3) RSSD-FSWT can analyze the LFO signal from multiple aspects. Due to full band time-frequency distribution analysis by FSWT, the dominant mode of LFO can be determined, and a 3D map expression of the LFO signal can be obtained. According to fine analysis of frequency slices based on FSWT, we can separate and extract LFO's modal components. Combining with the Hilbert transform, the parameters of the LFO mode components could be identified accurately.

Acknowledgments: This work has been supported by the National Natural Science Funds (51577023).

Author Contributions: Zhao Yan, Zhimin Li and Yonghui Nie checked and discussed the simulation results. Zhao Yan confirmed the series of simulation parameters and arranged and organized the entire simulation process. Zhimin Li and Yonghui Nie revised the paper. Zhimin Li and Yonghui Nie made many useful comments and simulation suggestions. In addition, all authors reviewed the manuscript.

Conflicts of Interest: The authors declare no conflict of interest.

References

1. Kishor, N.; Haarla, L.; Seppänen, J.; Mohanty, S.R. Fixed-order controller for reduced-order model for damping of power oscillation in wide area network. *Int. J. Electr. Power Energy Syst.* **2013**, *1*, 719–732. [CrossRef]

2. Pal, A.; Thorp, J.S.; Veda, S.S.; Centeno, V.A. Applying a robust control technique to damp low frequency oscillations in the WECC. *Int. J. Electr. Power Energy Syst.* **2013**, *1*, 638–645. [CrossRef]

3. Wang, T.; Pal, A.; Thorp, J.S.; Wang, Z. Multi-Polytope-Based Adaptive Robust Damping Control in Power Systems Using CART. *IEEE Trans. Power Syst.* **2015**, *30*, 1–10. [CrossRef]

4. Dosiek, L.; Zhou, N.; Pierre, J.W.; Huang, Z.; Trudnowski, D.J. Mode shape estimation algorithms under ambient conditions: A Comparative Review. *IEEE Trans. Power Syst.* **2013**, *2*, 779–787. [CrossRef]

5. Zhou, N.; Pierre, J.; Trudnowski, D. A stepwise regression method for estimating dominant electromechanical modes. *IEEE Trans. Power Syst.* **2012**, *27*, 1051–1059. [CrossRef]

6. Mandadi, K.; Kalyan, K.B. Identification of Inter-Area Oscillations Using Zolotarev Polynomial Based Filter Bank with Eigen Realization Algorithm. *IEEE Trans. Power Syst.* **2016**, *99*, 1–10. [CrossRef]

7. Yan, Z.; Miyamoto, A.; Jiang, Z. An overall theoretical description of frequency slice wavelet transform. *Mech. Syst. Signal Proc.* **2010**, *2*, 491–507. [CrossRef]

8. Yan, Z.; Miyamoto, A.; Jiang, Z. Frequency slice algorithm for modal signal separation and damping identifycation. *Comput. Struct.* **2011**, *1*, 14–26. [CrossRef]

9. Yan, Z.; Miyamoto, A.; Jiang, Z. Frequency slice wavelet transform for transient vibration response analysis. *Mech. Syst. Signal Proc.* **2009**, *5*, 1474–1489. [CrossRef]

10. Selesnick, I.W. Resonance-based signal decomposition: A New Sparsity-Enabled Signal Analysis Method. *Signal Proc.* **2011**, *12*, 2793–2809. [CrossRef]

11. Kundur, P. *Power System Stability and Control*; McGraw-Hill Professional: New York, NY, USA, 2005.

12. Selesnick, I.W. Wavelet Transform with Tunable Q-Factor. *IEEE Trans. Signal Proc.* **2011**, *8*, 3560–3575. [CrossRef]

13. Afonso, M.V.; Bioucas-Dias, J.M.; Figueiredo, M.A.T. Fast image recovery using variable splitting and constrained optimization. *IEEE Trans. Image Proc. A Publ. IEEE Signal Proc. Soc.* **2010**, *9*, 2345–2356. [CrossRef] [PubMed]

14. Duan, C.D.; Gao, Q. Noval fault diagnosis approach using time-frequency slice analysis and its application. *J. Vib. Shock* **2011**, *9*, 1–774.

15. Huang, N.E.; Shen, Z.; Long, S.R. A new view of nonlinear water waves: The Hilbert Spectrum. *Annu. Rev. Fluid Mech.* **1999**, *6*, 417–457. [CrossRef]

16. Li, T.Y.; Gao, L.; Zhao, Y. The analysis for low frequency oscillation based on HHT. *Proc. CSEE* **2006**, *14*, 24–30. (In Chinese)

17. Xiao, J.; Xie, X.; Zhixiang, H.U.; Han, Y. Improved prony method for online identification of low-frequency oscillations in power systems. *J. Tsinghua Univ.* **2004**, *7*, 883–887. (In Chinese).

18. Ni, J.M.; Shen, C.; Liu, F. Estimation of the electromechanical characteristics of power systems based on a revised stochastic subspace method and the stabilization diagram. *Sci. China Technol. Sci.* **2012**, *6*, 1677–1687. (In Chinese) [CrossRef]

19. Tian, L.I.; Yuan, M.Z.; Jun, L.; Li, J.Q.; Yuan, J.T.; Cai, G.W.; Qian, K. Method of modal parameter identification of power system low frequency oscillation based on EMD and SSI. *Power Syst. Prot. Control.* **2011**, *8*, 6–10. (In Chinese)

20. Kumaresan, R.; Tufts, D.W.; Scharf, L.L. A prony method for noisy data: Choosing the Signal Components and Selecting the Order in Exponential Signal Models. *Proc. IEEE* **1984**, *2*, 230–233. [CrossRef]

21. Li, D.H.; Cao, Y.J. An online identification method for power system low-frequency oscillation based on fuzzy filtering and prony algorithm. *Autom. Electr. Power Syst.* **2007**, *1*, 14–19.

22. Ghasemi, H.; Canizares, C.; Moshref, A. Oscillatory stability limit prediction using stochastic subspace identification. *IEEE Trans. Power Syst.* **2006**, *2*, 736–745. [CrossRef]

23. Li, T.; Yuan, M.; Xu, G.; Cai, G.; Liu, Z.; Bai, B. An inter-harmonics high-accuracy detection method based on stochastic subspace and stabilization diagram. *Autom. Electr. Power Syst.* **2011**, *20*, 50–54. (In Chinese)

24. Boudraa, A.O.; Cexus, J.C. Emd-based signal filtering. *IEEE Trans. Instrum. Meas.* **2007**, *6*, 2196–2202. [CrossRef]

energies

MDPI

Article

A Co-Simulation Framework for Power System Analysis

Seaseung Oh * and Suyong Chae

Energy Efficiency Research Division, Korea Institute of Energy Research, 152 Gajeong-ro, Yuseong-gu, Daejeon 34129, Korea; sychae@kier.re.kr
* Correspondence: shung@kier.re.kr; Tel.: +82-42-860-3012; Fax: +82-42-860-3544

Academic Editor: Ying-Yi Hong
Received: 16 November 2015; Accepted: 5 February 2016; Published: 25 February 2016

Abstract: Power system electromagnetic transient (EMT) simulation has been used to study the electromagnetic behavior of power system components. It generally comprises detailed models of the study area and an equivalent circuit which represents an external part of the study area. However, a detailed description of an external system that includes transmission or distribution system models is required to study the interaction among power system components because the number of high power converter based devices in a power grid have been increasing. Since detailed models of the system components are necessary to simulate a series of events such as cascading faults the computational burden of power system simulation has increased. Therefore a more effective and practical framework has been sought to handle this computational challenge. This paper proposes a co-simulation framework including a delay compensation algorithm to compensate the time delayed signals due to network segmentation and a fast and flexible simulation environment composed of non-real time power system EMT simulation on a general purpose computer with a multi core central processing unit (CPU), which is currently very popular owing to its performance. The proposed methods are applied to an AC/DC power system model.

Keywords: co-operative simulation; time-delay compensation; electromagnetic transient

1. Introduction

Since power electronic systems such as high voltage direct current (HVDC) and flexible alternating current transmission are widely used in power systems, the need for electromagnetic transient (EMT) studies to assess their effect on power systems has increased. Recently, the penetration level of distributed energy resources (DERs) such as photovoltaic, wind, fuel cell, has increased due to environmental issues and fossil fuel prices, which can result in uncertainties on both the supply and demand sides [1]. DERs use power electronics circuits to generate, convert and inject electric power into a distribution system. Therefore, an EMT simulation which includes both DERs and detailed power system component models is required to evaluate the effect of the increased uncertainties on a power grid and to develop appropriate countermeasures. However, to adequately simulate power systems using these detailed models, the integration time-step of the simulation must be sufficiently smaller than the time constant of the fastest EMT phenomena of the system. Because EMT simulation aims to solve both the nodal equations for the substituted network and the differential equations which describe the dynamic behavior of the detailed models [2,3], EMT simulation of a complex power system with a small fixed time-step would result in huge computational burden and would remarkably increase the computation time of the EMT simulation.

A high performance CPU can make the EMT simulation run faster. However, technical and economic constraints currently exist in terms of increasing the operating speed of CPUs. To overcome these constraints much research have been done on the simulation algorithms and techniques [4–7]. Most of these techniques are based on the fact that a power system can be divided into several subsystems and the subsystems can be simulated independently by the transmission line model (TLM) [8] which enables the decoupling of two subsystems by the wave propagation delay. This parallel simulation enables to distribute the computational load among the processing cores to more efficiently utilize the available computing power of a multi-core CPU. It is one of the co-simulation techniques which allows individual components to be simulated using different simulation tools which simultaneously execute and exchange information in a collaborative manner [9,10]. Unlike parallel processing [11,12], which divides a complex system into several subsystems and processes them in a single simulation system, co-simulation consists of several independent simulation programs and offers more simulation flexibility. The wave propagation delay which has been widely used to decouple two subsystems and allow each subsystem to be executed independently is the foundation of the co-simulation. However this time delay can be a bottleneck to enhance the overall accuracy and the stability of the co-simulation in some cases.

This paper proposes a co-simulation framework including a time-delay compensation algorithm which enables processing of time-consuming EMT simulation of complex power systems by two independent EMT simulations in a cooperative manner. The delay compensation method can enhance the accuracy and the stability of the EMT co-simulation and thus can alleviate the hardware and software limitations of the conventional EMT simulation on a general purpose personal computer. Unlike the message passing interface parallel programming, the proposed method can be implemented using commercial power system analysis software without modification of its source code. To evaluate the performance of the proposed method, examples of co-simulation of an AC/DC power system are presented and compared with the conventional simulation results.

2. EMT Co-Simulation Framework

The EMT co-simulation method generally comprises network partitioning, interfacing algorithm which integrates two decoupled subsystems. The network partitioning uses the reactive components, inductor or capacitor as transmission line to decouple two parts at the end of the line using the wave propagation delay. If a transmission line length is 15 km its propagation delay is about 50 μs. Therefore two subsystems which is connected with a 15 km long transmission line cannot see the change of the other side in 50 μs thus can be simulated independently. However the delay will be decreased proportional to the interface line length thus the time-step of the simulation should be decreased. The decreased time-step will increase the number of computation and thus total computation time.

The total simulation time depends on both the computation and data communication. The computation time for each subsystem is proportional to the size and complexity of the power systems being simulated. Because each pre-defined simulation sequence may not be completed before the data exchange among computing nodes in a non-real time simulation environment, some of the nodes idly wait for the data exchange, even in a parallel protocol, unless a prediction algorithm is applied [13]. Therefore, the need to reduce the computation time of each subsystem simulation and the total communication time between simulations is necessary to accelerate the co-simulation.

2.1. Network Partitioning

A network can be split into small subsystems by a network partitioning algorithm. The number of subsystems should be determined to minimize both communications between subsystems and the idle time of the simulation process since small subsystems take less time to solve system equations however increases the communication cost [14].

Since a power electronics circuit is generally very complex and has several high frequency switching devices which require high computational power to be simulated a power system which has power electronics circuits may be split into two subsystems at the interconnection point, as shown in Figure 1. The interconnection point will be the interface of the subsystems. The AC grid can be further partitioned to adjust the balance of the computational load of co-simulation. This characteristic-base partitioning can easily embrace the multi-time-step simulation algorithm and its loosely coupled connection increases reusability of a simulation model. It can also be easily applied in a commercial off-the-shelf graphical user interface (GUI)-based power system simulation program. This approach requires equivalent circuits that can fully simulate the dynamic behavior of each subsystem because the simulation accuracy depends on these equivalent circuits [15,16]. Power frequency based R-L-C networks have been used for the equivalent circuits. However, more sophisticated models are required if waveform distortion or phase imbalance occurs at the interface bus [17]. The circuits are modeled as frequency dependent using vector fitting [18,19].

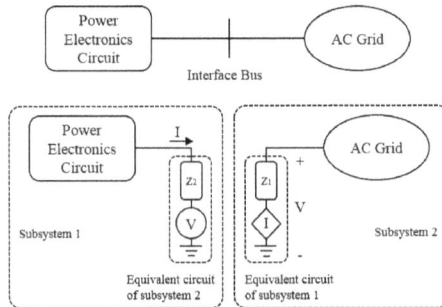

Figure 1. Network partitioning.

2.2. Interfacing

Much research on the interfacing method has been carried out for power hardware testing in simulation environments [20–22]. Two major approaches to split a network into subsystems are the travelling wave model [23], and the insertion of time-step delays [24].

The subsystems have been split using travelling wave transmission lines in order to decouple the subsystems and solve them independently. The simulation time-step should be set to an integer fraction of the propagation delay of a travelling wave thus partitioning small power systems which have no long transmission lines such as microgrids, industrial power systems, shipboard power systems with the propagation delay using the travelling wave model requires very short simulation time-step to decouple the subsystems. The short time-step naturally increases the number of computation and communication thus makes EMT simulation very slow.

The time-step delay insertion method adds artificial delay to decouple the subsystems. The artificial delay is usually implemented with a stubline which length has one simulation time-step for the wave propagation delay. This artificial delay is quite useful especially to split the small size power systems, however, it may accumulate phase drifts and make simulation inaccurate and unstable [14].

Interface algorithm defines how two decoupled subsystems are interconnected. In this paper ideal transformer model (ITM) [25] is adopted for interfacing subsystems. It generates accurate result and is easy to set up. The measured voltage and current values at the interface of the subsystems are to be exchanged between the subsystems through their equivalent circuits as in Figure 1.

An ideal communication protocol for subsystem interfacing can make co-simulation continuously run without any idle time between the simulation time-steps. To minimize the idle time of the co-simulation of power systems, the parallel protocol proposed in [26] is adopted. It is reasonable to initiate the interfacing sequence with transferring the voltage values first which is 1 in Figure 2 since all the power electronics systems start to operate after being fed enough voltage from a power grid except black starting.

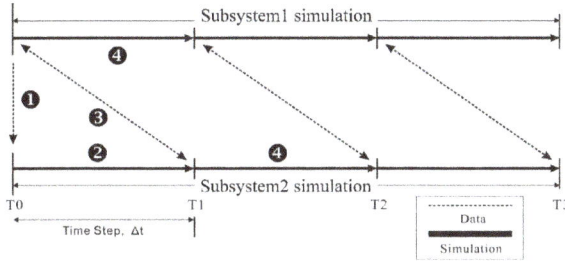

Figure 2. Parallel interface protocol.

In a serial protocol, only one simulation time segment runs at each time instance while the other one stays idle. However, using the parallel interface protocol shown in Figure 2, both simulation segments, labeled as 4, can be concurrently run, thus minimizing the idle time. To eliminate the idle time, two distinct simulation sequences, labeled as 4, must be simultaneously executed. However, achieving this simultaneous simulation is very difficult with the network partitioning in a GUI based simulation environment. In practice, in an EMT-to-EMT co-simulation environment, the total idle time can be slightly longer than the ideal case.

The actual data transfers between two subsystems are carried by a pipe which is one of the inter-process communication (IPC) mechanisms as shown in Figure 3. It is an information conduit and allows a process to read and write from its end. It can be used to transfer data between unrelated processes over a network. It can provide simple programming interfaces since the synchronization is built into the pipe mechanism itself.

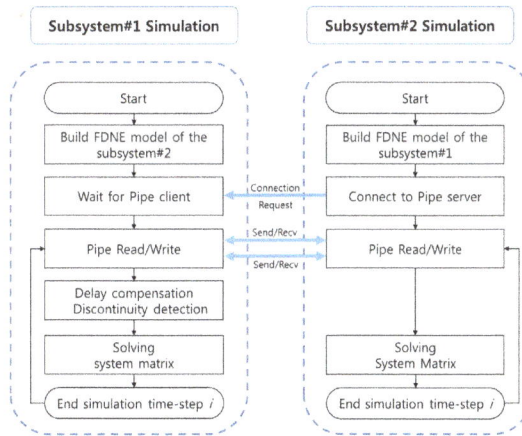

Figure 3. Inter-process communication (IPC) process of co-simulation.

2.3. Time-Step Delay Error

Network partitioning inherently introduces time delays in the network. The parallel interface protocol shown in Figure 2 introduces a time-step delay to the lower side of the co-simulation, which is usually intended for the grid. The simulation time-step which is conventionally set to 50 to 100 µs requires 15 to 30 km length transmission line. Without a proper interface transmission line the artificial delay which is imposed by the simulation time-step causes the phase error to the sinusoidal signal at the interface and should be compensated to maintain the accuracy of the simulation.

A differential equation, namely Equation (1), which describes the dynamic behavior of the power system components, can be converted into a simple algebraic equation, *i.e.*, Equation (3), by the trapezoidal rule:

$$\dot{x} = \mathbf{A}x + \mathbf{B}u \tag{1}$$

$$y = \mathbf{C}x + \mathbf{D}u \tag{2}$$

$$x(n) = ax(n-1) + bu(n-1) \tag{3}$$

$$y(n) = cx(n) + du(n) \tag{4}$$

Each side of the co-simulation processes shown in Figure 2 can be described by the following equations. Equations (5) and (6) describe the upper side shown in Figure 2, whereas Equations (7) and (8) represent the lower side. The $u_{bci}(n)$ term in Equation (9) represents the i_{th} interface value, which is exchanged between the subsystems. The errors which result from the delay in one subsystem would influence the simulation result of the other subsystem:

$$x_1(n) = a_1 x_1(n-1) + b_1 \hat{u}_1(n-1) \tag{5}$$

$$y_1(n) = c_1 x_1(n) + d_1 \hat{u}_1(n) \tag{6}$$

$$x_2(n) = a_2 x_2(n-1) + b_2 \hat{u}_2(n-1) \tag{7}$$

$$y_2(n) = c_2 x_2(n) + d_2 \hat{u}_2(n) \tag{8}$$

$$\hat{u}_1(n) = \begin{bmatrix} u_{11}(n) \\ \vdots \\ \hat{u}_{bci}(n) \\ u_{1i}(n) \end{bmatrix}, \ \hat{u}_2(n) = \begin{bmatrix} u_{21}(n) \\ \vdots \\ \hat{u}_{bci}(n-1) \\ u_{2j}(n) \end{bmatrix} \tag{9}$$

$$\varepsilon_1 = y(n) - y_1(n), \ \varepsilon_2 = y(n) - y_2(n) \tag{10}$$

The error due to the delay in the control signals is usually negligible. However, it can have a strong effect on the network solutions because the calculation of the power flow on the interface of the subsystems may be different at both sides of the subsystems due to the delay. In the case of a power flow with large effect on the grid, the errors cannot be neglected. The error in the delay should be properly compensated to enhance the accuracy of the entire EMT co-simulation. The error due to the delay varies with the size of the integration time-step and the frequency of the signals to be exchanged.

2.4. Data Prediction by Extrapolation

The delay error can be compensated using two time-step time shifts. The time shift of a signal can be obtained by predicting future data. Extrapolation is the process of estimating new data points outside a discrete set of known data points. Fundamentally, it is the same as the interpolation process, which constructs new points among known points. The interpolation algorithm can be applied as is to the estimation using the extended x_i range outside the known values. However, the interval should be kept as small as possible to maintain the estimation quality.

The most straightforward extrapolation method is the linear extrapolation, as expressed in Equation (11). It creates a tangent line at the end of the known data and extends it beyond that limit. It assumes that any slopes between two adjacent points are fixed, and the extrapolation interval is sufficiently small to satisfy the assumption because the error of the linear method is proportional to the square of the interval, *i.e.*, $O(h^2)$. As expressed in Equation (11), linear extrapolation assumes that the future data y_{k+1} are on the line which passes two adjacent past data y_{k-1} and y_k. The algorithm is straightforward, and the computational load is minimal. However, the assumption can fail with high frequency signal components in the EMT simulation; thus, the integration time-step should be limited to be short enough to maintain the accuracy of the co-simulation. This process results in the increase in the total number of computations and thus the computational load:

$$\hat{y}_{(k+1)} = y_{(k-1)} + \frac{\hat{x}_{(k+1)} - x_{(k-1)}}{x_{(k)} - x_{(k-1)}}(y_{(k)} - y_{(k-1)}) \tag{11}$$

A polynomial based algorithm shows much better performance than the linear method, as shown in Figure 4. The cubic spline method, whose error is expressed as $O(h^4)$, can reduce the interval to 13% compared with the linear method at the same level of summation of squared error (SSE).

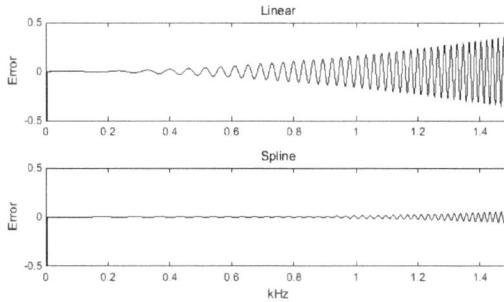

Figure 4. Prediction error comparison for a chirp signal.

The fundamental principle of the cubic spline method is to find a piecewise function which fits a given data set. The piecewise function consists of $n - 2$ third-degree polynomial, as expressed in Equation (12), with n points. The second derivative of each polynomial may be set to zero at the endpoints to satisfy the natural cubic spline condition. With the natural cubic spline condition $S_k(x)$ can be arranged as a tridiagonal system which is easy to solve to obtain the polynomial coefficients:

$$S_k(x) = \begin{cases} s_1(x) & (x_1 \leqslant x < x_2) \\ s_2(x) & (x_2 \leqslant x < x_2) \\ \quad \vdots \\ s_{n-1}(x) & (x_{n-1} \leqslant x < x_n) \end{cases} \tag{12}$$

$$s_i(x) = a_i(x - x_i)^3 + b_i(x - x_i)^2 + c_i(x - x_i) + d_i$$

The extrapolation would provide a minimum error using two properly selected parameters such as the number of given data points and the extrapolation interval. The interval should be set to a small value because it is directly related to the error of the cubic spline method. The iterative extrapolation based on a sliding window, as shown in Figure 5, can in effect reduce half of the interval to estimate the two step forward value.

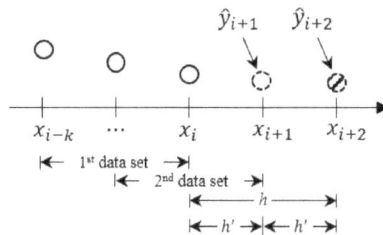

Figure 5. Iterative extrapolation of a two step forward estimation.

The other factor which affects the accuracy of the cubic spline extrapolation is the length of the data window because the extrapolation estimates the unknown values using the given data set. Figure 6 shows the SSEs of the cubic spline method with different numbers of data points to extrapolate a one-cycle sinusoidal signal. Given five data points, the sliding window method gives a minimum error in the two step forward estimation of the sinusoidal function.

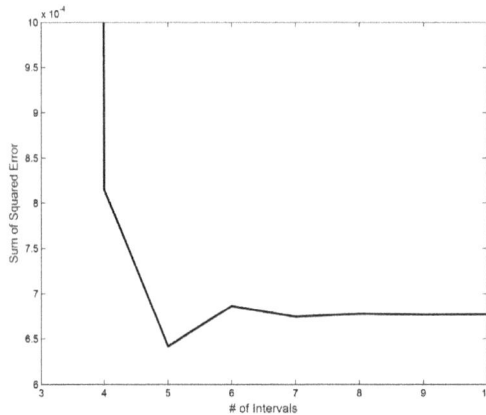

Figure 6. Summation of squared error (SSE) variation in the spline extrapolation.

2.5. Discontinuity Detection

Discontinuities such as voltage drops due to a line fault are common phenomena in power systems. Accurately predicting such discontinuities using a series of discrete data to compensate the delay is quite difficult. Prediction using a limited number of known points tends to induce numerical spikes or oscillation. The spline method is more vulnerable to discontinuity, and its effect would appear more severe than the linear method because it uses derivatives to solve the tridiagonal system equation. The magnitude of this phenomenon is proportional to that of the derivative at the discontinuity.

Although an oscillation would disappear within a few steps, it may affect the accuracy of the co-simulation. To avoid this numerical oscillation, the spline extrapolation should be turned off until the discontinuity moves out of the data window. The spline extrapolation would take effect again as soon as the discontinuity is out of the data window.

The first derivative of each x_i, which is one of the equations for determining the spline polynomial coefficients, is compared with the heuristically determined threshold value to detect discontinuities. Discontinuity detection basically uses the slope of two successive data points thus magnitude, frequency and size of the computational time-step affect the threshold. k_p is a heuristically determined constant which is generally in between 0.05 and 0.1 in 60 Hz system:

$$V_{threshold} = k_p \cdot Hz \cdot V_{mag} \tag{13}$$

More frequent and periodic discontinuities, such as power electronic switching noise, are assumed to be eliminated or attenuated by harmonic filters; thus, a threshold is set to detect large disturbances such as transmission line faults which cause severe line voltage drop. Figure 7 shows that the spline extrapolation with discontinuity detection (spline/DD) effectively handles the numerical oscillation, which is unacceptably high in the normal spline method.

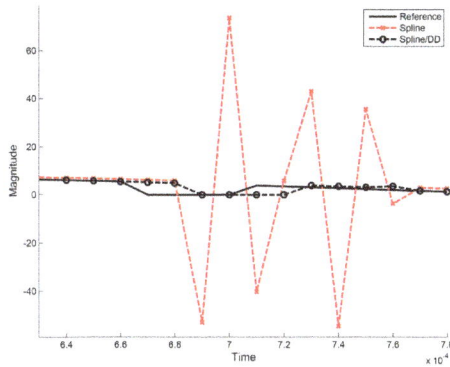

Figure 7. Discontinuity detection effect on the numerical oscillation.

Figure 8 shows the interface node voltage between the capacitor and inductor of a simple R-L-C circuit [27], which demonstrates the time-step delay and the compensation effect on the co-simulation.

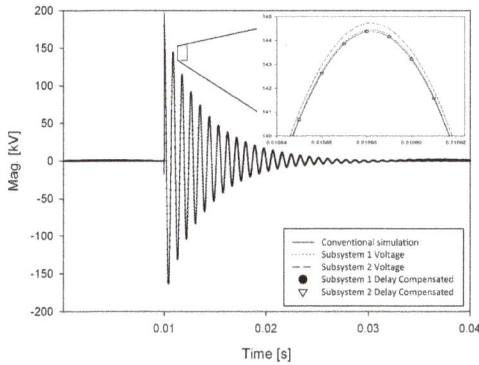

Figure 8. Interface node voltage during resistor, inductor and capacitor (R-L-C) energization.

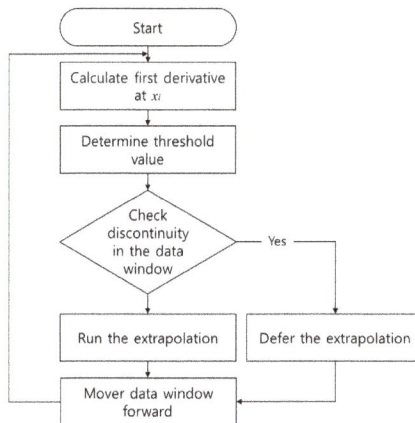

Figure 9. Extrapolation process of the discontinuity detection (DD) algorithm.

Both the dotted and dashed lines represent voltage waveforms, which are simulated with a 10 μs time-step, of the interface buses in each subsystem. The two voltage waveforms must be the same as the interface bus voltage waveform of the conventional simulation, which is represented by the solid line. However, phase and magnitude differences exist among the three voltage waveforms because the phase-shifted currents of subsystem 2, which result from the time-step delayed voltage signal generated by subsystem 1, are transferred back to subsystem 1 and input to the nodal equations. By using the proposed extrapolation method as Figure 9, the delayed voltage signal is successfully compensated and matches the reference signal.

3. Case Study

Two case studies which are an AC/DC system and an AC distribution system are used to verify the proposed framework. The cases are set up using PSCAD™. Each test system of the cases is divided into two subsystems. Each subsystem is run with two PSCAD programs concurrently which communicate each other via IPC.

3.1. AC/DC Power System

The AC/DC power system shown in Figure 10 is a simplified model of an island grid fed by a 300 MW HVDC system [28]. The selected information on the system is summarized in Table 1. A line commutated converter HVDC system continuously absorbs reactive power as part of the conversion process, and two synchronous condensers at the inverter bus supply reactive power to the HVDC system. The test system is partitioned into two subsystems along a fictitious bus at the inverter terminal. A frequency dependent network equivalent (FDNE) model based on vector fitting, which covers up to a 1 kHz frequency range, represents the AC equivalent circuit in the HVDC subsystem.

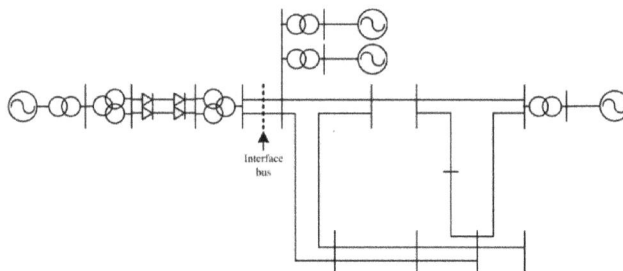

Figure 10. AC/DC test system.

Table 1. AC/DC test system parameters.

Specifications	HVDC System	AC System
Type	Line commutated converter	
Rate Voltage	DC ±180 kV	154 kV
Rate Current	849 A	-
Rating	300 MW	75 MVA synchronous generator
Number of circuits	2	-
Converter Transformer	3phase 3winding (YDY)	-
Reactive power compensation	70 MVA synchronous condensers	-
Harmonic Filter	DTF 27.5 MVA X 2 for 11, 13th harmonics HPF 27.5 MVA X 2 for 23, 25th harmonics	-

Figure 11 shows the comparison of three inverter bus AC voltage waveforms and their effect on the AC power injected by HVDC system. To show the delayed voltage signal effect on a simulation result more clearly the time-step is set to 100 μs.

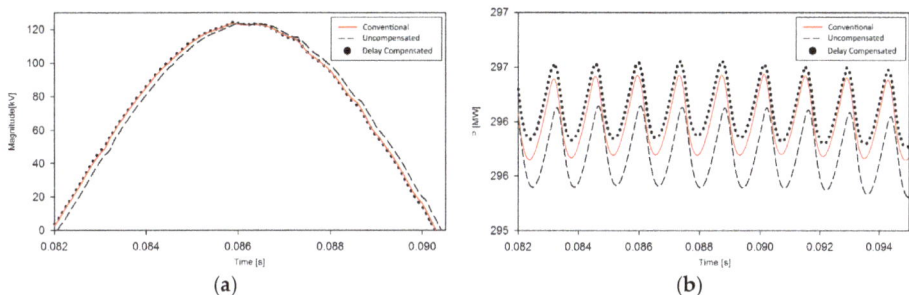

Figure 11. The effect of the compensation on (**a**) the voltage and (**b**) the high voltage direct current (HVDC) power.

As seen in Figure 11a a delayed voltage signal is well compensated by the proposed algorithm and reduces the error on the injected HVDC power to the AC system. The gap between the conventional simulation result and compensated co-simulation result is mainly due to the errors in the parameters in the FDNE model.

Figure 12 shows the grid frequency response to a bus fault. Each subsystem is simulated with a 10 μs time-step. At 0.1 s, a fault is applied to the inverter bus. The fault is successfully cleared after three cycles. Two real power values calculated at both sides of the interface bus deviate from the reference value from the conventional simulation because of the time delay, thus leading to inaccurate frequency response, as shown in Figure 12. The power deviation is successfully compensated for by the proposed extrapolation algorithm thus the results of the co-simulation match well with those of the conventional simulation.

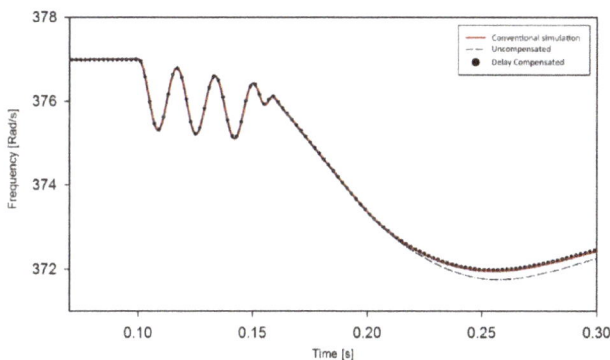

Figure 12. AC grid frequency during a bus fault at the inverter bus.

3.2. AC Distribution System

The AC distribution system is the IEEE 34 bus test feeder [29] (Figure 13). It is the complete data set of four radial distribution systems for analysis of unbalanced three-phase radial distribution feeders. It has been widely used as a test bench to evaluate the performance of newly developed technologies [30,31]. The selected information on the system specifications is summarized in Table 2.

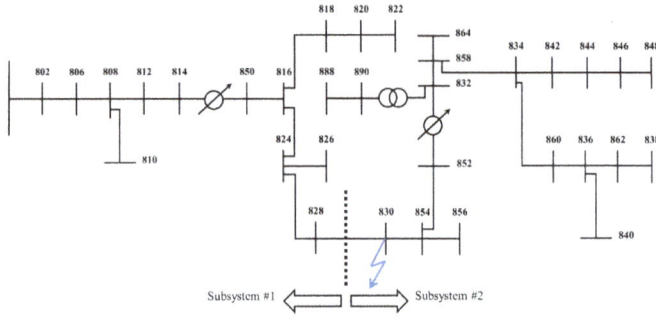

Figure 13. IEEE 34 bus test system.

Table 2. IEEE 34 bus test feeder configuration.

Type	Radial
Total Number of Bus	34
Frequency	60 Hz
Voltage	24.9 kV
Number of Loads	Spot: 6, Distributed: 19
Total Load	Active: 1769 kW, Reactive: 1044 kW
Transformer	24.9 kV/4.16 kV, D-Y, 500 kVA
Voltage Regulator	Y-Y with OLTC
Shunt Capacitor	750 kVA

It is divided into two subsystems at the sending end of the transmission line between the bus 828 and the 830 thus the subsystems have similar size and computational burden. The three phase ground fault is applied at the sending end of the transmission line between the bus 830 and the bus 854. The fault is applied at $t = 0.2$ s and cleared after three cycles.

As shown in Figure 14a the simulation result matches well to that of the conventional simulation which is shown in Figure 14b. Since two subsystems share one interface point the voltage and the current at the interface bus of each subsystem should be same as in Figure 15. The dashed line with square and the white circle which represent the interface bus voltage of the subsystem #1 and subsystem #2, respectively. The overshoot in the compensated signal appears to last longer than the reference signal due to the DD algorithm and the inaccuracy in the FDNE model parameters during the fault. However, after the overshoot the compensated signal quickly tracks the reference signal.

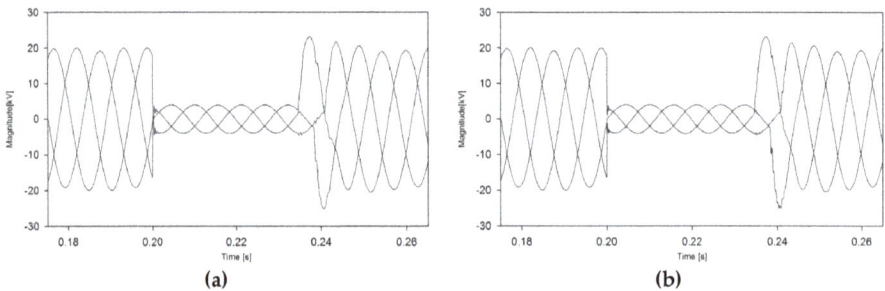

Figure 14. Three-phase ground fault simulation using (**a**) proposed framework and (**b**) conventional simulation.

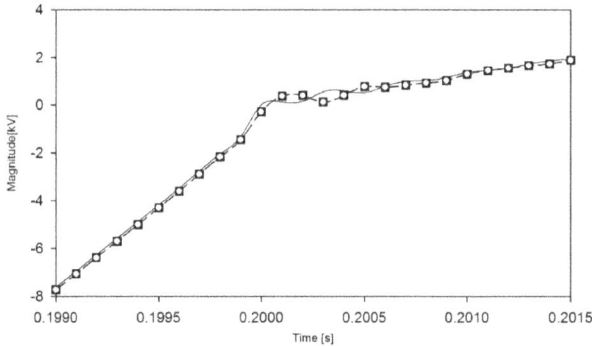

Figure 15. Zoomed view of the phase A voltage comparison at the fault.

4. Conclusions

The speed of the conventional EMT simulation using both single-core and multi-core CPUs is similar because the conventional EMT simulation can hardly utilize the maximum computing power of a multi-core CPU. Co-simulation requires two concurrent simulations for the different subsystems, and inter-process communications between them. If the processor doesn't handle this properly the two simulations will be executed in series, resulting in much longer computation time as is shown in Table 3. The proposed compensation algorithm effectively mitigates the negative impact of the time delay on co-operative EMT simulation thus enables to take advantage of the power of a multi-core CPU. It is a practical method as it can be applied to virtually all EMT simulation programs without source code modification.

Table 3. Simulation speed comparison.

Type	Conventional		Co-simulation	
Case	*Reference*		*Common time-step*	
CPU	Single core (3.2 GHz)	Multi core (2.66 GHz)	Single core (3.2 GHz)	Multi core (2.66 GHz)
Time-step	10 μs	10 μs	10 μs	10 μs
Computation time *	1	1.12	6.72	0.51

* normalized by the reference case.

Acknowledgments: This work (NRF-2015R1A2A2A01005497) was supported by Mid-career Researcher Program through NRF grant funded by the MEST and by the Power Generation & Electricity Delivery Core Technology Program of the Korea Institute of Energy Technology Evaluation and Planning (KETEP) granted financial resource from the Ministry of Trade, Industry & Energy, Republic of Korea (No. 20141010502280).

Author Contributions: Seaseung Oh contributed the simulation, analysis; Seaseung Oh and Suyong Chae wrote the paper.

Conflicts of Interest: The authors declare no conflict of interest.

Abbreviations

The following abbreviations are used in this manuscript:

CPU   Central processing unit
DD   Discontinuity detection
EMT   Electromagnetic transient
FDNE   Frequency dependent network equivalent
GUI   Graphical user interface
HVDC   High voltage direct current transmission
IPC   Interprocess communication

R-L-C   Resistor, Inductor and Capacitor

SSE   Sum squared error

References

1. Chowdhury, S.P.; Chowdhury, S.; Crossley, P.A. UK scenario of islanded operation of active distribution networks with renewable distributed generators. *Int. J. Electr. Power Energy Syst.* **2011**, *33*, 1251–1255. [CrossRef]
2. Dommel, H.W.; Meyer, W.S. Computation of electromagnetic transients. *Proc. IEEE* **1974**, *62*, 983–993. [CrossRef]
3. Dommel, H.W. *EMTP Theory Book*; Bonneville Power Admin: Portland, OR, USA, 1986.
4. Heffernan, M.D.; Turner, K.S.; Arrillaga, J.; Arnold, C.P. Computation of AC-DC System Disturbances—Part I. Interactive Coordination of Generator and Convertor Transient Models. *IEEE Trans. Power Syst.* **1981**, *11*, 4341–4348. [CrossRef]
5. Gao, F.; Kai, S. Multi-scale simulation of multi-machine power systems. *Int. J. Electr. Power Energy Syst.* **2009**, *31*, 538–545. [CrossRef]
6. Su, H.T.; Snider, L.A.; Chung, T.S.; Fang, D.Z. Recent advancements in electromagnetic and electromechanical hybrid simulation. In Proceedings of the International Conference on Power System Technology, Singapore, 21–24 November 2004; pp. 1479–1484.
7. Lefebvre, S.; Mahseredjan, J. Interfacing Techniques for Transient Stability and Electromagnetic Transient Programs. *IEEE Trans. Power Deliv.* **2009**, *24*, 2385–2395.
8. Hui, S.Y.R.; Fung, K.K.; Christopoulos, C. Decoupled simulation of DC-linked power electronic systems using transmission-line links. *IEEE Trans. Power Electron.* **1994**, *9*, 85–91. [CrossRef]
9. Chiocchio, T.; Leonard, R.; Work, Y.; Fang, R.; Steurer, M.; Monti, A.; Khan, J.; Ordonez, J.; Sloderbeck, M.; Woodruff, S.L. A co-simulation approach for real-time transient analysis of electro-thermal system interactions on board of future all-electric ships. In Proceedings of the 2007 Summer Computer Simulation Conference, San Diego, CA, USA, 16 July 2007.
10. Cécile, J.F.; Schoen, L.; Lapointe, V.; Abreu, A.; Bélanger, J. A Distributed Real-Time Framework for Dynamic Management of Heterogeneous Co-Simulations. Available online: http://www.rtlab.com/files/scs_article.pdf (accessed on 16 October 2015).
11. Tylavsky, D.J.; Bose, A.; Alvarado, F.; Betancourt, R.; Clements, K.; Heydt, G.T.; Huang, G.; Ilic, M.; La Scala, M.; Pai, M.A. Parallel Processing in Power Systems Computation. *IEEE Trans. Power Syst.* **1992**, *7*, 629–638.
12. Tomin, M.A.; De Rybel, T.; Marti, J.R. Extending the Multi-Area Thévenin Equivalents method for parallel solutions of bulk power systems. *Int. J. Electr. Power Energy Syst.* **2013**, *44*, 192–201. [CrossRef]
13. Su, H.T.; Chan, K.W.; Snider, L.A. Parallel interaction protocol for electromagnetic and electromechanical hybrid simulation. *IEEE Proc. Gener. Trans. Distrib.* **2005**, *152*, 406–414. [CrossRef]
14. Marti, J.R.; Linares, L.R.; Calvino, J.; Dommel, H.W.; Lin, J. OVNI: An object approach to real-time power system simulators. In Proceedings of the International Conference on Power System Technology, Beijing, China, 18–21 August 1998; pp. 977–981.
15. Sultan, M.; Reeve, J.; Adapa, R. Combined transient and dynamic analysis of HVDC and FACTS systems. *IEEE Trans. Power Deliv.* **1998**, *13*, 1271–1277. [CrossRef]
16. Wang, Y.P.; Watson, N.R. A benchmark test system for testing frequency dependent network equivalents for electromagnetic simulations. *Int. J. Electr. Power Energy Syst.* **2013**, *44*, 364–374. [CrossRef]
17. Annakkage, U.D.; Nair, N.-K.C.; Gole, A.M.; Dinavahi, V.; Noda, T.; Hassan, G.; Monti, A. Dynamic system equivalents: A survey of available techniques. *IEEE Trans. Power Deliv.* **2009**, *27*, 411–420. [CrossRef]
18. Gustavsen, B.; Semlyen, A. Rational approximation of frequency domain responses by vector fitting. *IEEE Trans. Power Deliv.* **1999**, *14*, 1052–1061. [CrossRef]
19. Gustavsen, B.; Mo, O. Interfacing convolution based linear models to an electromagnetic transients program. *Int. Conf. Power Syst. Trans.* **2007**, *1*, 4–7.

20. Zhu, W.; Pekarek, S.; Jatskevich, J.; Wasynczuk, O.; Delisle, D. A Model-in-the-Loop Interface to Emulate Source Dynamics in a Zonal DC Distribution System. *IEEE Trans. Power Electron.* **2005**, *20*, 438–445. [CrossRef]

21. Wu, X.; Lentijo, S.; Deshmuk, A.; Monti, A.; Ponci, F. Design and implementation of a power-hardware-in-the-loop interface: A nonlinear load case study. In Proceedings of IEEE Applied Power Electronics Conference and Exposition, Austin, TX, USA, 6–10 March 2005; Volume 2, pp. 1332–1338.

22. Ren, W.; Steurer, M.; Woodruff, S.; Andrus, M. Demonstrating the Power Hardware-in-the-Loop through Simulation of a Notional Destroyer-Class AllElectric Ship System during Crashback. In Proceedings of the ASNE Advanced Propulsion Symposium, Arlington, VA, USA, 30 October 2006.

23. Noda, T.; Sasaki, S. Algorithms for distributed computation of electromagnetic transients toward pc cluster based real-time simulations. In Proceedings of the International Conference on Power System Transients, New Orleans, LA, USA, 28 September 2003.

24. Dufour, C.; Paquin, J.-N.; Lapointe, V.; Be'langer, J.; Schoen, L. PC cluster-based real-time simulation of an 8-synchronous machine network with HVDC link using RT-LAB and TestDrive. In Proceedings of the 7th International Conference on Power Systems Transients, Lyon, France, 4–7 June 2007.

25. Ren, W.; Steurer, M.; Woodruff, S. Accuracy evaluation in power hardware-in-the-loop (PHIL) simulation center for advanced power systems. In Proceedings of the 2007 Summer Computer Simulation Conference, San Diego, CA, USA, 16 July 2007.

26. Fang, T.; Chengyan, Y.; Zhongxi, W.; Xiaoxin, Z. Realization of electromechanical transient and electromagnetic transient real time hybrid simulation in power system. In Proceedings of the IEEE/PES Transmission and Distribution Conference Exhibition, Dalian, China, 14 August 2005; pp. 1–6.

27. Belanger, J.; Snider, L.A.; Paquien, J.; Pirolli, C.; Li, W. A Modern and Open Real-Time Digital Simulator of Contemporary Power Systems. In Proceedings of the International Conference on Power Systems Transients (IPST 2009), Kyoto, Japan, 2 June 2009; pp. 2–6.

28. Jang, G.; Oh, S.; Hann, B.-M.; Kim, C.K. Novel reactive power compensation scheme for the Jeju-Haenam HVDC system. *IEEE Proc. Gener. Trans. Distrib.* **2005**, *152*, 514–520. [CrossRef]

29. Kersting, W.H. Radial distribution test feeders. In Proceedings of the 2001 Power Engineering Society Winter Meeting, Columbus, OH, USA, 31 January 2001; Volume 2, pp. 908–912.

30. Jang, S.; Kim, K. An islanding detection method for distributed generations using voltage unbalance and total harmonic distortion of current. *IEEE Trans. Power Deliv.* **2004**, *19*, 745–752. [CrossRef]

31. Xin, H.; Qu, Z.; John, S.; Ali, M. A Self-Organizing Strategy for Power Flow Control of Photovoltaic Generators in a Distribution Network. *IEEE Trans. Power Deliv.* **2011**, *26*, 1462–1473. [CrossRef]

energies

MDPI

Article

Field Experiments on 10 kV Switching Shunt Capacitor Banks Using Ordinary and Phase-Controlled Vacuum Circuit Breakers

Wenxia Sima [1], Mi Zou [1,*], Qing Yang [1], Ming Yang [1] and Licheng Li [2]

[1] State Key Laboratory of Power Transmission Equipment & System Security and New Technology, Chongqing University, Chongqing 400044, China; cqsmwx@cqu.edu.cn (W.S.); yangqing@cqu.edu.cn (Q.Y.); cqucee@cqu.edu.cn (M.Y.)

[2] School of Electric Power, South China University of Technology, Guangdong 510640, China; lilc@csg.cn

* Correspondence: zoumi@cqu.edu.cn; Tel.: +86-23-6511-1795 (ext. 8307); Fax: +86-23-6510-2442

Academic Editor: Ying-Yi Hong

Received: 17 December 2015; Accepted: 27 January 2016; Published: 30 January 2016

Abstract: During the switching on/off of shunt capacitor banks in substations, vacuum circuit breakers (VCBs) are required to switch off or to switch on the capacitive current. Therefore, the VCBs have to be operated under a harsh condition to ensure the reliability of the equipment. This study presents a complete comparison study of ordinary and phase-controlled VCBs on switching 10 kV shunt capacitor banks. An analytical analysis for switching 10 kV shunt capacitor banks is presented on the basis of a reduced circuit with an ungrounded neutral. A phase selection strategy for VCBs to switch 10 kV shunt capacitor banks is proposed. Switching on current waveforms and switching off overvoltage waveforms with, and without, phase selection were measured and discussed by field experiments in a 110 kV substation in Chongqing, China. Results show that the operation of phase-controlled VCBs for 10 kV switching shunt capacitor banks is stable, and phase-controlled VCBs can be used to implement the 10 kV switching on/off shunt capacitor banks to limit the transient overvoltage and overcurrent. The values of overvoltage and inrush current using phase-controlled VCBs are all below those with ordinary VCBs.

Keywords: vacuum circuit breakers (VCBs); phase selection; shunt capacitor banks; field tests

1. Introduction

Switching shunt capacitor banks is the most widespread method for the compensation of reactive power or for the stabilization of voltage for power quality reasons [1–6]. Circuit breakers are usually switched to adjust the reactive power capacity according to the load change. Capacitive switching is conducted with an almost daily frequency according to an inquiry performed by the CIGRE (International Council on Large Electric systems) Working Group 13.04 in 1999 [7]. About 60% of all capacitor banks are switched up to 300 times a year, and an additional 30% are switched up to 700 times a year. Capacitor-switching transients have caused many problems for the past few years [8–14]. Among these problems, the capacitors bearing switching overvoltage may be the most direct and severe one. Thus, studying the effect of capacitor-switching transients on capacitors is essential.

In China, some 220 kV power substations (especially in Hainan, an island in Southern China) or most 500 kV series compensation station do not directly offer power supply to the users, and their 10 kV busbars are not usually connected to the industry loads or the other loads. Substation transformer and reactive compensation equipment are connected to 10 kV busbars in most of these substations. According to the fault statistics report from Chongqing electric power corporation, the faults caused by 10 kV switching transients are always under the condition with no load or light loads. Thus, the condition of no load or light loads will be more severe to the safety of power system and equipment during the switching transients.

In a 10 kV ungrounded neutral system, switching shunt capacitor banks often caused severe insulation faults of shunt capacitor banks or dry type transformers in China in recent years [15,16]. For example, in our previous work on switching shunt capacitor banks and shunt reactors [15,16], 12 kV vacuum circuit breakers (VCBs) were used, and capacitor explosion occurred during our field experiments on switching 10 kV shunt capacitor banks using ordinary VCBs in Chongqing, China, as shown in Figure 1. After checking the burnt capacitor body, we found that the failure was caused by a breakdown between the two terminals of the capacitors. Overvoltage was deemed as the main reason of this accident. Therefore, based on our previous study and experience in this field, in a 10 kV ungrounded system, we proposed that the VCBs with phase selection was a potential and alternative method to limit the overvoltage and overcurrent caused by switching shunt capacitor banks.

Figure 1. An explosion involving switching 10 kV shunt capacitor banks with vacuum circuit breakers (VCBs). (**a**) Testing site; and (**b**) burnt capacitors.

Even though the transient duration is usually very short, switching shunt capacitor banks produces harmful transients which can reduce the life of the devices [17,18]. Conventionally, pre-insertion resistors or inductors, and R-C snubber circuits, have been applied to limit the switching transients during opening or closing shunt capacitor banks [19,20]. These methods may have reduced or limited the switching overvoltage and transients to some extent. However, pre-insertion resistors will dissipate energy and release a large amount of heat, and the operation of pre-insertion inductors may cause electric sparks and bring about resonance. Pre-insertion resistors or inductors will also result in increasing equipment costs and may not provide appreciable transient reduction [18,21]. The method of R-C snubber circuits may bring about severe harmonic distortion [21,22]. Thus, an alternative, more economical, and effective approach to reduce and limit the switching overvoltage and transients is phase-controlled switching. The significant advantages of phase-controlled switching include limitation of closing inrush currents, suppression of switching overvoltages, lower rate of power equipment failures and less maintenance of frequently used circuit breakers [17,19,23]. CIGRE Working Group A3-07 presented a series of studies on the development and application of VCBs with phase selection [17,24]. According to their statistical data, circuit breakers with phase selection have been widely used in 26.4–800 kV power systems, and about 64% of all circuit breakers with phase selection are applied to switch shunt capacitor banks. Many theoretical and simulation studies on the overvoltage and insulation failures caused by switching shunt capacitor banks have been reported [25–28]. However, only a few studies have been done on the comparison and application of ordinary and phase-controlled VCBs in substation by field tests. For this reason, field measurements of switching 10 kV shunt capacitor banks have been conducted with the use of both ordinary VCBs and phase-controlled VCBs to compare the opening overvoltage and closing current caused by 10 kV switching shunt capacitor banks.

The main contribution of this study is to investigate and demonstrate the potential effects and advantages of phase-controlled VCBs in 10 kV switching transients instead of 110 kV system, when compared with the ordinary VCBs. The rest of this paper is organized as follows. In Section 2, the overvoltage and the overcurrent of switching 10 kV Y-connected shunt capacitor banks are analyzed by analytical method based on a simplified circuit. In Section 3, a phase-selecting control strategy for VCBs to switch on or to switch off capacitor banks is proposed. In Section 4, six typical cases of field experiments are compared and discussed. Lastly, the conclusion is presented in Section 5.

2. Analytical Analysis

The analytical analysis of switching shunt capacitor banks is studied. Figure 2a illustrates the switching shunt capacitor bank circuit for an ungrounded power system with Y-connected shunt capacitor banks. $U_{sa}(t)$, $U_{sb}(t)$, and $U_{sc}(t)$ are the power sources. R is the internal resistance of source. K_a, K_b, and K_c are the circuit breakers. L is the series inductance, and C are the shunt capacitor banks. In a 10 kV ungrounded neutral system, the typical values of these parameters are as follows: $U_{sa}(t) = 8.165\sin(100\pi t)$ kV, $U_{sb}(t) = 8.165\sin(100\pi t + 120°)$ kV, $U_{sc}(t) = 8.165\sin(100\pi t - 120°)$ kV, $R = 0.2\ \Omega$, $C = 27.14\ \mu F$, $L = 9.90$ mH. The reduced circuit in Figure 2b can be obtained with the assumption of simultaneous switching on/off for the three-phase circuit breakers.

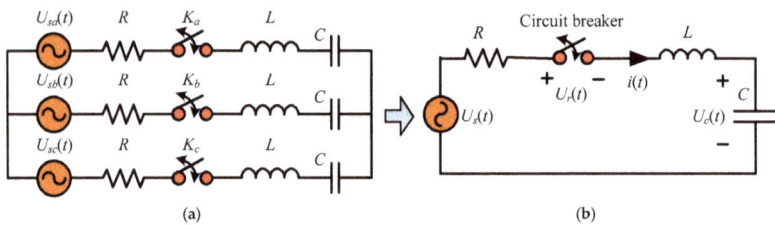

Figure 2. Circuit of switching shunt capacitor banks. (**a**) Symmetrical three-phase circuit; and (**b**) reduced circuit.

2.1. Closing Operation

According to the principle of Kirchhoff's voltage law and initial conditions, Equation (1) can be obtained as:

$$LC\frac{d^2U_c(t)}{dt^2} + RC\frac{dU_c(t)}{dt} + U_c(t) = U_s(t) \tag{1}$$

where $U_s(t) = U_{sm}\sin(\omega t + \theta_c)$, ω is the angular frequency of source, θ_c is the closing phase angle, and U_{sm} is the source voltage amplitude.

With $\theta_c = 90°$, the solutions for Equation (1) are given as follows:

$$U_c(t) = U_{cm}\cos(\omega t) + (U_0 - U_{cm})\cos(\omega_0 t) \tag{2}$$

$$i_c(t) = C\frac{dU_c(t)}{dt} = -U_{cm}\omega C\sin(\omega t) - \frac{(U_0 - U_{cm})}{\sqrt{L/C}}\sin(\omega_0 t) \tag{3}$$

where U_0 is the value of the initial capacitor voltage, U_{cm} is the steady voltage of capacitor $U_{cm} = \dfrac{U_{sm}}{\omega C\sqrt{R^2 + (\omega L - 1/\omega C)^2}}$, $\omega_0 = 1/\sqrt{(LC)}$, and ω_0 is the oscillation frequency. The amplitude of the closing current and voltage are given as:

$$|U_{cmax}| = |2U_{cm} - U_0| \tag{4}$$

$$i_{cmax} = I_{cm}(1 + f_0/f) \qquad (5)$$

where I_{cm} is the steady current of capacitor banks $I_{cm} = \dfrac{U_{sm}}{\sqrt{R^2 + (\omega L - 1/\omega C)^2}}$, $f_0 = \omega_0/2\pi$, and $f_0 \geqslant 500$ Hz in the most serious case. Thus, the closing overvoltage is below $2.0U_{cm}$, and the closing current may reach $11\,I_{cm}$ in the most serious case.

With $\theta_c = 0°$, the amplitude of the closing current and voltage are given in the same way as:

$$U_{cmax} = U_{sm}(1 + f/f_0) \qquad (6)$$

$$|i_{cmax}| = |2I_{cm} - I_0| \qquad (7)$$

where I_0 is the value of the initial capacitor current. Therefore, closing at $\theta_c = 0°$ significantly decreases the closing current by comparing Equation (7) with Equation (5).

2.2. Opening Operation

When the moving and static contacts are separated, $U_c(t)$ is almost equal to $U_s(t)$ because the arc voltage and inductance voltage are very low ($j\omega L = 1/j\omega C \times 5\%$). The capacitor voltage will be maintained at U_{sm} after arc extinguishing if capacitor the discharge is ignored. The transient recovery voltage of the circuit breaker contacts will be as follows:

$$U_r(t) = U_s(t) - U_{sm} = U_{sm}\left[\sin(\omega t + \pi/2) - 1\right] \qquad (8)$$

$U_r(t)$ is equal to $-2U_{sm}$ after half a cycle ($\omega t = \pi$). If the dielectric recovery voltage is less than $2U_{sm}$ at this moment, the re-striking and subsequent high-frequency oscillation will occur. The high-frequency voltage and current of the capacitor are given as follows:

$$\begin{cases} U_c(t) = U_{sm} - 2U_{sm}\cos(\omega_0 t) \\ i_c(t) = C\dfrac{dU_c(t)}{dt} = 2\omega_0 C U_{sm}\sin(\omega_0 t) \end{cases} \qquad (9)$$

The overvoltage of the capacitor is equal to $3U_{sm}$ at $i_c(t)$ initially crossing zero during high-frequency oscillation. In the same moment, the overvoltage of capacitor will be maintained at $3U_{sm}$ if the arc extinguishes. After a half cycle, the transient recovery voltage of the contacts reaches $4U_{sm}$. If re-striking occurs again, the maximum value of $U_c(t)$ will reach $5U_{sm}$. With this analogy, the overvoltage of the capacitor becomes 3 p.u., 5 p.u., 7 p.u., and so on.

It should be noted that, however, in practical switching operation, the switching operation time cannot be completely consistent because of the switching time scatter of three-phase vacuum breakers. In a 10 kV ungrounded system, shunt capacitor banks are always in Y-type ungrounded neutral in China; random non-simultaneously switching generally will result in a shift of the voltage of the neutral point of the capacitor banks, higher overvoltage and overcurrent on capacitors and circuit breakers, and more complicated oscillation [29,30]. In particular, the overvoltage and overcurrent of the last switching pole will be affected by the switching angle, delay time and line-to-line voltage of the other two phases [29,30]. Thus, in 10 kV ungrounded system, the phase-controlled switching method is proposed in this study to limit the overvoltage and transients.

3. Phase-Selecting Control Strategy for 10 kV Ungrounded Capacitor Banks

The structure of phase-controlled VCBs used in this experiment (MDS2B-12) is presented. As depicted in Figure 3, the MDS2B-12 mainly consists of a separate three-pole permanent magnetic actuator VCB, a three-pole phase selection intelligent controller, a power supply, a potential transformer testing capacitance residual voltage, and a telecommunications system.

The strategies and processes of switching shunt capacitor banks based on the phase-controlled methodology are analyzed, as shown in Figures 4 and 5. To reduce the closing inrush current for ungrounded shunt capacitor banks, the strategy for closing capacitor banks is to close the two phases while their phase voltages are equal and then to close the third phase while its phase voltage crosses zero (5 ms later than the above two phases). Figure 4 shows the sequence of closing capacitor banks with phase selection. Taking the line voltage of phase A and phase B as reference, the breaker receives the closing signal at t_0, the breakers of phase A and phase B close at t_{ab}, and the breaker of phase C closes at t_c. t_{Ad}, t_{Bd}, and t_{Cd} are the closing delay times, respectively.

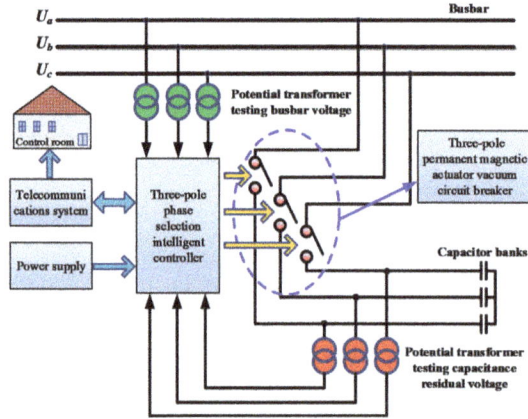

Figure 3. Structure chart of phase-controlled VCBs (MDS2B-12).

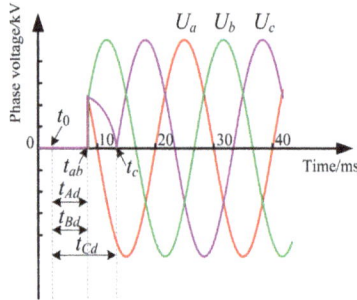

Figure 4. Sequence of closing capacitor banks with phase selection.

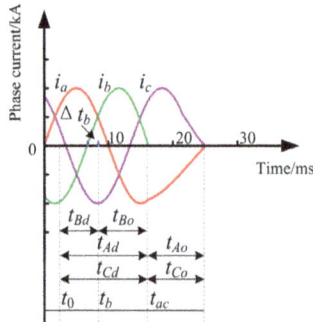

Figure 5. Sequence of opening capacitor banks with phase selection.

The breakers should start to open a little earlier than the phase current crosses zero to ensure sufficient circuit breaker fracture gap distance (Δt_b as shown in Figure 5). The strategy for opening capacitor banks is to open one phase and then to open the other two phases when the current of the first-pole-to-clear reaches zero (5 ms later than the first phase). Figure 5 shows the sequence of opening capacitor banks with phase selection. The breaker receives the opening signal at t_0, the breaker of phase B opens at t_b, and breakers of phase A and phase C opens at t_{ac}. t_{Ad}, t_{Bd}, and t_{Cd} are the opening delay times, respectively. t_{Ao}, t_{Bo}, and t_{Co} are the times between the moving contact starting to move and the moment the moving and static contacts are completely separated.

4. Field Test Results and Discussion

The field test of 10 kV switching shunt capacitor banks is conducted on the 10 kV side of a 110 kV substation in Chongqing, China. The short circuit capacity for the 10 kV system is 245 MVA, and the system short circuit reactance is 0.45 Ω. Figure 6 shows the site layout of the field experiment, and Figure 7 presents the corresponding test circuit. #653 represents the ordinary VCBs, and #65301, #65302, and #65303 represent the three groups of phase-controlled VCBs, respectively. The layout of the voltage measurement points (VMPs) and the current measurement points (CMPs) in the field test are also shown in Figure 7. The divider ratio of the capacitor voltage dividers is 350:1, and the frequency bandwidth of the capacitor dividers is from 50 Hz to 50 MHz. The current measurement range and the frequency bandwidth of the Rogowski coils are 0–6 kA and 30 Hz–1 MHz, respectively. The parameters of the capacitors and inductors are shown in Table 1.

During the operation of ordinary VCBs and phase-controlled VCBs switching on and off the corresponding shunt capacitor banks, the signals from the Rogowski coils and capacitor voltage dividers are transferred to the signal processing unit. The signals are sampled and stored through a multichannel high-speed frequency conversion data acquisition card (maximum sampling frequency of up to 40 MHz) of a computer.

Figure 6. Site layout of the field tests in a 110 kV substation.

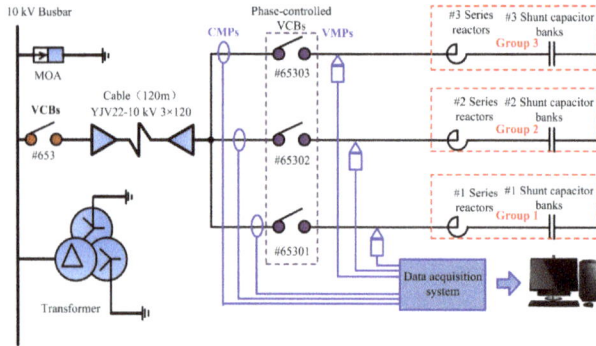

Figure 7. Test circuit of switching 10 kV shunt capacitor banks in a 110 kV substation.

Table 1. Parameters of the capacitors and reactors.

Parameters	Phase A	Phase B	Phase C	Total
Power frequency	50HZ	50HZ	50HZ	-
Rated power of capacitors	2004 kVAR	2004 kVAR	2004 kVAR	6012 kVAR
Number of capacitors	6	6	6	18
Number of groups	3	3	3	9
Number of capacitors in each group	2	2	2	6
Capacitance of each capacitor	27.14 μF	27.14 μF	27.14 μF	-
Rated voltage of each capacitor	$11/\sqrt{3}$ kV	$11/\sqrt{3}$ kV	$11/\sqrt{3}$ kV	-
Rated power of each capacitor	334 kVAR	334 kVAR	334 kVAR	-
Series reactor rate of shunt capacitor bank	5%	5%	5%	5%
Rated power of reactors	100.2 kVAR	100.2 kVAR	100.2 kVAR	300.6 kVAR
Number of reactors	3	3	3	9
Number of groups	3	3	3	9
Number of reactors in each group	1	1	1	3
Inductance of each reactor	9.90 mH	9.90 mH	9.90 mH	-
Rated voltage of each reactor	10 kV	10 kV	10 kV	-
Rated power of each reactor	33.4 kVAR	33.4 kVAR	33.4 kVAR	-
Quality factor (Q) of the reactor	50	50	50	-

The switching of the shunt capacitor banks are conducted 10 times by using phase-controlled VCBs (#65301), and are conducted 12 times by ordinary VCBs (#653). The operation of opening shunt capacitor banks is implemented after 5 min of closing the shunt capacitor banks. However, the operation of closing is conducted after 10 min of opening the shunt capacitor banks to fully discharge the capacitor banks. Six different typical cases are analyzed and discussed, as shown in Table 2 and Figures 8–13. It should be noted that, in Cases 2 and 5, the three groups of capacitor banks are simultaneously switched. The main reason of these two cases is to make a comparison with Cases 1 and 4, in which the shunt capacitors are also switched by ordinary VCBs, to investigate the impact of capacitor compensation capacity on the switching overvoltage, overcurrent, and transient.

Table 2. Typical cases of switching 10 kV shunt capacitor banks.

No.	Operation	Breaker	Group	Voltage Base Value (kV)	Current Base Value (A)
Case 1	Switching on	Without phase selection (#653)	1	8.165	155.56
Case 2	Switching on	Without phase selection (#653)	1–3	8.165	466.69
Case 3	Switching on	With phase selection (#65301)	1	8.165	155.56
Case 4	Switching off	Without phase selection (#653)	1	8.165	155.56
Case 5	Switching off	Without phase selection (#653)	1–3	8.165	466.69
Case 6	Switching off	With phase selection (#65301)	1	8.165	155.56

The peak line-to-neutral voltage of 10 kV power systems is as follows [16,31]:

$$U_{peak} = 10 \times \sqrt{2}/\sqrt{3} = 8.165 \text{ kV} \tag{10}$$

4.1. Case 1

Before the switching-on operations in Case 1, #65302 and #65303 are opened, and #65301 is closed. The circuit breaker of this switching operation is #653. Figure 8a shows the closing current waveform of Case 1 without phase selection. The maximum value of the closing current in this case occurs in phase B, and its value is 0.63 kA (4.04 p.u.). The duration of the closing shunt capacitor banks' current transient is about 110–120 ms, and the high-frequency oscillation frequency of the closing shunt capacitor banks' current is about 198 Hz.

Figure 8b illustrates corresponding closing voltage waveforms in Case 1. The maximum of the closing voltage is 10.08 kV (1.23 p.u.). The duration of the closing shunt capacitor banks' voltage transient is only 20 ms.

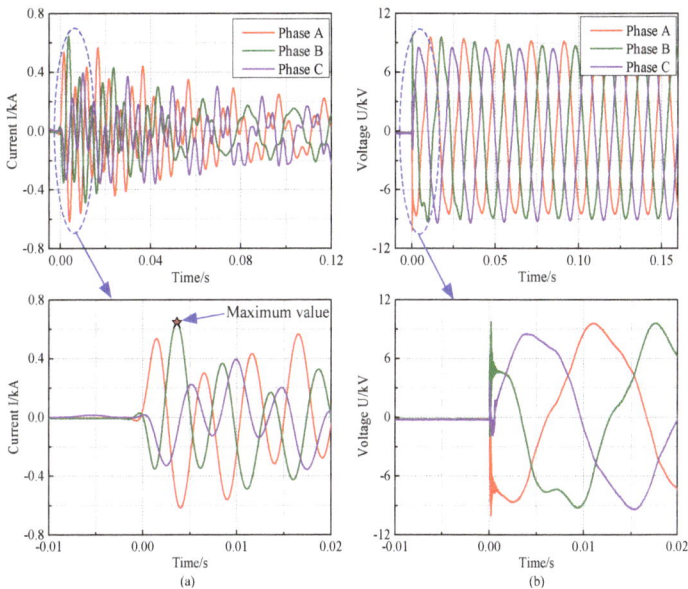

Figure 8. Waveforms of the closing current and capacitor voltage in Case 1. (**a**) Current waveform; and (**b**) voltage waveform.

4.2. Case 2

Before the switching-on operations in Case 2, #65301, #65302, and #65303 are all closed. The circuit breaker of this switching operation is also #653. Figure 9a shows the closing current waveform in Case 2 without phase selection but with three groups of shunt capacitor banks. The maximum value of the closing current in this case occurs in phase A, and its value is − 2.09 kA (4.49 p.u.). The duration of the closing shunt capacitor banks' current transient is about 90–100 ms, and the high-frequency oscillation frequency of the closing shunt capacitor banks' current is about 192 Hz in this case.

Figure 9b illustrates the corresponding closing voltage waveforms in Case 2. The maximum of the closing voltage also occurs in phase A, and its value is 14.25 kV (1.75 p.u.). The voltage transient reaches the steady state after about 20 ms.

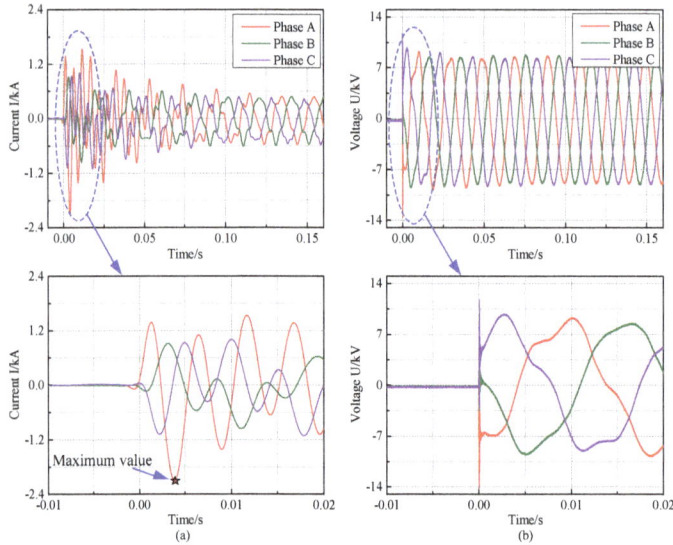

Figure 9. Waveforms of the closing current and capacitor voltage in Case 2. (**a**) Current waveform; and (**b**) voltage waveform.

4.3. Case 3

Before the switching-on operations in Case 3, #653 is closed, and #65302 and #65303 are opened. The circuit breaker of this switching operation is #65301. Figure 10a shows the closing current waveform in Case 3 with phase selection. The strategy of the closing shunt capacitor banks with phase selection is addressed in Section 2. The maximum value of the closing current in this case occurs in phase A, and its value is 0.33 kA (2.12 p.u.). The duration of the closing shunt capacitor banks' current transient is about 100–110 ms, and the high-frequency oscillation frequency of the closing shunt capacitor banks' current is about 206 Hz in this case.

Figure 10b illustrates the corresponding closing voltage waveforms. As shown in Figure 10b, phase A and phase B are closed at t_1 when their line voltage U_{ab} crosses zero, and then phase C is closed after 5 ms at t_2 when its phase voltage crosses zero. The transient voltage of phase C from t_1 to t_2 is equal to the neutral voltage of the shunt capacitor banks. Thus, the zero-crossing switching strategy is achieved. Almost no high-frequency voltage oscillation occurs in this case, and the corresponding maximum of the high-frequency oscillation voltage is very low (about 0.76 p.u.). The time error of the phase-controlled circuit breaker for closing the shunt capacitor banks is below ±0.2 ms.

4.4. Case 4

Before the switching-off operations in Case 4, #65302 and #65301 are opened, and #65301 is closed. The circuit breaker of the switching operation is #653. Figure 11a shows the opening current waveform in Case 4 without phase selection. The breaker contacts of phase B start to separate at t_3, and phase B is the first-pole-to-clear. The transient currents of phase A and phase C are always equal, but in the opposite direction, after the current of phase B decreases to zero at the moment of t_4, which is also the time of the contacts of phase A and phase C start to separate. The breaking arc duration of phase B is $\Delta t_1 = 1.7$ ms ($\Delta t_1 = t_4 - t_3$). After a quarter of a cycle (5 ms), the currents of phase A and phase C both reach zero at the moment of t_5. Thus, the phenomenon of the power frequency extinguishing arcing is observed. Statistical results according to our field tests show that the breaking arc duration of the first-pole-to-clear (phase B in this test) is almost 1.0–4.5 ms.

Figure 11b illustrates the corresponding opening voltage waveforms in Case 4. The maximum of the opening voltage in this case occurs in the first-pole-to-clear (phase B), and its value is 18.45 kV (2.26 p.u.). Residual charge of the capacitors remains more than 30 power frequency cycles after the capacitors are disconnected. Thus, the switching off operation is 10 min later.

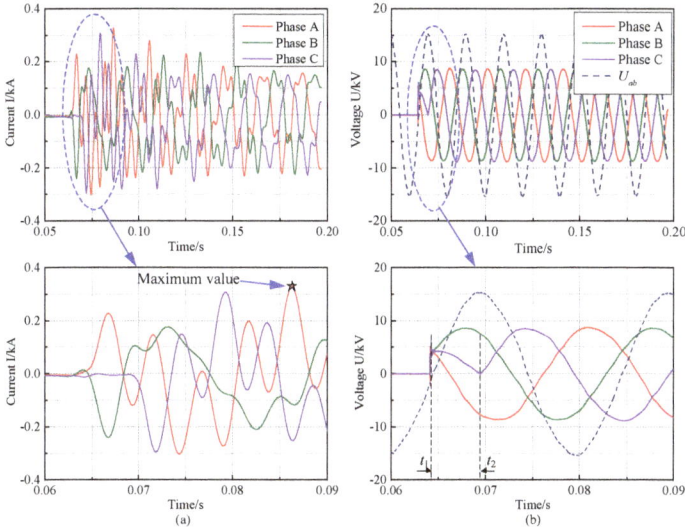

Figure 10. Waveforms of the closing current and capacitor voltage in Case 3. (**a**) Current waveform; and (**b**) voltage waveform.

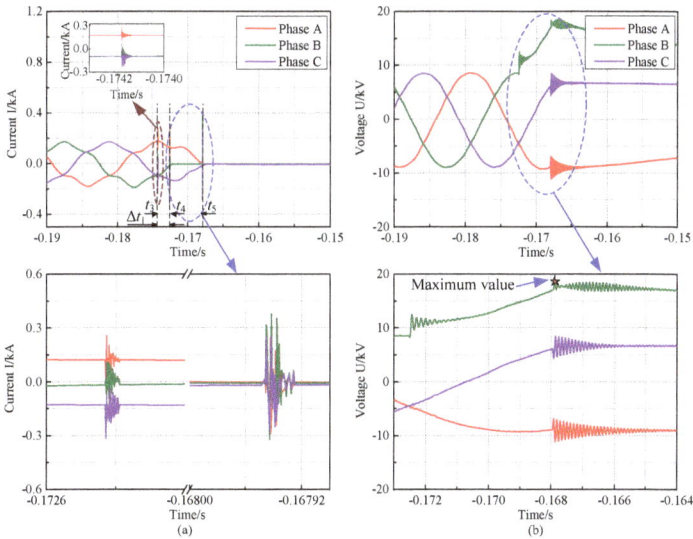

Figure 11. Waveforms of the opening current and capacitor voltage in Case 4. (**a**) Current waveform; and (**b**) voltage waveform.

4.5. Case 5

Before the switching-off operations in Case 5, #65301, #65302, and #65303 are all closed. The circuit breaker of this switching operation is still #653. Figure 12a shows the opening shunt capacitors banks' current waveform in Case 5 without phase selection but with three groups of shunt capacitor banks. The breaker contacts of phase B start to separate at t_6, and phase B is also the first-pole-to-clear. The breaking arc duration of phase B is $\Delta t_2 = 4.1$ ms ($\Delta t_2 = t_7 - t_6$). The currents of phases A and C are equal but opposite in direction from t_7 to t_8, during which the current of phase B reaches zero. After t_8, the currents of the three phases remain zero.

Figure 12b shows the opening voltage waveforms in Case 5. The maximum value of the opening voltage is -18.20 kV (2.23 p.u.), and it still occurs in the first-pole-to-clear (phase B) in this case.

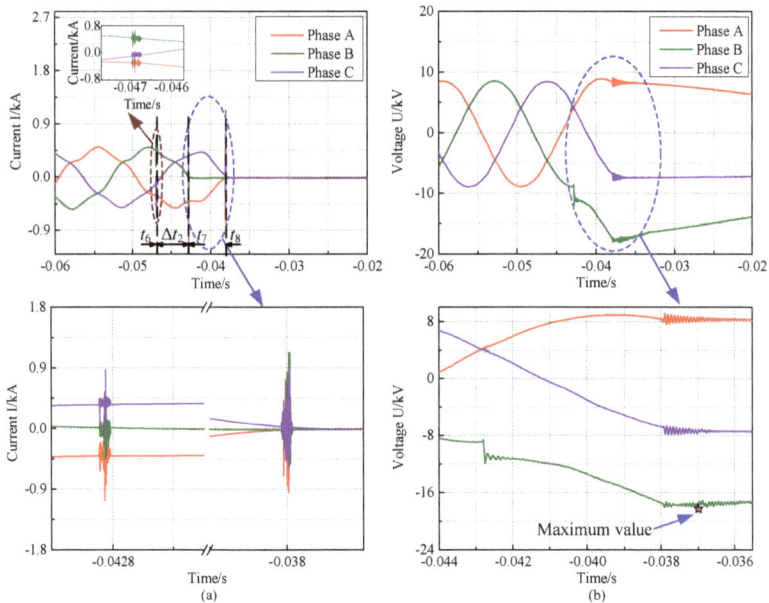

Figure 12. Waveforms of the opening current and capacitor voltage in Case 5. (**a**) Current waveform; and (**b**) voltage waveform.

4.6. Case 6

Before the switching-off operations in Case 6, #653 is closed, and #65302 and #65303 are opened. The circuit breaker of this switching operation is #65301. Figure 13a shows the opening current waveforms in Case 6 with phase selection. The strategy for opening shunt capacitor banks with phase selection is also addressed in Section 2. The breaker contacts of phase B start to separate 2.67 ms ($\Delta t_3 = 2.67$ms) later than its last current zero-crossing point at t_9, and phase B is the first-pole-to-clear. The transient currents of phase A and phase C are always equal, but in the opposite direction, after the current of phase B decreases to zero at the moment of t_{10}, which is also the time that the contacts of phase A and phase C start to separate. The breaking arc duration of phase B is $\Delta t_4 = 7.4$ ms ($\Delta t_4 = t_{10} - t_9$). After a quarter of a cycle (5 ms), the currents of phase A and phase C both reach zero at the moment of t_{11}. Therefore, the power frequency extinguishing arcing still occurs in this case. Statistical results according to our field tests indicate that the breaking arc duration of the first-pole-to-clear (phase B in this test) is about 7.5 ms, and the time error of the phase-controlled circuit breaker for opening shunt capacitor banks is below ±0.3 ms.

Figure 13b illustrates the opening voltage waveform in Case 6. The maximum of the opening voltage in this case occurs in the first-pole-to-clear (phase B), and its value is −12.83 kV (1.57 p.u.). The residual charge of capacitors remains more than 30 power frequency cycles after the capacitors are disconnected.

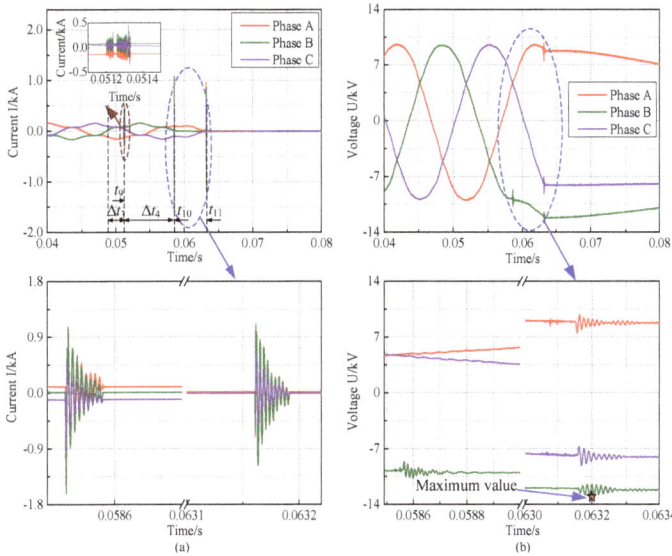

Figure 13. Waveforms of the opening current and capacitor voltage in Case 6. (**a**) Current waveform; and (**b**) voltage waveform.

5. Conclusions

Based on our previous work in this field [15,16], this study presents a complete comparison between ordinary VCBs and phase-controlled VCBs in the application of switching 10 kV shunt capacitor banks through field tests. The following conclusions are drawn:

» The overcurrent of closing 10 kV shunt capacitor banks was about 2.12 p.u. with phase selection, and it was far below than those of ordinary VCBs (4.04 p.u. for Case 1 and 4.49 p.u. for Case 2). Moreover, high-frequency voltage oscillation did not occur for switching on shunt capacitor banks when phase-controlled VCBs were used.

» The overvoltage of opening 10 kV shunt capacitor banks was about 1.57 p.u. with phase selection, and it was below those of ordinary VCBs (2.26 p.u. for Case 4 and 2.23 p.u. for Case 5).

» The arc duration of closing shunt capacitor banks without phase selection was about 1.0–4.5 ms. However, for the phase-controlled VCBs, the circuit breaker of the first-pole-to-clear was opened 2–3 ms later than its last current zero-crossing point, which result in an average of 7.5 ms arc duration of first-pole-to-clear.

» The time error of phase-controlled VCBs for opening and closing shunt capacitor banks was below ±0.3 ms. Furthermore, the higher arc duration increased the fracture gap distance of phase-controlled VCB contacts, and the probability of prestrike and re-ignition was reduced. This can contribute to keeping the power system safe and achieving steady operation.

Acknowledgments: This work was supported by the National Natural Science Foundation of China (51177182), the National Natural Science Foundation of China (51477018), the National Natural Science Foundation of China (51507019), and the Funds for Innovative Research Groups of China (51321063).

Author Contributions: Wenxia Sima contributed to the research idea and theoretical analysis of switching shunt capacitor banks. Mi Zou designed the experiments, and drafted the manuscript. Qing Yang, Ming Yang and Licheng Li worked on the data analysis and revision of the manuscript. All authors have read and approved the final manuscript.

Conflicts of Interest: The authors declare no conflict of interest.

References

1. Dullni, E.; Shang, W.K.; Gentsch, D. Switching of capacitive currents and the correlation of restrike and pre-ignition behavior. *IEEE Trans. Dielectr. Electr. Insul.* **2006**, *13*, 65–71. [CrossRef]
2. Glinkowski, M.; Greenwood, A.; Hill, J.; Mauro, R.; Varneckes, V. Capacitance switching with vacuum circuit-breakers—A comparative-evaluation. *IEEE Trans. Power Deliv.* **1991**, *6*, 1088–1095. [CrossRef]
3. Jones, R.A.; Fortson, H.S. Consideration of phase-to-phase surges in the application of capacitor banks. *IEEE Trans. Power Deliv.* **1986**, *1*, 240–244. [CrossRef]
4. Witte, J.F.; DeCesaro, F.P.; Mendis, S.R. Damaging long term overvoltages on industrial apacitor banks due to transformer energization inrush currents. *IEEE Trans. Ind. Appl.* **1994**, *30*, 1107–1115. [CrossRef]
5. Bruns, D.P.; Newcomb, G.R.; Miske, S.A.; Taylor, C.W.; Lee, G.E.; Edris, A.A. Shunt capacitor bank series group shorting (CAPS) design and application. *IEEE Trans. Power Deliv.* **2001**, *16*, 24–32. [CrossRef]
6. Anderson, E.; Karolak, J.; Wisniewski, J. Application of modern switching methods in reactive power compensation systems. *Prz. Elektrotechniczny* **2012**, *88*, 1–5.
7. Bonfanti, I. Shunt Capacitor Bank Switching, Stressesand Test methods (2nd part) (Cigre WG 13-04). *Electra* **1999**, *183*, 13–41.
8. Mcgranaghan, M.F.; Reid, W.E.; Law, S.W.; Gresham, D.W. Overvoltage protection of shunt-capacitor banks using MOV arresters. *IEEE Trans. Power Appar. Syst.* **1984**, *103*, 2326–2336. [CrossRef]
9. Pflanz, H.M.; Lester, G.N. Control of overvoltages on energizing capacitor banks. *IEEE Trans. Power Appar. Syst.* **1973**, *92*, 907–915. [CrossRef]
10. Boehne, E.W.; Low, S.S. Shunt capacitor energization with vacuum interrupters—A possible source of overvoltage. *IEEE Trans. Power Appar. Syst.* **1969**, *88*, 1424–1443. [CrossRef]
11. Surge Protective Devices Committee. Surge protection of high-voltage shunt capacitor banks on AC power-systems survey results and application considerations. *IEEE Trans. Power Deliv.* **1991**, *6*, 1065–1072.
12. Das, J.C. Analysis and control of large-shunt-capacitor-bank switching transients. *IEEE Trans. Power Appar. Syst.* **2005**, *41*, 1444–1451.
13. Zadeh, M.K.; Hinrichsen, V.; Smeets, R.; Lawall, A. Field emission currents in vacuum breakers after capacitive switching. *IEEE Trans. Dielectr. Electr. Insul.* **2011**, *18*, 910–917. [CrossRef]
14. Smeets, R.P.P.; Thielens, D.W.; Kerkenaar, R.W.P. The duration of arcing following late breakdown in vacuum circuit breakers. *IEEE Trans. Plasma Sci.* **2005**, *33*, 1582–1588. [CrossRef]
15. Yang, Q.; Ouyang, S.; Sima, W.X.; Xi, S.Y.; Kang, H.F.; Yang, M. Mechanism of overvoltage induced by fast switching on-off 10 kV shunt capacitors using vacuum circuit breakers. *High Volt. Eng.* **2014**, *40*, 3135–3140. (In Chinese)
16. Yang, Q.; Zhang, Z.H.; Sima, W.X.; Yang, M.; Wei, G.W. Field experiments on overvoltage caused by 12 kV vacuum circuit breakers switching shunt reactors. *IEEE Trans. Power Deliv.* **2015**. [CrossRef]
17. CIGRE (International Council on Large Electric systems) Working Group A3, 07. *Controlled Switching of HVAC Circuit Breaker: GUIDANCE for Further Applications Including Unloaded Transformer Switching, Load and Fault Interruption and Circuit-Breaker Updating*; CIGRE: Paris, France, 2004.
18. Mahurkar, T.M.; Murali, M. Suppression of capacitor switching transients using symmetrical structure transient limiter [SSTL] and its applications. In Proceedings of the International Conference on Computation of Power, Energy, Information and Communication, Chennai, India, 16–17 April 2014.
19. Filion, Y.; Coutu, A.; Isbister, R. Experience with controlled switching systems (CSS) used for shunt capacitor banks: planning, studies and testing accordingly with CIGRE A3-07 Working Group Guidelines. In Proceedings of the Quality and Security of Electric Power Delivery Systems Conference, Montreal, PQ, Canada, 8–10 October 2003.

20. Borghetti, A.; Napolitano, F.; Nucci, C.A.; Paolone, M.; Sultan, M.; Tripaldi, N. Transient recovery voltages in vacuum circuit breakers generated by the interruption of inrush currents of large motors. In Proceedings of the International Conference on Power Systems Transients (IPST2011), Delft, The Netherlands, 14–17 June 2011.
21. Ding, F.H.; Duan, X.Y.; Zou, J.Y.; Liao, M.F. Controlled switching of shunt capacitor banks with vacuum circuit breaker. In Proceedings of the International Symposium on Discharges and Electrical Insulation in Vacuum (ISDEIV 2006), Matsue, Janpan, 25–29 September 2006.
22. Ning, D.; Yonggang, G.; Jingsheng, Z.; Jirong, N.; Shuhua, Y.; Guozheng, X. Protections of overvoltages caused by 40.5 kV vacuum circuit breakers switching off shunt reactors. *High Volt. Eng.* **2010**, *36*, 345–349. (In Chinese)
23. Roseburg, T.; Tziouvaras, D.; Pope, J. Controlled Switching of HVAC Circuit Breakers: Application Examples and Benefits. In Proceedings of the 61st Annual Conference for Protective Relay Engineers, College Station, TX, USA, 1–3 April 2008.
24. Reid, J.F.; Tong, Y.K.; Waldron, M.A. *Controlled Switching Issues and the National Grid Company's Experience of Switching Shunt Capacitor Banks and Shunt Reactor*; August Session; CIGRE: Paris, France, 1998; pp. 223–231.
25. Jones, S.; Gardner, K.; Brennan, G. *Switchgear issue in deregulated electricity industries in Australia and New Zealand*; August Session; CIGRE: Paris, France, 2000; pp. 1–6.
26. Nordin, R.; Holm, A.; Norberg, P. *Ten Years of Experience with Controlled Circuit Breaker Switching in the Swedish Regional Network*; August Session; CIGRE: Paris, France, 2002; pp. 21–23.
27. Li, D. Study on forepart re-strike rate of 35kV vacuum circuit breaker in the closing-opening process of capacitor group. *High Volt. Eng.* **2002**, *28*, 22–23. (In Chinese)
28. Yan, X.L.; Li, Z.B.; Wang, C.Y.; Liu, B.Y.; Wang, H. Analysis on characteristics of phase-selectable circuit breaker for switching capacitor banks in UHV power transmission project. *Power Syst. Technol.* **2014**, *38*, 1772–1778.
29. Liu, S.H. Analysis and simulation study on overvoltage of shunt power capacitor. Master's thesis, South China University of Technology, Guangdong, China, 2011. (In Chinese)
30. Zhong, X. Study on the transition process of closing shunt capacitors in 10 kV power system. Master's Thesis, Chongqing University, Chongqing, China, 2010. (In Chinese)
31. *Insulation Co-Ordination—Part I: Definitions, Principles and Rules*; IEC Std. 60071-1:2006; IEC (the International Electrotechnical Commission): Geneva, Switzerland, 2006.

energies

MDPI

Article

Multi-Objective Demand Response Model Considering the Probabilistic Characteristic of Price Elastic Load

Shengchun Yang [1,2,*], Dan Zeng [2], Hongfa Ding [1], Jianguo Yao [2], Ke Wang [2] and Yaping Li [2]

[1] School of Electrical & Electronic Engineering, Huazhong University of Science and Technology, Wuhan 430074, Hubei, China; dinghongfa@sina.com

[2] China Electric Power Research Institute-Nanjing Branch, Nanjing 210003, Jiangsu, China; zengdan@epri.sgcc.com.cn (D.Z.); yaojianguo@epri.sgcc.com.cn (J.Y.); wangke@epri.sgcc.com.cn (K.W.); liyaping@epri.sgcc.com.cn (Y.L.)

* Correspondence: yangshengchun@epri.sgcc.com.cn; Tel.: +86-135-0516-2439

Academic Editor: Ying-Yi Hong

Received: 30 November 2015; Accepted: 18 January 2016; Published: 27 January 2016

Abstract: Demand response (DR) programs provide an effective approach for dealing with the challenge of wind power output fluctuations. Given that uncertain DR, such as price elastic load (PEL), plays an important role, the uncertainty of demand response behavior must be studied. In this paper, a multi-objective stochastic optimization problem of PEL is proposed on the basis of the analysis of the relationship between price elasticity and probabilistic characteristic, which is about stochastic demand models for consumer loads. The analysis aims to improve the capability of accommodating wind output uncertainty. In our approach, the relationship between the amount of demand response and interaction efficiency is developed by actively participating in power grid interaction. The probabilistic representation and uncertainty range of the PEL demand response amount are formulated differently compared with those of previous research. Based on the aforementioned findings, a stochastic optimization model with the combined uncertainties from the wind power output and the demand response scenario is proposed. The proposed model analyzes the demand response behavior of PEL by maximizing the electricity consumption satisfaction and interaction benefit satisfaction of PEL. Finally, a case simulation on the provincial power grid with a 151-bus system verifies the effectiveness and feasibility of the proposed mechanism and models.

Keywords: price elastic load (PEL); demand response; uncertainty; electricity consumption satisfaction (ECS); interaction benefit satisfaction (IBS); stochastic optimization

1. Introduction

Wind power is one of the fastest growing and cheapest renewable energy sources. In Reference [1], wind power is expected to account for 50% of the world's clean energy by 2030. However, wind power and other renewable energies are often variable, intermittent, anti-peaking, and difficult to dispatch. In traditional economic dispatch, generation follows the change of load. Therefore, wind power is unsuitable for the operation of a system with a high wind power penetration. The coordinated interactions among power sources, power grid, and loads are studied to address the challenges by optimizing the allocation of resources, such as traditional unit commitment methods and demand response (DR) programs [2,3].

In recent years, many studies [4–7] have indicated that the uncertainty and forecast errors of wind power have a significant effect on dispatch. Moreover, some studies have recently concluded that a power system can accommodate considerable wind power with high reliability by considering various

DR dispatch programs, *i.e.*, DR can be integrated as dispatchable resources that can eliminate wind power output randomness [8–11]. Reference [12] studied a stochastic unit commitment model for assessing the effects of the large-scale integration of renewable energy sources and deferrable demand in power systems in terms of reserve requirements. In Reference [13], for an electric market with high wind power penetration, a new two-step design approach of forward electricity markets containing DR programs is designed.

In general, DR has the potential to accommodate the uncertainty of wind power output. DR can also benefit consumers [14], ancillary services [15], and even all involved market parties [16]. In this field, deterministic analysis methods for DR behavior have been widely studied [17–19]. However, the actual DR is uncertain because of various reasons, including lack of attention, latency in communication, and change in consumption behaviors [20–24]. In the current study, we focus on the uncertainty of the demand response behavior of price elastic load (PEL). In Reference [22], the uncertain region of the price elasticity demand curve varies within a given range. Similarly, the actual price elasticity demand curve is uncertain in nature [23]. This finding indicates that the actual response from consumers in real time can be different from the forecasted values.

The demand response of PEL should be modeled with an uncertain price elasticity demand curve through the preceding analysis. In this paper, a new methodology for analyzing the relationship between the price elasticity and probabilistic characterization of PEL is proposed. Probabilistic characteristics can reflect the uncertainty and subjectivity of demand response behavior under an imprecise price elasticity demand curve. The price elasticity coefficient has a strong leading effect on the error between the real demand response amount and expectation of demand response amount. This study aims to show that the demand curve of PEL can vary within an uncertain range represented by probabilistic mathematical expression. In the aforementioned research, the electricity consumption patterns of end users change via time-varying prices. At the same time, ensuring the satisfaction of electricity customers is an important precondition. Hence, the objective of our optimization model can maximize the electricity consumption satisfaction (ECS) and interaction benefit satisfaction (IBS) of PEL.

The remaining part of this paper is organized as follows. In Section 2, the relationship between the amount of demand response and the efficiency of interactive response is presented. In Section 3, the probabilistic representation and uncertainty range of the PEL demand response are established. In Section 4, both the ECS and IBS of PEL are considered the objectives of stochastic optimization to solve uncertainties from wind power output. A multi-objective optimization method is also developed to solve the problem. In Section 5, a case study is provided and associated simulation results are analyzed. In Section 6, this paper is concluded with a summary of our contributions and conclusions.

2. Stochastic DR Characteristic

This section may be divided by subheadings. It should provide a concise and precise description of the experimental results, their interpretation, as well as the experimental conclusions that can be drawn.

2.1. Demand Response to Balance Wind Power Fluctuations

The anticipated response of PELs can be determined by the static method on the basis of the distributed slack buses [25]. Similarly, we use the static method to determine the allocation of stochastic demand response. Before calculating the probabilistic flow, power imbalance can be allocated to different PELs. Thereafter, the expected demand response amount of each PEL is obtained. Finally, the price signal can be given to each PEL.

The balanced system shows a power imbalance when source-side power fluctuations occur. The imbalance is mitigated by the demand response of PEL. In particular, PEL can increase electricity consumption when wind power is higher than expected. On the contrary, PEL can decrease electricity consumption when wind power output is lower than expected. Finally, the power imbalance of the system can be expressed as follows Equation (1):

$$\sum_{i=1}^{N} \Delta P_{Gi} + \Delta P_{Loss} + \sum_{i=1}^{N} \Delta P_{l,i} = 0 \qquad (1)$$

where, ΔP_{Gi} and $\Delta P_{l,i}$ are the active power and active load at node i respectively, N is the number of buses, and ΔP_{Loss} is the change of the transmission loss in the system. The grid change is usually small in a short time. Therefore, ΔP_{Loss} is negligible. Thus, the interaction benefit of all PELs can be calculated as follows:

$$\sum_{i=1}^{N} \Delta P_{l,i} = -\sum_{i=1}^{N} \Delta P_{Gi} \qquad (2)$$

2.2. Probabilistic Characterization of PELs

In the actual demand response process, the price elasticity demand curve is an uncertain issue. We propose a new methodology for analyzing the relationship between the price elasticity and probabilistic characterization of PEL. This methodology emphatically analyzes the effect of different price elasticity coefficients on the error between the real demand response amount and the expectation.

The price elasticity of demand response refers to the sensitivity of demand to price variation [15], which can be expressed as follows:

$$\alpha = \frac{\partial P}{\partial c} = \frac{c_0}{P_0} \times \frac{dP}{dc} \qquad (3)$$

where, α is price elasticity, which represents the sensitivity of electricity demand (P) with respect to the change of price (c). P_0, c_0 are, respectively, the initial electricity demand and initial price.

The relationship between load i ($P_{l,i}$) and its price (c_i) can be defined as follows:

$$P_{l,i} = P(c_i) \qquad (4)$$

We then describe how the explicit formulation of Equation (4) is obtained. According to the PEL defined with price elasticity, a higher price related to the less electricity consumption of PEL. In Reference [15], the relationship between the active power of PEL i ($P_{l,i}$) and the price (c_i) is linear and can be expressed as follows:

$$P_{l,i} = \alpha_i c_i + \beta_i \quad P_{l,i} \in \left[P_{l,imin}, P_{l,imax} \right] , \qquad (5)$$

where, the coefficients $\alpha_i < 0$ and $\beta_i < 0$.

Then, the PEL demand response curves with uncertainties can be described in Figure 1.

Figure 1 illustrates the price elasticity coefficient differs with different PELs. This observation indicates that the degree of PEL sensitivity to price is different. A small value of $|\alpha|$ means low sensitivity to price while a large value of $|\alpha|$ means high sensitivity to price. The price elasticity coefficient of PEL i is smaller than that of PEL j in Figure 1, *i.e.*, $|\alpha_i| > |\alpha_j|$.

The probability distribution of PEL can be represented in the form of $(P_{i0} + \mu_{\Delta P_i}, \delta_{\Delta P_i})$. When the price elasticity coefficient ($|\alpha|$) is larger, price has a strong leading effect on demand response behavior and the error ($\delta_{\Delta P_i}$) between the PEL real demand response amount and expectation is smaller. On the contrary, a smaller price elasticity coefficient means greater uncertainty in demand response behavior because of the greater change in electricity consumption.

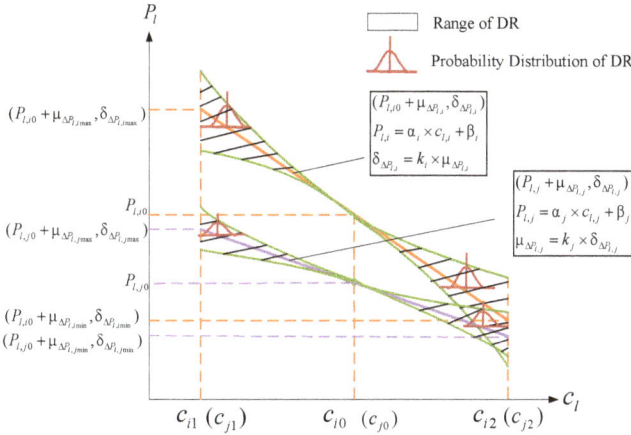

Figure 1. Demand response uncertainty curve of PELs.

The distribution coefficient of the PEL stochastic parameter is defined as k. The standard deviation and average value meet the following:

$$\delta_{\Delta P_l} = k \times \mu_{\Delta P_l}. \tag{6}$$

On the basis of the preceding analysis, we determine that a smaller price elasticity coefficient ($|\alpha|$) means a smaller stochastic parameter (k) and that a larger $|\alpha|$ means a larger k; where $k_i > k_j$. Thus, the uncertain demand response range of PEL i is larger than that of PEL j (Figure 1).

3. PEL Interaction Benefit Model

When PELs participate in grid interactions under the pricing mechanism of electricity markets, their electricity consumption will change with the variation in the relationship of price and power. Thus, electricity expenditure and customer satisfaction are affected.

In general, the relationship between the interaction benefit (C_i) of PEL and its amount of demand response ($\Delta P_{l,i}$) can be defined as follows:

$$C_i = f\left(\Delta P_{l,i}\right). \tag{7}$$

The interaction benefit of PEL (C_i) can be defined as the changes in the electricity costs of customers.

$$C_i = P_{l,i0} \times c_{i0} - P_{l,i} \times c_i. \tag{8}$$

Equation (5) can be transformed into Equation (9):

$$c_i = \frac{1}{\alpha_i} P_{l,i} - \frac{\beta_i}{\alpha_i}. \tag{9}$$

We can obtain Equation (10) by substituting Equation (9) into Equation (8):

$$C_i = -\frac{1}{\alpha_i} \times P_{l,i}{}^2 + \frac{\beta_i}{\alpha_i} \times P_{l,i} + P_{l,i0} \times c_{i0}. \tag{10}$$

In this paper, the relationship between $P_{l,i}$ and $\Delta P_{l,i}$ can be expressed as follows:

$$P_{l,i} = P_{l,i0} + \Delta P_{l,i}. \tag{11}$$

The relationship between C_i and $\Delta P_{l,i}$ can be described as Equation (12) by substituting Equation (11) into Equation (10):

$$C_i = -\frac{1}{\alpha_i} \times \Delta P_{l,i}{}^2 + \frac{\beta_i - 2P_{L,i0}}{\alpha_i} \times \Delta P_{l,i}. \tag{12}$$

4. Stochastic DR Model of PEL

4.1. Demand Response Satisfaction of PEL

The demand response satisfaction of PEL is considered the optimization objective in the demand side. Demand response satisfaction can be measured in two ways: one is electricity consumption satisfaction (ECS), and the other is interaction benefit satisfaction (IBS). In this paper, we define the ECS of PEL and the IBS of PEL as the index to measure the variation of electricity consumption manner and the index to measure the variation of customer electricity costs, respectively.

Before demand response occurs, the electricity consumption manner of PEL should be arranged. At this time, ECS reaches the maximum point. After demand response occurs, PEL changes the electricity consumption manner to pursue the maximum demand response interaction benefit. Thereafter, the PEL comfort of electricity consumption is changed.

The ECS of PEL is established in the response of load and its original level:

$$\eta_i = 1 - |\Delta P_{l,i}| / P_{l,i}, \tag{13}$$

where η_i represents the ECS of PEL i, and ECS submits to $\eta_i \in (0,1]$. ECS reaches the maximum (its value is 1.0) when the electricity consumption manner of PEL is not changed.

Combined with the model of the load response interaction benefit described in Section 2, the IBS of PEL is expressed as follows:

$$\begin{aligned}
\varepsilon_i &= 1 + \frac{P_{l,i0} \times c_{i,0} - P_{l,i} \times c_i}{P_{l,i0} \times c_{i,0}} \\
&= 1 + \frac{-\dfrac{1}{\alpha_i} \times \Delta P_{l,i}{}^2 + \dfrac{\beta_i - 2P_{L,i0}}{\alpha_i} \times \Delta P_{l,i}}{P_{l,i0} \times c_{i,0}} \cdot
\end{aligned} \tag{14}$$

4.2. Probabilistic Demand Response Model

4.2.1. Objective Function

The maximum demand response satisfaction of PEL is considered to be the objective in the load side. Thereafter, by considering the probabilistic demand response of PEL, we maximize the expected demand response satisfaction of various PELs:

$$\begin{cases} \max & E\left(\eta_1^t\right), \cdots, E\left(\eta_i^t\right), \cdots, E\left(\eta_{N_{PL}}^t\right) \\ \max & E\left(\varepsilon_1^t\right), \cdots, E\left(\varepsilon_i^t\right), \cdots, E\left(\varepsilon_{N_{PL}}^t\right) \end{cases} , \forall t = 1, 2, \cdots 24, \tag{15}$$

where, $E\left(*\right)$ represents the mathematical operator of expectation at time t hour.

The objective function is a multi-objective optimization problem. The objective function weighting is introduced to transform the problem into a single-objective optimization problem. Thus, the demand response satisfaction of PEL should be the weighted average number of ECS and IBS. Thereafter, the demand response satisfaction of PEL i is transformed as follows:

$$\max \quad E\left(f_i^t\right) = E\left(\lambda_{i1}\eta_i^t + \lambda_{i2}\varepsilon_i^t\right), \tag{16}$$

$$\lambda_{i1} + \lambda_{i2} = 1, \tag{17}$$

where, λ_{i1} represents the weight of IBS and λ_{i2} represents the weight of ECS. With regard to different PELs, weights can be set as different values to reflect that the degrees of attention to IBS and ECS are different for various PELs.

The following can be obtained by substituting Equations (13) and (14) into Equation (18):

$$E\left(f_i^t\right) = E\left(\begin{array}{c} \lambda_{i,1} \times \left(1 - \dfrac{\Delta P_{l,i}^t}{P_{l,i0}^t}\right) \\[4mm] +\lambda_{i,2} \times \left(1 + \dfrac{-\dfrac{1}{\alpha_i} \times \Delta P_{l,i}^{t\,2} + \dfrac{\beta_i - 2P_{l,i0}}{\alpha_i} \times \Delta P_{l,i}^t}{P_{l,i0}^t \times c_{i,0}^t}\right) \end{array}\right). \tag{18}$$

$$= \lambda_{i,1} \times \left(1 - \frac{E\left(\Delta P_{l,i}^t\right)}{P_{l,i0}^t}\right)$$

$$+\lambda_{i,2} \times \left(1 + \frac{-E\left(\Delta P_{l,i}^{t\,2}\right) + \left(\beta_i - 2P_{l,i0}^t\right) \times E\left(\Delta P_{l,i}^t\right)}{\alpha_i \times P_{l,i0}^t \times c_{i,0}^t}\right)$$

The objective function includes $E(\Delta P_{l,i}^{t\,2})$ and $E(\Delta P_{l,i}^t)$, which are expressed as follows:

$$E\left(\Delta P_{l,i}^{t\,2}\right) = \int_{-\infty}^{+\infty} (\Delta P_{l,i}^{t\,2} \times f(\Delta P_{l,i}^t))d\Delta P_{l,i}^t, \tag{19}$$

$$E(\Delta P_{l,i}^t) = \int_{-\infty}^{+\infty} (\Delta P_{l,i}^t \times f(\Delta P_{l,i}^t))d\Delta P_{l,i}^t, \tag{20}$$

According to probability theory, if $\begin{bmatrix} a, & b \end{bmatrix}$ is the range of $\Delta P_{l,i}^t$, then $\Delta P_{l,i}^t$ follows a normal distribution $N(\mu_{\Delta P_{l,i}^t}, \delta_{\Delta P_{l,i}^t}^2)$. If $\Delta P_{l,i}^t$ submits to $\begin{bmatrix} \mu - 3\delta, & \mu + 3\delta \end{bmatrix} \in \begin{bmatrix} a, & b \end{bmatrix}$, the confidence is 99.7% and can be approximated to one. Thereafter, we convert Equations (19) and (20) to Equations (21) and (22), respectively:

$$E(\Delta P_{l,i}^{t\,2}) = \mu_{\Delta P_{l,i}^t}^2 + \delta_{\Delta P_{l,i}^t}^2, \tag{21}$$

$$E(\Delta P_{l,i}^t) = \mu_{\Delta P_{l,i}^t}. \tag{22}$$

4.2.2. Equality Constraints

According to the description in Section 3, the demand response of PEL should meet the following two conditions:

- Power balance constraints

When wind power fluctuation causes system power imbalance, the fluctuations are absolutely eliminated by the demand response of PEL:

$$\Delta P_{\Sigma L}^t = \sum_{i=1}^{N_{PL}} E(\Delta P_{l,i}^t), \tag{23}$$

where $\Delta P_{\Sigma L}^t$ is the power imbalance caused by wind power output fluctuations at time t hour.

- Stochastic Constraint

The actual demand response is uncertain. The relationship between standard deviation ($\delta_{\Delta P_{l,i}^t}$) and the average value ($\mu_{\Delta P_{l,i}^t}$) of PEL i is expressed as follows:

$$\delta_{\Delta P_{l,i}^t} = k_i \times \mu_{\Delta P_{l,i}^t}. \tag{24}$$

The distribution coefficient of the PEL i stochastic parameter k_i is strongly affected by the price elasticity coefficient ($|\alpha_i|$).

4.2.3. Inequality Constraints

When load increases, demand response constraints are expressed as follows:

$$\begin{aligned}\Delta P_{l,i}^{t\,-} &\geqslant 0 \\ \Delta P_{l,i}^{t\,+} &\leqslant P_{l,imax}^t - P_{l,i0}^t\end{aligned}. \tag{25}$$

When load decreases, the demand response constraints are expressed as follows Equation (26):

$$\begin{aligned}\Delta P_{l,i}^{t\,-} &\geqslant P_{l,imin}^t - P_{l,i0}^t \\ \Delta P_{l,i}^{t\,+} &\leqslant 0\end{aligned}, \tag{26}$$

where ($\Delta P_{l,i}^{t\,-}, \Delta P_{l,i}^{t\,+}$) is PEL response random fluctuation range.

4.3. Solution Methodology

We introduce the weight of the objective function to solve the multi-objective optimization problem, and Equation (16) can be transformed to Equation (27):

$$\max \sum_i^{N_{rl}} v_i \times E\left(f_i^t\right). \tag{27}$$

Finally, the objective function switches to a non-linear optimization problem. We use the particle swarm optimization (PSO) [26] to solve this model.

5. Case Study

5.1. Data and Assumptions

The provincial power grid, which is a main network with 220 kV and 330 kV in Northwest China, contains 151 buses, 252 lines, and 43 generators. The total installed capacity is 24 GW. 11 buses are wind-driven generator with 4 GW capacity. The day-ahead forecasted wind output power and the total load of PEL are predicted in Figure 2. Moreover, a one-hour-ahead forecasted wind power is assumed to have deviation (Figure 2).

Figure 2 illustrates that the one-hour-ahead forecasted wind power fluctuates by the day-ahead forecasted wind output power. When the real-time wind power output is at the day-ahead power value, the power of system is in balance. The power imbalance can be calculated when the wind power output fluctuates. If the value is positive, a power surplus occurs. Moreover, a negative value shows a lack of generator power. Decreasing the load power consumption is needed in this situation.

Assuming that, there will be sufficient PEL to balance the variation in wind. Eight load buses are selected as PELs. The parameters of the price response curve and the weight values of the demand response satisfaction of PELs are presented in Table 1.

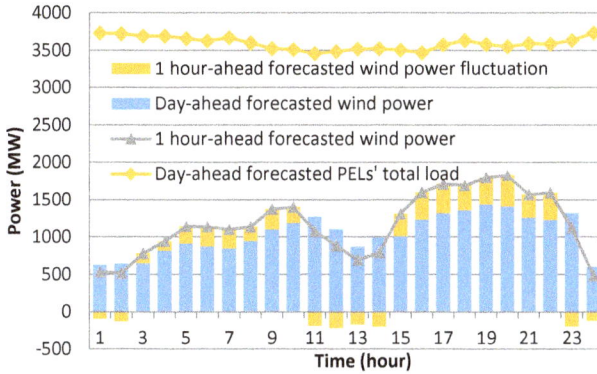

Figure 2. Forecasted Wind Power and PEL Load.

Table 1. Weight Values of PEL Response Satisfaction.

PEL	α_i	β_i	λ_{i1}	λ_{i2}
1	−0.559	9.3403	0.7	0.3
2	−0.509	8.3973	0.7	0.3
3	−0.506	5.3461	0.6	0.4
4	−0.556	4.1522	0.6	0.4
5	−0.107	3.2877	0.5	0.5
6	−0.117	3.2574	0.5	0.5
7	−0.306	2.9123	0.4	0.6
8	−0.336	2.9642	0.4	0.6

The price elasticity coefficient α_i and β_i of each PEL bus can be calculated as Table 1. We then set the corresponding distribution coefficient of probabilistic parameter k_i.

We can also know that the price elasticity coefficients satisfy $|\alpha_4| > |\alpha_1| > |\alpha_3| > |\alpha_2| > |\alpha_8| > |\alpha_7| > |\alpha_6| > |\alpha_5|$, thus indicating that the sensitivity to price of PEL 5 is the smallest and that of PEL 4 is the largest. On the basis of the analysis in Section 4.2, a smaller price elasticity coefficient $|\alpha|$ leads to a stronger response leading role, and a smaller response deviation than expected. Thus, k_i decreases. On the contrary, a larger $|\alpha|$ corresponds to the greater uncertainty of demand response. Thus, k_i increases. Therefore, $k_4 > k_1 > k_3 > k_2 > k_8 > k_7 > k_6 > k_5$. In this paper, the following are set: $k_4 = 0.25$, $k_1 = 0.2$, $k_3 = 0.18$, $k_2 = 0.15$, $k_8 = 0.12$, $k_7 = 0.1$, $k_6 = 0.08$, and $k_5 = 0.05$.

The day-ahead forecasted PEL nodal prices are assumed, as shown in Figure 3.

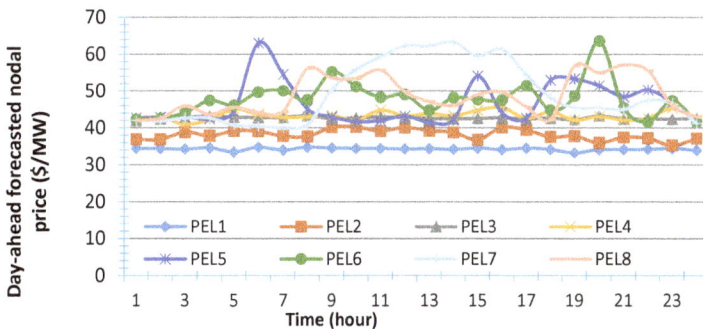

Figure 3. Day-ahead forecasted PEL nodal prices.

5.2. Relationship between Wind Power Fluctuation and Demand Response Amount

By using the simulation that considers the parameters shown in Figures 2 and 3 and Table 1, the calculation results of PEL demand response amounts are shown in Figure 4, the one-hour-ahead nodal prices are shown in the Figure 5, and demand response satisfactions are shown in Figures 6 and 7.

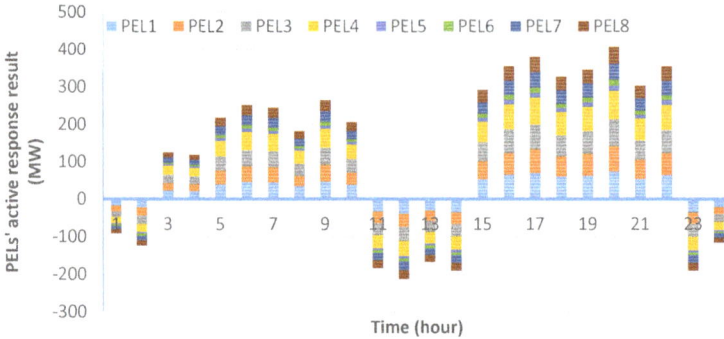

Figure 4. Demand response amounts of PEL.

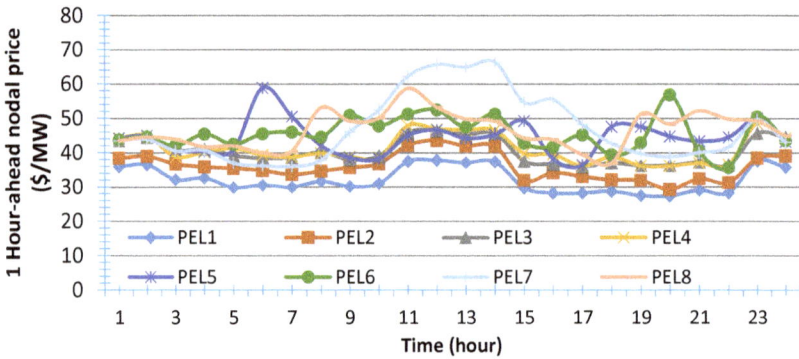

Figure 5. One-hour-ahead PEL nodal prices.

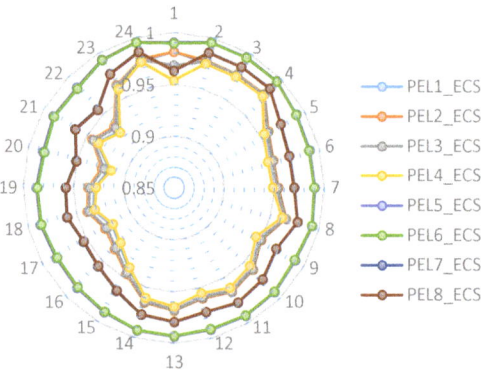

Figure 6. Electricity consumption satisfaction (ECS) of PELs.

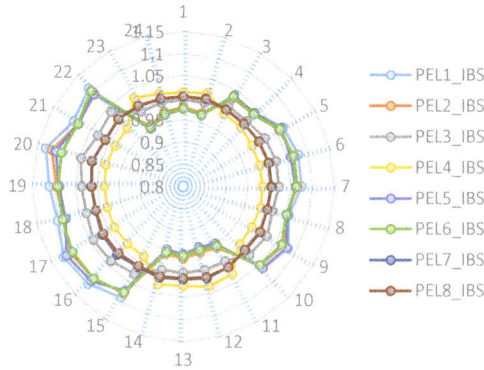

Figure 7. Interaction benefit satisfaction (IBS) of PELs.

Figure 4 illustrates that PELs show the overall responses to wind fluctuation. The comparison of Figures 3 and 5 indicates that each one-hour-ahead PEL nodal price at time 11:00–15:00 is higher than the day-ahead nodal price. Furthermore, each PEL decreases the load power consumption because the wind fluctuation is negative and the generator power is lacking. Figures 6 and 7 show the demand response satisfaction of PEL. A higher demand response at PEL 1–4 leads to a smaller ECS is, but the greater IBS and demand response overall satisfaction are. The reason is that IBS factor and ECS factor of PEL 1, 2 and PEL 3, 4 are (0.7, 0.3) and (0.6, 0.4), respectively. In particular, the factor of IBS is equal or greater than the factor of ECS factor.

5.3. Price Elasticity Affecting Demand Response Amount

Given that wind power output fluctuation is set as Figure 2, the price elasticity of PEL 2 is changed to $(-0.409, 8.3236)$, and other parameters of PELs are the same (Table 1). The demand response amounts of different price sensitivities are studied.

Under this scenario, the calculation results of demand response amount are shown in Figure 8, and demand response satisfaction is shown in Figures 9 and 10.

Figure 8 illustrates that the expected demand response amount of PEL 2 is decreased with decrease of price elasticity coefficient. Figures 9 and 10 illustrate that ECS increases when price elasticity coefficient decreases oppositely, thereby decreasing IBS. The reason is that the decrease of demand response expectation indicates that the change amount of electricity consumption manner decreases, thereby increasing ECS and decreasing IBS. These results are consistent with the analysis of the preceding conclusion.

Figure 8. Demand response amount of PEL 2 with changing α.

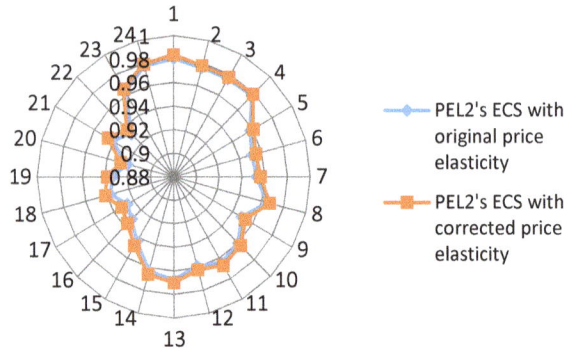

Figure 9. Electricity consumption satisfaction (ECS) of PEL 2 with changing α.

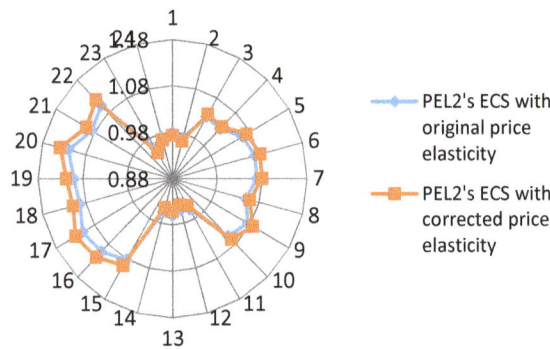

Figure 10. Interaction benefit satisfaction (IBS) of PEL 2 with changing α.

5.4. Effect of PEL Probabilistic Characterization on Demand Response Amount

If the wind power output fluctuation is set as Figure 2, the distribution coefficient k of the stochastic demand response of PEL 2 is changed to 0.3. We study the effect PEL probabilistic characterization on demand response. Other PEL parameters are the same, as shown in Table 1.

In this scenario, the demand response amount of PEL 2 is illustrated in Figure 11, and the results of demand response satisfaction are shown in Figures 12 and 13.

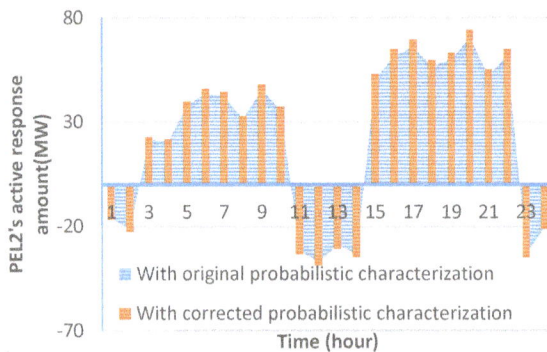

Figure 11. Demand response Amount of PEL2 with Changing k.

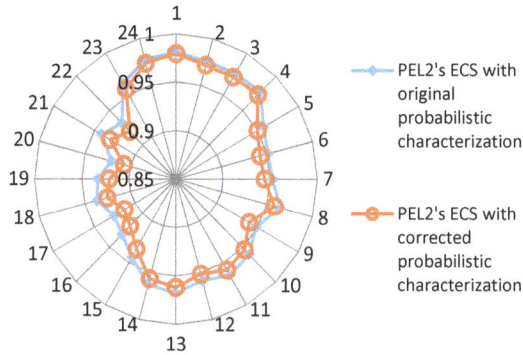

Figure 12. Electricity consumption satisfaction (ECS) of PEL 2 with changing *k*.

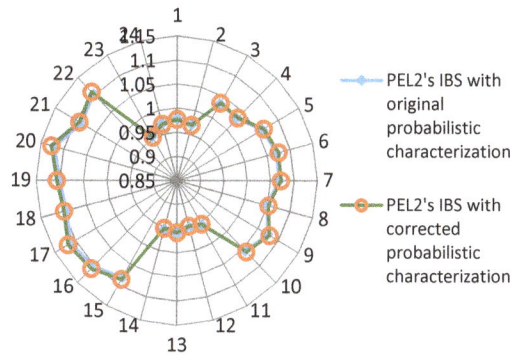

Figure 13. Interaction benefit satisfaction (IBS) of PEL 2 with changing *k*.

Figure 11 illustrates that the demand response amount expectation of PEL 2 increases with increasing uncertain factors. We can also draw the following two conclusions: first, ECS decreases while IBS increases with increasing uncertain factors; and second, Figures 8 and 11 have different changing trends because the price sensitivity and price elasticity coefficient are large for a large distribution coefficient of PEL stochastic demand response. Hence, a large *k* or high |*α*| corresponds to the similar trend of demand response amount. This result demonstrates the discussion about the relationship between price elasticity coefficient and distribution coefficient of PEL's stochastic demand response in Section 2.2. Thus, under a power shortage situation, the load has a small price elasticity coefficient, which has few demand response uncertainties, and can be chosen to participate in the interaction.

5.5. Computing Performance

To solve the problem, the stochastic constraints are transformed into deterministic constraints. The solution methodology used in this paper is the particle swarm optimization (PSO). All numerical simulations are coded in MATLAB. With iterations of PSO set to 300, the running time of each case on a provincial power system with 151 buses is approximately 2.5 min on a 2.4 GHz Windows-based laptop with 8 Gb of RAM. The simulation is fully able to meet the requirement of the hourly process.

6. Conclusions

In this paper, modeling the stochastic demand response behavior of PEL is proposed. The proposed model analyzes the demand response behavior of PEL by maximizing the electricity consumption satisfaction (ECS) and interaction benefit satisfaction (IBS) of PEL. In the proposed model, the uncertainties in wind power variability are considered. Meanwhile, the uncertainties in the stochastic process of PEL demand response to the power grid are included. The confidence intervals are introduced to transform this problem to a deterministic optimization problem. This problem is solved by the particle swarm optimization (PSO) method. The main contributions are as follows:

(1) The output of the uncertain model contains abundant probability information. It provides practical information on how PELs actively respond to the power grid integrated with wind power, thus decreasing the effects caused by the response deviation;

(2) The relationship between the elasticity coefficient of PELs and the distribution coefficient of the stochastic demand response is elaborated. The increasing elasticity coefficient of PELs, *i.e.*, decreasing flexibility, leads to a large distribution coefficient of stochastic demand;

(3) Choosing PELs with small sensitivity to the price elasticity coefficient into the interaction with the power grid reduce the uncertainty and enhance reliability,

(4) Proper choice of the distribution coefficient of ECS for PELs increases the comprehensive satisfaction of demand responses,

(5) The proposed model enables demand response resources to respond to wind power variability. It also contributes to mitigating power imbalance, and consideration of the interaction profit with the power grid is presented; and

(6) This approach is applicable in hourly real-time pricing models, and also in day-ahead pricing models.

Acknowledgments: This work was supported by National Natural Science Foundation of China (No. 51407165) and State Grid Corporation of China (DZ71-14-042, DZN17201300197 and DZ71-15-039).

Author Contributions: Shengchun Yang and Dan Zeng designed the research, carried out data analysis and prepared the manuscript. Hongfa Ding and Jianguo Yao proposed the research topic, provided technical support. Ke Wang and Yaping Li revised the paper. All authors discussed the results and contributed to the writing of the manuscript.

Conflicts of Interest: The authors declare no conflict of interest.

Abbreviations

The following abbreviations are used in this manuscript:

DR	Demand response
PEL	Price elastic load
ECS	Electricity consumption satisfaction
IBS	Interaction benefit satisfaction

References

1. Mark, Z.J.; Cristina, L.A. Saturation Wind Power Potential and Its Implications for Wind Energy. 25 September 2012. Available online: http://www.pnas.org/content/109/39/15679.full (accessed on 15 August 2014).
2. Yao, J.G.; Yang, S.C.; Wang, K.; Yang, Z.; Song, X. Concept and Research Framework of Smart Grid "Source-Grid-Load" Interactive Operation and Control. *Autom. Electr. Power Syst.* **2012**, *36*, 1–6.
3. Wang, Q.; Guan, Y.; Wang, J. A chance-constrained two-stage stochastic program for unit commitment with uncertain wind power output. *IEEE Trans. Power Syst.* **2012**, *27*, 206–215. [CrossRef]
4. Bouffard, F.; Galiana, F. Stochastic security for operations planning with significant wind power generation. *IEEE Trans. Power Syst.* **2008**, *23*, 306–316. [CrossRef]
5. Wang, J.; Botterud, A.; Miranda, V.; Monteiro, C.; Sheble, G. Impact of wind power forecasting on unit commitment and dispatch. In Proceedings of the 8th International Workshop Large-Scale Integration of Wind Power into Power Systems, Bremen, Germany, 14–15 October 2009.

6. Ruiz, P.; Philbrick, C.; Zak, E.; Cheung, K.; Sauer, P. Uncertainty management in the unit commitment problem. *IEEE Trans. Power Syst.* **2009**, *24*, 642–651. [CrossRef]

7. Wang, J.; Shahidehpour, M.; Li, Z. Security-constrained unit commitment with volatile wind power generation. *IEEE Trans. Power Syst.* **2008**, *23*, 1319–1327. [CrossRef]

8. Zeng, B.; Zhang, J.; Yang, X.; Wang, J.; Dong, J.; Zhang, Y. Integrated Planning for Transition to Low-Carbon Distribution System with Renewable Energy Generation and Demand Response. *IEEE Trans. Power Syst.* **2014**, *29*, 1153–1165. [CrossRef]

9. Nikzad, M.; Mozafari, B. Reliability assessment of incentive- and priced-based demand response programs in restructured power systems. *Int. J. Electr. Power Energy Syst.* **2014**, *56*, 83–96. [CrossRef]

10. Jia, W.; Kang, C.; Chen, Q. Analysis on demand-side interactive response capability for power system dispatch in a smart grid framework. *Electr. Power Syst. Res.* **2012**, *90*, 11–17. [CrossRef]

11. Andreas, G.V.; Pandelis, N.B. Demand Response in a Real-Time Balancing Market Clearing with Pay-as-Bid Pricing. *IEEE Trans. Smart Grid* **2013**, *4*, 1966–1975.

12. Papavasiliou, A.; Oren, S.S. Large-Scale Integration of Deferrable Demand and Renewable Energy Sources. *IEEE Trans. Power Syst.* **2014**, *29*, 489–499. [CrossRef]

13. Wang, J.; Kennedy, S.W.; Kirtley, J.L. Optimization of Forward Electricity Markets Considering Wind Generation and Demand Response. *IEEE Trans. Smart Grid* **2014**, *5*, 1254–1261. [CrossRef]

14. Li, S.; Zhang, D.; Roget, A.B.; O'Neill, Z. Integrating Home Energy Simulation and Dynamic Electricity Price for Demand Response Study. *IEEE Trans. Smart Grid* **2014**, *5*, 779–788. [CrossRef]

15. Ma, O.; Alkadi, N.; Cappers, P.; Denholm, P.; Dudley, J.; Goli, S.; Hummon, M.; Kiliccote, S.; Macdonald, J.; Matson, N.; *et al.* Demand Response for Ancillary Services. *IEEE Trans. Smart Grid* **2013**, *4*, 1988–1995. [CrossRef]

16. Nguyen, D.T.; Negnevitsky, M.; de Groot, M. Market-Based Demand Response Scheduling in a Deregulated Environment. *IEEE Trans. Smart Grid* **2013**, *4*, 1948–1956. [CrossRef]

17. Su, C.; Kirschen, D. Quantifying the effect of demand response on electricity markets. *IEEE Trans. Power Syst.* **2009**, *24*, 1199–1207.

18. Kirschen, D. Demand-side view of electricity markets. *IEEE Trans. Power Syst* **2003**, *18*, 520–527. [CrossRef]

19. Federal Energy Regulatory Commission. Wholesale Competition in Regions with Organized Electric Markets: FERC's Advanced Notice of Proposed Rulemaking, 22 February 2007. Available online: http://www.kirkland.com/siteFiles/Publications/C430B16C519842DE1AEB2623F7DE21D6.pdf (accessed on 12 March 2015).

20. Mortensen, R.E.; Haggerty, K.P. A Stochastic Computer Model for Heating and Cooling Loads. *IEEE Trans. Power Syst.* **1998**, *3*, 1213–1219. [CrossRef]

21. Molina-García, A.; Kessler, M.; Fuentes, J.A.; Gómez-Lázaro, E. Probabilistic Characterization of Thermostatically Controlled Loads to Model the Impact of Demand Response Programs. *IEEE Trans. Power Syst.* **2011**, *26*, 241–251. [CrossRef]

22. Sun, Y.; Elizondo, M.; Lu, S.; Fuller, J.C. The Impact of Uncertain Physical Parameters on HVAC Demand Response. *IEEE Trans. Smart Grid* **2014**, *5*, 916–923. [CrossRef]

23. Zhao, C.; Wang, J.; Watson, J.; Guan, Y. Multi-Stage Robust Unit Commitment Considering Wind and Demand Response Uncertainties. *IEEE Trans. Power Syst.* **2003**, *28*, 2708–2719. [CrossRef]

24. Navid-Azarbaijani, N. *Load Model and Control of Residential Appliances*; McGill University: Montreal, QC, Canada, 1996; p. 81.

25. Zhang, B.M.; Wang, S.Y.; Xiang, N.D. A linear recursive bad data identification method with real-time application to power system state estimation. *IEEE Trans. Power Syst.* **1992**, *7*, 1378–1385. [CrossRef]

26. Wu, Y.-C.; Debs, A.S.; Marsten, R.E. A direct nonlinear predictor-corrector primal-dual interior point algorithm for optimal power flows. *IEEE Trans. Power Syst.* **1994**, *9*, 876–883.

energies

MDPI

Article

A Study on Maximum Wind Power Penetration Limit in Island Power System Considering High-Voltage Direct Current Interconnections

Minhan Yoon [1], Yong-Tae Yoon [1] and Gilsoo Jang [2],*

[1] Department of Electrical & Computer Engineering, Seoul National University, Gwanakero 599, Gwanakegu, Seoul 151-742, Korea; minhan.yoon@gmail.com (M.Y.); ytyoon@snu.ac.kr (Y.-T.Y.)
[2] School of Electrical Engineering, Korea University, Anam-dong, Sungbuk-gu, Seoul 136-713, Korea
* Correspondence: gjang@korea.ac.kr; Tel.: +82-2-3290-3246; Fax: +82-2-3290-3692

Academic Editor: Ying-Yi Hong
Received: 7 September 2015; Accepted: 7 December 2015; Published: 17 December 2015

Abstract: The variability and uncontrollability of wind power increases the difficulty for a power system operator to implement a wind power system with a high penetration rate. These are more serious factors to consider in small and isolated power systems since the system has small operating reserves and inertia to secure frequency and voltage. Typically, this difficulty can be reduced by interconnection with another robust power system using a controllable transmission system such as a high-voltage direct current (HVDC) system. However, the reliability and stability constraints of a power system has to be performed according to the HVDC system implementation. In this paper, the method for calculation of maximum wind power penetration in an island supplied by a HVDC power system is presented, and the operational strategy of a HVDC system is proposed to secure the power system reliability and stability. The case study is performed for the Jeju Island power system in the Korean smart grid demonstration area.

Keywords: high-voltage direct current (HVDC); wind power; power system stability; power system reliability

1. Introduction

As a renewable energy source, wind power has many benefits from both the economic and environmental perspective [1,2]. This positive fact is currently being accelerated by the notable cost reduction of wind turbine generators [3]. Subsequently, several countries such as Germany, United States, and China are promoting wind power implementation to increase the penetration rate. The number of annual new wind power installations in Europe has increased steadily and the total installed capacity exceeded 32% (117 GW) of total power generation capacity in 2013 [4]. The growth of wind power installation in the US is also significant, and total wind power capacity has reached 61 GW [5]. In particular, the wind power integration rate in Texas reached 25% in 2014 [6]. This type of high wind power penetration rate is common in areas with robust interconnections with other large-scale power systems, e.g., Denmark and Texas.

However, the characteristics of wind power resource—variability and uncontrollability—create problems to a power system operator when scheduling a generation dispatch or establishing a grid operation strategy to maintain power balance in the system [7]. Especially, it is more difficult to implement wind power in small isolated power systems for that reason. Research institutions around the world have published various technical reports and guidelines on this topic [4,8–11]. Riso National Laboratory in Denmark published guidelines to implement wind power into isolated power systems by introducing operational engineering design and assessment methods [8]. The U.S. National Renewable

Energy Laboratory (NREL) reported the impact of industry structure on high wind penetration potential, and the advantages and difficulties in wind power accommodation [9].

In addition, studies on promoting wind power penetration have been performed. A grid code for operation with wind penetration, the impact on a system, and wind speed forecasting, are discussed in the IEEE power and energy society [10]. Conseil International des Grands Réseaux Électriques (CIGRE) established the B4-62 working group to study the connection of wind farms with weak AC networks in 2013 [11]. Similarly, the power limitations of a wind farm on an island system using energy yield evaluation was introduced in [12]. From the perspective of a power system, estimation of the frequency deviation and the power quality according to wind power penetration was performed by McGill University [13]. In addition, an economic approach was studied by modeling thermal generator dispatches and wind power curves to determine the optimal wind power generation capacity [14]. To mitigate the problems related to high wind power penetration, energy storage systems can be utilized for wind power smoothing [15]. There are efforts to solve the wind power fluctuation problem by system operation or control strategies [16,17]. Similarly, the interconnection with a robust power system can be a solution to promote penetration level [8]. In recent works, the stochastic and probabilistic methods are implemented to estimate maximum wind power penetration [18,19]. Furthermore, the implementation of the grid-scale energy storage system is considered to promote the wind power penetration [20].

To link systems such as wind farms to another system, there are two options from a broad point of view [21–25]: a HVAC system has advantages in transmission loss or competitive investment cost and compared to a HVDC system, it doesn't require complicated control systems [21]. However, the break-even distance of the HVDC implementation is shorter in the case of underground or undersea transmission to connect isolated systems or offshore wind farms due to the cable capacitive effect of AC transmission [22,23]. Several wind farm construction projects including the BorWin and DolWin ones in Germany have applied HVDC to connect offshore wind farms [24,25]. Two types of HVDCs are applicable at present: a traditional DC transmission technology, commonly known as a line commutated converter (LCC) HVDC and a voltage-source converter (VSC) HVDC. LCC technology is based on the firing angle of the thyristor, which results in the phase of the AC current of the converter always lagging the voltage. To compensate this lagging current, reactive power has to be supplied. Generally, a reactive power of approximately 50%–55% of the DC transmission capacity must be supplied from the AC grid for LCC HVDC. This fact concerning the reactive power consumption, can cause more difficulties as the HVDC transmission capacity increases [26,27]. VSC HVDC, which has reactive power control capability, is frequently used to connect the wind power generators [28]. However, it is necessary to consider both properties in terms of the power system operation.

Jeju Island in Korea is a system that has the characteristics mentioned above. According to the 6th power system planning of MKE (a government-affiliated organization of Korea), the load level of Jeju Island is expected to increase until 2022. Additionally, a LCC HVDC system was installed to link with the mainland and the installation of a large-scale wind power generator is planned. An impact analysis of wind farm implementation on Jeju Island was performed by Jeju University [29]. In addition, a study on the maximum instantaneous wind power penetration on Jeju Island was performed by a Korea power exchange (KPX) project [30,31].

Moreover, there must be research on the analysis of the mutual influence between HVDC operation and wind power penetration level on this isolated island. In this paper, the analytical modeling of power system generation including LCC/VSC HVDC and wind farms of an island system has been performed to study the proper amount of maximum wind power generation. The characteristics of the wind generation and HVDC system are formulated as a function to be reflected in the proposed algorithm. The maximum wind power penetration limit estimation algorithm includes power system reliability and stability criteria. Furthermore, the effect of the type of interconnection on penetration rate has been checked. The proposed method can be utilized for the assessment of an appropriate wind power penetration in power system operation and planning, even as real-time.

2. Problem Formulation

Due to the fluctuating output characteristics of wind generation, the generating units are key-role components on the determination of the maximum wind power generation in isolated weak power systems. As shown in Figure 1, in this paper the generating units are categorized as thermal generator, wind power generator and HVDC system. Each of components are dependent on the power system operation conditions. The operator of this mixed generation power system has to consider appropriate power system operation criteria to secure mutual system reliability and stability.

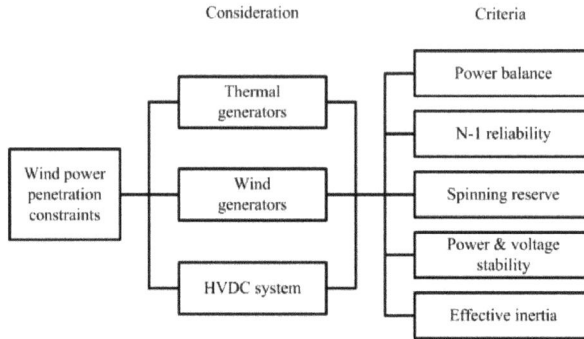

Figure 1. Criteria of the wind power generation constraints.

From the viewpoint of a power system operation, it is important to consider not only steady state analysis but also an $N - 1$ contingency situation. A generation outage in thermal, wind generators or a HVDC system is able to cause a severe frequency drop or voltage/power instability. According to these constraints, at a steady state or in an $N - 1$ contingency situation, the maximum wind power generation capacity must be limited. In this study, the categorized analysis for a large scale wind power penetrated in an island power system is presented. As shown in Figure 1, there are differences in an analysis criteria depending on the presence or type of HVDC system. The hierarchical structure of the study of wind power generation calculation is as follows:

Figure 2 shows the algorithm for the maximum wind power penetration limit estimation considering power system reliability and stability constraints. The objective function is to minimize the summation of generations from thermal generators and HVDC system in Equation (8). As a result, calculation of the proper wind power penetration limit to match the power demand can be performed instantly by using a mathematical model. The skeleton of the algorithm is based on a power balance equation considering the reliability/stability constraints.

The algorithm starts with the power flow data assimilation required to analyze the maximum wind power penetration limit. The control variables are the generating unit capacity $P_{G,i}$ and P_{DC}, the constraints of each control variable range are explained in Sections 2.1 and 2.2. The dynamic constraint branch considers the spinning reserve of the system to cover the amount of the wind power fluctuation. The ramp-up/ramp-down spinning reserve of each of the thermal generating units is smaller than the operation margin in the range. For a detailed explanation of the discriminant readers can refer to Section 2.5. The technical minimum is the result from minimum/maximum output capacity constraints of the components. The constraints are based on the power flow equation to match the power balance without any other stability and reliability constraints. The stability constraints are mainly related to determine the HVDC operation point within the stable operation range. The method analyzing the stable operation point considering power voltage stability and effective inertia is described in Sections 2.6 and 2.7. To calculate the maximum available power (MAP) of HVDC, the optimization method genetic algorithm (GA) is applied.

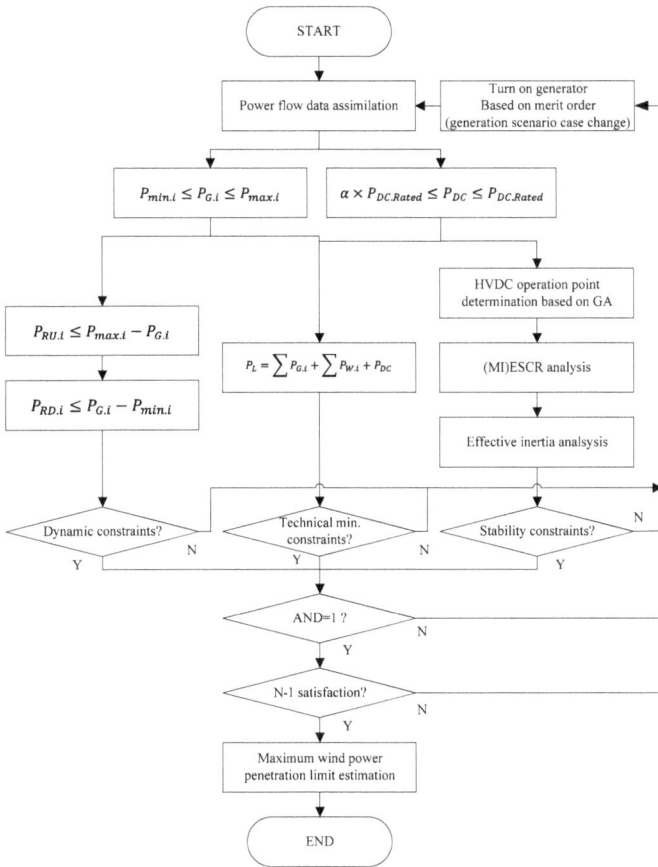

Figure 2. The algorithm for the maximum wind power penetration limit estimation.

In case that all stability constraints are satisfied (AND = 1), the condition to check the system reliability, $N - 1$ contingency is applied to the system as described in Section 2.4. In case of an unsatisfactory condition, we turn on an additional generator to improve the system strength and process the iteration loop. Finally, the result of the maximum wind power penetration limit estimation can be acquired.

2.1. Thermal Generator Constraints

Each thermal generator is in the range of normal operational constraints. The total ramp-up/down power capacity of a system depends on the up/down ramping rate of the generators. The ramp-up/down capacity contributes to maintain power balance when a contingency occurs. A more detailed analysis of spinning reserve is mentioned in Section 2.4:

$$P_{\text{min}.i} \leqslant P_{\text{G}.i} \leqslant P_{\text{max}.i} \tag{1}$$

Power ramp-up capacity:

$$P_{\text{RU}.i} = UR_i \times t \tag{2}$$

$$P_{\text{RU}.i} \leqslant P_{\text{max}.i} - P_{\text{G}.i} \tag{3}$$

Power ramp-down capacity:

$$P_{RD.i} = UD_i \times t \tag{4}$$

$$P_{RD.i} \leqslant P_{G.i} - P_{min.i} \tag{5}$$

2.2. High-Voltage Direct Current Operation Range

As shown in Figure 3, the amount of HVDC power transfer from positive to negative is flexible if a function that reverses power flow is supported. A VSC HVDC system typically has a function that allows current flow in the opposite direction to the capacity.

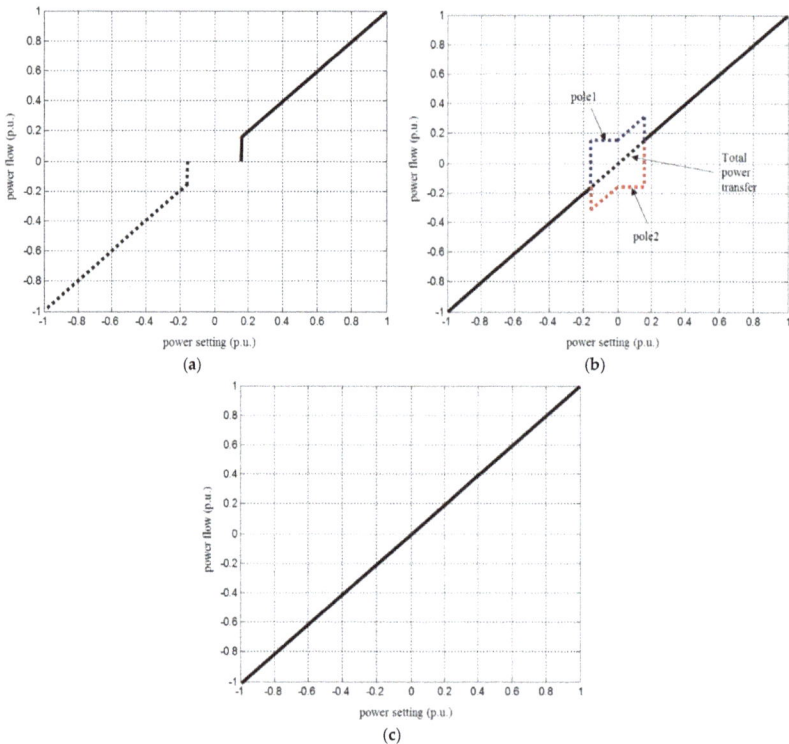

Figure 3. Power flow *versus* power setting depends on: (**a**) unidirectional line commutated converter (LCC) high-voltage direct current (HVDC); (**b**) bidirectional LCC HVDC; and (**c**) voltage-source converter (VSC) HVDC.

In that case, the minimum transferrable power capacity coefficient (τ) is −1 and the permissible DC power transfer range is from − 1 p.u. to 1 p.u., according to Equation (6):

$$\tau \times P_{DC.rated} \leqslant P_{DC} \leqslant P_{DC.rated}$$

$$\tau = \begin{cases} -1 & \text{VSC and LCC with power reversal} \\ 0.13 & \text{LCC without power reversal} \end{cases} \tag{6}$$

On the other hand, this is not simple for an LCC HVDC system owing to the requirement of reversing the voltage polarity [26]. The minimum transferrable capacity of an LCC HVDC system is 0.1–0.2 p.u. under normal conditions; however, reversal of the power flow is possible on LCC HVDC with simultaneous bidirectional power flow control or bipolar power flow control [22]. As an analysis of the Jeju Island case, a coefficient (α) of 0.13 is selected for unidirectional LCC study, then, the P_{DC} range is from 0.13 p.u. to 1 p.u. as presented in Section 3.2.

2.3. Power Balance Constraints

2.3.1. Real Power Criteria

According to this analysis, the maximum wind power penetration level can be calculated in terms of supply and demand, as shown in Equation (7). The wind power fluctuation must be compensated by controllable thermal generators and an HVDC system to satisfy the demands of the load:

$$P_{L} = \sum P_{G.i} + \sum P_{W.i} + P_{DC} \tag{7}$$

As presented in Equation (8), the thermal generators and HVDC systems are required to be minimum generation to maximize the wind power generation capacity by satisfying power balance equation. If it is possible, an HVDC system is required to be operated as maximum power reversal mode as -1 p.u. Thermal generators must be operated at minimum generation capacity while maintaining a stable system condition:

$$P_{W.max} = P_{L} - \min \left(\sum P_{G.i} + P_{DC} \right) \tag{8}$$

2.3.2. Reactive Power Criteria

Wind Power Side

The previous type of wind generator, which has an induction turbine, consumes reactive power in the motoring/generating regions by the slip phenomenon. PF control capacitors compensate when the slip effect occurs. This capacitor is able to generate reactive power when the wind generators are off. However, the latest doubly-fed induction generator (DFIG), doubly-fed asynchronous generator (DFAG) type and full converter type wind generators have the capability to control reactive power using a power converter during operation [32]. As a result, the wind generators can be assumed to be operated as almost a unity power factor and this issue does not need to be considered in terms of the local reactive power.

High-Voltage Direct Current Side

The surplus reactive power supply from shunt elements (Q_c) that compensate the reactive power consumption of an LCC HVDC system (Q_d) affects the bus voltage with HVDC system blocking. A large short circuit capacity of the system or a sufficient amount of reactive power compensation is required to respond instantaneously. VSC HVDC is more applicable to weak ac system in that sense, having a reactive power control capability [28]. In this study, must-run generators are set to supply short-circuit capacity. This will be noted in Section 2.5:

$$Q_c \approx Q_d = \frac{1}{4} V_{d0} I_d \frac{\sin 2\gamma - \sin 2\left(\gamma + \mu\right) + 2\mu}{\cos\alpha - \cos\gamma + \mu} \tag{9}$$

2.4. N − 1 Reliability

In a power system operation, an $N-1$ contingency event should be considered [33]. Therefore, the system operator should evaluate the system and maintain a margin for when there is an outage in the largest generators or a transmission line. In this study, the outage of large-scale wind farms is a factor that must be seriously considered. The power system including several generating units is required to satisfy the power balance from the most critical generation outage as shown in Equation (10):

$$P_{L} \leqslant \sum P_{G.i} + \sum P_{W.i} + P_{DC} - \max \left(P_{G.i}, P_{W}, P_{DC} \right) \tag{10}$$

143

However, wind power is uncontrollable and is excluded from a generation dispatch:

$$P_L \leqslant \sum P_{G,i} + P_{DC} - \max(P_{G,i}, P_W, P_{DC}) \tag{11}$$

In case of a bipolar or multiple monopolar HVDC system, each pole can be operated complementarily with the control function. Equation (11) has to be modified with the n-number of HVDC poles that can be operated independently. The maximum outage component in HVDC system is not whole system but a pole as shown in Equation (12):

$$P_L \leqslant \sum P_G + P_{DC} - \max(P_G, P_W, \frac{1}{n} \cdot P_{DC}) \tag{12}$$

2.5. Spinning Reserve

Spinning reserve refers to the generation capacity that can respond within 10 min for the outage of generation or a transmission line [34]. The ramp-up/ramp-down spinning reserve of thermal generating units are the summation of an each unit $P_{RU,i}$ and $P_{RD,i}$ as shown in Equations (13) and (14). The ramp-up/down rate of an HVDC system is definitely high to follow an order within 10 min [26]. Therefore, HVDC is able to contribute to spinning reserve on condition that upper margin to maximum operation capacity ($P_{DC.rated} - P_{DC}$) or lower margin to minimum ($P_{DC} - \alpha \times P_{DC.rated}$). The total spinning reserve (P_{USR}, P_{DSR}) has to be sufficiently above the required reserve capacity such as wind power fluctuation or other power outage ($\sigma_W \times P_W$).

In this study, the wind power fluctuation coefficient (σ_W) was defined as maximum wind power fluctuation ratio *versus* its capacity. The coefficient needs to be determined carefully depending on the system operation strategy and the spinning reserve planning. Typically, the maximum Jeju Island wind power fluctuation within 10 min is in the range of less than 20% [30]. In conclusion, the up/down spinning reserve of the system is calculated as follows:

Up spinning reserve constraints:

$$P_{USR} = \sum P_{RU,i} + (P_{DC.rated} - P_{DC})$$
$$P_{USR} \geqslant \sigma_W \times P_W \tag{13}$$

Down spinning reserve constraints:

$$P_{DSR} = \sum P_{RD,i} + (P_{DC} - \alpha \times P_{DC.rated})$$
$$P_{DSR} \geqslant \sigma_W \times P_W \tag{14}$$

2.6. Power and Voltage Stability

When there is a fault in a HVDC system or a severe fault in an AC system, the result can lead to the temporary blocking of HVDC converters. In the case of an LCC HVDC, shunt capacitors are installed for reactive power compensation of HVDC operation, which can cause temporary overvoltage (TOV) at a converter bus by converter blocking. A VSC HVDC system has the capability to support reactive power by AC voltage control, but this compensated amount will disappear by converter blocking. However, if the power system strength is sufficiently high, the temporary overvoltage problem can be alleviated. In this study, we assumed that the wind power system is connected with the voltage control capability, and the voltage support function of VSC is neglected. The MAP concept is widely utilized to analyze the power and voltage stability of a HVDC system [35–37]. Consideration of the commutation failure problem is essential to the planning of an LCC HVDC system. Therefore, the strength index is computed as a ratio of the short circuit capacity (SCC) to the DC power transfer amount, called the effective short circuit ratio (ESCR) and critical effective short circuit ratio (CESCR). Equation (15) is derived from converter equation and power flow equation based on a Jacobian matrix. For the procedure to derive Equation (15) readers are referred to [36,37]:

$$\frac{dP_d}{dI_d} = \frac{aU\cos(\gamma+\mu)\left[U^4ESCR^2 + b\left(U^2ESCR + Q_d\right) - Q_d^2 - P_{DC}^2\right]}{\det J} = 0$$

$$a = \frac{2U_d}{\cos\gamma + \cos(\gamma+\mu)} \tag{15}$$

$$b = 2\left(Q_d - P_{DC}\tan(\gamma+\mu)\right)$$

CESCR is the designated value to determine the MAP point, where dP_d/dI_d is zero. The ESCR index is required to have a value greater than the CESCR to secure power system stability with an LCC HVDC system. A MAP condition occurs when:

$$U^4ESCR^2 + b\left(U^2ESCR + Q_d\right) - Q_d^2 - P_{DC}^2 = 0$$

$$CESCR = ESCR, \quad \frac{dP_d}{dI_d} = 0 \tag{16}$$

CESCR is the ESCR value which satisfies the condition of Equation (16). Equation (17) can be derived by a quadratic formula since Equation (16) is a quadratic equation which has ESCR as unknown quantity:

$$CESCR = \frac{1}{U^2}\left[-Q_d + P_{DC}\cot\left(\frac{\pi}{4} + \frac{\gamma+\mu}{2}\right)\right] \tag{17}$$

To stably operate the HVDC system, the ESCR of the converter bus is required to be greater than the CESCR:

$$ESCR = \frac{SCC - Q_C}{P_{DC}} \geqslant CESCR \tag{18}$$

Assume the power system is in a flat condition, where U is the unity value. Equation (19) is able to be derived by substituting of Equation (17) into Equation (18):

$$SCC \geqslant \cot\left(\frac{\pi}{4} + \frac{\gamma+\mu}{2}\right) \cdot P_{DC}^2 - Q_d \cdot P_{DC} + Q_c \tag{19}$$

In addition, through the positive voltage sensitivity factor (VSF), the voltage stability of the system can be secured in a stable region when the ESCR is greater than the CESCR [35].

2.7. Effective Inertia Constant Constraints

The effective inertia constant is an index that represents the ability to sustain the power system frequency by the rotational inertia (H) of an AC system. The rotational mechanical inertia of the system is able to provide energy to maintain the electromotive force of the system during the loss of power generation. H_{DC} is the effective inertia constant of the DC power through the inverter *versus* the rotational inertia of the AC system [38]. In the case of a bipolar system, each pole has the ability to operate independently. However, the total transfer capacity of a DC system should be considered for the analysis of a severe case such as the contingency of a converter station bus or DC line fault.

General situations that cause a decrease in power system frequency are:

- HVDC commutation failure
- The faults in a HVDC system while sending and receiving to an end system
- DC line fault

In a HVDC infeed situation, the effective inertia of the HVDC system is as follows:

$$H_{DC} = \frac{\sum H_i \cdot S_{MVA.i}}{P_{DC}} \tag{20}$$

The frequency deviation of a machine is presented as:

$$\Delta f = \frac{\Delta P \cdot f_0 \cdot \Delta t}{2H} \tag{21}$$

Frequency change after HVDC drop, Equation (22), is derived from Equations (20) and (21):

$$\Delta f = \frac{\Delta P_{DC.pu} \cdot f_0 \cdot \Delta t}{2 \sum (H_i \cdot S_{MVA.i})/P_{DC}} \tag{22}$$

The power loss period (Δt) calculated from the summation of the fault duration (t_f) and recovery time (t_r). To maintain the frequency deviation within the specified range ($\Delta f / f_0$) from the power loss during the period (Δt), the system rotational mechanical inertia of ON state generators should be secured as shown in Equation (23). In other words, the DC power transfer capacity is required to be limited related to the rotational inertia of the AC system to sustain frequency stability:

$$\sum H_i \cdot S_{MVA.i} \geqslant \frac{P_{DC} \cdot \Delta P_{DC.pu} \cdot f_0 \cdot (\Delta t_f + \Delta t_r)}{2\Delta f} \tag{23}$$

3. Case Study

3.1. Introduction of the Jeju Island Power System

The proposed analysis and simulation were performed on the basis of Jeju Island power system data in 2013 of the 6th power system planning of MKE. The schematic diagram of Jeju Island is shown in Figure 4. The peak load level on Jeju Island is 800 MW, and the off-peak load level is 300 MW [31]. A LCC HVDC inverter station is installed in the Jeju C/S bus, and six thermal generators are being operated to supply power to meet the demand load. Hanlim C/C is normally not in operation because of the high generation cost and because the HVDC system and other generators are able to meet the demand load. For wind power generation, approximately 630 MW in contracts have been ordered for installation. The rated capacity, technical minimum, inertia, and ramp rate of generators in Jeju Island are shown in Table 1.

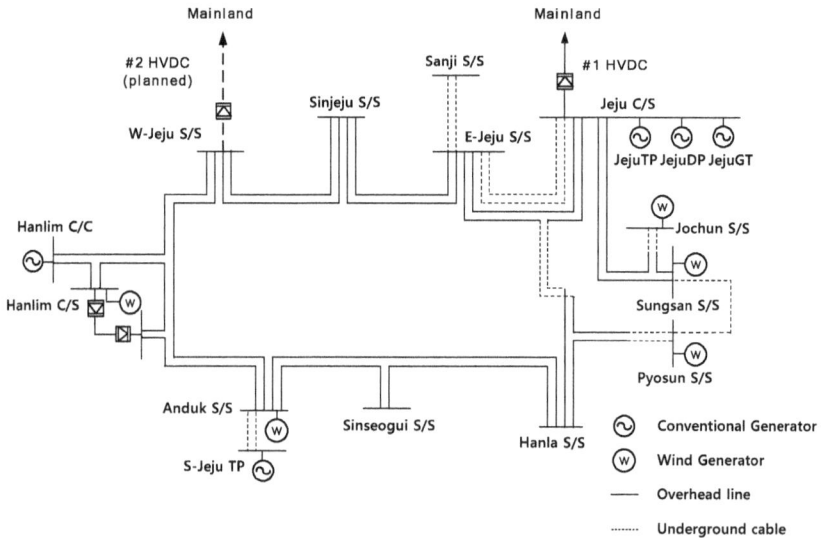

Figure 4. Jeju Island power system diagram.

Table 1. Generators in Jeju Island operation.

Generators	Pmax (MW)	Pmin (MW)	Inertia	Ramp Rate (MW/min)	Rating (MVA)
Jeju GT#1 (SC)	-	-	2.56	-	50
Jeju GT#2 (SC)	-	-	2.56	-	50
Jeju DP#1	40	26	6.71	2.0	44.96
Jeju DP#2	40	26	6.71	2.0	44.96
S-Jeju TP#3	100	50	5.93	5	130
S-Jeju TP#4	100	50	5.93	5	130
Jeju TP#2	75	42	5.4	1	97.06
Jeju TP#3	75	42	5.4	1	97.06
Hanlim CC	105	41	6	8.7	150

The basic information of the installed HVDC system, including the topology, rating, firing angle in normal operation, and transformer reactance, is presented in Table 2. A unidirectional 300-MW LCC HVDC system was installed in 1998. The second LCC HVDC is undergoing a system operational test to be implemented. In this study, the effectiveness of a single HVDC system for wind power penetration is analyzed, and the analysis of bidirectional LCC and VSC replacement has been performed as a supplement.

Table 2. HVDC in Jeju Island operation.

Specifications	HVDC Bipole 1	Unit
Power rating	150 × 2 (Bipoles)	MW
Voltage	180	kV
Current	840	A
Gamma firing angle (inverter)	27	Deg
Power reversal	Not available	-
% Impedance of transformer	12	%
Smoothing reactor	60	mH

3.2. Power System Operation and Analysis Criteria

The generator input sequence by the merit order to satisfy the power balance and other constraints in Jeju Island generators is shown in Figure 5 and each generator operation case are shown in Table 3.

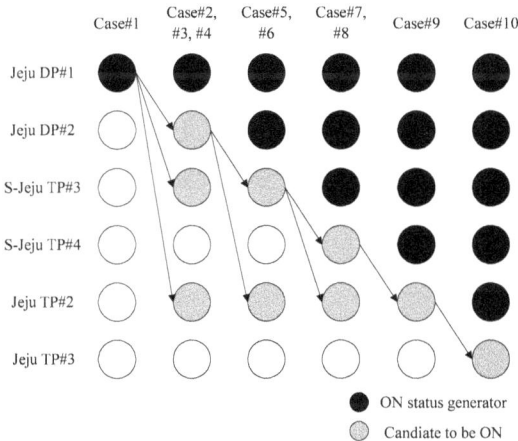

Figure 5. Jeju Island generation scenarios.

Table 3. Jeju Island generator in-service cases.

Case	Jeju DP#1	Jeju DP#2	S-Jeju TP#3	S-Jeju TP#4	Jeju TP#2	Jeju TP#3	Pmax (MW)	Pmin (MW)	Ramp Rate (MW/min)
#1	ON	-	-	-	-	-	40	26	2
#2	ON	ON	-	-	-	-	80	52	4
#3	ON	-	ON	-		-	140	76	7
#4	ON	-	-	-	ON	-	115	68	3
#5	ON	ON	ON	-		-	180	102	9
#6	ON	ON	-	-	ON	-	155	94	5
#7	ON	ON	ON	ON	-	-	280	152	14
#8	ON	ON	ON		ON	-	255	144	10
#9	ON	ON	ON	ON	ON	-	355	194	15
#10	ON	ON	ON	ON	ON	ON	430	236	16

As explained above, short circuit capacity and rotational inertia of AC system is essential to operate LCC HVDC. Therefore, must-run generators and an HVDC system are required to be operated at minimum capacity to ensure system reliability and stability. The results of analysis criteria based on the case set of generation units are presented in Table 4.

Table 4. System analysis results considering Jeju #1 HVDC transmission. Short circuit capacity: SCC; effective short circuit ratio: ESCR.

Case	Spinning Reserve (MW)				SCC (MVA)	ESCR	Inertia	H_{dc}
	w/o HVDC	With Unidirectional LCC	With Bidirectional LCC	With VSC				
#1	20	280	470	470	436.07	0.95	813.68	2.71
#2	40	300	490	490	581.64	1.44	1115.36	3.72
#3	70	330	520	520	918.91	2.56	1584.58	5.28
#4	30	290	480	480	838.13	2.29	1337.81	4.46
#5	90	350	540	540	1018.87	2.90	1886.26	6.29
#6	50	310	500	500	939.12	2.63	1639.49	5.46
#7	140	400	590	590	1335.39	3.95	2657.16	8.86
#8	100	360	550	550	1376.38	4.09	2410.39	8.03
#9	150	410	600	600	1692.71	5.14	3181.29	10.6
#10	160	420	610	610	2055.5	6.35	3705.41	12.35

For an $N-1$ contingency in the Jeju Island power system, 100 MVA S-Jeju TP is the largest power loss before HVDC construction. However, the largest power supplier is changed to the HVDC or the large-scale wind farms. As presented Sections 2.1 and 2.4 the controllable units of the Jeju system have to compensate the outage of the largest power generation. The ramp rate to change the power output of conventional generators are not sufficient as spinning reserve or standby generation for the contingency. Therefore, the HVDC system transfers power from the mainland to Jeju or in the opposite direction rapidly. The point of an HVDC system is to increase frequency stability significantly with a rapid response and to make the power system flexible for large-scale wind power penetration.

In power and voltage stability analysis, CESCR calculations are being utilized for the analysis of the stable operation of an HVDC region. The MPC of the 300-MW LCC HVDC operation considering full capacity is shown in Figure 6. The minimum limit of stability is analyzed in Case 2 (MAP is where DC current is 0.86 kA) when SCC of Jeju Island system is 581.6 MVA. In other words, Case 2 set of generation units is almost the boundary for stable operation region with the LCC HVDC system. However, $N-1$ contingency has to be considered, Case 5 is the optimal operation point considering and merit order and generator trip. In the case of a VSC HVDC system, there were no violation cases, and stable operation was possible with no generators, as explained in Section 2.6.

In system effective inertia analysis, the permissible maximum frequency decline by a more severe $N-1$ contingency in Korea, such as an entire HVDC block failure, is 5%, and the maximum AC fault duration time is 100 ms. In the case of an HVDC system recovery time within 100 ms, the H_{dc} index of the power system should be over 2.0 for the system stability for HVDC operation. The result of the analysis on effective inertia has less of an effect on the must-run generator constraints.

The frequency drop in Jeju Island caused by the largest generation loss can be sustained by two synchronous condensers and one generator unit in Case 1.

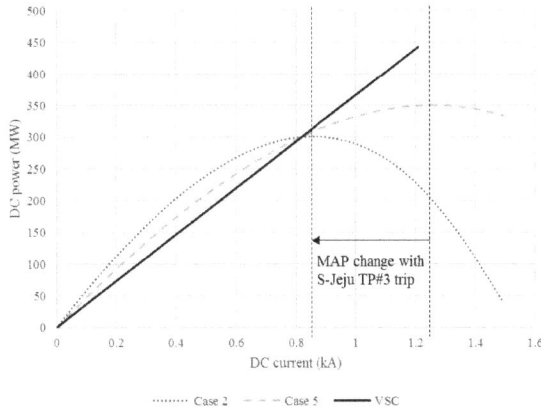

Figure 6. Maximum Power Curve of the Jeju system following different generation cases.

In addition, the amount of spinning reserve is being changed by the installed HVDC type, as explained in Sections 2.2 and 2.5. For a unidirectional LCC HVDC system, the contribution to the spinning reserve increased compared to the situation without any interconnections. A minimum available operation point of the Jeju—Haenam HVDC system is 40 MW and approximately 0.13 p.u., and the margin to the rated capacity is 260 MW. In case of bidirectional LCC and VSC HVDC, the spinning reserve can be increased to 450 MW due to the − 150 MW reversal power flow function. Furthermore, the PSS/e simulation has been performed to compare the variation of voltage and frequency between with and without the VSC HVDC situation in off-peak load.

3.3. Maximum Wind Power Penetration Limit According to the Load Level

After the analysis of system reliability and stability constraints, the maximum wind power generation based on generator constraints and the HVDC operation point can be studied. The maximum fluctuation coefficient (σ_W) of wind power generation in Jeju Island is 100%, considering a total loss of wind generation on the basis of the system operation rule of KPX [31]. As a result, the wind penetration limit according to the spinning reserve can be analyzed on the basis of Equations (13) and (14). According to the load level, the maximum wind power generation capacity is varied by the required generation mixture.

As shown in Figure 7a, only a small amount of wind power penetration is possible without any interconnections. The wind power penetration limit is mainly constrained by the technical minimum except the small part of dynamic constrained range by deficient spinning reserve. The unidirectional LCC HVDC system that is installed on Jeju Island for the current requires must-run generators to sustain the reliability and stability of the system. Especially, the limitation by stability constraint is calculated off-peak load level due to the weak AC power system characteristics. However, it is able to promote wind power penetration by a fast responsible capacity, as shown in Figure 7b. In the case of HVDC replacement for a bidirectional LCC or VSC HVDC system, the penetration level can be increased significantly, as shown in Figure 7c,d. The available wind power penetration rate based on Equation (23) of all cases is shown in Figure 8. The penetration rate is promoted by HVDC interconnection regardless of type. The significant gap between unidirectional and bidirectional LCC is caused by the permissible operation range described in Section 2.2. The difference of the penetration level exist between VSC and unidirectional LCC at off-peak load level. It is caused by that the stability constraint of the LCC HVDC leads to the more number of must-run thermal generators as explained above:

$$\text{Penetration Rate} = \frac{P_W}{P_W + \sum P_G + P_{DC}} \tag{24}$$

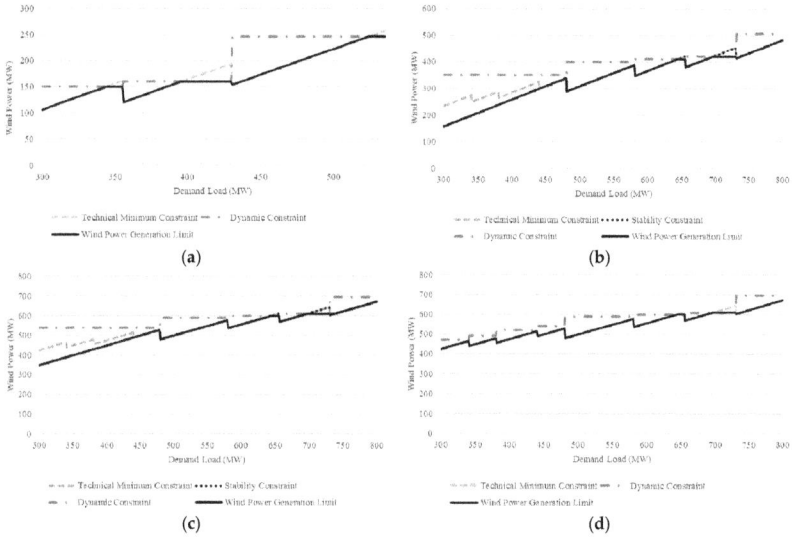

Figure 7. Maximum wind penetration in Jeju island, (a) w/o interconnection; (b) w/ unidirectional LCC HVDC; (c) w/ bidirectional LCC HVDC and (d) w/ VSC HVDC.

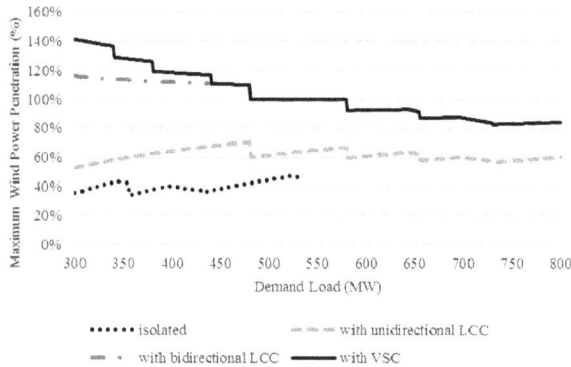

Figure 8. Wind power penetration rate according to HVDC interconnection.

4. Conclusions

It is a delicate problem for power system operators to implement an HVDC system or large-scale wind power generation in an island system. This paper describes an analytic model of power system operation including an LCC/VSC HVDC system to secure system reliability and stability, including $N − 1$ contingency analysis, power and voltage analysis, and effective inertia analysis. Subsequently, estimation was performed on the effect of each type of HVDC interconnection for the maximum wind power penetration limit considering the uncontrollable and variable characteristics of wind. The Jeju

Energies **2015**, *8*, 14244–14259

Island power system data and grid code of Korea was applied for the case study. As a result, there are significant differences depending on the types of HVDC in the must-run generator constraint to restrain system stability. Furthermore, a relation between the wind power penetration limit and the system constraints with a HVDC system was discovered. In addition to the above results, the superiority of the VSC HVDC system was ascertained for a wind power penetration level increase in an island power system. The analytic modeling of the maximum wind power penetration limit estimation depending on HVDC operation would be useful for an economic evaluation of HVDC system installation and the establishment of an optimal system operational strategy.

Acknowledgments: This work was supported by National Research Foundation of Korea (NRF) grant (No. NRF-2013R1A2A2A01067762) and Human Resources Development of KETEP Grant (No. 20134030200340) funded by the Korea government Ministry of Knowledge Economy.

Author Contributions: The authors have contributed equally in writing and revision of this manuscript.

Conflicts of Interest: The authors declare no conflict of interest.

Nomenclature

P_G	Total power generation of i-th generator
$P_{G.i}$	Power generation of i-th generator
$P_{min.i}$	Minimum active power generation limit of i-th generator
$P_{max.i}$	Maximum active power generation limit of i-th generator
UR_i	Ramp-up rate of i-th generator
UD_i	Ramp-down rate of i-th generator
t	Ramp up/down time
$P_{RU.i}$	Generation ramp-up capacity of i-th generator
$P_{RD.i}$	Generation ramp-down capacity of i-th generator
P_{DC}	HVDC power transfer
$P_{DC.rated}$	Rated capacity of HVDC
τ	Minimum HVDC operation point coefficient (p.u.)
P_{USR}	Up spinning reserve capacity
P_{DSR}	Down spinning reserve capacity
σ_W	Wind power fluctuation coefficient
U	AC terminal voltage
U_d	DC voltage
P_L	Demand load in the system
P_W	Total wind power generation in the system
$P_{W.i}$	Wind power generation of i-th wind farm
Q_d	Reactive power consumption of HVDC system
Q_c	Reactive power compensation by shunt element
V_{d0}	No-load DC voltage
I_d	DC current
α	Firing(delay) angle
μ	Overlap angle
γ	Extinction angle
SCC	Short circuit capacity at converter bus
H_i	Rotational inertia constant of i-th machine
$S_{MVA.i}$	MVA rating of i-th machine
Δf	Frequency deviation (p.u)
ΔP_i	Power mismatch (p.u)
f_0	System nominal frequency
Δt	Power loss period

Δt_f	Fault duration time
Δt_r	Power recovery time
CC	Combined cycle gas turbine power-plant
GT	Gas turbine power-plant
DP	Diesel engine power-plant
TP	Thermal power-plant

References

1. Kennedy, S. Wind power planning: Assessing long-term costs and benefits. *Energy Policy* **2005**, *33*, 1661–1675. [CrossRef]
2. Snyder, B.; Kaiser, M.J. Ecological and economic cost-benefit analysis of offshore wind energy. *Renew. Energy* **2008**, *34*, 1567–1578. [CrossRef]
3. *Wind Energy—The Facts: Costs & Prices*; The European Wind Energy Association (EWEA): Brussels, Belgium, 2009; Volume 2, pp. 94–110.
4. The European Wind Energy Association (EWEA). *Wind in Power: 2013*; European Statistics: Brussels, Belgium, 2014.
5. *2012 Wind Technologies Market Report*; U.S. Department of Energy: Washington, DC, USA, 2013.
6. *ERCOT Grid Operations Wind Integration Report*; ERCOT: Austin TX, USA, 2014.
7. Clausen, N.; Bindner, H.; Frandsen, S.; Hansen, J.C.; Hansen, L.H.; Lundsager, P. *Isolated Systems with Wind Power an Implementation Guideline*; Risø National Laboratory Technical Report Riso-R-1257 (EN); Risø National Laboratory: Roskilde, Denmark, 2001.
8. Weisser, D.; Garcia, R.S. Instantaneous wind energy penetration in isolated electricity grids: Concepts and review. *Renew. Energy* **2005**, *30*, 1299–1308. [CrossRef]
9. Milligan, M.; Kirby, B. *Impact of Electric Industry Structure on High Wind Penetration Potential*; National Renewable Energy Laboratory: Golden, CO, USA, 2009.
10. Peter, B.E.; Thomas, A.; Hans, A.; Paul, S.; Wilhelm, W.; JuanMa, R.G. System operation with high wind penetration. *IEEE Power Energy Mag.* **2005**, *3*, 65–74.
11. *Integration of Large Scale Wind Generation Using HVDC and Power Electronics*; Conseil International des Grands Réseaux Électriques Working Group B4.39: Paris, France, 2009.
12. Papathanassiou, S.A.; Boulaxis, N.G. Power limitations and energy yield evaluation for wind farms operating in island systems. *Renew. Energy* **2006**, *31*, 457–479. [CrossRef]
13. Luo, C.; Far, H.G.; Banakar, H.; Keung, P.; Ooi, B. Estimation of wind penetration as limited by frequency deviation. *IEEE Trans. Energy Convers.* **2007**, *22*, 783–791. [CrossRef]
14. Chen, C.-L.; Lee, T.-Y.; Jan, R.-M. Optimal wind-thermal coordination dispatch in isolated power systems with large integration of wind capacity. *Energy Convers. Manag.* **2006**, *47*, 3456–3472. [CrossRef]
15. Francisco, D.; Andreas, S.; Oriol, G.; Roberto, V. A review of energy storage technologies for wind power applications. *Renew. Sustain. Energy Rev.* **2012**, *16*, 2154–2171.
16. Shan, J.; Botterud, A.; Ryan, S.M. Impact of Demand Response on Thermal Generation Investment with High Wind Penetration. *IEEE Trans. Smart Grid* **2013**, *4*, 2374–2383.
17. Margaris, I.D.; Papathanassiou, S.A.; Hatziargyriou, N.D.; Hansen, A.-D.; Sorensen, P. Frequency Control in Autonomous Power Systems with High Wind Power Penetration. *IEEE Trans. Sustain. Energy* **2012**, *3*, 189–199. [CrossRef]
18. Nasrolahpour, E.; Ghasemi, H. A stochastic security constrained unit commitment model for reconfigurable networks with high wind power penetration. *Electr. Power Syst. Res.* **2015**, *121*, 341–350. [CrossRef]
19. Lyu, Y.; Liang, J.; Yan, J.; Yu, Z.; Sun, S.; Lu, G. On-line probabilistic dynamic security assessment considering large scale wind power penetration. In Proceedings of the International Conference on Power System Technology (POWERCON), Chengdu, China, 20–22 October 2014; pp. 2635–2641.
20. Liu, W.; Liu, L.; Xu, G.; Liang, F.; Yang, Y.; Zhang, W.; Wu, Y. A Novel Hybrid-Fuel Storage System of Compressed Air Energy for China. *Energies* **2014**, *7*, 4988–5010. [CrossRef]
21. Barberis Negra, N.; Todorovic, J.; Ackermann, T. Loss evaluation of HVAC and HVDC transmission solutions for large offshore wind farms. *Electr. Power Syst. Res.* **2006**, *76*, 916–927. [CrossRef]

22. Bresesti, P.; Kling, W.L.; Hendriks, R.L.; Vailati, R. HVDC Connection of Offshore Wind Farms to the Transmission System. *IEEE Trans. Energy Convers.* **2007**, *22*, 37–43. [CrossRef]

23. Wang, L.; Thi, M. Comparative Stability Analysis of Offshore Wind and Marine-Current Farms Feeding Into a Power Grid Using HVDC Links and HVAC Line. *IEEE Trans. Power Deliv.* **2013**, *28*, 2162–2171. [CrossRef]

24. Bahrman, M. *Offshore Wind Connections: HVDC for Offshore Grids*; UWIG Technical Workshop: Maui, HI, USA, 2011.

25. Chou, C.; Wu, Y.; Han, G.; Lee, C. Comparative Evaluation of the HVDC and HVAC Links Integrated in a Large Offshore Wind Farm—An Actual Case Study in Taiwan. *IEEE Trans. Ind. Appl.* **2012**, *48*, 1639–1648. [CrossRef]

26. Kim, C.; Sood, V.K.; Jang, G. *HVDC Transmission*; Wiley: Hoboken, NJ, USA, 2009.

27. Banakar, H.; Luo, C.; Ooi, B.T. Impacts of Wind Power Minute-to-Minute Variations on Power System Operation. *Electr. Power Syst. Res.* **2008**, *23*, 150–160. [CrossRef]

28. Zhang, L.; Harnefors, L.; Nee, H.P. Modeling and control of VSC-HVDC links connected to island systems. *IEEE Trans. Power Syst.* **2011**, *26*, 783–793. [CrossRef]

29. Kim, E.; Kim, J.; Kim, S.; Choi, J.; Lee, K.; Kim, H. Impact Analysis of Wind Farms in the Jeju Island Power System. *IEEE Syst. J.* **2012**, *6*, 134–139. [CrossRef]

30. Park, J.; Park, Y.; Moon, S. Instantaneous Wind Power Penetration in Jeju Island. In Proceedings of the IEEE Power Engineering Society General Meeting, Pittsburgh, PA, USA, 20–24 July 2008.

31. *A Long-Term View of Power System Operation*; Korea Power Exchange: Seoul, Korea, 2013; pp. 251–268.

32. Ackermann, T. *Wind Power in Power Systems*; John Wiley & Sons, Ltd.: Chichester, UK, 2005.

33. Kundur, P. *Power System Stability and Control*; McGraw-Hill Education: Columbus, OH, USA, 1994.

34. Western Electricity Coordinating Council. *Operating Reserves*; WECC Standard BAL-STD-002-0; Western Electricity Coordinating Council: Salt Lake City, UT, USA, 2007.

35. *On Voltage and Power Stability in AC/DC Systems*; Conseil International des Grands Réseaux Électriques Working Group 14.05: Paris, France, 2003.

36. Franken, B.; Andersson, G. Analysis of HVDC converters connected to weak AC systems. *IEEE Trans. Power Syst.* **1990**, *5*, 235–242. [CrossRef]

37. Denis, L.H.A.; Andersson, G. Voltage stability analysis of multi-infeed HVDC systems. *IEEE Trans. Power Deliv.* **1997**, *12*, 1309–1318.

38. *IEEE Guide for Planning DC Links Terminating at AC Locations Having Low Short-Circuit Capacities*; IEEE Standard 1204-1997; IEEE: Piscataway, NJ, USA, 1997.

Article

Designing an Incentive Contract Menu for Sustaining the Electricity Market

Ying Yu [1,*], Tongdan Jin [2] and Chunjie Zhong [1]

[1] School of Mechatronics Engineering and Automation, Shanghai University, Shanghai 200072, China; stephzcj@sina.cn

[2] Ingram School of Engineering, Texas State University, San Marcos, TX 78666, USA; tj17@txstate.edu

[*] Correspondence: squarey@shu.edu.cn; Tel.: +86-21-5633-1568; Fax: +86-21-6613-4021

Academic Editor: Ying-Yi Hong

Received: 14 October 2015; Accepted: 7 December 2015; Published: 16 December 2015

Abstract: This paper designs an incentive contract menu to achieve long-term stability for electricity prices in a day-ahead electricity market. A bi-level Stackelberg game model is proposed to search for the optimal incentive mechanism under a one-leader and multi-followers gaming framework. A multi-agent simulation platform was developed to investigate the effectiveness of the incentive mechanism using an independent system operator (ISO) and multiple power generating companies (GenCos). Further, a Q-learning approach was implemented to analyze and assess the response of GenCos to the incentive menu. Numerical examples are provided to demonstrate the effectiveness of the incentive contract.

Keywords: stackelberg game; Q-learning; multi-agent simulation; electricity market; incentive mechanism

1. Introduction

As the vertically integrated power industry evolves into a competitive market, electricity now can be treated as a commodity governed by demand and generation interactions. Competition among generating companies (GenCos) is highly encouraged in order to lower the energy price and benefit end consumers. However, when GenCos are given more flexibility to choose their bidding strategies, larger uncertainties are also brought into the electricity markets. Many factors that affect bidding strategies include the risk appetite of Gen Cos, price volatility of fuels, weather conditions, network congestion and overloading. These factors and their interactions may lead to larger price volatility in the deregulated power market.

Many efforts have been made to design and optimize the bidding strategies in the presence of uncertainty. Zhang *et al.* [1] proposed an efficient decision system based on a Lagrangian relaxation method to find the optimal bidding strategies of GenCos. Kian and Cruz [2] modeled the oligopolistic electricity market as a non-linear dynamical system and used dynamic game theory to develop bidding strategies for market participants. Swider and Weber [3] proposed a methodology that enables a strategically behaving bidder to maximize the revenue under price uncertainty. Centeno *et al.* [4] used a scenario tree to represent uncertain variables that may affect price formation, which include hydro inflows, power demand, and fuel prices. They also presented a model to analyze GenCos' medium-term strategic analysis. In [5], a dynamic bid model was used to simulate the bidding behaviors of the players and study the inter-relational effects of the players' behaviors and the market conditions on the bidding strategies of players over time. Li and Shi [6] proposed an agent-based model to study the strategic bidding in a day-ahead electricity market, and found that applying learning algorithms could help increase the net earnings of the market participants. Nojavan and Zare [7] proposed an information gap decision theory model to solve the optimal bidding strategy problem by incorporating the uncertainty of market price. Their case study further shows that risk-averse or risk-taking decisions could affect the expected profit and the bidding curve in the day-ahead electricity market. Qiu *et al.* [8] discussed the impacts of model deviations on the design of a GenCo's bidding strategies using the conjectural variation (CV) based methods, and further proposed a CV-based dynamic learning algorithm with data filtering to alleviate the influence of demand uncertainty. Kardakos *et al.* [9] point out that when making a bidding decision, a GenCo would take into accounts the behavior of its competitors as well as specific features and enacted rules of the electricity market. They further developed an optimal bidding strategy for a strategic generator in a transmission-constrained day-ahead electricity market. Other studies [10–12] emphasized that transmission constraints, volatile loads, market power exertions, and collusions may induce GenCos to bid higher prices than their true marginal costs, thereby aggravating the price volatility issue.

As concern for the sustainability of the power market increases, efforts also have been made to reduce the risk of price variation. Most studies focus on the employment of price-based demand response (DR) programs for the electricity users to control and reduce the peak-to-average load ratio [13–22]. For instance, Oh and Hildreth [13] proposed a novel decision model to determine whether or not to participate in the DR program, and further assessed the impact of the DR program on the market stability. Faria *et al.* [14] suggested that adequate tolls could motivate the potential market players to adopt the DR programs. Ghazvini *et al.* [15] showed that multi-objective decision-making is more realistic for retailers to optimize the resource schedule in a liberalized retail market. In [16], a two-stage stochastic programming model was formulated to hedge the financial losses in the retail electricity market. Zhong *et al.* [17] proposed a new type of DR program to improve social welfare by offering coupon incentives. The researchers in [18,19] handled the energy scheduling issue by optimizing the DR program in a smart grid environment. Yousefi *et al.* [20] proposed a dynamic model to simulate a time-based DR program, and used Q-learning methods to optimize the decisions for the market stakeholders. In [21–23], much more detailed reviews were provided for benefit analyses and applications of DR in a smart grid environment.

However, in electricity markets where the demand side is regulated, how to design and optimize GenCos' bidding strategies is treated as one of the most efficient ways to sustain the market price in the presence of uncertainty. Some studies have been made by proposing an incentive mechanism or contract for the GenCos to mitigate the risk of price variation caused by their subjective preferences during the bidding process [24–27]. Silva *et al.* [24] introduce an incentive compatibility mechanism, which is individually rational and feasible, to resolve the asymmetric information problem. Liu *et al.* [25] proposed an incentive electricity generation mechanism to control GenCos' market power and reduce the pollutant emissions using the signal transduction of game theory. Cai *et al.* [26] proposed a sealed dynamic price cap to prevent GenCos from exercising market power. Heine [27] performed a series of studies on the effectiveness of regulatory schemes in energy markets, and pointed out that potential improvements exist in contemporary systems when incentive-based regulations are appropriately implemented.

Although there is a large body of literature in bidding and incentive policy, most of the studies neglect assessment of the long-term effects of the incentive programs on the GenCos' learning behaviors. Besides, less attention is paid to the dynamic response of the GenCos to the volatile loads and incentive schemes. To maintain the market stability, it is highly desirable to understand the interplays between the incentive mechanism and the GenCos' adaptive responses to the variable market. This paper aims to fill this gap by proposing an incentive mechanism in a day-ahead power market to reduce price variance, and further assessing the subsequent long-term impacts of the incentive mechanism. To that end, a two-level Stackelberg gaming model was developed to analyze the bidding strategies of the market participants including one independent system operator (ISO) and several GenCos. An optimal menu of incentive contracts was derived under a one-leader and multi-followers game theoretic framework. Finally, a Stackelberg-based Q-learning approach was employed to assess the GenCos' response to the incentive-based generation mechanism.

The remainder of the paper is organized as follows: in Section 2, we introduce the menu of incentive contracts, and describe the workflow of the multi-agent game framework; in Section 3, we give a mathematical description of the problem, and present the details of the Stackelberg model; in Section 4, we use a Q-learning methodology to investigate the long-term effectiveness of the incentive contracts; in Section 5, numerical examples are provided to demonstrate the application and performance of the method; Section 6 concludes the paper.

2. Problem Statement

2.1. Description of the Menu of Incentive Contracts

A commercial agreement between the ISO and the GenCo is proposed, which defines a reward scheme in exchange for a consistent bidding behavior: the GenCo agreed to bid a reasonable power generation with a constant bidding curve during the contract period. Note that the reasonable power output should be larger than the regulated threshold of power output.

In general it is not efficient to design a uniform incentive contract with a constant threshold due to the fact that GenCos usually possess diverse bidding behaviors. It is also rather complex to design customized incentive contracts for all GenCos. One viable approach is to design a pertinent incentive menu comprised of key characteristic incentive contracts for certain target GenCos. These GenCos are selected as representatives from the entire group of power generators. Though the ISO cannot precisely predict the target GenCos' bidding information in a future bidding round, the customized incentive contracts still can be devised by incorporating the target GenCos' interest through inference of historical bidding data. For the target GenCo, its expected profit could be amplified if it complies with the incentive contract which takes into account its individual rational constraints and incentive compatibility constraints. Hence it is reasonable to assume that target GenCos would not reject the customized contract.

Though the incentive contracts in the menu are designed based on the individual rationality and incentive compatibility of the target GenCos, they also benefit non-target GenCos that are willing to accept the incentive contracts. For a non-target GenCo, the incentive contract would be appealing if the expected profit is higher by abiding with the agreement. To better illustrate how the menu of the incentive contracts works, some notations are given as follows.

2.1.1. Target GenCo and Contracted Generating Companies

The GenCos are termed the target GenCos if their individual rationality constraints and incentive compatibility constraints are considered so that they could be motivated to accept the incentive contract. We define A^0 as set of all possible combinations of target GenCos. For each $a^0 \in A^0$, $a^0 = (a_1^0, a_2^0, ..., a_m^0)$, where a_i^0 represents whether GenCo i is chosen as a target GenCo or not. Further, we define $I = \{k | a_k^0 = 1, k \in M\}$ as a set of target GenCos.

Note that a is a combination of strategies of GenCos, $a = (a_1, a_2, ..., a_m)$, where a_i represents whether GenCo i decides to accept the menu of the incentive contracts or not. If $a_i = 0$, it is "not", and $a_i \neq 0$ is "yes". If $a_i \neq 0$, $a_i = k$, $k \in I$, meaning GenCo k accepts the incentive contract and becomes the target GenCo. The GenCos with $a_i \neq 0$ are termed as contracted GenCos.

2.1.2. Bidding Curve and Market Clearing Price

For electricity transactions, The study in [28] shows that GenCos submit power output in MW (Megawatts) along with associated prices for one bid in both discrete form and continuous form. A bid in discrete form with three different blocks is shown in Table 1. If the power output level is below 30 MW, the price is 10 \$/MWh; If the power output level is between 30 MW and 60 MW, the price is 15 \$/MWh; If the power output level is between 60 MW and 90 MW, the price is 20 \$/MWh. Generally, this bid could also take a continuous form as shown in Figure 1. Without loss of generality, a continuous bid curve model is adopted in this paper. For GenCo i, its bidding curve at time t is in the form of $p_{it} = \alpha_{it} + \beta_{it}q_{it}$, where α_{it} and β_{it} are the bidding coefficients of GenCo i at time t. Here p_{it} and q_{it} respectively, represent the bidding price and the bidding power output of GenCo i at time t.

Table 1. Block bid.

Blocks	Price ($/MWh)	Power output level (MW)
Block 0	10	30
Block 1	15	60
Block 2	20	90

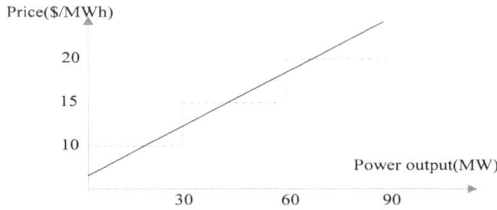

Figure 1. Block bid and continuous bid curves.

The market clearing price (MCP) is a uniform price shared by all GenCos, and the actual MCP depends on all GenCos' bidding behaviors. Assume the bidding curve of GenCo i is $p_{it} = \alpha_{it} + \beta_{it}q_{it}$, and the electricity demand at time t is D_t. The MCP at time t, which is denoted as p_t, can be obtained by solving the following power balance equation:

$$D_t = \sum_{i=1}^{m} q_{it} \tag{1}$$

$$p_t = \alpha_{it} + \beta_{it}q_{it}, \text{ for } i = 1, 2 \ldots, m \tag{2}$$

2.1.3. Menu of the Incentive Contracts

The menu of the incentive contracts is composed of multiple contracting terms in the form of $(\alpha_i, \beta_i, \pi_i)$, where α_i and β_i represent the thresholds of bidding coefficients respectively, and π_i is the relevant reward for meeting the incentive contract. Though each incentive contract is originally tailored to the rational and incentive-compatibility constraints of a certain target GenCo, it is also expected that these contracts are designed appropriately to motivate non-target GenCos to participate in the incentive program.

Assuming an incentive contract is customized for a target GenCo with bidding coefficients α_{i0} and β_{i0}, and the amount of the reward is π_{i0} by calculating the target GenCo's individual rationality and incentive compatibility conditions. This contract could be expressed by the triplet $(\alpha_{i0}, \beta_{i0}, \pi_{i0})$ which specifies the reward and the obligation associated with the contracted GenCo: if the dispatched power output of the contracted GenCo during the contract period is always greater than the required level, which is prescribed as $\underline{q} = (p_t - \alpha_{i0})/\beta_{i0}$ with p_t being the MCP at time t, a reward of π_{i0} would be received at the end of the contract period.

All sets of $(\alpha_i, \beta_i, \pi_i)$ are further incorporated into (AL, B, π), where AL, B, π are the vectors of α_i, β_i, π_i, respectively. Hence the menu of the incentive contracts could be concisely specified in the form of (AL, B, π).

2.2. Scenario-Based Approach

At certain time, unexpected events like hot weather, network congestion, and demand spikes may occur randomly, which causes load soaring and demand forecast errors. Scenario based approach [16,29,30] is often employed to address these types of uncontrollable events. These uncertain events are characterized by scenarios with corresponding probabilities. The scenarios considered in this paper include both normal scenario and bad scenario. In the latter, the load is 20% higher than the average demand. Probability of each scenario could be inferred from historical data and experiences. The Monte Carlo method is adopted to simulate both normal and bad scenarios.

2.3. The Workflow of Multi-Agent System

In this paper, a multi-agent system, adapted to a simulated context with multiple GenCos and one ISO, is proposed to study a day-ahead electricity market based on the proposed incentive menu of contracts. The multi-agent system was developed in a Java platform that was partly inherited from the Repast platform [31]. Some actions of GenCos were coded by Matlab, and then packaged and implanted into the multi-agent Java platform.

Figure 2 shows the flowchart of multi-agent system (MAS) scheduling procedure.

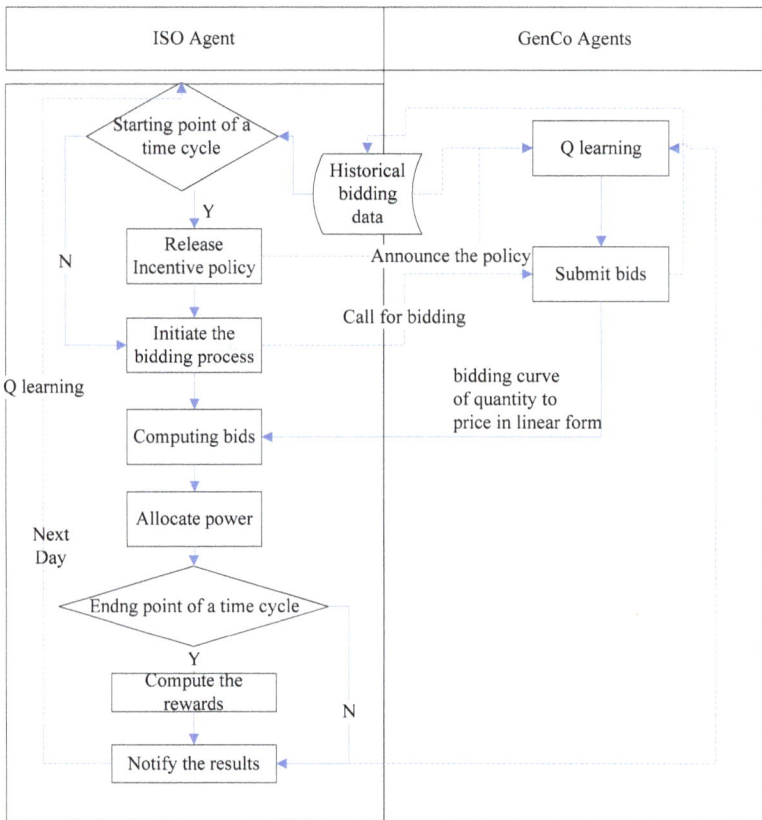

Figure 2. Flow chart of multi-agent system (MAS) scheduling mechanism (note: Y = yes and N = no). ISO: independent system operator.

At the beginning of a specified period $\tilde{t}$ that consists of T days (*i.e.*, a period may include several days, or several months), ISO announces the menu of the incentive contracts. GenCos, which act for their own interest, decide whether they accept the incentive contract or not by using the periodic Q-learning method. This decision-making system resembles the one-leader and multi-followers Stackelberg game where the ISO is the leader and the GenCos are the followers. An algorithm using the idea of the Stackelberg game, which is further illustrated in Section 4.3, is presented for the ISO to find the initial optimal menu of incentive contracts in the first period. In addition, a periodic Stackelberg-based Q-learning method, which is illustrated in Section 4.2, is proposed for the ISO to find the subsequent optimal menu of incentive contracts over the following periods.

At the beginning of a period, GenCo i decides whether or not to accept the incentive program, by using its periodic Q-learning method. In each day of the period, GenCo i chooses to place a high bid or a normal bid by using its daily Q-learning method, and submits its bid, taking the form of $p_{it} = \alpha_{it} + \beta_{it}q_{it}$. After the ISO receives all the bids from the participants, the relevant information is aggregated and stored in a central repository. Based on the estimated hourly electricity demand of the next day, the ISO decides the unified hourly MCP of the next day, and announces the hourly power output schedule of individual GenCos for the next day.

At the end of period $\tilde{t}$, the ISO computes the relevant rewards based on the bidding data retrieved from the central repository. If any GenCo's bidding data in the given period are constant, and always in alignment with certain contract in the menu of the incentive contracts, the GenCo would receive the relevant reward. During the repetitive bidding periods, both ISO and the GenCos improve their pricing policy and bidding strategies using the Q-learning algorithm.

3. Multi-Agent Stackelberg Game Model

3.1. Model Assumption and Description

Model assumptions are given as follows:

(1) To prevent GenCos reaping extra profits by modifying their bidding data to satisfy the incentive contract, it is stipulated that any GenCo using new bidding coefficients is not eligible to join the incentive program until after several rounds.

(2) At some time, due to the uncertainties in weather condition, network reliability and consumer behavior, unexpected demand spikes may occur, and the load may vary with a large degree of uncertainty. Probabilistic scenarios trees are adopted to accommodate the uncertain characteristics of the load profile. For instance, the electricity demand at time t is estimated to be 100 MW with probability of 0.8 for the normal demand scenario, and 150 MW with probability 0.2 for the high demand scenario, or bad scenario. Enumeration methods can be used to capture all possible scenarios if the problem size is not too large. Let Λ denote a set of uncertain scenarios, and λ_t denotes a realized scenario in Λ at time t. In addition, Λ^B is used to represent a set of bad scenarios.

(3) It is assumed that each GenCo within the MAS framework have two bidding options: either place a high bid (*i.e.*, $b_{i,t} = 1$) or place a normal bid (*i.e.*, $b_{i,t} = 0$), where $b_{i,t}$ is the bidding strategy of GenCo i at time t. The coefficients for different bidding options are defined as follows:

$$\alpha_{i,t} = \begin{cases} \alpha_i^c & b_{i,t} = 0 \\ \alpha_i^h & b_{i,t} = 1 \end{cases} \tag{3}$$

$$\beta_{i,t} = \begin{cases} \beta_i^c & b_{i,t} = 0 \\ \beta_i^h & b_{i,t} = 1 \end{cases} \tag{4}$$

where α_i^c, β_i^c are parameters of the normal bidding curve for GenCo i, and α_i^h, β_i^h are parameters of the corresponding high bidding curve. Obviously, if a GenCo has accepted an incentive contract, we have $b_{it} = 0$, $\alpha_{i,t} = \alpha_i^c$, $\beta_{i,t} = \beta_i^c$ for $t \in \tilde{t}$.

3.2. Single-Period Decision-Making Model of GenCo

Based on the given menu of incentive contracts, in each time period, a GenCo tries to maximize its profit by choosing the best bidding strategy as follows:

$$\max \quad \Pi_i (a_i) \tag{5}$$

For a GenCo who does not accept any incentive contract, its profit is given as:

$$\prod_i (a_i = 0) = \sum_{t \in \tilde{t}} \sum_{\lambda_t \in \Lambda} (Y \times \rho (\lambda_t)) \tag{6}$$

where $Y = \prod_i(\lambda_t, a_i = 0)$, and $\rho(\lambda_t)$ represents the probability of λ_t, and π_k is the reward specified in the incentive contract for target GenCo k.

It is usually difficult for a GenCo to know the actual bidding behavior of others, but it is reasonable to assume that the probability of its competitors' decision can be inferred from historical data. Hence $Y = \prod_i(\lambda_t, a_i = 0)$ can be obtained as:

$$Y = \sum_{t \in \tilde{t}, \lambda_t \in \Lambda} X \tag{7}$$

where:

$$
\begin{aligned}
X = \; & \text{pos}_i(b_t^{i,C}) \prod_{j \neq i} \text{pos}_j(b_{j,t}) \left(p\left(\lambda_t, b_t^{i,C}\right) K_{i,t} - c_{i1} K_{i,t} - 0.5 c_{i2} K_{i,t}{}^2 \right) \\
& + \text{pos}_i(b_t^{i,H}) \prod_{j \neq i} \text{pos}_j(b_{j,t}) \left(p\left(\lambda_t, b_t^{i,H}\right) K_{i,t} - c_{i1} K_{i,t} - 0.5 c_{i2} K_{i,t}{}^2 \right)
\end{aligned}
\tag{8}
$$

where $K_{i,t} = q_{i,t}(\lambda_t, b_{i,t})$ is the power output of GenCo i when its bidding action is $b_{i,t}$ in a scenario λ_t. c_{i1} and c_{i2} are cost coefficients of GenCo i. $b_t^{i,c} = \{b_{1,t}, b_{2,t}, b_{i-1,t}, 0, b_{i+1,t}, \ldots, b_{m,t}\}$ represent a bidding combination when GenCo i places a normal bid. Note that $\text{pos}_j(b_{j,t})$ is the probability for GenCo j to take action $b_{j,t}$. Here $p(\lambda_t, b_t(a))$ is the expected electricity price when the bidding combination of GenCos is $b_t(a)$ in scenario λ_t. Note that $p(\lambda_t, b_t(a))$ is $p(\lambda_t, b_t^{i,C})$ when GenCos' bidding action is $b_t^{i,C}$ in scenario λ_t, and is $p(\lambda_t, b_t^{i,H})$ when GenCos' bidding action is $b_t^{i,H}$ in scenario λ_t.

For a contracted GenCo who agrees on the acceptance of an incentive contract which is tailored to the target GenCo k, its profit could be calculated as follows:

$$\Pi_i(a_i = k) = \sum_{t \in \tilde{t}} \sum_{\lambda_t \in \Lambda_t} \rho(\lambda_t) \left(\Pi_i(\lambda_t, a_i = k) \right) + \pi_k \tag{9}$$

Assuming the incentive contract is prescribed as $(\alpha_k, \beta_k, \pi_k)$ triplet, we have:

$$K_{i,t} \geqslant \frac{p_t - \alpha_k}{\beta_k} \tag{10}$$

$$\text{pos}_i(b_t^{i,C}) = 1 \tag{11}$$

$$\text{pos}_i(b_t^{i,H}) = 0 \tag{12}$$

So $\Pi_i(\lambda_t, a_i = k)$ could be calculated as:

$$\Pi_i(\lambda_t, a_i = k) = \sum_{t \in \tilde{t}} \left(\prod_{j \neq i} \text{pos}_j(b_{j,t}) \left(p\left(\lambda_t, b_t^{i,C}\right) K_{i,t} - c_{i1} K_{i,t} - 0.5 c_{i2} K_{i,t}{}^2 \right) \right), k \in I \tag{13}$$

subject to:

$$K_{i,t} \geqslant \frac{p_t - \alpha_k}{\beta_k} \tag{14}$$

$$\text{pos}_i(b_t^{i,C}) = 1 \tag{15}$$

$$\text{pos}_i(b_t^{i,H}) = 0 \tag{16}$$

The optimization problem faced by a GenCo is how to choose an optimal bidding strategy such that its expected profit is maximized. Hence the incentive-compatibility constraint could be formulated as:

$$\text{IC}: a_i = \text{argmax}\{\Pi_i(a_i = 0), \Pi_i(a_i = k)\}, k \in I \tag{17}$$

If a GenCo accepts an incentive contract, its expected profit should be higher than the alternative. Thus the personal rationality constraint could be re-formulated as:

$$\text{PC}: \Pi_i(a_i = 0) < \Pi_i(a_i = k) + \pi_k, \quad k \in I \tag{18}$$

3.3. Optimization Problem of Independent System Operator

From the ISO's point of view, its goal is to design an optimal menu of incentive contracts such that the average MCP during the period remains at a relatively stable level, or the volatility of price in the worst scenarios could be mitigated, while the total electricity payment is minimized. To that end, it is necessary for the ISO to identify the optimal set of target GenCos (*i.e.*, a^0) as well as designing the incentive menu for attracting contracted GenCos, so that its objectives could be optimized. Since an incentive contract, which is specified in the triplet form of (α_i, β_i, π_i), is dependent upon a^0, how to target suitable GenCos is the key to designing an optimal incentive menu of contracts. Hence the ISO's initial decision is to choose optimal a^0, so as to minimize the total cost with certain price stability.

As the leader of the Stackelberg game, the ISO can analyze the response of the followers (*i.e.*, GenCos) so as to find the optimal decision variable a^0. A two-level programming model is proposed to facilitate ISO's decision-making. The sub-problem at the first level enables the ISO to minimize the total cost with price stability by finding an optimal value of a^0. The sub-problem at the second level can be treated as GenCos' reaction model upon the release of the menu of incentive contracts from the first level decision:

$$\min_{a^0 \in A^0} \left(C(a^0) \right) \tag{19}$$

$$\min_{a^0 \in A^0} \left((1 - \delta) \times \text{EP}(a^0) + \delta \times \text{BP}(a^0) \right) \tag{20}$$

subject to:

$$a = a(a^0) = (a_1(a^0), a_2(a^0), ..., a_m(a^0)) \triangleq (a_1, ..., a_i, ..., a_m) \tag{21}$$

$$C\left(a^0\right) = \sum_{t \in \tilde{t}} \sum_{\lambda_t \in \Lambda_t} \left(\rho\left(\lambda_t\right) P(\lambda_t, a) K_{i,t} \right) + \sum_{i=1}^{m} \pi\left(a_i\right) \tag{22}$$

$$\pi\left(a_i\right) = \begin{cases} \pi_k & a_i = k, k \in I \\ 0 & a_i = 0 \end{cases} \tag{23}$$

$$\text{EP}\left(a^0\right) = \sum_{t \in \tilde{t}} \sum_{\lambda_t \in \Lambda_t} \rho\left(\lambda_t\right) \times P\left(\lambda_t, a\right) \tag{24}$$

$$\text{BP}(a^0) = \sum_{t \in \tilde{t}} \sum_{\lambda_t \in \Lambda_t^B} \rho\left(\lambda_t\right) \times \left[P\left(\lambda_t, a\right) - \text{EP}^*\right]^2 \tag{25}$$

$$\sum_{i=1}^{m} K_{i,t} = D\left(\lambda_t\right) \tag{26}$$

$$P\left(\lambda_t, a\right) = \prod_{j \in M} \text{pos}_j(b_{j,t}) p\left(\lambda_t, b_t\left(a\right)\right) \tag{27}$$

$$p\left(\lambda_t, b_t\left(a\right)\right) = \alpha_{i,t} + \beta_{i,t} K_{i,t}, i \in M \tag{28}$$

$$b_t(a) = (b_{1,t}(a), b_{2,t}(a), \cdots, b_{m,t}(a)) \tag{29}$$

$$b_{i,t}(a_i) = \begin{cases} 0 & a_i > 0 \\ 1 \quad \text{or} \quad 0 & a_i = 0 \end{cases} \tag{30}$$

where a_i is obtained by solving follows:

$$\text{IC} : a_i = \text{argmax}\left\{\Pi_i\left(a_i\left(a^0\right)\right)\right\} \tag{31}$$

$$\text{s.t. PC} : \Pi_i(a_i = 0) < \Pi_i(a_i = k) + \pi_k, \quad k \in I \tag{32}$$

$$K_{i,t} \geqslant \frac{p_t - \alpha_k}{\beta_k} \text{ for } a_i > 0 \tag{33}$$

$$\text{pos}_i(b_t^{i,C}) = 1 \text{ for } a_i > 0 \tag{34}$$

$$\text{pos}_i(b_t^{i,H}) = 0 \text{ for } a_i > 0 \tag{35}$$

where $C(a^0)$ is the total power purchasing cost when the combination of the target GenCos is a^0, and δ is a balance parameter. $EP(a^0)$ is the expected electricity price when the combination of the target GenCos is a^0, and EP^* is the best expected price. $BP(a^0)$ is the variance of mean price *versus* EP^* when the combination of the target GenCos is a^0. Here $P(\lambda_t, a)$ is the expected MCP when the combination of contracted GenCos' is a in scenario λ_t. The MCP is $p(\lambda_t, b_t(a))$ when the bidding behavior of GenCos is $b_t(a)$ in scenario λ_t.

As shown in Equation (19), one of the ISO's objectives is to minimize the electricity payment. Equation (20) is another objective of the ISO, that contains dual goals: Firstly, minimizing the common price in the contract period; and secondly minimizing the volatility of price. Both goals are combined by a balance factor. Equation (21) indicates that a is also decided by a^0. Equations (22) and (23) are the mathematical descriptions of the cost and the reward, respectively. Equation (24) calculates the average price in one period under multiple scenarios. Equation (25) calculates the variation of mean price *versus* EP^* in one period in multiple scenarios. Equation (26) ensures that the electricity demand is always satisfied. Equation (27) computes the average price by multiplying the price for certain bid combination in a specified scenario with its occurrence possibility. Equations (28)–(30) provide the mathematical descriptions for $p(\lambda_t, b_t(a))$, $b_{t(a)}$, $b_{i,t}(a_i)$, respectively. Equation (31) represents the GenCos' objective which is also their incentive-compatibility constraint with a_i being the decision variable for GenCo i. Equation (32) gives the personal rational constraint of the GenCos who is willing to accept an incentive contract. Finally, Equations (33)–(35) defines the constraints of contracted GenCos including power output capacity of contracted GenCos, and the possibilities of contracted GenCos to place high bids or normal bids.

A multi-objective optimization can be solved by turning it into a single objective model through appropriately assigning weight to each objective function. Using a weight w to combine the two objectives in Equations (19) and (20), the ISO's decision model could be further expressed as:

$$\max_{a^0 \in A^0} J = w\left(\frac{C_{\max} - C(a^0)}{C_{\max} - C_{\min}}\right) + (1-w)\left(\frac{EPM_{\max} - EPM(a^0)}{EPM_{\max} - EPM_{\min}}\right) \tag{36}$$

subject to:

$$EPM(a^0) = \left((1-\delta) \times EP\left(a^0\right) + \delta \times BP(a^0)\right) \tag{37}$$

$$a = a(a^0) = (a_1(a^0), a_2(a^0), ..., a_m(a^0)) \triangleq (a_1, ..., a_i, ..., a_m) \tag{38}$$

$$C\left(a^0\right) = \sum_{t \in \tilde{t}} \sum_{\lambda_t \in \Lambda_t} \left(\rho(\lambda_t) P(\lambda_t, a) K_{i,t}\right) + \sum_{i=1}^{m} \pi(a_i) \tag{39}$$

$$\pi\left(a_i\right) = \begin{cases} \pi_k & a_i = k, k \in I \\ 0 & a_i = 0 \end{cases} \tag{40}$$

$$EP\left(a^0\right) = \sum_{t \in \tilde{t}} \sum_{\lambda_t \in \Lambda_t} \rho\left(\lambda_t\right) \times P\left(\lambda_t, a\right) \tag{41}$$

$$BP(a^0) = \sum_{t \in \tilde{t}} \sum_{\lambda_t \in \Lambda_t^B} \rho\left(\lambda_t\right) \times \left[P\left(\lambda_t, a\right) - EP^*\right]^2 \tag{42}$$

$$\sum_{i=1}^{m} K_{i,t} = D\left(\lambda_t\right) \tag{43}$$

$$P\left(\lambda_t, a\right) = \prod_{j \in M} pos_j(b_{j,t}) p\left(\lambda_t, b_t\left(a\right)\right) \tag{44}$$

$$p\left(\lambda_t, b_t\left(a\right)\right) = \alpha_{i,t} + \beta_{i,t} K_{i,t}, i \in M \tag{45}$$

$$b_t\left(a\right) = \left(b_{1,t}\left(a\right), b_{2,t}\left(a\right), \cdots, b_{m,t}\left(a\right)\right) \tag{46}$$

$$b_{i,t}\left(a_i\right) = \begin{cases} 0 & a_i > 0 \\ 1 \quad or \quad 0 & a_i = 0 \end{cases} \tag{47}$$

where a_i is obtained by solving follows:

$$IC : a_i = \operatorname{argmax}\left\{\Pi_i\left(a_i\left(a^0\right)\right)\right\} \tag{48}$$

$$\text{s.t. PC} : \Pi_i(a_i = 0) < \Pi_i(a_i = k) + \pi_k, \quad k \in I \tag{49}$$

$$K_{i,t} \geq \frac{p_t - \alpha_k}{\beta_k} \text{ for } a_i > 0 \tag{50}$$

$$pos_i(b_t^{i,C}) = 1 \text{ for } a_i > 0 \tag{51}$$

$$pos_i(b_t^{i,H}) = 0 \text{ for } a_i > 0 \tag{52}$$

where C_{max} is the maximum available $C(a^0)$; C_{min} is the minimum available $C(a^0)$; $EPM(a^0)$ is a balance between price minimization and price variation minimization when the decision variable is a^0. EPM_{max} is the maximum available $EPM(a^0)$; and EPM_{min} is the minimum available $EPM(a^0)$. Equations (38)–(52) are the same with Equations (21)–(35).

4. Q-Learning for Agents' Optimal Decision Making

Each agent interacts in the volatile market environment due to the uncertain load and lack of precise knowledge of its competitors. It is imperative for the agents to evolve their actions through the learning of repeated bidding processes. Q-learning is one of the reinforcement learning methods, and could guide the agents to improve the performance of their decision making over time. In each period, an agent perceives the state of the market environment, and takes certain actions based on its perception and past experience, which result in a new state. This sequential learning process would reinforce its subsequent actions. Quite a few studies have been done on Q-learning, and its application in the electricity market has been reported. For instance, Rahimiyan and Mashhadi [32] propose a fuzzy Q-learning method to model the GenCos' strategic bidding behavior in a competitive market condition, and find that GenCos could accumulate more profit by using fuzzy Q-learning. Naghibi-Sistani *et al.* [33] developed a modified reinforcement learning based on temperature variation, and applied it to the electricity market to determine the GenCos' optimal strategies. Attempts also have been made to combine the Q-learning with Nash-Stackelberg games for reaching a long-run equilibrium. Haddad *et al.* [34] incorporate a Nash-Stackelberg fuzzy Q-learning into a hierarchical and distributed learning framework for decision-making, with which mobile users are guided to enter the equilibrium state that optimizes the utilities of all the network participants.

In this paper, Q-learning methods are adopted by ISO and GenCos for the making decisions. Different learning algorithms are designed for ISO and GenCos because they have different goals. For GenCos, both a periodic Q-learning method and a daily Q-learning method are applied to the bidding decision process. For ISO, a Stackelberg-based Q-learning is adopted to design the menu of incentive contracts in each period.

4.1. Periodic and Daily Q-Learning Methods for Generating Companies

At the starting point of a period, a GenCo decides whether the incentive contract should be accepted or declined. In each day of the period, the GenCo should choose to place a high bid or place a normal bid. Especially, if a contracted GenCo decides to place a high bid, the reward at the end of the period, would be cancelled. To calculate the potential reward, a multi-step Q-learning method is adopted by the GenCo to decide its bidding strategy in daily basis. Two Q-learning methods are proposed for the GenCo's periodic and daily decision making. The state, actions, reward and Q-value function are defined as follows:

4.1.1. State Identification

State $s_{\tilde{t}}$ is defined for GenCo's Q-learning method for a period, and it is composed of values of all possible average electricity prices over one period.

State s_t denote the states for GenCo's Q-learning method in each day, and it is composed of values of all possible average electricity prices over one day.

4.1.2. Action Selection

Let a discrete set of actions $a_{i,\tilde{t}} = \{0, k\}, k \in I$, denote the action selection of GenCo i at the starting point of a period for GenCo's Q-learning method for a period. When $a_{i,\tilde{t}} = 0$, GenCo i chooses not to accept the incentive contract over period $\tilde{t}$. When $a_{i,\tilde{t}} = k, k \in I$, GenCo i accepts the incentive contract which is tailored to the target GenCo k over period $\tilde{t}$.

$a_{i,t}$ denotes the action selection of GenCo i in each day for GenCo's Q-learning method for a day, and its value is 0 or 1. When $a_{i,t} = 1$, GenCo i adopts a normal bidding strategy; When $a_{i,t} = 0$, GenCo i adopts a high bidding strategy.

When $a_{i,\tilde{t}} \neq 0$, in each day of the period $\tilde{t}$, GenCo i places a normal bidding strategy. So for a contracted GenCo, $a_{i,t} = 1$, with a high probability.

4.1.3. Reward Calculation

The periodic reward function for Q-learning method over period $\tilde{t}$ is defined as:

$$r(s_{\tilde{t}}, a_{i,\tilde{t}}) = \sum_{t \in \tilde{t}} \Pi_{i,t}(a_{i,t}) + R(a_{i,\tilde{t}}) \tag{53}$$

Equation (53) represents the reward assigned to action $a_{i,\tilde{t}}$ from the old state $s_{\tilde{t}}$. If $a_{i,\tilde{t}} = 0$, which means that the menu of incentive contracts is not accepted over period $\tilde{t}$, $R(a_{i,\tilde{t}}) = 0$. If $a_{i,\tilde{t}} = k, k \in I$, which means that GenCo i accepts the incentive contract which is tailored to the target GenCo k over period $\tilde{t}$, an amount of $R(a_{i,\tilde{t}})$ is received as the reward for meeting the incentive contract. The reward would further influence the periodic Q-value which guides the GenCo to determines the next action as whether or not to accept the incentive menu.

Every day the GenCo agent evaluates the current state, and chooses the best action that optimizes its objectives. Then the current state evolves to the new state, with a transition probability, and the agent receives a reward.

The reward *r* for daily Q-learning is made up of two parts. One is the direct profit subject to all the GenCo agents' bidding behaviors, loads, and cost of the GenCo agent. The second is a portion of the expected reward if the GenCo accepts the incentive contract. If a GenCo agent accepts the incentive menu and fulfills the contractual obligations in the contract period, a reward would be obtained at the end of the period, so the reward is a delayed reward. A multi-step reward function, which captures the characteristic of the delayed reward, is defined to describe GenCo agent's daily Q-learning as follows:

$$r(s_t, a_{i,t}) = \Pi_{i,t}(a_{i,t}) + R(a_{i,t}) \tag{54}$$

subject to:

$$R(a_{i,t}) = \begin{cases} \Gamma & \text{for } \prod_{t \in \tilde{t}} a_{i,t} = 1 \\ 0 & \text{for } \prod_{t \in \tilde{t}} a_{i,t} = 0 \end{cases} \tag{55}$$

where,

$$\begin{aligned}
\Gamma &= T_s \frac{1}{T} \pi\left(a_{i,t}\right) + (T - T_s) \left(\varphi \frac{1}{T} \pi\left(a_{i,t}\right) + \varphi^2 \frac{1}{T} \pi\left(a_{i,t}\right) + \ldots + \varphi^{T-T_s} \frac{1}{T} \pi\left(a_{i,t}\right) \right) \\
&= \frac{1}{T} \pi(a_{i,t}) \left[T_s + (T - T_s) \sum_{i=1}^{T-T_s} \varphi^i \right], (T_s = 1, 2, \ldots, T)
\end{aligned} \tag{56}$$

where φ is a discount factor, and $\pi(a_{i,t})$ is the reward for GenCo *i* to meet the incentive contract terms at time *t*, and *T* is the total number of days in a contract period, and T_s is the number of the days elapsed in the period.

4.1.4. Q-Value Update

By Q-learning, using the Bellman optimality in Equations (57) and (58), each GenCo agent tries to find the optimal action to maximize the Q-value of each state in a long run.

The periodic Q-value function defined for GenCo *i* over period $\tilde{t}$ is given as follows:

$$Q_{\tilde{t}+1}(s_{\tilde{t}}, a_{i,\tilde{t}}) = Q_{\tilde{t}}(s_{\tilde{t}}, a_{i,\tilde{t}}) + \ell_{\tilde{t}} \left[r(s_{\tilde{t}}, a_{i,\tilde{t}}) + \gamma_{\tilde{t}} \max_{a_{i,\tilde{t}+1}} Q(s_{\tilde{t}+1}, a_{i,\tilde{t}+1}) - Q_{\tilde{t}}(s_{\tilde{t}}, a_{i,\tilde{t}}) \right] \tag{57}$$

where $\ell_{\tilde{t}}$ is a positive learning rate at period $\tilde{t}$, and $\gamma_{\tilde{t}}$ is a discount parameter at period $\tilde{t}$.

These action-state value functions $Q_{\tilde{t}+1}(s_{\tilde{t}}, a_{i,\tilde{t}})$ ($i = 1, \ldots m$), which are greatly affected by the reward function as illustrated in Equation (53), determine the GenCo agents' most suitable actions for the next run. That is, if the Q-value for accepting the incentive menu is less than the Q-value for not accepting it, the GenCo agent would not take the action of accepting the incentive menu. Conversely, it would. The daily Q-value function defined for GenCo *i* at each day is given as follows:

$$Q_{t+1}(s_t, a_{i,t}) = Q_t(s_t, a_{i,t}) + \ell_t \left[r(s_t, a_{i,t}) + \gamma_t \max_{a_{i,t+1}} Q(s_{t+1}, a_{i,t+1}) - Q_t(s_t, a_{i,t}) \right] \tag{58}$$

where ℓ_t is a positive learning rate in day *t*, and γ_t is a discount parameter in day *t*.

4.2. Q-Learning for the Leader of the Stackelberg Game (Independent System Operator)

4.2.1. State Identification

State $\vec{s}_{\tilde{t}} = \{(s_{\tilde{t}}, a^0)\}$ for ISO's Q-learning is composed of two state variables, one is the values of all possible average electricity prices during that period, and the other is the decision variable for menu of the incentive contracts.

4.2.2. Action Selection

The set of all possible combinations of target GenCos, or A^0 is defined as the set of action selection of ISO agent. The ISO takes the action at each step, or at the starting point of each period. $(a^0)_{\tilde{t}}$ denotes the action selection of ISO in period $\tilde{t}$.

4.2.3. Reward Calculation

Reward function $r\left(s_{\tilde{t}}, (a^0)_{\tilde{t}}\right)$ is given by:

$$r\left(s_{\tilde{t}}, (a^0)_{\tilde{t}}\right) = w\left(\frac{C_{max} - C\left((a^0)_{\tilde{t}}\right)}{C_{max} - C_{min}}\right) + (1-w)\left(\frac{EPM_{max} - EPM\left((a^0)_{\tilde{t}}\right)}{EPM_{max} - EPM_{min}}\right) \tag{59}$$

$$\text{s.t. } a = a\left[(a^0)_{\tilde{t}}\right] = \left[a_{1,\tilde{t}}\left(a^0\right)_{\tilde{t}}, a_{2,\tilde{t}}\left(a^0\right)_{\tilde{t}}, \ldots, a_{m,\tilde{t}}\left(a^0\right)_{\tilde{t}}\right] \triangleq \left(a_{1,\tilde{t}}, a_{2,\tilde{t}}, \ldots, a_{m,\tilde{t}}\right) \tag{60}$$

$$a_{i,\tilde{t}} = \operatorname{argmax}(Q_{\tilde{t}+1}(s_{\tilde{t}}, a_{i,\tilde{t}})) \tag{61}$$

Equations (59)–(61) illustrates that $a_{i,\tilde{t}}$, which is the GenCo i's action in period t, depends on its Q-learning, and so the reward of the ISO is obtained by using a Stackelberg-based Q-learning method.

4.2.4. Q-Value Update

As the leader of the Stackelberg game, Q-learning algorithm for ISO is given as follows:

$$Q^0_{\tilde{t}+1}\left(s_{\tilde{t}}, (a^0)_{\tilde{t}}\right) = Q^0_{\tilde{t}}\left(s_{\tilde{t}}, (a^0)_{\tilde{t}}\right) + \ell_{\tilde{t}}\left[r_{\tilde{t}}\left(s_{\tilde{t}}, (a^0)_{\tilde{t}}\right) + \gamma \max_{(a^0)_{\tilde{t}+1}} Q^0\left(s_{\tilde{t}+1}, (a^0)_{\tilde{t}+1}\right) - Q^0_{\tilde{t}}\left(s_{\tilde{t}}, (a^0)_{\tilde{t}}\right)\right] \tag{62}$$

4.3. Solution Methodology for Independent System Operator's Initial Q Value

For problems with multiple decision variables, chaos search is more capable of hill-climbing and escaping from the local optima than the random search [35]. Hence a chaos optimization algorithm is proposed to solve the problem
The detailed procedure of the algorithm is given as follows:

Step 1: set initial parameters incorporating bidding coefficients of GenCos and its power capacity.
Step 2: set $v = 1$.
Step 3: generate a non-zero chaos variable η_{v+1} using cube mapping method as shown below:

$$\eta_{v+1} = 4\eta_v^3 - 3\eta_v \tag{63}$$

Step 4: decoding the chaos variables into a binary variable which represents a value for the sets of target GenCos.
Step 5: calculate the tailoring values of $(\alpha_i, \beta_i, \pi_i)$ for all target GenCos using Equation (18).
Step 6: for the designed menu of the incentive contracts, check each GenCo's optimal reaction by solving Equation (17).
Step 7: calculate the corresponding objective value and the state $\vec{s}_{\tilde{t}}$ for ISO's Q-learning. The latter includes the mean electricity price during the period, and the menu of the incentive contracts. If the obtained objective value is larger than the existing one, substitute the existing one.
Step 8: substitute the chaos variables into Equation (64) to yield new chaos variables:

$$x_v = c_v - \left[d_v \eta_{v+1}\right] \tag{64}$$

where c_v and d_v are two constant vectors, and $\left[d_v \eta_{v+1}\right]$ is the integr part of $d_v \eta_{v+1}$.
Step 9: Set $v = v + 1, k = k + 1$.
Step 10: If $v > v_{max}$, stop searching, else go to Step 4.

5. Simulations and Analysis

The simulation is performed in a day-ahead electricity market with the participation of five GenCos. The original probability for a GenCo to bid high or normal is 0.5. Electricity demand at each hour varies between 170 MW and 230 MW. The probability of high-demand scenarios, which are also termed as bad scenarios, is less than 0.2. The length of a contract could be several days, or a couple of months. For computational convenience, firstly it is assumed that each period consists of seven days or one week.

Parameters of GenCos' bidding curves are listed in Table 2, and the GenCos' cost parameters are listed in Table 3, and the weights of the two objectives are listed in Table 4. Three cases are investigated for the comparative analysis.

Case 1: no menu of incentive contracts or Q-learning.

Case 2: menu of incentive contracts without Q-learning in one period. Eight sub-cases are further analyzed, and the comparative results are listed in Tables 5 and 6.

Case 3: menu of incentive contracts with Q-learning in multiple periods. Note that load demand over the multiple periods varies between 170 MW and 230 MW.

Table 2. Parameters of GenCos' bidding curves. (Unit for $\alpha_{i,t}{}^c$, $\alpha_{i,t}{}^h$: $/MW per hour, unit for $\beta_{i,t}{}^c$, $\beta_{i,t}{}^h$: $/(MW)^2$ per hour).

GenCo No.	Case 1, Cases 2.1–2.5, Case 3				Cases 2.6–2.8			
	$\alpha_{i,t}{}^c$	$\beta_{i,t}{}^c$	$\alpha_{i,t}{}^h$	$\beta_{i,t}{}^h$	$\alpha_{i,t}{}^c$	$\beta_{i,t}{}^c$	$\alpha_{i,t}{}^h$	$\beta_{i,t}{}^h$
1	10	0.5	10.5	0.525	10	0.5	10.5	0.525
2	11	0.8	12	0.84	11	0.8	12	0.84
3	8	0.6	8.4	0.63	8	0.9	8.4	0.945
4	15	0.5	15.75	0.525	15	0.5	15.75	0.525
5	20	0.9	21	0.945	20	0.6	21	0.63

Table 3. GenCos' cost parameters. (Unit for c_{i1}: $/MW, unit for c_{i2}: $/(MW)^2$).

GenCo No.	Cases 2.1–2.4 and 2.6–2.8		Case 2.5	
	c_{i1}	c_{i2}	c_{i1}	c_{i2}
1	10	0.5	5	0.25
2	11	0.55	6	0.35
3	8	0.5	4	0.25
4	15	0.8	7	0.4
5	20	0.9	10	0.45

Table 4. Objectives of Cases 2.1–2.8.

Cases	Cost	EPM ($0.5 \times$ EP $+ 0.5 \times$ BP)
2.1, 2.5, 2.6	1	0
2.2, 2.7	0	1
2.3, 2.8, 3	0.5	0.5
2.4	Without menu of incentive contracts	

Table 5. Comparative results for Cases 2.1–2.3 and 2.5 (unit: $,). (Note: Y = saying "Yes" to offer of the incentive contract menu and N = saying "No" to the offer).

Items	Case 2.1					Cases 2.2 and 2.3					Case 2.5				
Target GenCos	1	0	0	0	1	1	0	0	1	1	0	1	0	1	1
GenCos' response	Y	Y	Y	N	Y	Y	Y	Y	Y	Y	Y	Y	Y	Y	Y
Expected reward	208,750					543,530					123,070				
Expected cost saving (compared with Case 2.4)	597,830					512,120					932,580				
Expected price (EP)	37.17					36.94					36.94				
BP (mean price variance in bad scenarios)	2.78					2.53					2.53				
EPM ($0.5 \times$ EP $+ 0.5 \times$ BP)	19.97					19.74					19.74				

Table 6. Comparative results for Cases 2.6–2.8 (unit: $, N/A = Not Applicable).

Items		Case 2.6					Cases 2.7 and 2.8				
Target GenCos		0	0	1	0	1	0	0	1	1	1
GenCos' response		Y	Y	Y	N	Y	Y	Y	Y	Y	Y
Expected reward		215,980					581,130				
Expected cost saving (compared with Case 2.9)		607,780					497,270				
EP		38					36.94				
BP (meanprice variance in bad scenarios)		2.78					2.53				
EPM ($0.5 \times$ EP $+ 0.5 \times$ BP)		20.39					20.15				
Threshold for GenCo's power output	1	6558					6477				
	2	7145					7091				
	3	7145					7091				
	4	N/A					9852				
	5	6558					6477				

Case 2.1 aims at minimizing cost. Cost in Case 2.1 is less than that in Cases 2.2 and 2.3, but EP and BP in Case 2.1 are larger compared with that in Cases 2.1 and 2.3.

In Case 2.5, though it also aims at minimizing cost, as GenCos have small cost coefficient, they could gain more profits compared with Cases 2.1–2.4, and so GenCos in Case 2.5 prefer to make normal bids since they could obtain more power output and hence more profit, and so both cost and EPM could be minimized.

Since GenCo 3 has higher bidding coefficients in Cases 2.6–2.8, it has more influence on MCP than in Cases 2.1–2.5. Hence GenCo 3 is more likely to be the target GenCo in Cases 2.6–2.8 than in Cases 2.1–2.5.

Figures 3 and 4 show the simulation results of variations in price and cost across 112 days in Cases 1 and 3.

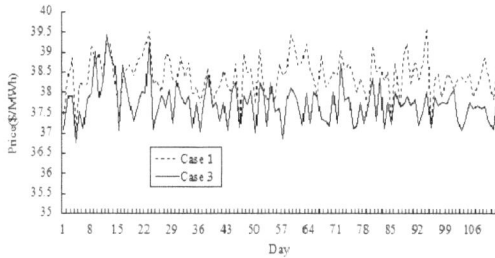

Figure 3. Comparative results of price variation for Cases 1 and 3 (for 112 days).

It can be seen that the variations of electricity price and cost could be reduced in the long term provided that the incentive contract is adopted. It could be seen that in the early phases, the effect of the incentive contract is not obvious as the daily price and daily electricity purchase cost do not significantly decrease in Case 3. However, as time evolves, GenCos can enhance their bidding experience through learning from past bidding processes and realize that accepting the incentive contract could help improve their profitability. They become more interested in participating in the incentive program. As a result, the electricity price is kept at a low and stable level.

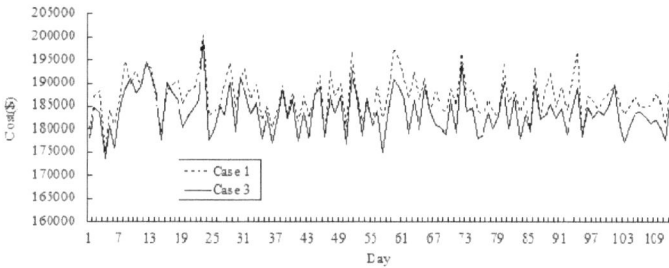

Figure 4. Comparative results of cost variation for Cases 1 and 3 (for 112 days).

Figures 5 and 6 show the simulation results of the variations of price and the cost across 112 days in Cases 2 and 3. It could be seen that in the early periods, the cost in Case 3 may be higher than that in Case 2 over certain number of days, and the price in Case 3 is higher than that in Case 2. As GenCos and ISO accumulate more bidding experiences by Q-learning, optimum decisions in Case 3 could be made by both players, and the cost and the price could be reduced compared with that in Case 2.

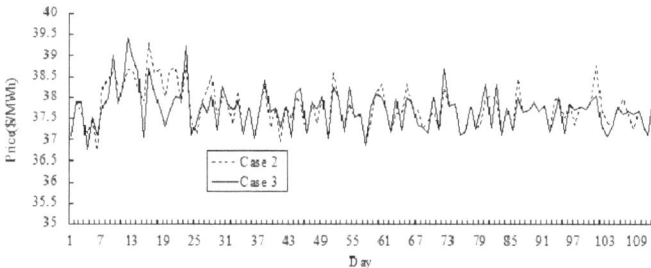

Figure 5. Comparative results of price variation for Cases 2 and 3 (for 112 days).

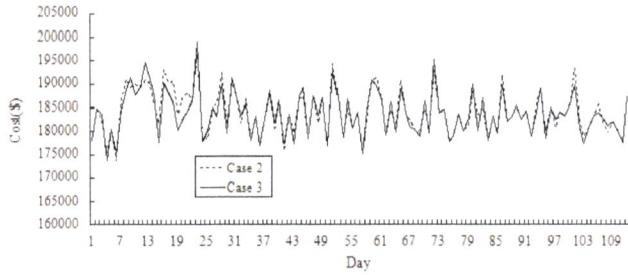

Figure 6. Comparative results of cost variation for Cases 2 and 3 (for 112 days).

Figures 7 and 8 show the comparative results of average price variation and average cost variation for three cases, respectively. Based on Case 3, it could be seen that both the cost and the price could be reduced and remain stable in a long run.

Extending the length of the contract period to 14 days, the comparative results for the duration of 224 days (*i.e.*, 16 periods) are shown in Figure 9, and it could be seen that price variation in the electricity market with incentive mechanism is less than the market without incentive mechanism. In fact, the price variance in the former market is 0.242 *versus* 0.270 in the latter, and the average price in the former market is 38.33 *versus* 38.50 in the latter (price unit is $/MWh).

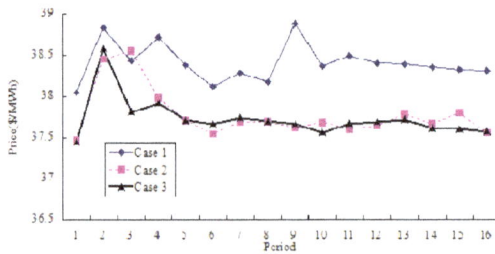

Figure 7. Comparative results of average price variation for 3 Cases (for 16 periods).

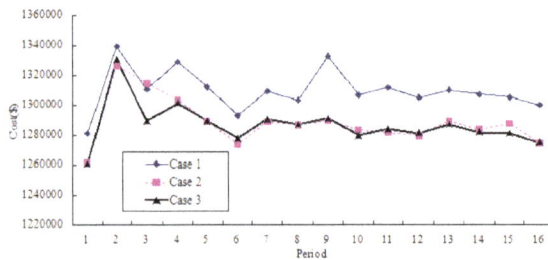

Figure 8. Comparative results of average cost variation (for 16 periods).

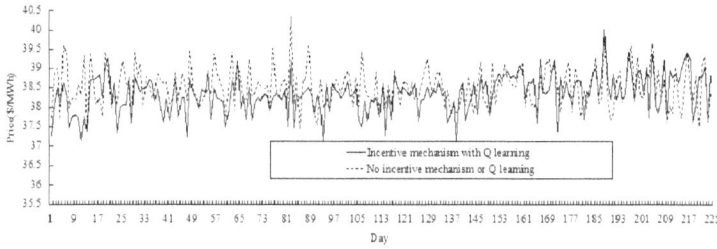

Figure 9. Comparative results of price variation (for 224 days).

Extending the single contract period to 28 days, the comparative results for 448 days (*i.e.*, 16 periods) are shown in Figure 10. It could be seen that price variation in the electricity market with incentive mechanism is less than the market without incentive mechanism. In fact, the price variance in the former market is 0.252 *versus* 0.310 in the latter, and the average price in the former market is 37.86 *versus* 38.43 in the latter (price unit is $/MWh).

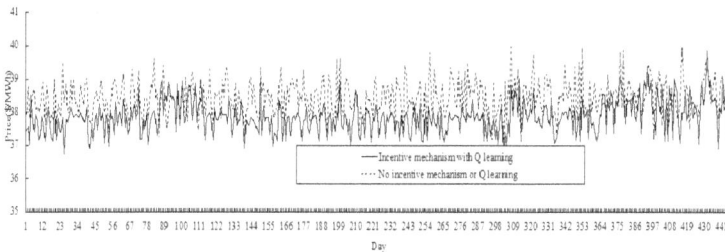

Figure 10. Comparative results of price variation (for 448 days).

6. Conclusions

In this paper a menu of incentive contracts is presented in a Stackelberg game model, aiming at seeking an incentive bidding mechanism with which the electricity price could be kept at a low and stable level. To ensure market equilibrium, a Stackelberg game based Q-learning is proposed for the ISO to analyze the responses of GenCos to the market as well as searching for the optimal menu of incentive contracts. For GenCos, a periodic Q-learning method is adopted to determine whether the incentivized menu should be accepted or not. In addition, a multi-step Q-learning method is adopted by the GenCo to decide its daily bidding policy. Based on the multi-agent platform, the long-term effectiveness of the incentive program is validated up to 14 months using simulation methods. Numerical results show that an incentivized menu, which is suitably designed by the ISO with the perspective of a central planner, could lead to desirable bidding behavior of GenCos, and hence guarantees the market sustainability. When multiple types of Q-learning methods are adopted by the ISO and the GenCos for decision makings, both the electricity price and purchasing cost could be reduced. Hence a desirable trade-off between the price variation and the purchasing cost could be reached at equilibrium. Future efforts could be directed to analyzing GenCos' reactions to the menu of incentive contracts under different risk preferences or generation uncertainties with wind and solar power integration.

Acknowledgments: This research is supported by National Natural Science Funds of China (Grant No. 71201097) and Action plan for scientific and technological innovation Program of Science and Technology Commission Foundation of Shanghai (Grant No. 15511109700). We would like to thank the anonymous reviewers and the editor for their valuable time and constructive comments for the improvement of the original manuscript.

Author Contributions: This paper designs an incentive contract menu to achieve long-term stability for electricity prices in a day-ahead electricity market. A bi-level Stackelberg game model is proposed to search for the optimal incentive mechanism under a one-leader and multi-followers gaming framework. A multi-agent simulation platform was developed to investigate the effectiveness of the incentive mechanism using an independent system operator (ISO) and multiple power generating companies (GenCos). Further, a Q-learning approach was implemented to analyze and assess the responses of GenCos to the incentive menu.

Conflicts of Interest: The authors declare no conflict of interest.

Notations

N1. Set Parameters

A^0	Set of all possible a^0, which denotes a combination of target GenCos.
M	Set of all serial numbers of GenCos.
I	Set of target GenCos, $I = \{k \mid a_k^0 = 1, k \in M\}$.
Λ	Set of uncertain scenarios during a bidding period.
Λ^B	Set of all bad scenarios.
(AL, B, π)	Data set for the menu of incentive contracts.

N2. Decision Variables

a_i^0	Whether a GenCo is chosen as the target GenCo. $a_i^0 = 1$ is "yes" and 0 is "not".
a^0	$a^0 = (a_1^0, a_2^0, ..., a_m^0), a^0 \in A^0$.
a_i	*Whether GenCo i accepts the incentive menu. $a_i = 0$ means "not", and $a_i \neq 0$ means "yes". Moreover, if $a_i \neq 0$, $a_i = k$, $k \in I$, meaning GenCo i accepts the incentive contract with GenCo k as the target GenCo.*
a	$i \in M$.

N3. Model Parameters

$C(a^0)$	Total electricity purchasing cost when the combination of the target GenCos is a^0.
$\pi(a_i)$	Award received by GenCo i.
t	Time interval in one period.
$\tilde{t}$	Length of contract period.
λ_t	A certain scenario in Λ at time t.
$\rho(\lambda_t)$	Probability of λ_t.
$EP(a^0)$	Expected electricity price when the combination of the target GenCos is a^0.
EP^*	The best expected price.
δ	Balance parameter.
w	Weight of the objective functions.
$BP(a^0)$	Variance of mean price in bad scenarios *versus* EP^* when the set of the target GenCos is a^0.
EPM	A representative symbol of the model objective which combines the expected price (EP) and robustness of the price (BP) with a balance factor.
$P(\lambda_t, a)$	In scenario λ_t, the expected market price when the set of the GenCos' options for the accepted incentive contract is a.
m	Number of GenCos.
$K_{i,t}$	Power output of GenCo i when the bidding combination of GenCos is $b_t(a)$ in scenario λ_t.
α_{it}, β_{it}	Parameters of GenCo i's bidding curve at time t.

b_{it}	Bidding strategy of GenCo i at time t. $b_{it} = 1$ means a high bidding; and $b_{it} = 0$ implies a normal bidding.
$pos_i(b_{it})$	Probability for GenCo i to make a bid b_{it} at time t.
$\Pi_i(a_i)$	Expected profit of GenCo i when its contract decision is a_i.
$\Pi_i(\lambda_t, a_i)$	Expected profit for GenCo i in scenario λ_t. When $a_i = 0$, GenCo i accepts the menu of incentive contracts; when $a_i \neq 0$ or $a_i = k$, GenCo i does not accept the incentive contract which is tailored to the target GenCo k ($k = a_i$).
$b_t^{i,c}$	$b_t^{i,c} = \{b_{1t}, b_{2t}, b_{i-1,t}, 0, b_{i+1,t}, ..., b_{mt}\}$.
$b_t^{i,H}$	$b_t^{i,H} = \{b_{1t}, b_{2t}, b_{i-1,t}, 1, b_{i+1,t}, ..., b_{mt}\}$.
$b_t(a)$	Combination of GenCos' bidding strategy with $b_t(a) = (b_{1,t}(a_1), b_{2,t}(a_2), \cdots, b_{m,t}(a_m))$.
$p(\lambda_t, b_t(a))$	The expected electricity price when the bidding combination of GenCos is $b_t(a)$. The value of $b_t(a)$ could be $b_t^{i,c}$ or $b_t^{i,H}$.
$D(\lambda_t)$	Electricity demand in scenario λ_t.
c_{i1}, c_{i2}	Cost coefficients of GenCo i.
$(\alpha_i, \beta_i, \pi_i)$	Parameters of an incentive contract. Note α_i and β_i represent the bidding coefficients of a target GenCo, and π_i denotes per-period reward.
p_{it}	Bidding price of GenCo i at time t.
p_t	MCP at time t.
q_{it}	Bidding power output of GenCo i at time t.
D_t	Electricity demand at time t.
α_i^c, β_i^c	Parameters of the normal bidding curve for GenCo i.
α_i^h, β_i^h	Parameters of the high bidding curve for GenCo i.

N4. Q-Learning Parameters

$s_{\tilde{t}}$	State identification for GenCo's Q-learning method in a period.
s_t	State identification for GenCo's Q-learning method in each day.
$a_{i,\tilde{t}}$	Periodical action selection of GenCo i at the starting point of a period for GenCo's Q-learning method.
$a_{i,t}$	Daily action selection of GenCo i in each day for GenCo's Q-learning method.
$r(s_{i,\tilde{t}}, a_{i,\tilde{t}})$	Periodical reward function for Q-learning method.
$R(a_{i,\tilde{t}})$	Reward obtained by GenCo i over period $\tilde{t}$.
$r(s_{i,t}, a_{i,t})$	Daily reward function for Q-learning method.
$R(a_{i,t})$	Reward obtained by GenCo i over period at a day.
φ	Discount factor.
T	Number of days in the contract period.
T_s	Number of days elapsed over a contract period $\tilde{t}$.
$Q_{\tilde{t}+1}(s_{\tilde{t}}, a_{i,\tilde{t}})$	Periodical Q-value function defined for GenCo i.
$Q_{t+1}(s_t, a_{i,t})$	Daily Q-value function defined for GenCo i.
$\ell_{\tilde{t}}$	Positive learning rate for periodical Q-learning function.
ℓ_t	Positive learning rate for daily Q-learning function.
$\gamma_{\tilde{t}}$	Discount parameter for periodical Q-learning function.
γ_t	Discount parameter for daily Q-learning function.
$\overrightarrow{s}_{\tilde{t}} = \{(s_{\tilde{t}}, a^0)\}$	State identification for ISO's Q-learning function.
$(a^0)_{\tilde{t}}$	Periodic action selection of ISO.
$r(s_{\tilde{t}}, (a^0)_{\tilde{t}})$	Periodic reward calculation for ISO's Q-learning function.
$Q^0_{\tilde{t}+1}(s_{\tilde{t}}, (a^0)_{\tilde{t}})$	Periodic Q-value function defined for ISO.

N5. Algorithms Parameters

v	An iteration number.
η_v	Non-zero chaos variable.
c_v, d_v	Constant vectors.
v_{max}	Max iteration times.

References

1. Zhang, D.; Wang, Y.; Luh, P.B. Optimization based bidding strategies in the deregulated market. *IEEE Trans. Power Syst.* **2000**, *15*, 981–986. [CrossRef]
2. Kian, A.R.; Cruz, J.B. Bidding strategies in dynamic electricity markets. *Decis. Support Syst.* **2005**, *40*, 543–551. [CrossRef]
3. Swider, D.J.; Weber, C. Bidding under price uncertainty in multi-unit pay-as-bid procurement auctions for power systems reserve. *Eur. J. Oper. Res.* **2007**, *181*, 1297–1308. [CrossRef]
4. Centeno, E.; Renese, J.; Barquin, J. Strategic analysis of electricity markets under uncertainty: A conjectured-price-response approach. *IEEE Trans. Power Syst.* **2007**, *22*, 423–432. [CrossRef]
5. Sahraei-Ardakani, M.; Rahimi-Kian, A. A dynamic replicator model of the players' bid in an oligopolistic electricity market. *Electr. Power Syst. Res.* **2009**, *79*, 781–788. [CrossRef]
6. Li, G.; Shi, J. Agent-based modeling for trading wind power with uncertainty in the day-ahead wholesale electricity markets of single-sided auctions. *Appl. Energy* **2012**, *99*, 13–22. [CrossRef]
7. Nojavan, S.; Zare, K. Risk-based optimal bidding strategy of generation company in day-head electricity market using information gap decision theory. *Int. J. Electr. Power Energy Syst.* **2013**, *48*, 83–92. [CrossRef]
8. Qiu, Z.; Gui, N.; Deconick, G. Analysis of equilibrium-oriented bidding strategies with inaccurate electricity market models. *Int. J. Electr. Power Energy Syst.* **2013**, *46*, 306–314. [CrossRef]
9. Kardakos, E.G.; Simoglou, C.K.; Bakirtzis, A.G. Optimal bidding strategy in transmission-constrained electricity markets. *Electr. Power Syst. Res.* **2014**, *109*, 141–149. [CrossRef]
10. Anderson, E.J.; Cau, T.D.H. Implicit collusion and individual market power in electricity markets. *Eur. J. Oper. Res.* **2011**, *211*, 403–414. [CrossRef]
11. Nam, Y.W.; Yoon, Y.T.; Hur, D.; Park, J.; Kim, S. Effects of long-term contracts on firms exercising market power in transmission constrained electricity markets. *Electr. Power Syst. Res.* **2006**, *76*, 435–444. [CrossRef]
12. David, A.K.; Wem, F.S. Market power in electricity supply. *IEEE Trans. Energy Convers.* **2001**, *16*, 352–360. [CrossRef]
13. Oh, S.; Hildreth, A.J. Decisions on energy demand response option contracts in smart grids based on activity-based costing and stochastic programming. *Energies* **2013**, *6*, 425–443. [CrossRef]
14. Faria, P.; Vale, Z.; Baptista, J. Demand response programs design and use considering intensive penetration of distributed generation. *Energies* **2015**, *9*, 6230–6246. [CrossRef]
15. Ghazvini, M.A.F.; Soares, J.; Horta, N.; Neves, R.; Castro, R.; Vale, Z. A multi-objective model for scheduling of short-term incentive-based demand response programs offered by electricity retailers. *Appl. Energy* **2015**, *151*, 102–118. [CrossRef]
16. Ghazvini, M.A.F.; Faria, P.; Ramos, S.; Morais, H.; Vale, Z. Incentive-based demand response programs designed by asset-light electricity providers for the day-ahead market. *Energy* **2015**, *82*, 786–799. [CrossRef]
17. Zhong, H.; Xie, L.; Xia, Q. Coupon incentive-based demand response: Theory and case study. *IEEE Trans. Power Syst.* **2013**, *28*, 1266–1276. [CrossRef]
18. Fakhrazari, A.; Vakilzadian, H.; Choobineh, F.F. Optimal energy scheduling for a smart entity. *IEEE Trans. Smart Grid* **2014**, *5*, 2919–2928. [CrossRef]
19. Christopher, O.A.; Wang, L. Smart charging and appliance scheduling approaches to demand side management. *Int. J. Electr. Power Energy Syst.* **2014**, *57*, 232–240.
20. Yousefi, S.; Moghaddam, M.P.; Majd, V.J. Optimal real time pricing in an agent-based retail market using a comprehensive demand response model. *Energy* **2011**, *36*, 5716–5727. [CrossRef]
21. Shariatazadeh, F.; Mandal, P.; Srivastava, A.K. Demand response for sustainable energy systems: A review, application and implementation strategy. *Renew. Sustain. Energy Rev.* **2015**, *45*, 343–350. [CrossRef]

22. Gu, W.; Yu, H.; Liu, W.; Zhu, J.; Xu, X. Demand response and economic dispatch of power systems considering large-scale plug-in hybrid electric vehicles/electric vehicles (PHEVs/EVs): A review. *Energies* **2013**, *6*, 4394–4417. [CrossRef]

23. Bradley, P.; Leach, M.; Torriti, J. A review of the costs and benefits of demand response for electricity in the UK. *Energy Policy* **2013**, *52*, 312–327. [CrossRef]

24. Silva, C.; Wollenberg, B.F.; Zheng, C.Z. Application of mechanism design to electric power markets. *IEEE Trans. Power Syst.* **2001**, *16*, 1–8. [CrossRef]

25. Liu, Z.; Zhang, X.; Lieu, J.; Li, X.; He, J. Research on incentive bidding mechanism to coordinate the electric power and emission-reduction of the generator. *Int. J. Electr. Power Energy Syst.* **2010**, *32*, 946–955. [CrossRef]

26. Cai, X.; Li, C.; Lu, Y. Price cap mechanism for electricity market based on constraints of incentive compatibility and balance accounts. *Power Syst. Technol.* **2011**, *35*, 143–148.

27. Heine, K. Inside the black box: Incentive regulation and incentive channeling on energy markets. *J. Manag. Gov.* **2013**, *17*, 157–186. [CrossRef]

28. Weber, J.D.; Overbye, T.J. A two-level optimization problem for analysis of market bidding strategies. In Proceedings of the IEEE Power Engineering Society Summer Meeting, Edmonton, AB, Canada, 18–22 July 1999; Volume 2, pp. 682–687.

29. Lei, W.; Shahidehpour, M.; Zuyi, L. Comparison of scenario-based and interval optimization approaches to stochastic SCUC. *IEEE Trans. Power Syst.* **2012**, *27*, 913–921.

30. Wang, B.; Yang, X.; Li, Q. Bad-scenario set risk-resisting robust scheduling model. *Acta Autom. Sin.* **2012**, *38*, 270–278. [CrossRef]

31. North, M.J.; Collier, N.T.; Vos, J.R. Experiences creating three implementations of the repast agent modeling toolkit. *ACM Trans. Model. Comput. Simul.* **2006**, *16*, 1–25. [CrossRef]

32. Rahimiyan, M.; Mashhadi, H.R. An adaptive Q-learning algorithm developed for agent-based computational modeling of electricity market. *IEEE Trans. Syst. Man Cybern. C Appl. Rev.* **2010**, *40*, 547–556. [CrossRef]

33. Naghibi-Sistani, M.B.; Akbarzadeh-Tootoonchi, M.R.; Bayaz, M.H.J.D.; Rajabi-Mashhadi, H. Application of Q-learning with temperature variation for bidding strategies in market based power systems. *Energy Convers. Manag.* **2006**, *47*, 1529–1538. [CrossRef]

34. Haddad, M.; Altmann, Z.; Elayoubi, S.E.; Altaman, E. A Nash-Stackelberg fuzzy Q-learning decision approach in heterogeneous cognitive networks. In Proceedings of the IEEE Global Telecommunications Conference, Miami, FL, USA, 6–10 December 2010.

35. Zuo, X.Q.; Fan, Y.S. A chaos search immune algorithm with its application to neuro-fuzzy controller design. *Chaos Solitons Fractals* **2006**, *30*, 94–109. [CrossRef]

energies

MDPI

Article

Voltage Control Method Using Distributed Generators Based on a Multi-Agent System

Hyun-Koo Kang [1], Il-Yop Chung [2,*] and Seung-Il Moon [3]

[1] Korea Electric Power Research Institute (KEPRI), Korea Electric Power Company (KEPCO), 105 Munji-Ro, Yuseong-Gu, Daejeon 34056, Korea; khyun9@kepco.co.kr

[2] School of Electrical Engineering, Kookmin University, Seoul 136-702, Korea

[3] School of Electrical Engineering and Computer Scirnce, Seoul National University, Gwanak-ro, Gwanak-gu, Seoul 151-744, Korea; moonsi@plaza.snu.ac.kr

* Correspondence: chung@kookmin.ac.kr; Tel.: +82-2-910-4702; Fax: +82-2-910-4449

Academic Editor: Ying-Yi Hong

Received: 21 September 2015; Accepted: 3 December 2015; Published: 11 December 2015

Abstract: This paper presents a voltage control method using multiple distributed generators (DGs) based on a multi-agent system framework. The output controller of each DG is represented as a DG agent, and each voltage-monitoring device is represented as a monitoring agent. These agents cooperate to accomplish voltage regulation through a coordinating agent or moderator. The moderator uses the reactive power sensitivities and margins to determine the voltage control contributions of each DG. A fuzzy inference system (FIS) is employed by the moderator to manage the decision-making process. An FIS scheme is developed and optimized to enhance the efficiency of the proposed voltage control process using particle swarm optimization. A simple distribution system with four voltage-controllable DGs is modeled, and an FIS moderator is implemented to control the system. Simulated data show that the proposed voltage control process is able to maintain the system within the operating voltage limits. Furthermore, the results were similar to those obtained using optimal power flow calculations, even though little information on the power system was required and no power flow calculations were implemented.

Keywords: distributed generation (DG); fuzzy inference system (FIS); multi-agent system (MAS); particle swarm optimization (PSO); reactive power control; voltage control

1. Introduction

In modern power systems, the integration of distributed generators (DGs), including micro-gas turbines, renewable energy resources and battery energy storage systems, has seen much recent development in response to economic, environmental and political interests [1–3]. The use of DGs has potential benefits in numerous aspects of power system operation; for example, they may lead to improvements in system efficiency, power quality and reliability [4–6]. However, the implementation of a grid with a large number of DGs also complicates power system control and management, because DGs may change the direction of power flow locally in distribution systems and may disturb the conventional operation schemes, including voltage management and protection [7–9].

Voltage regulation is a significant issue in the planning and operation of power distribution systems. The objective of voltage regulation is to supply electricity within a suitable voltage range to all power consumers and to ensure system stability and safety for electrical devices. The voltage range is typically ±5% of the rated voltage [10]. To maintain a voltage level within these limits, voltage regulators, including on-load tap changers, shunt capacitors and step-voltage regulators, are operated at the substation or in the middle of the distribution feeder. In general, these voltage regulators employ the line drop compensation (LDC) method, which depends on the load level, to supervise the voltages in conventional distribution systems [11]. Using LDC, the line voltage drop from the regulator-installed point to the regulating point can readily be estimated by measuring the bypass load current. However, in distribution systems that include multiple DGs, the LDC has limited application for voltage management, because the system voltage and load current vary not only with the load demand, but also with the output power of the DGs [12–14].

There have been a number of research works of voltage regulation for distribution systems, including DGs [14–17]. A common concept of these works has been using the compensation ability of DGs and operating multiple DGs in an autonomous and distributed manner. The application of a multi-agent system (MAS) to power engineering has emerged over the past decade [18,19]. MAS is a system comprising two or more agents that have decision-making capabilities and can communicate with other agents. In particular, MAS is appropriate for establishing autonomous and distributed control systems and for providing coordination and cooperation with agent-based devices, which may include an inverter controller of a DG [20].

Voltage control has local control characteristics; then it follows that MAS is a suitable framework to realize a voltage regulation strategy using various voltage compensation devices. There have been a number of papers of voltage control based on MAS [19,21–23]. The work in [19] describes a theoretical multi-agent secondary voltage control scheme. Voltage compensation using the reactive power output of DGs based on MAS has been proposed [21]. These studies described the voltage regulation problem using MAS; however, they focused on multi-agent-based voltage compensation rather than on optimal dispatch. On the other hand, numerous studies have used fuzzy inference systems (FISs) to solve power system problems, including voltage control and network reconfiguration [24–26].

The power distribution system is inherently area distributed, and then, it is too expensive to implement a centralized energy management system. In other words, the achievement of voltage regulation in a distributed system is not easy to accomplish in a centralized manner. Then, in this paper, a distributed intelligent voltage control method is proposed to achieve voltage regulation in a distributed manner that is based on a multi-agent-based voltage control participation and decision-making process. Voltage regulation is achieved by reactive power control of multiple DGs, where each is modeled as a DG agent. These DG agents can cooperate to accomplish voltage regulation, where coordination is achieved using a moderator. Here, DGs involved in the proposed voltage control are inverter-based ones so as to control its reactive power. Additionally, the proposed voltage regulation method can be achieved not only with inverter-based DGs, but also with any other reactive power compensator, such as STATCOM. These reactive power sources can be utilized to regulate voltage in the distribution network. Especially, inverter-based power electronic devices are very useful for voltage control, since they can inject their output immediately. The moderator determines the voltage control contributions of each DG based on the reactive power sensitivities and margins collected from each DG agent. FIS is used in this decision-making process and is optimized using a particle swarm optimization (PSO) method to enhance the efficiency of the voltage control process. Meanwhile, the stability proof of FIS, PSO and MAS are omitted in this paper, since they are already well-known theories and fully validated in the fields [27–30]. The remainder of this paper is organized as follows. Section 2 describes the MAS-based voltage control process. The modeling of the moderator using FIS is described in Section 3, and the optimization thereof using PSO is described in Section 4. Finally, in Section 5, a simulation model configuration is discussed that is used to assess the effectiveness of the proposed voltage control method.

2. Voltage Regulation Using Multi-Agent System

2.1. Multi-Agent System for Voltage Regulation

We propose an MAS-based voltage regulation method that maintains voltages within a specified range in a distribution system with multiple DGs. Using the proposed method, the bus voltages in the distribution system can be controlled by coordinating the reactive power output of the DGs.

Figure 1 shows the basic structure of the MAS platform. Each agent links to an electrical node of the distribution network or an output controller of a DG. There are two principle issues for the design and implementation of an MAS [21]. One is defining and classifying the role of the agent, and the other is determining how multiple agents will cooperate to achieve the global objective.

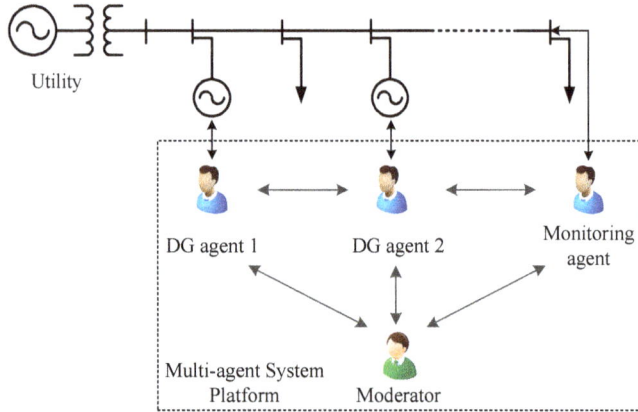

Figure 1. Schematic diagram illustrating the structure of the multi-agent system (MAS).

Three types of agent were defined: monitoring agents, DG agents and the moderator. Monitoring agents may be located at specific nodes to report on the voltage levels at strategic points in the power system. These agents monitor the local node voltage and can transmit warnings if voltage disturbances occur and request voltage compensation to restore the normal voltage levels. The nodes located farthest from the substation are strong candidates for being designated the optimal locations of the monitoring agents, because this voltage level is typically the lowest of the network. DG agents control the reactive power output of each DG to regulate the bus voltage. These DG agents must participate in the MAS-based decision-making process.

The moderator coordinates the DG agents to accomplish voltage compensation. It coordinates multiple agents to achieve a global goal, which here is voltage regulation. Contract net protocol (CNP) is widely used to coordinate multiple agents and was adopted to implement our MAS-based decision-making process. In determining the reactive power output of each DG, the moderator considers the status of the DG and the system performance. The objective of the moderator is to minimize the sum of the additional reactive power outputs by considering the reactive power margins of each DG, while maintaining the system within the operating limits. If a given DG has high reactive power output sensitivity and margin, then its participation rate in voltage regulation will be greater. The moderator is described in more detail in Section 3.

2.2. Voltage Control Process Based on Multi-Agent System

The voltage control process is conducted through CNP-based communication. CNP is a protocol that is used in MAS for sharing problem information and distributing tasks to solve the problem. To achieve this CNP-based decision-making process for the proposed voltage regulation method, the communication capability between agents should ensure that it can be done within a desired time. However, voltage regulation in a distribution system is generally a process that takes place within several seconds and minutes. Then, here, the communication time between agents would not affect its regulation result considering the baud rate for CNP. From a previous work related to this CNP-based communication, all decision-making processes could be completed within 0.5 s within a ZigBee environment for wireless personal area networking [31].

CNP is typically composed of five stages: recognition, announcement, bidding, awarding and expediting [32,33]. As a consequence, the voltage control process is implemented here in five stages, as summarized in Figure 2, and is described as follows.

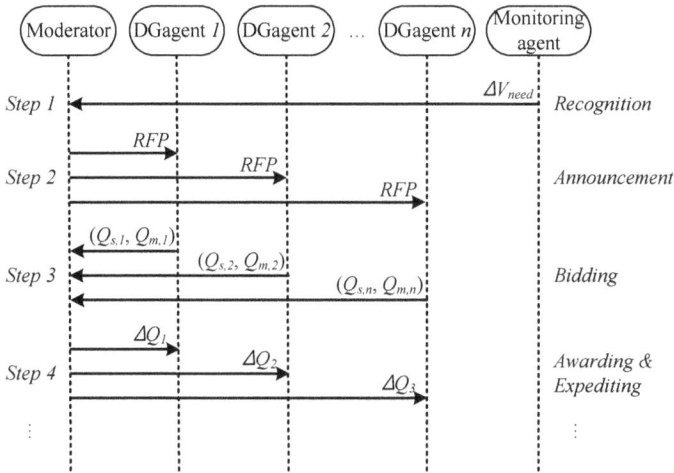

Figure 2. A summary of the voltage control process.

1. Recognition: In the normal state, monitoring agents measure a local node voltage. If the node voltage moves outside the predetermined range (typically ±5% of the rated voltage), the monitoring agent informs the moderator with a request message for voltage compensation. Therefore, the voltage deviation, ΔV_d, that requires restoration to the normal state is included in this request message.
2. Announcement: Once the request message for voltage compensation has been received by the moderator, it requests voltage control proposals from all DG agents. The bidding process begins with the issuance of a request for proposal (RFP).
3. Bidding: DG agents respond to this RFP by sending a proposal if they are able to participate in voltage control. In this proposal, two items of power system data are included: the reactive power sensitivity, Q_s, and the reactive power margin, Q_m. Definitions of these quantities and the derivative methods employed to make the proposals are detailed in the following subsection.
4. Awarding: The moderator reviews the proposals and decides how much of the reactive power is to be controlled by each DG for restoring the voltage error. That is to say, the moderator assigns tasks to the DG agents. These tasks are allocated as an increase or decrease in the reactive power at one or more DGs.
5. Expediting: Finally, once the task for increasing or decreasing the reactive power of DG has been received by each DG agent, it completes the assigned task by dispatching the reactive power reference to its own DG unit.

Strictly speaking, the reactive power sensitivity is nonlinear. Since it is assumed to be linear in the proposed voltage control process, the initial voltage problem may not be mitigated completely. In this case, a new voltage control process will be initiated, with Step 1 being repeated because one or more monitoring agents detect a voltage deviation. The process is therefore repeated until ΔV_d is within the predefined operating limits of the system.

2.3. Bidding

The bidding information is the reactive power sensitivity to the bus voltage, Q_s, and the reactive power margin of the DG, Q_m; these criteria are used in the moderator's decision-making model. This subsection presents the definitions of these quantities and how DG agents derive them.

The reactive power sensitivity corresponds to an incremental increase in the violated node voltage for an incremental injected reactive power. This can be determined from a simple power flow equation as follows:

$$\begin{bmatrix} \Delta\theta \\ \Delta V \end{bmatrix} = [J]^{-1} \begin{bmatrix} \Delta P \\ \Delta Q \end{bmatrix} \tag{1}$$

where J is the Jacobian matrix corresponding to the active and reactive powers from the bus voltage angles and magnitudes. If we assume that the injected active power does not change, we can rewrite Equation (1) as:

$$\begin{bmatrix} \Delta\theta \\ \Delta V \end{bmatrix} = [J]^{-1} \begin{bmatrix} 0 \\ \Delta Q \end{bmatrix} = \begin{bmatrix} X_{11} & X_{12} \\ X_{21} & X_{22} \end{bmatrix} \begin{bmatrix} 0 \\ \Delta Q \end{bmatrix} \tag{2}$$

From this expression, the relationship between the bus voltages and injected reactive power can be determined, *i.e.*,

$$\Delta V = X_{22} \times \Delta Q = Y \times \Delta Q \tag{3}$$

We can divide this equation into generator bus and load bus parts, as follows:

$$\begin{bmatrix} \Delta V_g \\ \Delta V_l \end{bmatrix} = \begin{bmatrix} Y_{11} & Y_{12} \\ Y_{21} & Y_{22} \end{bmatrix} \begin{bmatrix} \Delta Q_g \\ \Delta Q_l \end{bmatrix} \tag{4}$$

For voltage control by reactive power injection at the DGs, the change in the reactive power at the load bus can be assumed to be zero. We can then obtain a relationship between the load bus voltage and the reactive power injected by the DG bus, *i.e.*,

$$\Delta V_l = Y_{21} \times \Delta Q_g \tag{5}$$

and the reactive power sensitivity can be defined as:

$$Q_s = \frac{\partial V_l}{\partial Q_g} = Y_{21} \tag{6}$$

If there are p load buses and q DG buses in the distribution network, we can rewrite Equation (6) as follows:

$$\mathbf{Q_s} = \begin{bmatrix} \beta_{11} & \beta_{12} & \cdots & \beta_{1q} \\ \beta_{21} & \beta_{22} & \cdots & \beta_{2q} \\ \vdots & \vdots & \ddots & \vdots \\ \beta_{p1} & \beta_{p2} & \cdots & \beta_{pq} \end{bmatrix} \tag{7}$$

When a voltage violation occurs at the i-th load bus, the reactive power sensitivity of each DG is given by:

$$Q_{s,1} = \beta_{i1}, \ Q_{s,2} = \beta_{i2}, \ \dots, \ Q_{s,q} = \beta_{iq} \tag{8}$$

If each DG agent acquires Q_s, as given in Equation (8), then the injected active and reactive power, as well as the network topology should be known in order to establish the power flow equation. In the proposed MAS, the injected power at the DG buses can be obtained through

MAS-based communication channels among the agents. However, it is not straightforward to obtain the information describing the load buses without additional devices, such as measurement units and communication lines. In other words, to know the injected power of all buses in the distribution network, a comprehensive monitoring system is required, which may not be practical due to environmental and economic limitations. Additionally, the method described so far cannot achieve the required voltage regulation because it is essentially for distributed control.

However, reactive power sensitivity calculation in a distributed manner can be achieved by autonomous data exchange between DG agents and monitoring agents. For example, as illustrated in Figure 3, the j-th DG agent can calculate its own sensitivity factor by monitoring the voltage change, ΔV_p, which is caused by increasing or decreasing the reactive power output of DG, ΔQ_j [21]. Here, V_p indicates the bus voltage measured at the monitoring agent node. Otherwise, the j-th DG agent may assume that the voltage change at bus p is the same as that of the local bus j in the special case of a radially-structured distribution feeder. The reactive power sensitivity corresponding to an incremental increase in the voltage-violated bus p for an incremental reactive power injection by j-th DG can be expressed as:

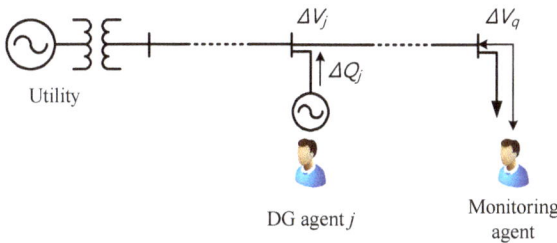

Figure 3. An example showing the method to calculate Q_s.

$$Q_{s,j} = \Delta V_p / \Delta Q_j \approx \Delta V_j / \Delta Q_j. \tag{9}$$

The reactive power margin is also an important factor in the voltage control method, and that of the j-th DG, $Q_{m,j}$, can be obtained by monitoring its reactive power output, $Q_{output,j}$, *i.e.*,

$$Q_{m,j} = Q_{rated,j} - Q_{output,j}. \tag{10}$$

3. Moderator Design Using a Fuzzy Inference System

In the proposed voltage control strategy, the moderator determines the reactive power output of each DG based on the bidding information. In other words, the reactive power sensitivities and margins are the criteria for the moderator to determine the participation rate of DGs to regulate the bus voltage. The decision-making model of the moderator must reflect the main concept of the proposed voltage control strategy. Therefore, the higher the Q_s and Q_m of a given DG are, the more likely it is to participate in voltage regulation. However, it is nontrivial to determine the participation rate of DGs based on the two uncorrelated criteria Q_s and Q_m. To deal with this problem, we introduce a fuzzy logic approach. Furthermore, to make the decision-making model effective, the moderator employs the fuzzy inference system (FIS), which is an artificial intelligence technique and can reflect the views of the expert or operator [34,35]. Negnevitsky As shown in Figure 4, the FIS of the moderator determines the participation factor, α, based on the bidding information and then assigns the increase or decrease in the reactive power to the DGs. We used the Mamdani-style FIS to determine the participation factor of each DG. This method is the most commonly-used fuzzy inference technique and consists of three main stages, as shown in Figure 4: fuzzification, the rule-based inference engine and defuzzification [35].

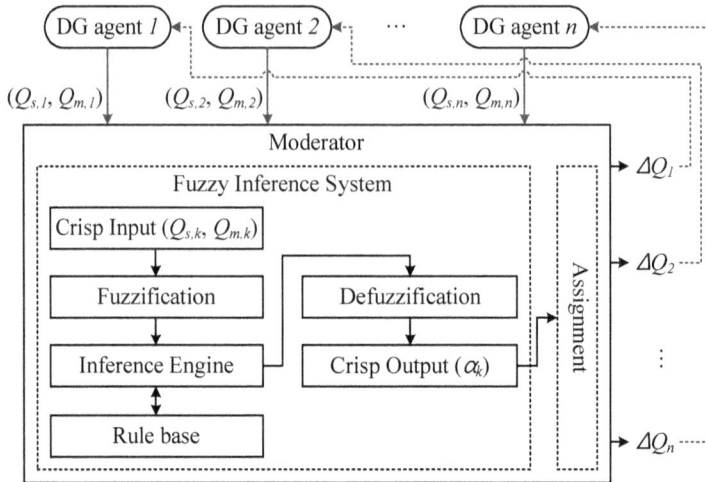

Figure 4. Configuration of the moderator.

3.1. Fuzzification

Fuzzification is the first stage in determining the degree to which each measurement (*i.e.*, crisp) input belongs to each of the appropriate fuzzy sets. For example, the reactive power sensitivity represents the voltage change in response to injected reactive power and is a crisp variable. However, in fuzzy set theory, we must fuzzify these numerical data against the membership degree of the appropriate linguistic fuzzy set [35].

All inputs are fuzzified using a membership function defined with three fuzzy sets, as shown in Figure 5a: S for small; M for medium; and L for large. Triangular and trapezoidal membership functions are used here, as these provide an adequate representation of the desired response of the system and significantly simplify the computational process.

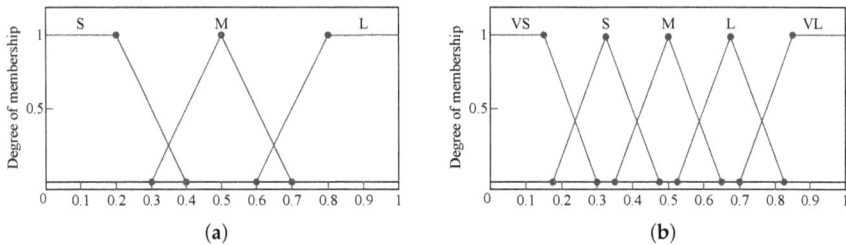

Figure 5. Membership functions. (**a**) Two inputs and (**b**) An output.

The two inputs to the FIS, Q_s and Q_m, are delivered from the DG agents to the moderator in numerical (crisp) values. To obtain the degree of membership in fuzzification, these crisp inputs are limited to the range [0,1], and so, Q_s is normalized as follows:

$$\hat{Q}_{s,j} = \frac{Q_{s,j}}{\max_i Q_{s,i}} \tag{11}$$

where the subscripts *i* and *j* indicate the index of the DG units. This normalization is unnecessary for the crisp values of Q_m because these are represented per unit. The degree of membership of the input variables $\mu_{\hat{Q}_s}$ and $\mu_{\hat{Q}_m}$ is determined from the values corresponding to the relevant fuzzy sets.

3.2. Rule-Based Inference Engine

The fuzzy membership inputs of the inference engine are evaluated by the fuzzy rules. A typical fuzzy rule has the form "if (antecedent), then (consequent)." Here, the antecedent is a fuzzy membership input, and the consequent is the participation factor in voltage control. The membership function of the consequent is defined with five fuzzy sets, as shown in Figure 5b: VS for very small; S for small; M for medium; L for large; and VL for very large.

Table 1 shows nine fuzzy rules applied to our inference engine. For example, one of the fuzzy rules for our study is "if (Q_s is large and Q_m is large), then (α is very large)." These rules are based on the voltage control strategy, which gives a DG with high reactive power sensitivity and margin a high participation factor for voltage compensation.

Table 1. Fuzzy rules for inference engine.

Q_m \ Q_s	S	M	L
S	VS	S	M
M	S	M	L
L	M	L	VL

NOTE: S=Small; M=Medium;
L=Large; V=Very.

3.3. Defuzzification

Defuzzification is the final stage of the FIS and is the process for calculating crisp data from the output of the fuzzy sets determined by the inference engine. The most popular defuzzification method is the centroid technique, and it is applied to obtain participation factors of each DG. The centroid defuzzification method, which is the process for obtaining the mathematical center of gravity (COG), can be expressed as follows:

$$COG = \frac{\int \mu_\alpha(x) \times x\, dx}{\int \mu_\alpha(x)\, dx} = \alpha \qquad (12)$$

3.4. Output of the Moderator

The moderator determines the reactive power control of each DG based on the value of α that is the output of the FIS. To assign reactive power control to the DGs, these participation factors should be normalized, *i.e.*,

$$\hat{\alpha}_k = \frac{\alpha_k}{\sum_i \alpha_i} \qquad (13)$$

The moderator finally assigns the tasks to the DGs to compensate for the voltage deviation so that we have:

$$\Delta Q_k = \frac{\hat{\alpha}_k}{Q_{s,k}} \times \Delta V_d \qquad (14)$$

4. Optimization of the Fuzzy Inference System Using Particle Swarm Optimization

The performance of the proposed voltage control system is determined by the output of the FIS, which is strongly affected by the membership functions of the input/output fuzzy sets and the fuzzy rules of the inference engine. To enhance the decision-making performance of the system, the membership functions used in the FIS may be optimized using particle swarm optimization (PSO). In this section, the PSO algorithm is briefly introduced, and the optimization process of the FIS is described.

4.1. Particle Swarm Optimization

Particle swarm optimization is an evolutionary computation technique originally proposed by Kennedy and Eberthart in 1995, and it is widely used in optimization problems [36]. In the PSO algorithm, multiple "particles," which represent potential solutions, are randomly moved in the search space. This iterative search process is carried out to find the optimum point corresponding to a fitness (objective) function. This optimization method exhibits a low rate of becoming trapped in local optima, as it offers a randomized and stochastic search process in an uncertain domain.

Each particle is iteratively moved in the search space to determine the optimum arrangement. During each iteration, the position information is updated, with the optimum arrangement of those trials labeled as the particle best, P_b. The particles share knowledge about the best position to determine the global optimum. The next positions are dependent on the "velocity" of the particles, which is determined based on the moving inertia, its particle best, Pb, and the global best, Gb. The PSO algorithm can be represented as:

$$V_i^{k+1} = w \times V_i^k + c_1 \times r_1 \times (Pb_i^k - X_i^k) + c_2 \times r_2 \times (Gb^k - X_i^k) \tag{15}$$

and:

$$X_i^{k+1} = X_i^k + V_i^{k+1} \tag{16}$$

where i and k are the particle and iteration indices, respectively; and X and V indicate the n-dimensional position and velocity vectors, respectively. The velocity constants c_1 and c_2 are generally set to two, and the random variables r_1 and r_2, which are in [0,1], determine stochastically the velocity of the next iteration. The inertia weight w changes from iteration to iteration as:

$$w = w_{max} - \frac{(w_{max} - w_{min}) \times k}{N} \tag{17}$$

where N is the total number of iterations; and w_{max} and w_{min} are the initial and final inertia weights. Typically, these are set to $w_{max} = 0.9$ and $w_{min} = 0.4$.

4.2. Optimization of the Fuzzy Inference System

The objective of the optimization of the FIS is to make the output of the FIS correspond to our voltage control strategy. The coordination strategy of the moderator assigns a high participation rate in voltage compensation to a DG that has high reactive power sensitivity and margin. In other words, the objective of voltage control is to minimize the sum of the additional reactive power injection of each DG while securing the control margin. The cost function of the i-th DG is defined as:

$$C_i = \frac{1}{(Q_{rated,i} - Q_{output,i})^2} \times (\Delta Q_i)^2 \tag{18}$$

where $Q_{rated,i}$ is the rated reactive power; $Q_{output,i}$ is the reactive output power and ΔQ_i is the change in the reactive power required for voltage control.

Using this cost function, the optimal point to coordinate multiple DGs can be determined by solving the following problem:

$$\min F = \sum_{i}^{N_{DG}} C_i(\mathbf{x}),$$ (19)

$$\text{subject to} \quad 0 \leq \mathbf{x} \leq 1, \text{for all elements}$$

$$V_n = 0.95$$

where N_{DG} is the total number of DGs that link to a DG agent and participate in voltage control and V_n indicates the last bus voltage. The vector $\mathbf{x}$ is a 27-dimensional vector, the elements of which determine the membership functions of the input/output fuzzy sets, as shown with dotted points in Figure 5: each input is determined by seven points, and an output is determined by 13 points.

To establish the optimal FIS for the target distribution system, we chose *cn* voltage problem cases randomly and induced the FIS of the moderator via the PSO optimization, as shown in Figure 6. The particles in the swarm were defined as described above by the 27-dimensional vector, $\mathbf{x}$. The basic form of the fitness function of the PSO was similar to that in Equation (19), except that it was modified to the form of the average of *cn* cases.

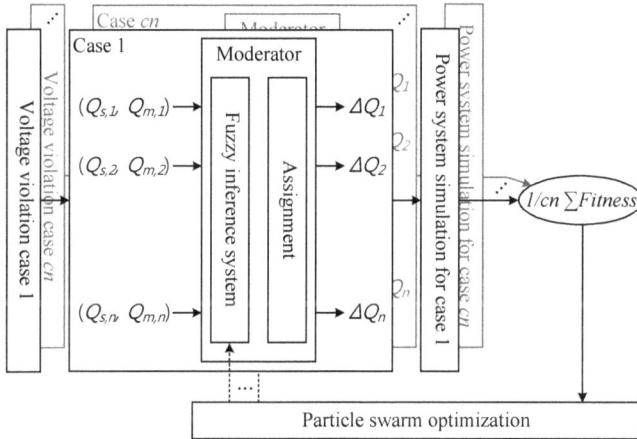

Figure 6. The PSO optimization process of the fuzzy inference system (FIS).

The optimized FIS was determined from the final output of the PSO algorithm, which was oriented only to that target distribution system. Therefore, to apply the proposed MAS-based voltage control method to a distribution system, the development of an FIS oriented to that system should be carried out.

5. Case Studies

5.1. Test System and Simulation Conditions

The test system shown in Figure 7 was developed to validate the proposed MAS-based voltage control method. This test distribution system was composed of four DGs and five aggregated loads. We assume that voltage control can be accomplished by reactive power control at the DGs and that the DG agents participate in the MAS-based voltage control process. The rated active and reactive power outputs of all DGs were set to 10 kW and 4 kVar, respectively. The rated active and reactive power consumption of each of the aggregated loads was 20 kW and 10 kVar, respectively. The line

185

impedances were $0.0242 + j0.0194\Omega$ for Z_1 and $0.0605 + j0.0486\Omega$ for Z_2. For simplicity, the voltage of Bus 1 was maintained at 1 p.u. by the on-load tap changer (OLTC) of the substation. The monitoring agents can be placed at the node where the system operator wants to regulate the voltage, *i.e.*, the node that requests voltage compensation from the moderator when a voltage problem occurs. In our case studies, the objective of voltage regulation was to keep the minimum voltage of this distribution system more than 0.95 p.u.; the monitoring agent was therefore placed at the last node.

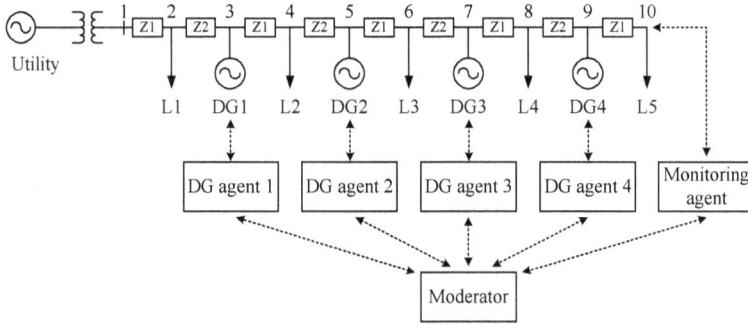

Figure 7. The test system to simulate the voltage control method.

5.2. Simulation Results

First, the FIS of the moderator oriented to the target system was developed using the PSO algorithm. To establish an optimized FIS of the moderator, we randomly chose 100 power system conditions of the target test system. These conditions were limited to voltage problem cases where the voltage of the last node was less than 0.95 p.u. We then randomly generated 30 particles and determined the membership functions of the fuzzy sets after 500 iterations of the PSO algorithm. Figure 8 shows the fitness of the global best particle during the optimization. The best among the initial 30 particles had a fitness of 1.8161, whereas we finally obtained a particle with a fitness of 1.1645 at the end of the optimization. Each fuzzy set obtained following the PSO optimization is shown in Figure 9. The membership functions for VS and S in the output fuzzy set cannot be seen in the figure, as they are aligned with the *x*-axis. It follows that a DG that only belongs to that area does not contribute to voltage compensation. To clarify the output of the FIS rule-based evaluation, a three-dimensional plot of the fuzzy rules described in Section 3.2 is shown in Figure 10.

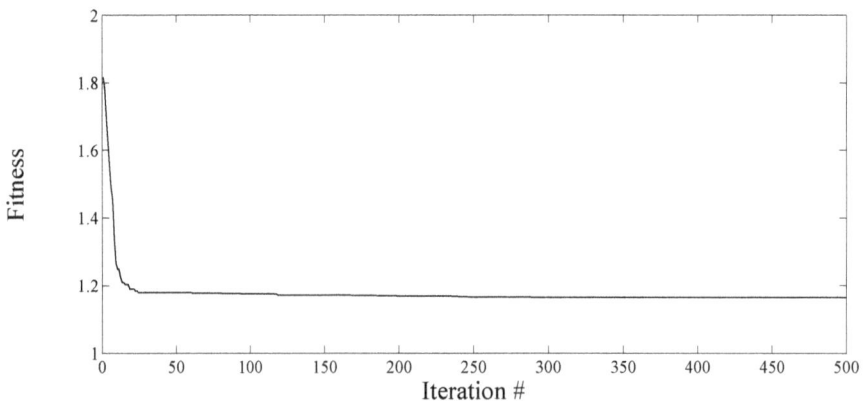

Figure 8. The global best particle's fitness over the PSO iteration.

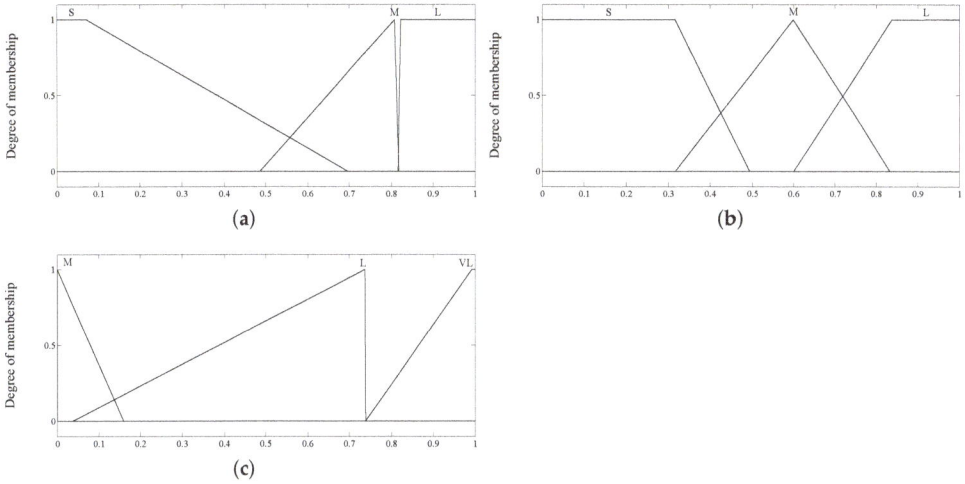

Figure 9. Optimized fuzzy sets of the moderator's FIS. (**a**) Membership functions for input Q_s; (**b**) Membership functions for input Q_m; (**c**) Membership functions for output α.

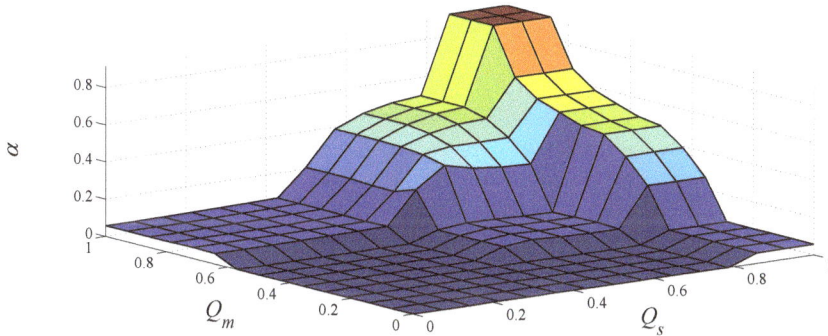

Figure 10. Three-dimensional plot for rule-based evaluation.

Following the development of the decision-making model of the moderator, we studied two voltage violation cases to verify the effectiveness of the proposed voltage control method.

The first case shows the general voltage control process when the voltage problem occurs due to a load increase. Figure 11a shows the initial operating conditions for Case 1: the demand power of the loads and the output power of DGs are given in the form of $P + jQ$ in kW and kVar. In the initial state, the voltage of the last bus was 0.9613 p.u. (normal state). The load demand of the last bus was then increased to $18 + j4$. This increase in the load resulted in a drop in the voltage at the last bus to 0.9451 p.u. (below the limit). The monitoring agent then requested voltage compensation of $\Delta V_d = 0.0049$ p.u. from the moderator.

After receiving this request, the moderator issued RFPs to DG agents. The DG agents then bid for that proposal by providing the data on Q_s and Q_m, the values of which are listed in Table 2. The moderator determined the additional reactive power output of each DG based on these bidding data; the values of the inputs and outputs to the moderator are also listed in Table 2.

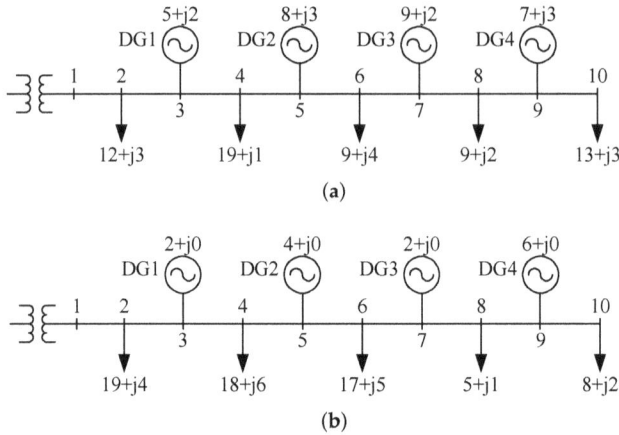

Figure 11. Initial condition of the test system. (**a**) Case 1; (**b**) Case 2.

Table 2. The simulation results for Case 1.

Unit	Q_s	Q_m	$\hat{Q}_s$	α	$\hat{\alpha}$	ΔQ (kVar)
DG agent 1	0.5034	0.75	0.2473	0.0578	0.0566	0.5454
DG agent 2	1.0022	0.50	0.4972	0.0479	0.0469	0.2247
DG agent 3	1.5245	0.75	0.7489	0.4347	0.4254	1.3547
DG agent 4	2.0358	0.50	1	0.4816	0.4712	1.1238

Figure 12a shows the variation in the voltage profile. The abnormal state of the last bus voltage (red-circle line) was restored to the normal state (blue-star line). Additionally, the result of optimal power flow (OPF) calculation (described in the Appendix) is shown by the black square line. We find that the performance of the proposed voltage control method is similar to that of OPF. This shows that our method has good performance, even though it has little information on the power system and no power flow calculations were implemented.

The second case shows how the voltage can be recovered to within the normal range when the voltage problem cannot be resolved using a single voltage control process. In this case, the initial condition was assumed to be as shown in Figure 11b, where the last bus voltage was significantly below the normal limit, at 0.9355 p.u. This voltage problem cannot be resolved through one voltage control process, as the voltage compensation determined by Equation (14) is not sufficient. This mismatch results from the nonlinearity of the power system, and another voltage control process is required to resolve the initial voltage problem. Table 3 lists the simulation results for the first and the second iterations of the process.

In the first part of the process, all of the DGs had the same (and sufficient) reactive power margin, because they were operating in the unit power factor. Therefore, the coordination criteria to distribute the voltage compensation were mostly related to the reactive power sensitivity of the DGs. The result of bidding information and the increased reactive power output of each DG are shown in Table 3. After this first process, the last bus voltage changed to 0.9497 p.u., which was not sufficient for our operating strategy. The monitoring agent was, therefore, once again requested to compensate for a voltage deviation.

(a)

(b)

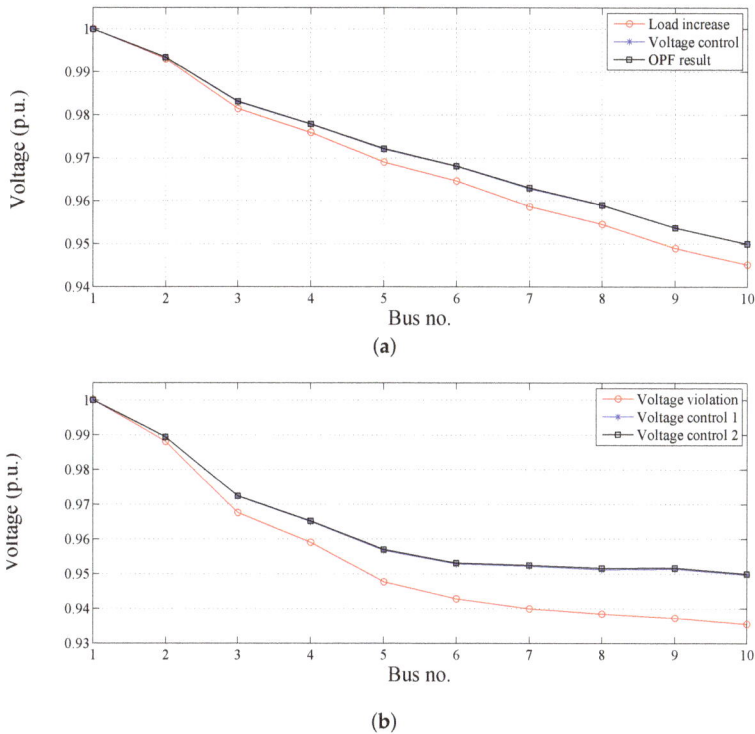

Figure 12. Voltage profiles: (**a**) For Case 1; and (**b**) For Case 2.

Table 3. The simulation results for Case 2.

Unit	Q_s	Q_m	$\hat{Q}_s$	α	$\hat{\alpha}$	ΔQ (kVar)
First process						
DG agent 1	0.5145	1	0.2483	0.0591	0.0350	0.9842
DG agent 2	1.0383	1	0.5011	0.2153	0.1276	1.7777
DG agent 3	1.5570	1	0.7515	0.4967	0.2944	2.7351
DG agent 4	2.0718	1	1	0.9162	0.5430	3.7918
Second process						
DG agent 1	0.5006	0.7539	0.2533	0.0576	0.3327	0.2141
DG agent 2	0.9994	0.5556	0.5057	0.0591	0.3415	0.1101
DG agent 3	1.4877	0.3162	0.7527	0	0	0
DG agent 4	1.9765	0.0521	1	0.0564	0.3258	0.0531

In the second part of the process, the reactive power margins were recalculated based on the new reactive power output. This took account of the reactive power sensitivity and margin of each unit. As shown in Table 3, the participation factors of DG Agents 1 and 4 were similar, even though the inputs of the FIS were significantly different. This is because the reactive power sensitivity of DG1 was smaller, but its reactive power margin was larger than that of DG4. The final reactive power outputs of DG1, DG2, DG3 and DG4 were 1.1983 kVar, 1.8878 kVar, 2.7351 kVar and 3.8449 kVar, respectively. Following this second voltage control stage, the last bus voltage was restored to within the limits. The initial voltage profile and those following voltage control process Steps 1 and 2 are shown in Figure 12b.

6. Conclusions

We have described a voltage control method using multiple DGs based on MAS to solve voltage problems in power systems in a distributed manner. Monitoring agents, DG agents and a moderator were employed to establish voltage regulation. These agents were able to cooperate to achieve voltage regulation. The local voltage was monitored at strategic locations; and alarm signals were sent to the moderator if a voltage problem was detected, and voltage compensation was requested. The DG agents then participated in a bidding process based on CNP, and the moderator coordinated the agents to achieve voltage control by dispatching the reactive power output to each DG. The moderator determined the reactive power output of each DG based on the DG status and the system performance.

The moderator used FIS to coordinate the voltage control response of multiple DGs, and the input and output fuzzy sets of the FIS were optimized to improve the decision-making process using PSO. The objective of this optimization was to minimize the sum of the additional reactive power injected by each DG by considering the reactive power margin of the DGs. We find that this voltage control process was able to regulate the voltages in the system and that the performance was similar to that obtained using OPF, even though little information on the power system was required and no power flow calculations were implemented.

Acknowledgments: This work was supported by the Korea Institute of Energy Technology Evaluation and Planning (KETEP), granted financial resource from the Ministry of Trade, Industry & Energy, Republic of Korea (No. 20151210200080) and by the National Research Foundation of Korea grant funded by the Korea government (NRF-2015R1C1A1A01054635).

Author Contributions: Hyun-Koo Kang proposed and developed the main idea of the paper and also compiled the manuscript. Il-Yop Chung advised to apply multi-agent based control algorithms to the paper. Seung-Il Moon advised to improve the simulation results of the paper.

Conflicts of Interest: The authors declare no conflict of interest.

Appendix

Optimal Power Flow Using Sequential Quadratic Programming

To validate the performance of the proposed voltage control method, the optimal power flow (OPF) problem is formulated using sequential quadratic programming (SQP). The objectives of our problem are to minimize the reactive power cost function defined as Equation (18) while maintaining the last bus voltage at the predetermined reference, which is 0.95 in this case. The optimization problem has a similar form as Equation (19), and it can be represented as:

$$\min f = \sum_i^{N_{DG}} C_i(\Delta Q_i) = \sum_i^{N_{DG}} \left(\frac{\Delta Q_i}{Q_{rated,i} - \hat{Q}_i} \right)^2$$

$$= \sum_i^{N_{DG}} \left(\frac{1}{Q_{m,i}} \times \Delta Q_i \right)^2, \tag{A1}$$

$$\text{subject to} \quad 0 \le \Delta Q_i \le Q_{m,i}, \text{ for all } i$$

$$V_n = 0.95$$

where ΔQ_i indicates the increased reactive power output of the i-th DG for restoring the last bus voltage to 0.95. $Q_{rated,i}$, $Q_{m,i}$ and $\hat{Q}_i$ represent the rated reactive power, the reactive power output margin and the current reactive power output of the i-th DG, respectively, and their relationship is $Q_{m,i} = Q_{rated,i} - \hat{Q}_i$.

The general matrix form of the QP is as follows.

$$\min_{x} \quad \frac{1}{2}\mathbf{X}^T\mathbf{H}\mathbf{X} + \mathbf{F}^T\mathbf{X} \tag{A2}$$
$$\text{subject to} \quad \mathbf{A}\mathbf{X} \leq \mathbf{B}$$
$$\mathbf{A_{eq}}\mathbf{X} = \mathbf{B_{eq}}$$
$$\mathbf{LB} \leq \mathbf{X} \leq \mathbf{UB}$$

x indicates the control variable matrix that is represented as $[\Delta Q_1, \Delta Q_2, \cdots, \Delta Q_{N_{DG}}]^T$.

References

1. Pepermans, G.; Driesen, J.; Haeseldonckx, D.; Belmans, R.; D'haeseleer, W. Distributed generation: Definition, benefits and issues. *Energy Policy* **2005**, *33*, 787–798.
2. Ackermann, T.; Andersson, G.; Söder, L. Distributed generation: A definition. *Electr. Power Syst. Res.* **2001**, *57*, 195–204.
3. Dugan, R.C.; Mcdermott, T.E. Distributed generation. *IEEE Ind. Appl. Mag.* **2002**, *8*, 19–25.
4. Dugan, R.C.; McDermott, T.E.; Bal, G.J. Planning for distributed generation. *IEEE Ind. Appl. Mag.* **2001**, *7*, 80–88.
5. Mao, Y.; Miu, K. Switch placement to improve system reliability for radial distribution systems with distributed generation. *IEEE Trans. Power Syst.* **2003**, *18*, 1346–1352.
6. Delfino, B. Modeling of the integration of distributed generation into the electrical system. In Proceedings of the Power Engineering Society Summer Meeting, Chicago, IL, USA, 25–25 July 2002; Volume 1, pp. 170–175.
7. Ochoa, L.F.; Padilha-Feltrin, A.; Harrison, G.P. Evaluating distributed generation impacts with a multiobjective index. *IEEE Trans. Power Deliv.* **2006**, *21*, 1452–1458.
8. Walling, R.; Saint, R.; Dugan, R.C.; Burke, J.; Kojovic, L.A. Summary of distributed resources impact on power delivery systems. *IEEE Trans. Power Deliv.* **2008**, *23*, 1636–1644.
9. Barker, P.P.; De Mello, R.W. Determining the impact of distributed generation on power systems. I. Radial distribution systems. In Proceedings of the Power Engineering Society Summer Meeting, Seattle, WA, USA, 16–20 July 2000; Volume 3, pp. 1645–1656.
10. Std, ANSI C84. 1-2011. American National Standard For Electric Power Systems and Equipment-Voltage Ratings (60 Hertz). 2011. Avaliable online: https://www.nema.org/Standards/Complimentary Documents/Contents-and-Scope-ANSI-C84-1-2011.pdf (accessed on 20 September 2015).
11. Kersting, W.H. *Distribution System Modeling and Analysis*; Chemical Rubber Company (CRC) Press: Boca Raton, FL, USA, 2012.
12. Dai, C.; Baghzouz, Y. On the voltage profile of distribution feeders with distributed generation. In Proceedings of the Power Engineering Society General Meeting, Toronto, ON, Canada, 13–17 July 2003; Volume 2.
13. Masters, C. Voltage rise: The big issue when connecting embedded generation to long 11 kV overhead lines. *Power Eng. J.* **2002**, *16*, 5–12.
14. Carvalho, P.; Correia, P.F.; Ferreira, L.A. Distributed reactive power generation control for voltage rise mitigation in distribution networks. *IEEE Trans. Power Syst.* **2008**, *23*, 766–772.
15. Hashim, T.T.; Mohamed, A.; Shareef, H. *A Review on Voltage Control Methods for Active Distribution Networks*; Przeglad Elektrotechniczny (Electrical Review): Warszawa, Poland, 2012; pp. 304–312. Available online: http://www.red.pe.org.pl/articles/2012/6/71.pdf (accessed on 20 September 2015).
16. Tonkoski, R.; Lopes, L.A.; El-Fouly, T.H. Coordinated active power curtailment of grid connected PV inverters for overvoltage prevention. *IEEE Trans. Sustain. Energy* **2011**, *2*, 139–147.
17. Viawan, F.; Karlsson, D. Combined local and remote voltage and reactive power control in the presence of induction machine distributed generation. *IEEE Trans. Power Syst.* **2007**, *22*, 2003–2012.
18. McArthur, S.D.; Davidson, E.M.; Catterson, V.M.; Dimeas, A.L.; Hatziargyriou, N.D.; Ponci, F.; Funabashi, T. Multi-agent systems for power engineering applications—Part I: Concepts, approaches, and technical challenges. *IEEE Trans. Power Syst.* **2007**, *22*, 1743–1752.

19. Rehtanz, C. *Autonomous Systems and Intelligent Agents in Power System Control and Operation*; Springer-Verlag: Berlin/Heidelberg, Germany, 2003.

20. Wooldridge, M. *An Introduction to Multiagent Systems*; John Wiley & Sons: Chichester, West Sussex, UK, 2009.

21. Baran, M.E.; El-Markabi, I.M. A multiagent-based dispatching scheme for distributed generators for voltage support on distribution feeders. *IEEE Trans. Power Syst.* **2007**, *22*, 52–59.

22. Elkhatib, M.E.; El-Shatshat, R.; Salama, M. Novel coordinated voltage control for smart distribution networks with DG. *IEEE Trans. Smart Grid* **2011**, *2*, 598–605.

23. Farag, H.E.; El-Saadany, E.F. A novel cooperative protocol for distributed voltage control in active distribution systems. *IEEE Trans. Power Syst.* **2013**, *28*, 1645–1656.

24. Spatti, D.H.; Da Silva, I.N.; Usida, W.F.; Flauzino, R.A. Real-time voltage regulation in power distribution system using fuzzy control. *IEEE Trans. Power Deliv.* **2010**, *25*, 1112–1123.

25. Liang, R.H.; Wang, Y.S. Fuzzy-based reactive power and voltage control in a distribution system. *IEEE Trans. Power Deliv.* **2003**, *18*, 610–618.

26. Miranda, V.; Moreira, A.; Pereira, J. An improved fuzzy inference system for voltage/VAR control. *IEEE Trans. Power Syst.* **2007**, *22*, 2013–2020.

27. Rigatos, G.; Siano, P.; Yousef, H. Adaptive fuzzy control based on output feedback for synchronization of distributed power generators. In Proceedings of the 3rd International Symposium on Energy Chanllenfes and Mechanics (ECM 2015), Aberdeen, UK, 7–9 July 2015.

28. Tomescu, M.L.; Preitl, S.; Precup, R.E.; Tar, J.K. Stability analysis method for fuzzy control systems dedicated controlling nonlinear processes. *Acta Polytech. Hung.* **2007**, *4*, 127–141.

29. Liu, J.; Liu, H.; Shen, W. Stability analysis of particle swarm optimization. In *Advanced Intelligent Computing Theories and Applications. With Aspects of Artificial Intelligence*; Springer-Verlag: Berlin/Heidelberg, Germany, 2007; pp. 781–790.

30. Moreau, L. Stability of multiagent systems with time-dependent communication links. *IEEE Trans. Autom. Control* **2005**, *50*, 169–182.

31. Oh, S.; Yoo, C.; Chung, I.; Won, D. Hardware-in-the-loop simulation of distributed intelligent energy management system for microgrids. *Energies* **2013**, *6*, 3263–3283.

32. Smith, R.G. The contract net protocol: High-level communication and control in a distributed problem solver. *IEEE Trans. Comput.* **1980**, *29*, 1104–1113.

33. Alibhai, Z. *What is Contract Net Interaction Protocol?* Intelligent Robotics and Manufacturing Systems (IRMS) Laboratory, Simon Fraser University: Vancouver, BC, Canada, 2003.

34. Ross, T.J. *Fuzzy Logic with Engineering Applications*, 3rd ed.; John Wiley & Sons: Chichester, West Sussex, UK, 2009.

35. Negnevitsky, M. *Artificial Intelligence: A Guide to Intelligent Systems*; Pearson Education: Harlow, Essex, UK, 2005.

36. Kennedy, J. Eberhart, R. Particle swarm optimization. In Proceedings of the IEEE International Conference on Neural Network, Piscataway, NJ, USA, 1995; pp. 1942–1948.

![energies logo] *energies*

MDPI

Article

A Two-Stage Optimal Network Reconfiguration Approach for Minimizing Energy Loss of Distribution Networks Using Particle Swarm Optimization Algorithm

Wei-Tzer Huang [1,*], Tsai-Hsiang Chen [2], Hong-Ting Chen [1], Jhih-Siang Yang [2], Kuo-Lung Lian [2], Yung-Ruei Chang [3], Yih-Der Lee [3] and Yuan-Hsiang Ho [3]

[1] Department of Industrial Education and Technology, National Changhua University of Education, No. 2, Shida Road, Changhua 500, Taiwan; edchen1991@gmail.com

[2] Department of Electrical Engineering, National Taiwan University of Science and Technology, No.43, Section 4, Keelung Road, Da'an District, Taipei City 106, Taiwan; thchen@mail.ntust.edu.tw (T.-H.C.); m10307106@mail.ntust.edu.tw (J.-S.Y.); ryanlian@mail.ntust.edu.tw (K.-L.L.)

[3] The Institute of Nuclear Energy Research, 1000 Wenhua Road, Jiaan Village, Longtan District, Taoyuan City 325, Taiwan; raymond@iner.gov.tw (Y.-R.C.); ydlee@iner.gov.tw (Y.-D.L.); twingo_ho@iner.gov.tw (Y.-H.H.)

* Correspondence: vichuang@cc.ncue.edu.tw; Tel.: +886-4-723-2105 (ext. 7264); Fax: +886-4-721-1287

Academic Editor: Neville Watson

Received: 7 October 2015; Accepted: 1 December 2015; Published: 5 December 2015

Abstract: This study aimed to minimize energy losses in traditional distribution networks and microgrids through a network reconfiguration and phase balancing approach. To address this problem, an algorithm composed of a multi-objective function and operation constraints is proposed. Network connection matrices based on graph theory and the backward/forward sweep method are used to analyze power flow. A minimizing energy loss approach is developed for network reconfiguration and phase balancing, and the particle swarm optimization (PSO) algorithm is adopted to solve this optimal combination problem. The proposed approach is tested on the IEEE 37-bus test system and the first outdoor microgrid test bed established by the Institute of Nuclear Energy Research (INER) in Taiwan. Simulation results demonstrate that the proposed two-stage approach can be applied in network reconfiguration to minimize energy loss.

Keywords: energy loss; distribution network; microgrid; network reconfiguration; phase balancing; particle swarm optimization (PSO); connection matrices

---

1. Introduction

The major function of traditional passive distribution networks is to distribute electrical power to customers. Because voltage levels in such networks are relatively lower and their total length is longer compared with transmission networks, reducing power losses in distribution networks is vital; furthermore, the annual energy loss of the power system will be diminished. At present, many distribution energy resources (DERs) are connected to distribution networks. Distribution networks have become active networks called microgrids. Microgrids consist of DERs and loads. DERs include renewable and nonrenewable generation units, as well as storage devices, such as photovoltaic systems, wind turbines, fuel cells, microturbines, diesel engines, battery banks, and supercapacitors, among other [1–3]. Microgrids can be operated under grid-tied and islanding modes through a static switch at the common coupling point between the main power grid and the microgrid [4]. In the grid-tied operation mode, the microgrid may act as a load or source at any time in terms of the main power grid. The islanding operation mode must be operated autonomously based on the power balance principle to maintain constant voltage and frequency. Numerous renewable energy units are used in microgrids. Thus, CO_2 emissions are reduced and global warming is prevented. Constructing microgrids in industrial parks, campuses, shopping malls, off-shore islands, and remote districts is worthwhile because of the all the aforementioned advantages.

The system planning, designing, operating, and controlling of microgrids is more complex compared with traditional passive distribution systems. Consequently, an energy management system (EMS) is essential in the system operation stage in microgrids [5–7]. To increase operating efficiency, the network reconfiguration and phase balancing approach, which is one of the functions in EMS, has been adopted to minimize power loss and improve voltage quality. Merlin and Back [8] used a spanning tree structure to model a distribution system. The obtained solution results were independent from the initial status of the switches; however, their algorithm was very time-consuming. Civanlar *et al.* [9] proposed a branch-exchange method to minimize the number of switching operations; however, this approach is not systematic and can only reduce power loss. Jeon *et al.* [10] presented a simulated annealing algorithm for network reconfiguration; this algorithm was easy to code but required considerable computation time in large-scale systems. Venkatesh and Ranjan [11] proposed an approach that used an evolutionary programming with fuzzy adaptation as a solution technique; however, as a system grew larger, this method became increasingly complex. Hamdoui *et al.* [12] used the ant colony approach algorithm to identify the optimal combination of feeders with different natures to find a new network topology. This method is highly efficient and convergence definitely occurs; however, the length of time required to achieve convergence remains uncertain.

In this work, a population-based stochastic optimization technique that adopts the particle swarm optimization (PSO) algorithm is used to search for the solutions of the proposed two-stage approach, which is to solve the optimal network reconfiguration at the first stage and phase balancing at second stage. This paper is divided into four sections: Section 1 presents the introduction; Section 2 reviews network reconfiguration and phase balancing algorithms, and then describes the proposed two-stage optimal network reconfiguration problem and its formulation; Section 3 demonstrates and discusses the simulation results; and Section 4 concludes the paper.

2. Problem Formulation

In this section, the network reconfiguration and phase balancing problems will be explained in details, and then the PSO algorithm and power flow solution technique for solving this problem will also be described; finally, a multi-objective function will be derived for the proposed two-stage approach.

2.1. Describes Network Reconfiguration Problem

Most distribution networks exhibit a radial configuration from the distribution substation to the customers. Sectionalizing switches and tie switches are installed in these systems to consider normal and abnormal operations. Under normal conditions, the sectionalizing switches are typically closed and the tie switches are generally open. Nevertheless, the network can be changed by performing switching actions for the best network topology to increase system performance. This process is called reconfiguration. Through network reconfiguration, power losses are reduced, load distribution becomes uniform, and overloading is avoided. System reliability is enhanced after a fault occurs.

A combinatorial problem arises because of switching actions. Therefore, when the number of switches is high, the possibility of reconfiguration increases. The most common approaches to solve this problem in network reconfiguration can be classified as follows:

- Mathematical optimization methods,
- Heuristic methods,
- Artificial intelligence methods.

These methods each have advantages and disadvantages. Based on literature reviews, these techniques can effectively address network reconfiguration problems. Solving a network reconfiguration problem involves two components: (1) the objective function and the system operating constraints; (2) the power flow algorithm. The common objective function is power loss minimization, and the constraints are the upper and lower limits of bus voltages, the ampere capacity of the conductor, and feasible network topology. The power flow algorithms must suit the characteristics of distribution networks with high R/X ratio, short distance between two connected buses, and unbalanced load distributions and system structure.

2.2. Describes Phase Balancing Problem

Distribution networks are inherently unbalanced, due the single-phase three-wire, three-phase three-wire, and three-phase four-wire connections of distribution transformers widely used in distribution networks to serve various loads; besides, feeder arrangements that are not completely three-phase four-wire (three-wire), two-phase or single-phase arrangements are usually adopted in laterals or sub-laterals. Moreover, the electricity consumption of customers is random. Consequently, these factors cause the three-phase currents in a distribution feeder to be unbalanced, and then result in three-phase voltage unbalance. The voltage and current unbalances are the dominators resulting in extra power losses, increasing the current in the neutral line of three-phase four-wire distribution networks, and so on. Figure 1 shows the six connection schemes of individual phase loads for three-phase buses, the individual phase loads can be derived by the integrated models of distribution transformers with theirs loads [13]; therefore, there are six possible connection schemes at the three-phase bus, and similarly, two connection schemes at a two-phase bus, and one connection scheme at a single-phase bus. To improve the phase voltage and current unbalance conditions, the common approach is to derive the combinations of suitable phase connections between the individual phase load and a primary feeder; this is called phase balancing. However, electrical distribution engineers usually use the conventional trial and error approach, and it is time consuming to achieve an acceptable result; ultimately it fails to solve for the phase balancing problem. Fortunately, some effective optimal algorithms for phase balancing were proposed to improve the unbalance and to increase system operation efficiency, such as the mixed-integer programming method [14], genetic algorithm [15], phase balancing algorithm considering time-varying load patterns [16], expert systems [17], *etc.* According to the literature reviews, these approaches are able to effectively address phase balancing problems. The objective function, system operating constraints, and the power flow algorithm are essential to solve this problem.

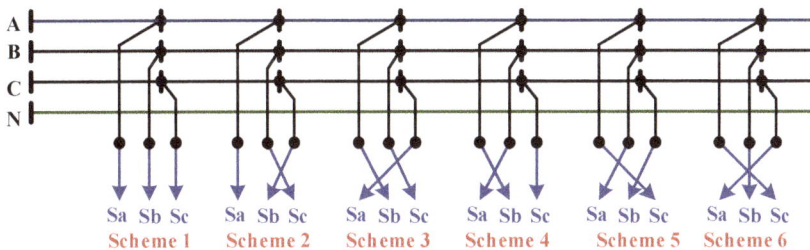

Figure 1. Six connection schemes for three-phase buses.

2.3. Particle Swarm Optimization Algorithm

In this paper, the PSO algorithm is used to solve the two-stage optimal network reconfiguration problem, it was introduced by Kennedy and Eberhart [18,19] in 1995. This algorithm is a population-based optimal search technique describing certain social behaviors of animals, such as fish schooling or bird flocking. PSO simulates the population behavior that combines the cognition-only model and the social-only model, as shown in Equations (1) and (2), respectively. The cognition-only model searches for the individual best solutions as the local best (*pbest*) and changes particle position and velocity to move in a multi-dimensional space until the convergence constraints are reached. In the social-only model, the *pbest* and global best (*gbest*) are compared to update the *gbest* and change particle position and velocity. The combination of *pbest* and *gbest* in PSO allows the particle to adjust rapidly and correctly, which results in fast convergence using Equations (3)–(5):

$$V_n^{k+1} = V_n^k + c_1 \times rand_1 \times (pbest_n^k - s_n^k) \tag{1}$$

$$V_n^{k+1} = V_n^k + c_2 \times rand_2 \times (gbest^k - s_n^k) \tag{2}$$

$$V_n^{k+1} = w \times V_n^k + c_1 \times rand_1 \times (pbest_n^k - s_n^k) + c_2 \times rand_2 \times (gbest^k - s_n^k) \tag{3}$$

$$s_n^{k+1} = s_n^k + v_n^{k+1} \tag{4}$$

$$w = w_{\max} - (w_{\max} - w_{\min}) \times \frac{k}{k_{\max}} \tag{5}$$

where $k_{\max}$ is the maximum iteration, n is the particle number, V_n^k is the velocity of particle n at the kth iteration, s_n^k is the kth position of particle n, c_1 and c_2 are learning factors, $rand_1$ and $rand_2$ are random numbers between 0 and 1, $pbest_n^k$ is the best value of particle n at the kth iteration, and $gbest^k$ is the global best value at the kth iteration. w, $w_{\max}$, and $w_{\min}$ are acceleration coefficients, maximum weighting values, and minimum weighting values, respectively.

2.4. Power Flow Algorithm

The network reconfiguration and phase balancing problems must be solved by the power flow algorithm. Two common frame of reference-based power flow algorithms are used in distribution networks: the Gauss and Newton–Raphson algorithms based on bus frame of reference are common techniques used for power flow solutions [20–24]; besides, other algorithms based on branch frame of reference were adopted for solving unbalanced power flows [25–28]. Graph theory and the backward/forward sweep method [25,26] were applied in the proposed power flow algorithm. Graph theory is a systematic approach to build incidence matrices that correspond to network topologies. The incidence matrices used in the proposed algorithm is the *A* matrix, which is the element-bus incidence matrix, and the *K* matrix, which is branch–path incidence matrix. Based on these matrices, the bus-injection to branch-current (BIBC) matrix and the branch-current to bus-voltage (BCBV) matrix can be established according to various system structures. Furthermore, BIBC and BCBV matrices are adopted in the power flow algorithm. The power flow solution procedure is described as follows:

Step 1: Build the *A* matrix. The *K* matrix can be derived using Equation (6). Establish the BIBC matrix using Equation (7), as follows:

$$K = [A^{-1}]^t \tag{6}$$

$$[BIBC] = -K \tag{7}$$

Step 2: Transpose the BIBC matrix and add the primitive line impedance into the corresponding non-zero element position to derive the BCBV matrix.

Step 3: Compute the equivalent bus injection current at each bus connected to the source or load using Equation (8) as follows:

$$I_i^k = \left(\frac{P_i + Q_i}{V_i^k}\right)^* \tag{8}$$

Step 4: Calculate the voltage derivation of each bus using Equation (9):

$$[\Delta V^k] = [BIBC][BCBV][I^k] \tag{9}$$

Step 5: Update the bus voltage using Equation (10), where $V_{\text{no\_load}}$ is the no-load voltage at each bus, that is:

$$[V^{k+1}] = [V_{\text{no\_load}}][\Delta V^k] \tag{10}$$

Step 6: Check whether convergence is achieved using Equation (11). If convergence is not achieved, then proceed to *step 3*; otherwise, end the solution procedure. ε is the maximum toleration, that is:

$$\max_i\left(\left|I_i^{k+1}\right| - \left|I_i^k\right|\right) > \varepsilon \tag{11}$$

2.5. Description of the Objective Function

Up to now, few literatures on this subject have simultaneously solved for both network reconfiguration and phase balancing problems, although some works have solved these two problems individually. It is time consuming and even more divergent to deal with them at the same time, especially for large scale systems. Consequently, in this paper, the problem is divided into two sub-problems, which are network reconfiguration and phase balancing; Figure 2 depicts the proposed two-stage optimal network reconfiguration approach. In general, due to the fact that power loss reduction by network reconfiguration is much better than phase balancing, and besides, because of the combinations of switch status are relatively less than those of phase arrangements of the loads at each bus, the network reconfiguration optimization is therefore chosen as a first stage for considerable power loss reduction, and then the power loss and voltage unbalance will be improved by phase balancing in the second stage. In order to rigidly consider the unbalanced characteristic of distribution networks, the negative and zero voltage factors and daily energy loss are included in the proposed multi-objective function, which is explained as follows.

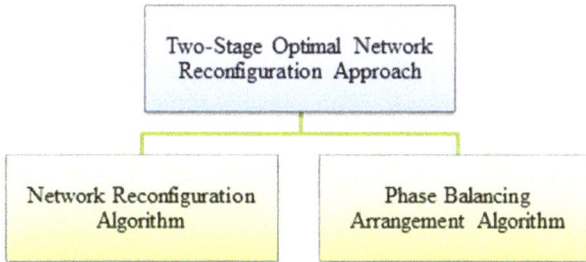

Figure 2. The schematic diagram of the proposed approach.

2.5.1. Three-Phase Voltage Unbalance

The zero- and negative sequence voltage factors, are defined as the zero- and negative sequence voltage component divided by the positive-sequence voltage component, respectively. Only positive-sequence component exists in a three-phase balanced voltage, but otherwise the zero- and negative components exist in a three-phase unbalanced voltage. These two factors can clearly explain the extra power loss and derated operation of motors, instead of the other definitions which only consider the voltage magnitude differences between each phase. In this paper, the total zero- and negative sequence voltage factors are expressed as Equations (12) and (13), respectively:

$$TD_0 = \sqrt{\frac{1}{n}\sum_{i=1}^{n}\left(\frac{|V_{0,i}|}{|V_{1,i}|}\right)^2} \tag{12}$$

$$TD_2 = \sqrt{\frac{1}{n}\sum_{i=1}^{n}\left(\frac{|V_{2,i}|}{|V_{1,i}|}\right)^2} \tag{13}$$

where the zero- positive- and negative sequence voltage components at bus i are $V_{0,i}$, $V_{1,i}$ and $V_{2,i}$ and n denotes the bus number.

2.5.2. Energy Loss

To reduce power loss is vital for increasing system operating efficiency. The ratio between the average load demand (P_{avg}) and maximum load demand (P_{peak}) in a period of time (T) is the definition of load factor (LF) as shown in Equation (14), where $p(t)$ is the instantaneous power. The typical value of LF in a distribution system is between 30% and 70% [29], so the LF is set as 62.68% in this paper. The maximum load demands can be derived from the measured daily load curve. Furthermore, the loss factor (LSF) is defined as the ratio between the average power loss ($P_{avg,loss}$) and maximum power loss ($P_{peak,loss}$) in a period of time (T) as shown in Equation (15), where $p_{loss}(t)$ is the instantaneous power loss, and the LSF is set as 49.11% in this paper.

In this paper, the maximum load of each bus is used for power flow simulation, and the peak power loss can be computed by Equation (16); moreover, the average power loss is calculated by the peak power loss product the LSF, and thus the daily energy loss (kWh) can be calculated by Equation (17). Once the daily energy loss is obtained, the seasonal and annual energy losses can be calculated by Equations (18) and (19), respectively.

$$LF = \frac{P_{avg}}{P_{peak}} \times 100\% = \frac{\frac{1}{T}\int_0^T p(t)dt}{P_{peak}} \times 100\% \tag{14}$$

$$LSF = \frac{P_{avg,loss}}{P_{peak,loss}} \times 100\% = \frac{\frac{1}{T}\int_0^T p_{loss}(t)dt}{P_{peak,loss}} \times 100\% \tag{15}$$

$$P_{peak,loss} = \sum_{j=1}^{m} \sum_{ph \in \{abc\}} \left|I_{peak,j}^{ph}\right|^2 \times R_j^{ph} \tag{16}$$

$$E_{daily,loss} = P_{peak,loss} \times LSF \times 24 \times 10^{-3} \tag{17}$$

$$E_{season,loss} = \sum_{weekday=1}^{w} E_{daily,loss}^{weekday} + \sum_{holiday=1}^{h} E_{daily,loss}^{holiday} \tag{18}$$

$$E_{annual,loss} = \sum_{season=1}^{4} E_{season,loss} \tag{19}$$

2.5.3. Multi-Objective Function

The goal in this paper is to minimize energy losses and improve the voltage profile in distribution networks. Consequently, the multi-objective function can be formulated as Equation (20) by combining daily energy loss and voltage unbalance factors:

$$f = w_1 \cdot \frac{E_{daily,loss} - E_{daily,loss}^{min}}{E_{daily,loss}^{max} - E_{daily,loss}^{min}} + w_2 \cdot \frac{TD_o - TD_o^{min}}{TD_o^{max} - TD_o^{min}} + w_3 \cdot \frac{TD_2 - TD_2^{min}}{TD_2^{max} - TD_2^{min}} \tag{20}$$

which is subject to:

$$w_1 + w_2 + w_3 = 1 \tag{21}$$

$$P_{i+1} = P_i - r_i I_i^2 - P_{Li+1} \tag{22}$$

$$Q_{i+1} = Q_i - x_i I_i^2 - Q_{Li+1} \tag{23}$$

$$V_{Li} \leq V_i \leq V_{Ui} \tag{24}$$

$$D_{0,i} = \frac{|V_{0,i}|}{|V_{1,i}|} \leq D_0^{max.} \tag{25}$$

$$D_{2,i} = \frac{|V_{2,i}|}{|V_{1,i}|} \leq D_2^{max.} \tag{26}$$

$$g \in G \tag{27}$$

In Equation (20), because the daily energy loss and voltage unbalance factors are with distinct units and the numerical values between them are quite different, normalization of the individual item between 0 and 1 is essential for multi-objective optimization. w_i is an adjustable weighting factor depend on the requirement; besides, $E_{daily,loss}^{max}$ and $E_{daily,loss}^{min}$ represent the maximum and minimum values of daily energy loss of the particles in a swarm; similarly, the same meanings of TD_0^{max}, TD_0^{min}, TD_2^{max}, and TD_2^{min} in Equation (20).

Equations (22) and (23) represent the power balance equations in radial networks, L represents the number of lines and I_j denotes the current of the jth line. Meanwhile, P_i and Q_i denote the real and reactive power flow out of bus i, respectively; r_i and x_i are the resistance and reactance between bus i and $i + 1$; Li represents the line current between bus i and $i + 1$; in Equation (24), V_i, V_{Ui}, and V_{Li} denote the voltage at bus i and its upper and lower limits, respectively. In Equations (25) and (26), $D_{0,i}$ and $D_{2,i}$ represent the zero- and negative sequence voltage factors at bus i, and $D_0^{max.}$ and $D_2^{max.}$ are the specified maximum values of zero- and negative sequence voltage factors, respectively. In Equation (27), g is the network topology; and G represents the sets of radial topologies, which cannot be closed-loop and islanding topologies, the A matrix that is the element-bus incidence matrix can be used to check the network topology, if the determinant of A equals 1 or -1 and then it is the radial topology; otherwise, if the determinant of A equals 0 and then it is not a radial topology.

The PSO algorithm is applied to solve the proposed two-stage optimal network reconfiguration and phase balancing approach, whereby the A matrix is built according to switch status and the phase connection arrangement of the load at each bus must be transferred to the PSO algorithm to have the corresponding network topology and individual complex power at each bus. The power flow algorithm is used to execute the specified network topology and then the value of the proposed function will obtained for each particle. The detailed solution procedure is illustrated in Figure 3.

199

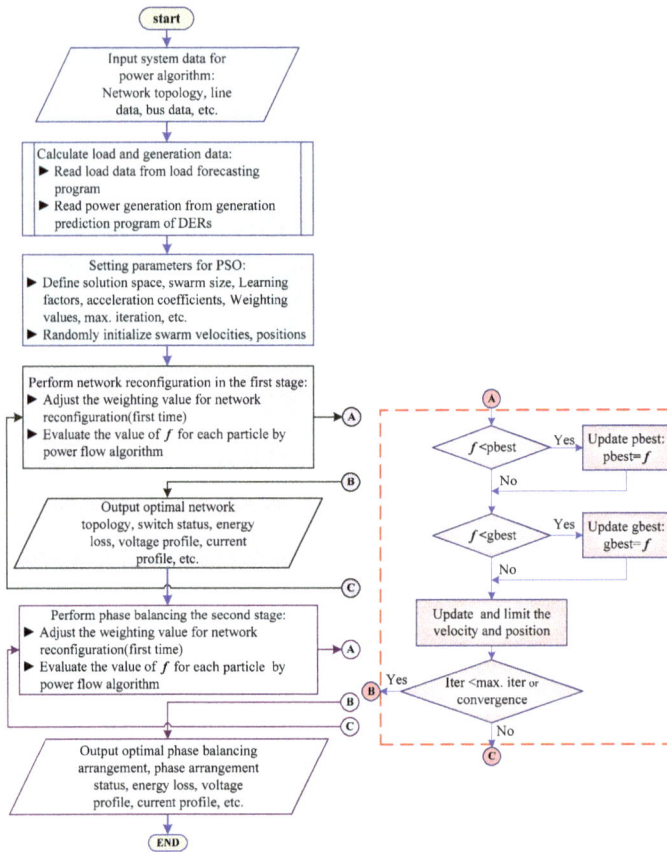

Figure 3. The flow chart of the solution procedure.

3. Numerical Results

In this section, the IEEE 37-bus test system and the microgrid of the Institute of Nuclear Energy Research (INER) in Taiwan were used as sample systems to verify the effectiveness of the proposed approach. The IEEE 37-bus test system is a traditional distribution system whose line data and bus data are shown in [30]. It is a three-phase unbalance passive network that is only connected with loads. The INER microgrid is an active network with both DERs and loads. The simulation results are discussed in the following subsections.

3.1. IEEE 37-Bus Test System

In this case, the related parameters of PSO are shown in Table 1; besides, the individual phase loads at each bus are shown in Table 2.

Table 1. Parameters of particle swarm optimization (PSO) of IEEE 37-bus test system.

Stage	Parameter							
	Particle	Max. iteration	c_1	c_2	w_1	w_2	w_3
First stage	100	200	2	2	1	0	0
Second stage	500	200	2	2	0.7	0.15	0.15

Table 2. Individual phase loads before and after phase balancing arrangement at each bus of IEEE 37-bus test system.

Bus	Loads	Before phase balancing					
		Phase A		Phase B		Phase C	
Bus number	Phase type	*P* (kW)	*Q* (kvar)	*P* (kW)	*Q* (kvar)	*P* (kW)	*Q* (kvar)
701	ABC	**224.54**	**181.07**	144.03	72.02	279.58	70.99
712	AC	33.24	42.85	0	0	54.21	−1.7
713	AC	33.24	42.85	0	0	54.21	−1.7
714	ABC	**10.84**	**−0.34**	20.07	8.21	8.18	10.65
718	AB	54.21	−1.7	33.24	42.85	0	0
720	AC	33.24	42.85	0	0	54.21	−1.7
722	ABC	8.18	10.65	90.36	−0.69	67.09	72.34
724	BC	0	0	27.11	−0.21	16.1	21.81
725	BC	0	0	27.11	−0.21	16.1	21.81
727	AC	16.1	21.81	0	0	27.11	−0.21
728	ABC	42	21	42	21	42	21
729	AB	27.11	−0.21	16.1	21.81	0	0
730	AC	33.24	42.85	0	0	54.21	−1.7
731	BC	0	0	54.21	−1.7	33.24	42.85
732	AC	**16.1**	**21.81**	0	0	27.11	−0.21
733	AB	**54.21**	**−1.7**	33.24	42.85	0	0
734	AC	16.1	21.81	0	0	27.11	−0.21
735	AC	33.24	42.85	0	0	54.21	−1.7
736	BC	0	0	27.11	−0.21	16.1	21.81
737	AB	90.36	−0.68	53.37	72.7	0	0
738	AB	81.06	−1.13	42.56	64.92	0	0
740	AC	**33.24**	**42.85**	0	0	**54.21**	**−1.7**
741	AC	16.1	21.81	0	0	27.11	−0.21
742	ABC	**5.16**	**−0.04**	57.27	2.45	33.24	42.85
744	AB	**27.11**	**−0.21**	16.1	21.81	0	0
Total		**888.62**	**551.05**	**683.88**	**367.6**	**945.33**	**315.07**

Bus	Loads	After phase balancing					
		Phase A		Phase B		Phase C	
Bus number	Phase type	*P* (kW)	*Q* (kvar)	*P* (kW)	*Q* (kvar)	*P* (kW)	*Q* (kvar)
701	ABC	**144.03**	**72.02**	279.58	70.99	224.54	181.07
712	AC	33.24	42.85	0	0	54.21	−1.7
713	AC	33.24	42.85	0	0	54.21	−1.7
714	ABC	**20.07**	**8.21**	10.84	−0.34	8.18	10.65
718	AB	54.21	−1.7	33.24	42.85	0	0
720	AC	33.24	42.85	0	0	54.21	−1.7
722	ABC	8.18	10.65	67.09	72.34	90.36	−0.69
724	BC	0	0	16.1	21.81	27.11	−0.21
725	BC	0	0	27.11	−0.21	16.1	21.81
727	AC	16.1	21.81	0	0	27.11	−0.21
728	ABC	42	21	42	21	42	21
729	AB	27.11	−0.21	16.1	21.81	0	0
730	AC	33.24	42.85	0	0	54.21	−1.7
731	BC	0	0	54.21	−1.7	33.24	42.85
732	AC	**27.11**	**−0.21**	0	0	**16.1**	**21.81**
733	AB	**33.24**	**42.85**	54.21	−1.7	0	0
734	AC	16.1	21.81	0	0	27.11	−0.21
735	AC	33.24	42.85	0	0	54.21	−1.7
736	BC	0	0	27.11	−0.21	16.1	21.81
737	AB	90.36	−0.68	53.37	72.7	0	0
738	AB	81.06	−1.13	42.56	64.92	0	0
740	AC	**54.21**	**−1.7**	0	0	**33.24**	**42.85**
741	AC	16.1	21.81	0	0	27.11	−0.21
742	ABC	**57.27**	**2.45**	33.24	42.85	5.16	−0.04
744	AB	**16.1**	**21.81**	27.11	−0.21	0	0
Total		**869.45**	**453.04**	**783.87**	**426.9**	**864.51**	**353.78**

Figure 4a shows the IEEE 37-bus test system with three tie switches and 33 sectionalizing switches, which are modified by the authors for this study. The simulation result of the optimal network topology that uses the proposed approach for the first stage network reconfiguration is illustrated in Figure 4b. In the figure, three tie switches between buses 701 and 722, 727 and 732, and 741 and 735 are closed and three sectionalizing switches between buses 704 and 720, 708 and 732, and 711 and 741 are opened.

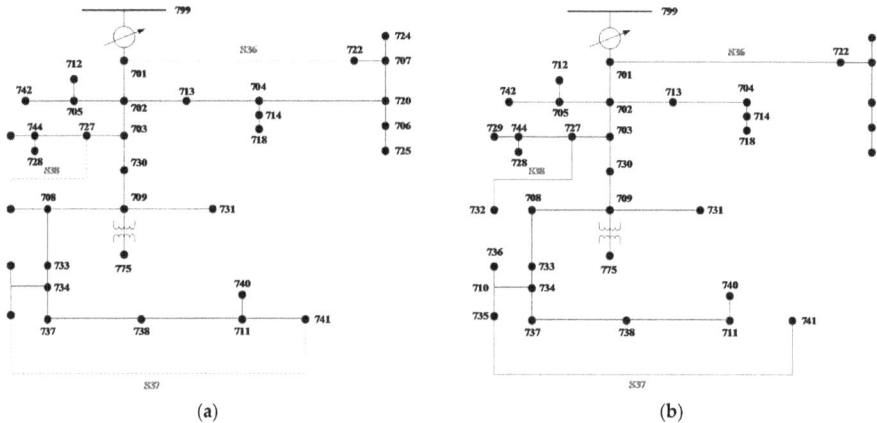

Figure 4. IEEE 37-bus test system: (**a**) before reconfiguration; (**b**) after the first stage reconfiguration.

The trend of convergence of the proposed method of the first and second stage are shown in Figure 5a,b, respectively. The multi-objective function f from initial values to the global optimum values of the first and second stage are at the 3th and 29th iteration, respectively. The new phase connections of individual phase loads are listed in Table 2 after second stage phase balancing algorithm, the simulation result depicts that the three-phase complex powers are more balanced than before phase arrangement.

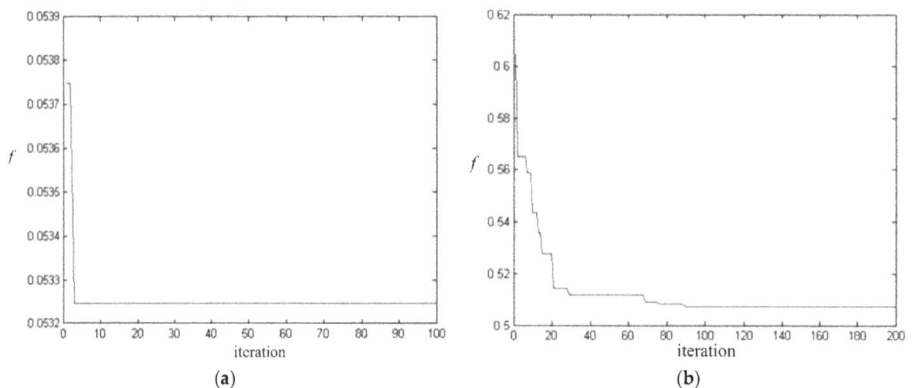

Figure 5. Convergence characteristics of proposed method of the IEEE 37-bus test system: (**a**) the first stage; (**b**) the second stage.

Figure 6 indicates the simulation result of the three-phase bus voltage profile. The voltage drop was decreased after the two-stage optimization approach; therefore, the voltage profile after optimization was better than that before optimization.

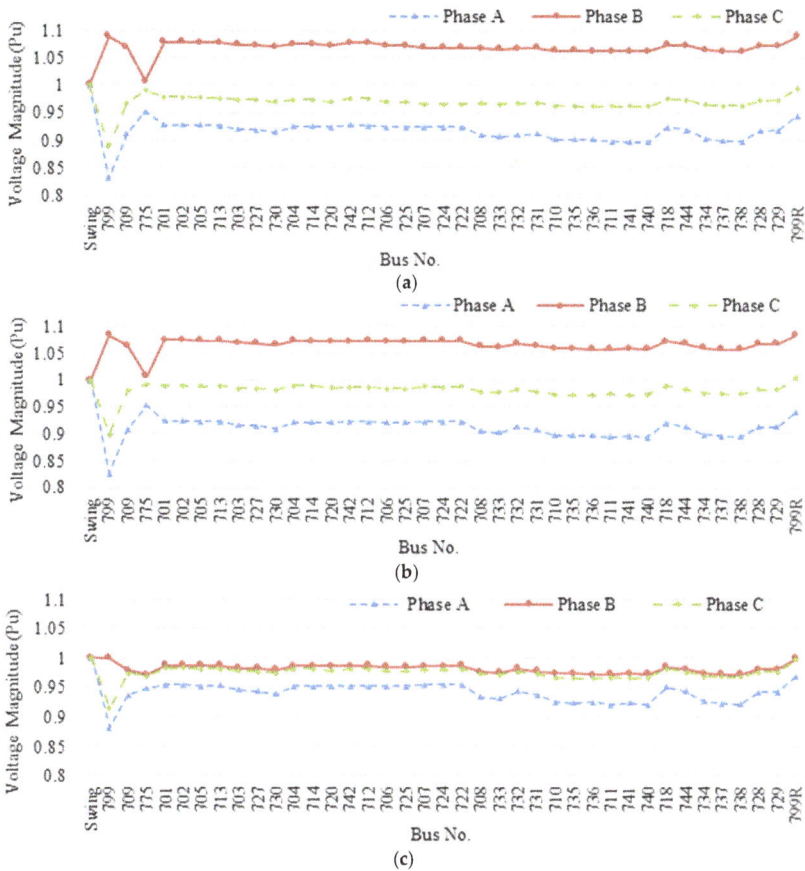

Figure 6. Simulation result of the bus voltage of the IEEE 37-bus test system: (**a**) before optimization; (**b**) after first stage; and (**c**) after second stage.

Moreover, Figure 7a shows that the zero-sequence voltage unbalance factor after optimization was better than that before optimization; similarly, the negative-sequence voltage unbalance factor after optimization was better than that before optimization, as shown in Figure 7b.

In addition, the simulation results of daily energy loss before and after optimization shown in Figure 8 indicated that the daily energy loss in each line section varied because the line flow was changed and the daily energy losses were 1739 kWh, 1668 kWh and 1648 kWh, respectively. Evidently, daily energy loss was reduced after optimization. Based on these numerical results, the proposed two-stage optimal network reconfiguration approach effectively improved voltage profile, reduced energy losses, and increased operation efficiency under normal operating conditions.

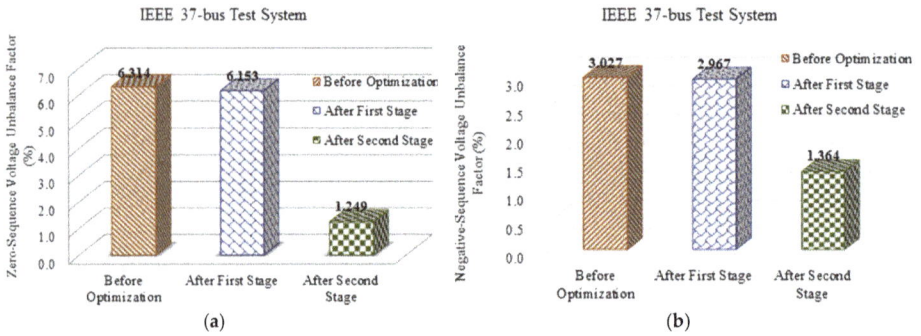

Figure 7. Simulation result of the voltage unbalance factors of the IEEE 37-bus test system: (**a**) zero-sequence voltage unbalance factor; and (**b**) negative-sequence voltage unbalance factor.

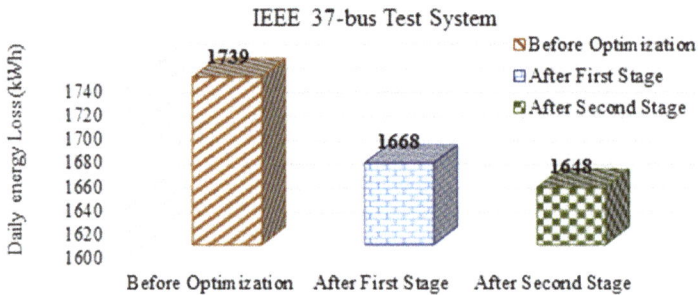

Figure 8. Simulation result of the daily energy loss of the IEEE 37-bus test system.

3.2. Institute of Nuclear Energy Research Microgrid

The first outdoor microgrid test bed was developed by INER in Taiwan. This system consists of three zones with DERs and loads and includes a tie switcher and 11 sectionalizing switches, as shown in Figure 9. For example, zone 1 comprises 21 units of 1.5 kW high concentrator photovoltaic, one 65 kW microturbine unit, a 60 kWh battery bank, and a lumped load in an office building (Building 048). The line data of the INER microgrid for the simulation is provided in Table 3. Although this is a sample network topology, the solution can be derived via a brute force search. Our proposed algorithm is a systematic approach that can be applied in a complex network topology. Thus, the effectiveness of the proposed approach can be verified using this sample system by comparing the results of the proposed approach with that of the brute force search method.

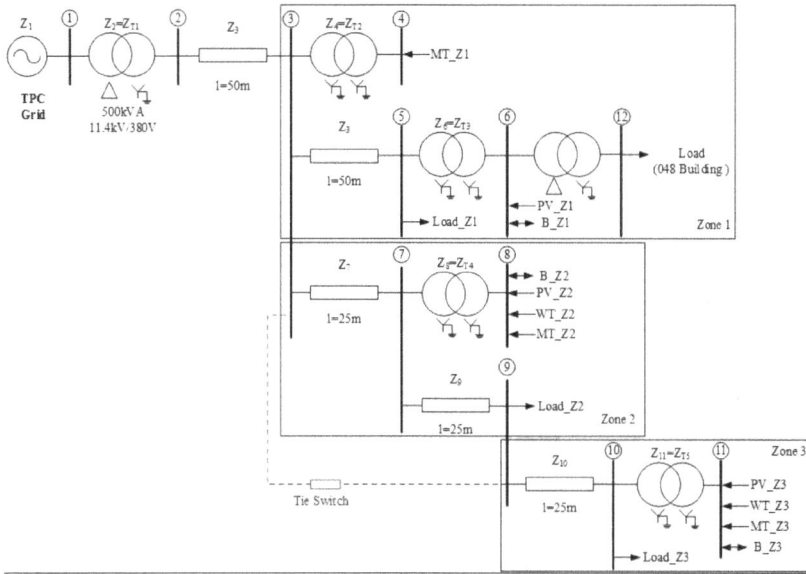

Figure 9. Single line diagram of the Institute of Nuclear Energy Research (INER) microgrid.

Table 3. Line data of the INER microgrid.

From bus	To bus	Line resistance (pu)	Line reactance (pu)	Z (%)	Distance (m)	Transformer rating (kV)	Transformer capacity (kVA)	X/R
1	2	-	-	3.85	-	11.4/0.38	500	8.02
2	3	0.2918	0.354	-	50	-	-	-
3	4	-	-	2	-	0.38/0.48	100	8
3	5	0.2918	0.354	-	50	-	-	-
5	6	-	-	4	-	0.38/0.38	150	8
3	7	0.2918	0.354	-	25	-	-	-
7	8	-	-	8	-	0.38/0.38	400	8
7	9	0.2918	0.354	-	25	-	-	-
9	10	0.2918	0.354	-	25	-	-	-
10	11	-	-	-	-	0.38/0.38	150	8
3	9	0.2918	0.354	-	25	0	-	-
6	12	-	-	4	-	0.38/0.208	150	8

Figure 10 illustrates the proposed approach applying in the optimal network reconfiguration function in EMS of the INER microgrid. In this case, the related parameters of PSO are similar to those of the IEEE 37-bus, and the differences are the particle number is 50, and the maximum iteration is 100; besides, the individual phase loads at each bus of a weekday in summer are shown in Table 4. The simulation result indicated that the tie switcher was closed and a sectionalizing switcher between buses 3 and 7 was opened. This outcome is the same as that in the brute force search method. The convergence speed is very fast due to the small number of combinations; the multi-objective function f of the two stages is convergent at the 20th and 21th iteration, respectively. The new phase connections of individual phase loads are shown in Table 4; the simulation result demonstrates that the three-phase complex powers are more balanced than before phase arrangement. Figure 11 shows the simulation result of the three-phase bus voltage profile of a weekday in summer. After the two-stage optimization approach, the voltage profile was better than that before optimization. Figure 12 indicates the simulation result of the voltage unbalance factors after optimization was better than that before optimization.

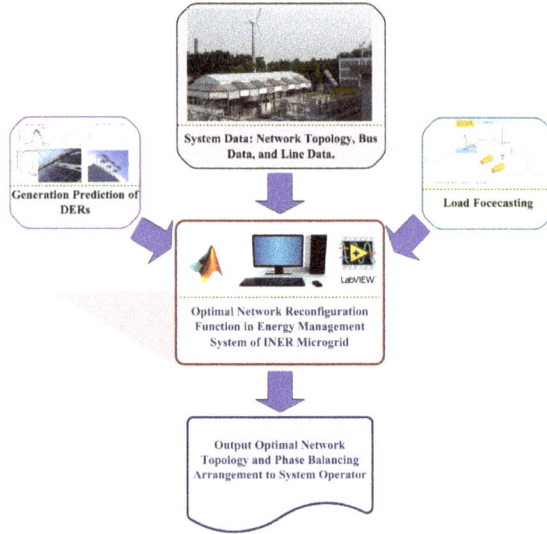

Figure 10. Optimal network reconfiguration function in energy management system (EMS) of the INER microgrid.

Table 4. Individual phase loads of a weekday in summer before and after phase balancing at each bus of the INER microgrid.

Bus	Loads	Before phase balancing					
		Phase A		Phase B		Phase C	
Bus number	Phase type	P (kW)	Q (kvar)	P (kW)	Q (kvar)	P (kW)	Q (kvar)
2	ABC	0.5342	0.3106	0.5342	0.3106	0.5342	0.3106
3	ABC	0.0694	0.0404	0.0694	0.0404	0.0694	0.0404
5	ABC	24	4	19.8	3.3	16.2	2.7
6	ABC	−18.5647	0.0932	−18.5647	0.0932	−18.5647	0.0932
7	ABC	0.4274	0.2485	0.4274	0.2485	0.4274	0.2485
8	ABC	−6.5333	0	−6.5333	0	−6.5333	0
9	ABC	24	0	19.8	0	16.2	0
10	ABC	12.1603	2.0932	10.0603	1.7432	8.2603	1.4432
11	ABC	−3.2667	0	−3.2667	0	−3.2667	0
12	ABC	14.5001	0.1118	11.9626	0.0922	9.7876	0.0755
Total		47.3267	6.8977	34.2892	5.8281	23.1142	4.9114

Bus	Loads	After phase balancing					
		Phase A		Phase B		Phase C	
Bus number	Phase type	P (kW)	Q (kvar)	P (kW)	Q (kvar)	P (kW)	Q (kvar)
2	ABC	0.5342	0.3106	0.5342	0.3106	0.5342	0.3106
3	ABC	0.0694	0.0404	0.0694	0.0404	0.0694	0.0404
5	ABC	19.8	3.3	24	4	16.2	2.7
6	ABC	−18.5647	0.0932	−18.5647	0.0932	−18.5647	0.0932
7	ABC	0.4274	0.2485	0.4274	0.2485	0.4274	0.2485
8	ABC	−6.5333	0	−6.5333	0	−6.5333	0
9	ABC	19.8	0	16.2	0	24	0
10	ABC	10.0603	1.7432	12.1603	2.0932	8.2603	1.4432
11	ABC	−3.2667	0	−3.2667	0	−3.2667	0
12	ABC	14.5001	0.1118	9.7876	0.0755	11.9626	0.0922
Total		36.8267	5.8477	34.8142	6.8614	33.0892	4.9281

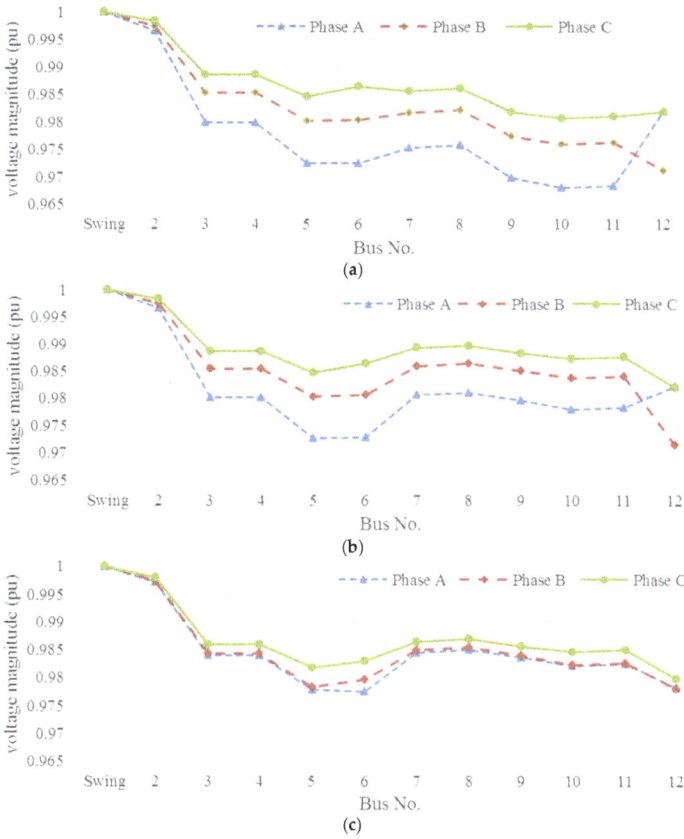

Figure 11. Simulation result of the bus voltage of a weekday in summer of the INER microgrid: (**a**) before optimization; (**b**) after first stage; and (**c**) after second stage.

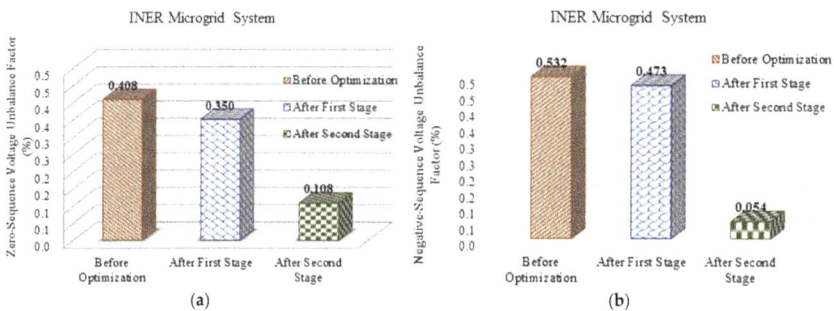

Figure 12. Simulation result of the Voltage Unbalance Factors of a weekday in summer of the INER microgrid: (**a**) zero-sequence voltage unbalance factor; (**b**) negative-sequence voltage unbalance factor.

Furthermore, the daily energy losses of the four seasons are all reduced after reconfiguration as shown in Figure 13. Based on the numerical results, the proposed algorithm was proven to be a feasible approach to improve voltage quality, reduce energy loss, and increase efficiency under normal operating conditions.

Energies **2015**, *8*, 13894–13910

INER Microgrid System

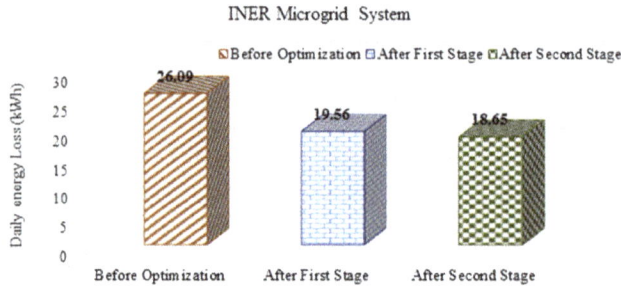

Figure 13. Simulation result of the daily energy loss of a weekday in summer of the INER microgrid.

4. Conclusions

A two-stage optimal approach, which is composed of network reconfiguration and phase balancing algorithms that applies a graph theory-based power flow solution technique, has been developed in this study. PSO exhibits self-learning capability to obtain the most optimal solution, and the graph theory-based power flow algorithm can easily establish the network topology using incidence matrices according to different system structures. The IEEE 37-bus system and the INER microgrid have been used as sample systems to verify the effectiveness of the proposed approach. The numerical results demonstrate that this approach can improve voltage profiles, reduce energy losses, and increase efficiency under normal operating conditions. The developed algorithm can be applied in traditional distribution networks with or without DERs and microgrids to improve system operation performance.

Acknowledgments: The authors are grateful for financial support from the Ministry of Science and Technology, Taiwan, under Grant MOST 104-3113-E-042A-004-CC2.

Author Contributions: The paper was a collaborative effort between the authors. All authors have read and approved the final manuscript.

Conflicts of Interest: The authors declare no conflict of interest.

References

1. Lasseter, R.H.; Asano, H.; Iravani, R.; Marnay, C. Microgrids. *IEEE Power Energy Mag.* **2007**, *5*, 78–94.
2. Lasseter, R.H. Smart distribution: Coupled microgrids. *IEEE Power Energy Mag.* **2011**, *99*, 1074–1082. [CrossRef]
3. Hong, Y.-Y.; Lai, Y.-M.; Chang, Y.-R.; Lee, Y.-D.; Liu, P.-W. Optimizing Capacities of Distributed Generation and Energy Storage in a Small Autonomous Power System Considering Uncertainty in Renewables. *Energies* **2015**, *8*, 2473–2492. [CrossRef]
4. Huang, W.-T.; Yao, K.-C.; Wu, C.-C. Using the Direct Search Method for Optimal Dispatch of Distributed Generation in a Medium-Voltage Microgrid. *Energies* **2014**, *7*, 8355–8373. [CrossRef]
5. Kim, H.-M.; Lim, Y.; Kinoshita, T. An intelligent multiagent system for autonomous microgrid operation. *Energies* **2012**, *5*, 3347–3362. [CrossRef]
6. Kim, H.-M.; Kinoshita, T.; Shin, M.-C. A multiagent system for autonomous operation of islanded microgrids based on a power market environment. *Energies* **2010**, *3*, 1972–1990. [CrossRef]
7. Yoo, C.-H.; Chung, I.-Y.; Lee, H.-J.; Hong, S.-S. Intelligent control of battery energy storage for multi-agent based microgrid energy management. *Energies* **2013**, *6*, 4956–4979. [CrossRef]
8. Merlin, A.; Back, H. Search for a minimal-loss operating spanning tree configuration in an urban power distribution system. In Proceedings of the 5th Power System Computation Conference (PSCC), Cambridge, UK, 1–5 September 1975; pp. 1–18.
9. Civanlar, S.; Grainger, J.J.; Yin, H.; Lee, S.S.H. Distribution reconfiguration for loss reduction. *IEEE Trans. Power Deliv.* **1988**, *3*, 1217–1223. [CrossRef]

10. Jeon, Y.J.; Kim, J.C.; Kim, J.O.; Shin, J.R.; Lee, K.Y. An efficient simulated annealing algorithm for network reconfiguration in large-scale distribution systems. *IEEE Trans. Power Deliv.* **2002**, *17*, 1070–1078. [CrossRef]

11. Venkatesh, B.; Ranjan, R. Optimal radial distribution system reconfiguration using fuzzy adaptation of evolutionary programming. *Int. J. Electr. Power Energy Syst.* **2003**, *25*, 775–780. [CrossRef]

12. Hamdoui, H.; Hadjeri, S.; Zeblah, A. A new constructive method for electric power system reconfiguration using ant colony. *Leonardo Electron. J. Pract. Tech.* **2008**, *12*, 49–60.

13. Chen, T.H.; Chang, Y.L. Integrated Models of Distribution Transformers and Their Loads for Three-phase Power Flow Analyses. *IEEE Trans. Power Deliv.* **1996**, *11*, 507–513. [CrossRef]

14. Zhu, J.; Chow, M.Y.; Zhang, G.F. Phase balancing using mixed-integer programming. *IEEE Trans. Power Syst.* **1998**, *13*, 1487–1492.

15. Chen, T.H.; Cherng, J.T. Optimal phase arrangement of distribution transformers connected a primary feeder for system unbalance improvement and loss reduction using a genetic algorithm. *IEEE Trans. Power Syst.* **2000**, *15*, 994–1000. [CrossRef]

16. Murat, D.; Robot, P.B.; Jeferey, C.T.; Richard, S. Simultaneous Phase Balancing at Substations and Switches with Time-Varying Load Patterns. *IEEE Trans. Power Syst.* **2001**, *16*, 922–928.

17. Lin, C.H.; Chen, C.S.; Chuang, H.J. An Expert System for Three-Phase Balancing of Distribution Feeders. *IEEE Trans. Power Syst.* **2008**, *23*, 1488–1496.

18. Kennedy, J.; Eberhart, R.C. Particle swarm optimization. In Proceedings of the IEEE International Conference on Neural Networks, Perth, Australia, 27 November–1 December 1995; Volume 4, pp. 1942–1948.

19. Eberhart, R.C.; Kennedy, J. A new optimizer using particle swarm theory. In Proceedings of the IEEE International Symposium on Micro Machine and Human Science, Nagoya, Japan, 4–6 October 1995; pp. 39–43.

20. Chen, T.H.; Chen, M.S.; Inoue, T.; Kotas, P.; Chebli, E.A. Three-phase cogenerator and transformer models for distribution system analysis. *IEEE Trans. Power Deliv.* **1991**, *6*, 1671–1681. [CrossRef]

21. Vieira, J.C.M.; Freitas, W.; Morelato, A. Phase-decoupled method for three-phase power-flow analysis of unbalanced distribution systems. *IEE Proc. Gener. Transm. Distrib.* **2004**, *151*, 568–574. [CrossRef]

22. Marinho, J.M.T.; Taranto, G. A Hybrid Three-Phase Single-Phase Power Flow Formulation. *IEEE Trans. Power Syst.* **2008**, *23*, 1063–1070. [CrossRef]

23. Kersting, W.H. Radial distribution test feeders. *IEEE Trans. Power Syst.* **1991**, *6*, 975–985. [CrossRef]

24. Dugan, R.C.; Santoso, S. An example of 3-phase transformer modeling for distribution system analysis. In Proceedings of the IEEE PES Transmission Distribution Conference and Exposition, Dallas, TX, USA, 7–12 September 2003; Volume 3, pp. 1028–1032.

25. Chen, T.H.; Yang, N.C. Three-phase power-flow by direct ZBR method for unbalanced radial distribution systems. *IET Gener. Transm. Distrib.* **2009**, *3*, 903–910. [CrossRef]

26. Teng, J.H. A network-topology based three: Phase load flow for distribution systems. *Proc. Natl. Sci. Counc. ROC A Phys. Sci. Eng.* **2000**, *24*, 259–264.

27. Teng, J.H. A direct approach for distribution system load flow solution. *IEEE Trans. Power Deliv.* **2003**, *18*, 882–887. [CrossRef]

28. Elsaiah, S.; Benidris, M.; Mitra, J. A three-phase power flow solution method for unbalanced distribution networks. In Proceedings of the North American Power Symposium, Boston, MA, USA, 4–6 August 2011; pp. 1–8.

29. Oliveira, M.E.; Boson, D.F.A.; Padilha-Feltrin, A. A statistical analysis of loss factor to determine the energy losses. In Proceedings of the IEEE/PES Transmission and Distribution Conference and Exposition, Bogota, Colombia, 13–15 August 2008; pp. 1–6.

30. Distribution Test Feeders. Available online: http://ewh.ieee.org/soc/pes/dsacom/testfeeders/index.html (accessed on 5 October 2015).

Article

Comparison between IEEE and CIGRE Thermal Behaviour Standards and Measured Temperature on a 132-kV Overhead Power Line

Alberto Arroyo [1,*], Pablo Castro [1], Raquel Martinez [1], Mario Manana [1], Alfredo Madrazo [1], Ramón Lecuna [1] and Antonio Gonzalez [2]

[1]   Electrical and Energy department, University of Cantabria, Av. Los Castros S/N, Santander 39005, Spain; pablo.castro@unican.es (P.C.); rakelmt1987@hotmail.com (R.M.); mario.manana@unican.es (M.M.); alfredo.madrazo@unican.es (A.M.); ramon.lecuna@unican.es (R.L.)

[2]   Viesgo, Santander 39011, Spain; antonio.gonzalez@viesgo.com

*   Correspondence: arroyoa@unican.es; Tel.: +34-942-201-371; Fax: +34-942-201-385

Academic Editor: Ying-Yi Hong

Received: 6 July 2015; Accepted: 19 November 2015; Published: 2 December 2015

Abstract: This paper presents the steady and dynamic thermal balances of an overhead power line proposed by CIGRE (Technical Brochure 601, 2014) and IEEE (Std.738, 2012) standards. The estimated temperatures calculated by the standards are compared with the averaged conductor temperature obtained every 8 min during a year. The conductor is a LA 280 Hawk type, used in a 132-kV overhead line. The steady and dynamic state comparison shows that the number of cases with deviations to conductor temperatures higher than 5 °C decreases from around 20% to 15% when the dynamic analysis is used. As some of the most critical variables are magnitude and direction of the wind speed, ambient temperature and solar radiation, their influence on the conductor temperature is studied. Both standards give similar results with slight differences due to the different way to calculate the solar radiation and convection. Considering the wind, both standards provide better results for the estimated conductor temperature as the wind speed increases and the angle with the line is closer to 90°. In addition, if the theoretical radiation is replaced by that measured with the pyranometer, the number of samples with deviations higher than 5 °C is reduced from around 15% to 5%.

Keywords: thermal rating; ampacity; overhead line temperature; weather parameters; real-time monitoring

---

1. Introduction

Electricity distribution networks are increasingly affected by new operation scenarios that make integration more complex. Some of these factors are electricity market liberalization and the integration of a large number of renewable installations [1]. As a result, line congestions are increasing, resulting in problems for both the distribution company, which is not capable of absorbing all of the energy generated, resulting in a decrease in efficiency, and the generation company because it will be requested to limit production and, in some cases, to stop it. These scenarios produce great inefficiencies in the system from both energy and environmental aspects because the generation of clean energy is limited to avoid problems in the distribution lines.

The basic solution is to increase the distribution and transmission line capacity, which can be performed in several ways. The most obvious way is to build new lines to reinforce the network. However, this solution is constrained by the high costs and legal difficulties of building new lines [2]. Because of the unviability of the first proposal, electrical line operators are focusing on solutions based on the modification of existing lines and an increase in their capacity.

Increasing the capacity of overhead power lines is currently one of the important areas of research due to a good balance between the results obtained and the costs involved. There are different techniques to increase this capacity: determine meteorological conditions by means of deterministic [3] or probabilistic [4] methods, up to the newest innovations in smart grids and line parameters real-time monitoring: temperature, sag, tilt, power, current and weather conditions [5–7].

In the case of wind farm integration into the grid, monitoring weather conditions in real time can be very useful to obtain a win-win situation [8,9]. Strong winds increase wind farm production. At the same time, they cool down the conductors of the distribution lines near the farm. This cooling effect allows the grid to be overloaded when it is most needed.

2. Ampacity, Conductor Temperature and Dynamic Calibration of Overhead Lines

The notion of ampacity appeared as a result of research on increasing power line capacity, and it is defined as the maximum amount of electrical current a conductor can continuously carry before sustaining deterioration. Ampacity is limited by several factors: the conductor structure and design, the surrounding environmental conditions and the operating conditions of the line.

Ampacity can be used as a static or dynamic value [10–12]. Static ampacity always assumes the most constrained conditions for the conductor and its environment. This condition gives very conservative values and low efficiency grids.

On the other hand, dynamic ampacity considers the variability of the grid and its surroundings (ambient temperature, solar radiation, wind, *etc.*). Thus, if the different conductor cooling and heating processes are measured in real time, the maximum instantaneous practical current can be measured (dynamic ampacity) without reaching the maximum thermal rating [13,14]. This is why dynamic ampacity is considered to be a more efficient control parameter of the power grid than static ampacity.

Working parameters should be measured or estimated by different methods (deterministic or probabilistic methods) to calculate the ampacity. International Council on Large Electric Systems (in French: Conseil International des Grands Reseaux Electriques, CIGRE) [15] and Institute of Electrical and Electronics Engineers (IEEE) [16] have standards in which the algorithms to estimate the ampacity and the temperature of the conductor are described.

3. Thermal Balance of Overhead Lines Calculation Methods

Both algorithms (CIGRE and IEEE) are based on the thermal balance between the gained and lost heat in the conductor due to the load and environmental conditions [17]. They suggest two ways to estimate the conductor temperature of an overhead power line. The first way uses steady state conditions to calculate the conductor temperature while the second way estimates the temperature in a dynamic balance taking into account the conductor thermal inertia.

The basic thermal balance used in steady state conditions is:

$$q_c + q_r = q_s + q_j + q_m \tag{1}$$

where q_c is the cooling due to convection, q_r is the cooling due to the radiation to the surroundings, q_s is the heating due to the solar radiation, q_j is the heating due to the Joule effect and q_m is the heating due to the magnetic effect.

If the thermal inertia of the conductor is considered, the following dynamic thermal balance is used instead:

$$\tag{2}$$

where m is the mass per unit length, c the specific heat capacity and T_c the theoretical conductor temperature.

The main similarities and differences between both algorithms are [18]:

- Both methods consider the weather conditions, including wind speed and direction, ambient temperature and solar radiation, but they use different approaches to calculate the thermal balance.

- Solar heating is calculated by considering the sun's position depending on the hour and day of the year. CIGRE uses a more complex algorithm including the direct, diffuse and reflected radiation.
- Convective cooling is approached by CIGRE using Morgan correlations based on Nusselt number and by IEEE using McAdams correlations based on Reynolds number.

Focusing on CIGRE and IEEE standards and in the guide for selection of weather parameters for bare overhead conductor ratings of CIGRE [19], the variables that should be measured or estimated are the ambient temperature (T_a), solar radiation (Q_s), wind speed (u_w), wind direction (ϕ_w) and the current of the conductor (I_c^{TMS}). This paper shows the measured conductor surface temperature (T_c^{TMS}) and compares it with the temperature estimated by the standards (T_c^{CIGRE} & T_c^{IEEE}).

On the one hand, the steady state balance is used and the conductor temperature $T_{c,ss}$ is obtained from the solution of Equation (1). On the other hand, the dynamic state balance is calculated by Equation (**??**), tracking the conductor temperature $T_{c,ds}$ using a time step $dt = 1$ s (Figure 1).

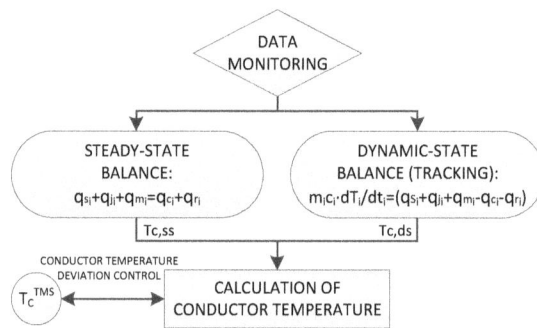

Figure 1. Conductor heat balance flow chart.

The values of the parameters to calculate the temperature are measured by a meteorological station placed in the tower (ambient temperature, humidity, wind speed and direction and solar radiation). Measured solar radiation is used to compare it with that estimated by the standards and to show the error made by the standards due to the estimated solar radiation use. The conductor temperature calculated for each set of data is then compared with the value measured by a temperature measurement sensor (TMS) placed in the overhead line and close to the meteorological station. This TMS is also used to measure the conductor current needed to calculate q_j. All data from meteorological stations and TMS are obtained every second and used to calculate their average values in periods of 8 min.

The evaluation of the optimal place for the location of the weather station has been carried out using both historical data and a meso-scale (convection-permitting) model called HIRLAM (High-Resolution Limited Area Model) that is widely used in Europe for numerical weather prediction. The HIRLAM model had a resolution of 0.05, which means a data grid of 4 km. The resolution of the micro-climatic study was reduced to 500 m by using bi-cubic interpolation. The model also included the surface roughness of the terrain provided by the database CORINE Land Cover. The results provided by the micro-climatic study defined critical points in terms of their ability to cool the cable.

4. Results for a Specific Overhead Line

To study the influence of each variable on the thermal balance of the algorithms, real time data of the ambient and conductor temperature, humidity, wind speed and direction and sun radiation were averaged every 8 min during an entire year—from September 2013 to September 2014— in a 132-kV overhead line with a LA 280 Hawk type conductor [20] located in northern Spain (Figure 2a).

Table 1 describes the variables and the equipment used to measure them. The meteorological station is placed in the electricity tower and the TMS attached to the conductor (Figure 2b).

With the set of values generated, the steady and dynamic thermal states and the associated conductor temperatures according to CIGRE ($T_{c,ss}^{CIGRE}$ and $T_{c,ds}^{CIGRE}$) and IEEE ($T_{c,ss}^{IEEE}$ and $T_{c,ds}^{IEEE}$) are calculated and compared with the conductor temperature measured by the TMS (T_c^{TMS}). A large amount of data was processed, and a statistical approach is used to study the individual influence of the variables.

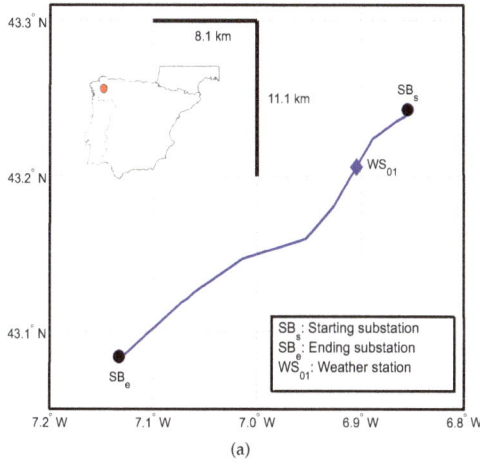

(a)

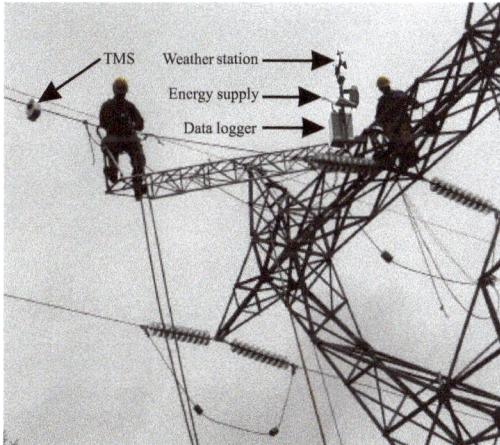

(b)

Figure 2. Description of the line and the system components. (**a**) 132 kV overhead transmission line located in northern Spain; (**b**) System components of the conductor temperature and meteorological data monitoring at the tower.

Table 1. Technical data of the measuring equipment.

Measurement	Measuring Equipment
Conductor Temperature (T_c^{TMS})	TMS Accuracy: 0–120 °C
Conductor current (I_c^{TMS})	TMS Accuracy: 100–1500 A
Solar Radiation (Q_s)	Pyranometer. Accuracy: 0–1100 W/m^2 ±0.5%
Wind Speed (u_w)	Vane Anemometer. Accuracy: 0–60 m/s ±0.3 m/s
Wind Angle Relative Direction (ϕ_w)	Vane Anemometer. Accuracy: 0–360°±2°
Ambient Temperature (T_a)	Thermometer. Accuracy: (−20)–80 °C ±0.3 °C
Humidity	Hygrometer. Accuracy: 0%–100% ±3%

Figure 3a,b and Table 2 provide information regarding the frequency and cumulative frequency of the deviation between the estimated and measured temperatures. Both standards are in good agreement for steady and dynamic balances and underestimate the measured temperature T_c^{TMS} in a 15% of cases.

The steady state assumption does not take into account the thermal inertia of the conductor materials and, thus, it can not model the transition between the set of values. This fact generates peaks in the estimated conductor temperature, which do not, in reality, exist. These mistakes are corrected if the dynamic balance is used, Equation (**??**). For instance, Table 2 indicates that the number of samples with deviations to conductor temperature lower than 5 °C increases from around 80% to 85% when the dynamic analysis is used.

Table 2. Cumulative frequency of differences between temperatures obtained using CIGRE ($T_{c,ss}^{CIGRE}$ and $T_{c,ds}^{CIGRE}$) and IEEE ($T_{c,ss}^{IEEE}$ and $T_{c,ds}^{IEEE}$) standards and T_c^{TMS} for an entire year.

Deviation Temperature (° C)	CIGRE S.S. Cum.Freq.[1] (%)	IEEE S.S. Cum.Freq.[2] (%)	CIGRE D.S. Cum.Freq.[3] (%)	IEEE D.S. Cum.Freq.[4] (%)
−5	0.00	0.00	0.00	0.00
−4	0.01	0.01	0.01	0.00
−3	0.16	0.12	0.14	0.04
−2	0.94	0.87	1.04	0.75
−1	3.93	3.81	4.08	3.68
0	14.77	13.54	15.58	13.70
1	35.34	33.41	39.74	37.37
2	51.05	49.57	56.97	55.67
3	63.37	62.65	69.81	69.20
4	72.89	72.70	78.70	79.14
5	79.62	80.12	84.87	86.02
6	84.51	85.30	89.37	90.59
7	87.76	88.91	92.35	93.56
8	90.22	91.51	94.37	95.76
9	92.38	93.30	95.95	97.24
.....	.....	.....	.....	.....
25	99.93	99.96	100.00	100.00

(1) Steady state cumulative frequency with CIGRE.
(2) Steady state cumulative frequency with IEEE.
(3) Dynamic state cumulative frequency with CIGRE.
(4) Dynamic state cumulative frequency with IEEE.

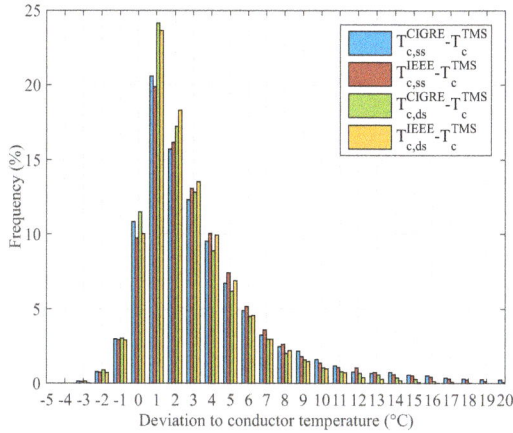

(a)

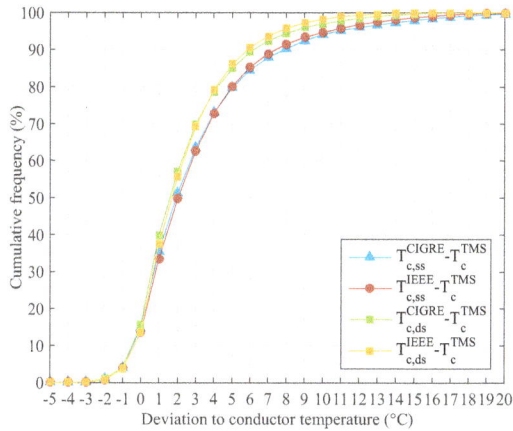

(b)

Figure 3. Frequency and cumulative frequency of differences between temperatures obtained using CIGRE ($T_{c,ss}^{CIGRE}$ and $T_{c,ds}^{CIGRE}$) and IEEE ($T_{c,ss}^{IEEE}$ and $T_{c,ds}^{IEEE}$) standards and the measured conductor temperature (T_c^{TMS}) for an entire year. (**a**) Frequency; (**b**) Cumulative frequency.

As an example, a representative day (30 August 2014) is shown in Figure 4. Figure 4a shows the deviation between the conductor temperature estimated by CIGRE and IEEE steady state balance ($T_{c,ss}^{CIGRE}$ & $T_{c,ss}^{IEEE}$) and the measured temperature (T_c^{TMS}). Figure 4b shows the same deviation for the dynamic state balance. Finally, the measured weather parameters are also represented in Figure 4c.

Comparing Figure 4a,b ,some differences between steady and dynamic balance can be observed. First of all, the dynamic balance models the transient states obtaining smoother curves with less deviations to the conductor temperature giving a better fit than the steady state. Secondly, the consideration of the thermal inertia of the conductor materials makes the slope of the dynamic curves closer to the slope of the T_c^{TMS} curve.

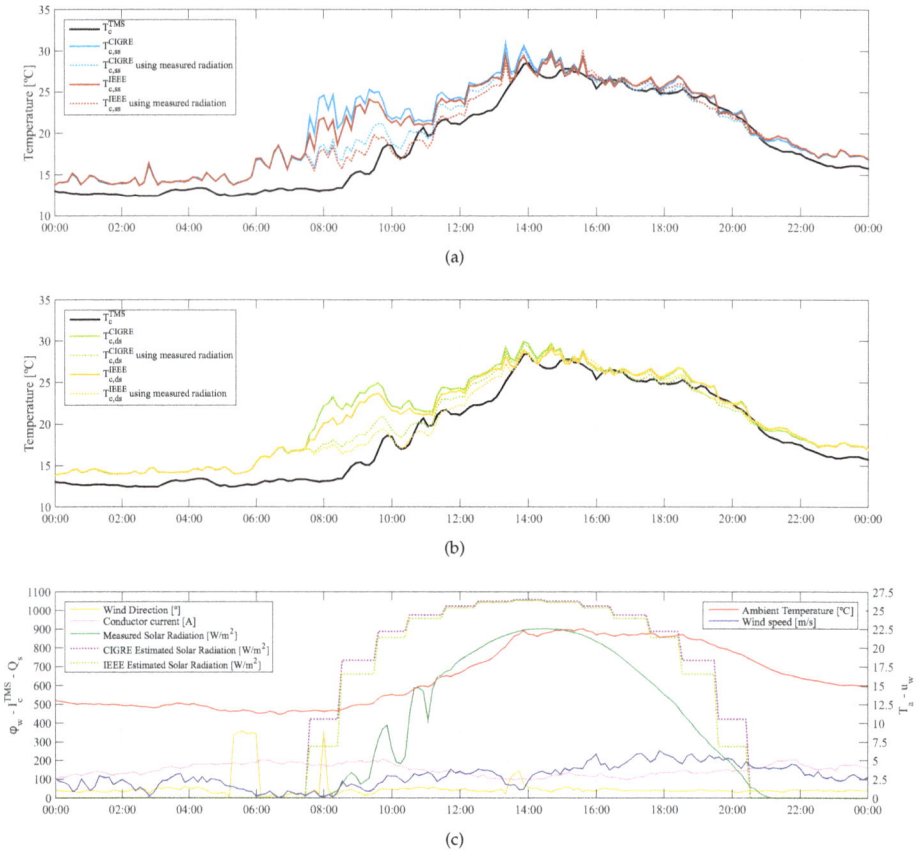

Figure 4. Comparison of conductor temperature obtained using IEEE ($T_{c,ss}^{IEEE}$ and $T_{c,ds}^{IEEE}$) and CIGRE ($T_{c,ss}^{CIGRE}$ and $T_{c,ds}^{CIGRE}$) standards with the measured conductor temperature (T_c^{TMS}) for a single day. (**a**) Steady state balance (30 August 2014); (**b**) Dynamic state balance (30 August 2014); (**c**) Weather conditions (30 August 2014).

From these figures, one can conclude that CIGRE and IEEE estimated temperatures differ more when the influence of radiation is appreciable (from 8:00 to 21:00). These differences between standards are due to the distinct ways to calculate the solar heat gain. CIGRE estimates the direct, diffuse and reflected radiation while IEEE only includes the direct radiation. This is the reason why the CIGRE estimated radiation is higher than the IEEE estimated one, as shown in Figure 4c. This effect makes the CIGRE estimated temperature to be higher than the IEEE estimated one. In addition, a systematic overstimation of the conductor temperature appears in both models when there is no solar radiation (*i.e.*, at night). This deviation, around 2 °C , might be due to the radiative cooling calculation. The equation used to evaluate this effect considers the ground and sky temperature to be equal to the ambient temperature [15,16] but during clear nights this assumption obtains worse estimated conductor temperatures because of radiation to deep space [19].

Figure 4a,b also show the error made if the estimated radiation is used instead of the one measured by the pyranometer. Temperatures obtained using the measured radiation fit better with the conductor temperature T_c^{TMS}. Additionally, the frequency and cumulative frequency of the deviation using estimated and measured radiation are plotted in Figure 5. The correction made using the measured radiation is clearly shown. The number of samples with deviations higher than 5 °C decreases from 15% to 5%. However, the number of samples which underestimate the conductor temperature increases 10% (from 15% to 25%). This makes the use of the measured radiation recommendable, but the increase of the underestimated values should also be taken into account.

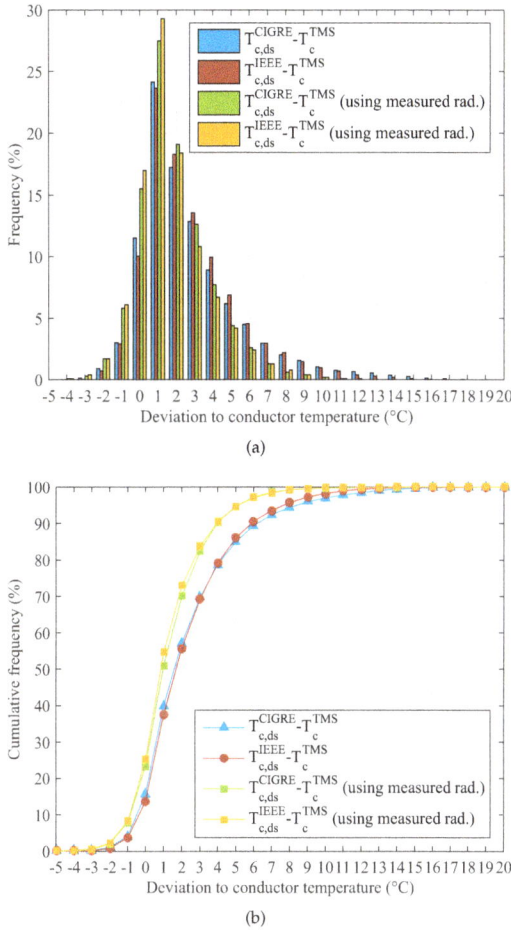

Figure 5. Frequency and cumulative frequency of differences between temperatures obtained using CIGRE ($T_{c,ds}^{CIGRE}$) and IEEE ($T_{c,ds}^{IEEE}$) standards and T_c^{TMS} for an entire year, with estimated and measured radiation. (**a**) Frequency; (**b**) Cumulative frequency.

As the dynamic thermal balance provides a better estimated temperature, the dynamic method will be used to study the influence of the wind on the estimated temperature. This influence is reported in previous studies [21] and the wind seems to be the most critical variable for the difference between the estimated and measured temperatures. In Figure 6a, it can be seen that as the wind speed decreases, this difference increases. If the influence of the other variables are minimized by selecting only the cases without solar radiation ($Q_s = 0\ W/m^2$), low radiation losses q_r ($T_c^{TMS} - T_a < 2\ °C$) and low current ($I_c^{TMS} < 200\ A$, the LA-280 maximum current to 80 °C is 600 A), the wind speed influence is clearer, as seen in Figure 6b.

As reported in the standards, the overestimation of the conductor temperature at low speeds is due to the difficulty of having accurate equations to model the convective effect. This fact can make the estimated temperature even 20 °C higher than the measured one.

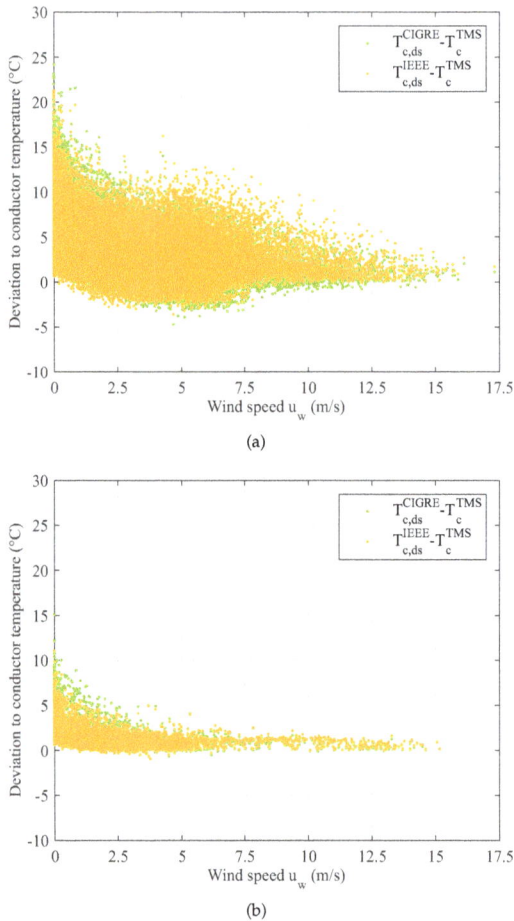

Figure 6. Deviation of conductor temperature obtained using CIGRE ($T_{c,ds}^{CIGRE}$) and IEEE ($T_{c,ds}^{IEEE}$) standards to the measured conductor temperature (T_c^{TMS}) vs. wind speed for an entire year. (a) For all data; (b) $Q_s = 0\ W/m^2$, $T_c^{TMS} - T_a < 2°C$ and $I_c^{TMS} < 200\ A$.

Going deeper into the influence of the wind is to consider how the deviation to conductor temperature is modified by the wind direction. If the temperature deviation is plotted against the angle between the wind and axis of the conductor ϕ_w (Figure 7), it can be seen that the lower the wind angle, the higher the deviation is, *i.e.*, in cases with wind blowing parallel to the conductor, standards generally overestimate the conductor temperature.

Finally, Table 3 shows the cumulative frequency of the deviation to conductor temperature lower than 5 °C obtained by the different methods. On the one hand, including the thermal inertia of the conductor materials improves the accuracy of the estimated temperature around 5%. On the other hand, replacing the theoretical radiation by the measured one continues improving the accuracy. In this case, 94.7% of the samples have a deviation lower than 5 °C.

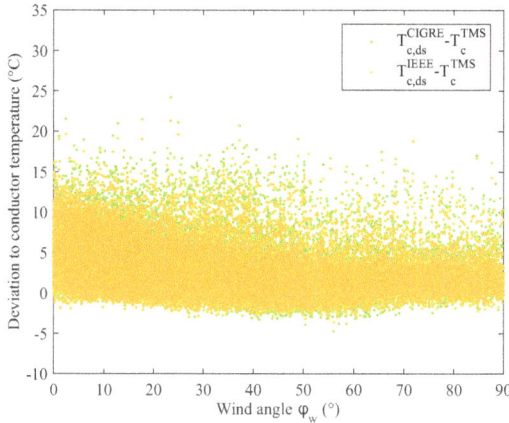

Figure 7. Deviation of conductor temperature obtained using IEEE ($T_{c,ds}^{IEEE}$) and CIGRE ($T_{c,ds}^{CIGRE}$) standards to the measured conductor temperature (T_c^{TMS}) *vs.* the angle between the wind and the axis of the conductor (ϕ_w).

Table 3. Cumulative frequencies of deviation to conductor temperature lower than 5 °C for the studied cases.

Cumulative Frequency	CIGRE S.S.	IEEE S.S.	CIGRE D.S.	IEEE D.S.	CIGRE D.S. (Using Measured Rad.)	IEEE D.S. (Using Measured Rad.)
(%)	79.6	80.1	84.9	86.0	94.7	94.7

5. Conclusions

This paper presents the steady and dynamic thermal balances of an overhead power line proposed by CIGRE [15] and IEEE [16] standards. The estimated temperatures calculated by the standards are compared with the averaged conductor temperature obtained every 8 min during an entire year. The conductor is a LA 280 Hawk type, used in a 132-kV overhead line and located in northern Spain.

A good monitoring system of the weather conditions surrounding power lines provides very important information to control the conductor temperature. The evaluation of the optimal place for the location of the weather station has been carried out using both historical data and a meso-scale model. The results provided by the micro-climatic study defined critical points in terms of their ability to cool the cable.

Regarding the type of heat balance, the dynamic method gives a better approach to the conductor temperature. The steady and dynamic state comparison shows that the number of cases with deviations to conductor temperature higher than 5 °C decreases from around 20% to 15% when the dynamic analysis is used (Table 2).

As some of the most critical variables for the IEEE and CIGRE thermal balances are speed and direction of the wind, ambient temperature and solar radiation, their influence on the conductor temperature is studied. Both standards give very similar results with slight differences due to the different way to calculate the solar radiation gain and the convection losses.

CIGRE estimates the direct, diffuse and reflected radiation while IEEE only includes the direct radiation. Focusing on a single day (Figure 4), the estimated temperatures present more differences when the influence of the radiation is appreciable, making the CIGRE estimated temperature to be higher than the IEEE estimated one. Worth noting also is the significant difference between the estimated and the measured temperature if there are large deviations between the estimated and the measured solar radiation (Figure 4c). If the measured radiation on site is used instead of the theoretical one suggested by the standards, the deviation to conductor temperature can also be decreased (Figure 5b). For example, using the estimated radiation, 15% of the samples present deviations higher than 5 °C, while using the measured radiation this percentage decreases to 5%.

Considering the wind, both standards provide better results for the estimated conductor temperature as the wind speed increases (Figure 6a,b) and the angle with the line is closer to 90° (Figure 7), giving the maximum deviation to the measured temperature for low wind speeds and quasi-parallel flows. As reported in the standards, the overestimation of the conductor temperature at low speeds is due to the difficulty of having accurate equations to model the convective effect. This fact can make the estimated temperature to be even 20 °C higher than the measured one.

In conclusion, as the algorithms and the input data are improved, from steady state analysis with estimated radiation to dynamic balance with measured radiation, the accuracy of the estimated temperature can increase up to 15% (Table 3).

Acknowledgments: This work was supported by the Spanish Government under the R+D initiative INNPACTO with reference IPT-2011-1447-920000 and Spanish R+D initiative with reference ENE2013-42720-R. The authors would also like to acknowledge Viesgo for its support.

Author Contributions: Alberto Arroyo, Pablo Castro, Raquel Martinez, Mario Manana, Alfredo Madrazo, Ramón Lecuna and Antonio Gonzalez contributed to this paper. Alberto Arroyo and Pablo Bernardo: definition of the methodology, Raquel Martinez, Mario Manana and Antonio Gonzalez: test execution. Alfredo Madrazo and Ramon Lecuna: review.

Conflicts of Interest: The authors declare no conflict of interest.

Abbreviations

ϕ_w: angle between wind and axis of conductor (°).
c: specific heat capacity (J/kg °C).
I_c^{TMS}: conductor measured current (A).
m: conductor mass per unit length (kg/m).
q_c: convective cooling (W/m).
q_r: radiative cooling (W/m).
q_s: solar radiative heating (W/m).
q_j: joule heating (W/m).
q_m: magnetic heating (W/m).
Q_s: solar radiation (W/m^2).
T_a: ambient air temperature (°C).
T_c: theoretical conductor temperature (°C).
$T_{c,ss}^{CIGRE}$: steady state conductor temperature estimated by CIGRE (°C).
$T_{c,ds}^{CIGRE}$: dynamic state conductor temperature estimated by CIGRE (°C).
$T_{c,ss}^{IEEE}$: steady state conductor temperature estimated by IEEE (°C).
$T_{c,ds}^{IEEE}$: dynamic state conductor temperature estimated by IEEE (°C).
T_c^{TMS}: measured conductor temperature (°C).
u_w: wind speed (m/s).

References

1. Nykamp, S.; Molderink, A.; Hurink, J.; Smit, J. Statistics for PV, wind and biomass generators and their impact on distribution grid planning. *Energy* **2012**, *45*, 924–932.
2. Jorge, R.S.; Hertwich, E.G. Environmental evaluation of power transmission in Norway. *Appl. Energy* **2013**, *101*, 513–520.
3. Hall, J.F.; Deb, A.K. Prediction of overhead transmission line ampacity by stochastic and deterministic models. *IEEE Trans. Power Deliv.* **1988**, *3*, 789–800.

Energies **2015**, *8*, 13660–13671

4. Reding, J.L. A method for determining probability based allowable current ratings for BPA's transmission lines. *IEEE Trans. Power Deliv.* **1994**, *9*, 153–161.

5. Pytlak, P.; Musilek, P. Modelling precipitation cooling of overhead conductors. *Electr. Power Syst. Res.* **2011**, *81*, 2147–2154.

6. Cho, J.; Kim, J. H.; Lee, H.J.; Kim, J.Y.; Song, I.K.; Choi, J.H. Development and improvement of an intelligent cable monitoring system for underground distribution networks using distributed temperature sensing. *Energies* **2014**, *7*, 1076–1094.

7. Holyk, C.; Liess, H.D.; Grondel, S.; Kanbach, H.; Loos, F. Simulation and measurement of the steady-state temperature in multi-core cables. *Electr. Power Syst. Res.* **2014**, *116*, 54–66.

8. Hosek, J. Dynamic thermal rating of power transmission lines and renewable resources. In Proceedings of the ES1002 Workshop, Paris, France, 22–23 March 2011; pp. 1–3.

9. Heckenbergerova, J.; Hosek, J. Dynamic thermal rating of power transmission lines related to wind energy integration. In Proceedings of the 11th International Conference on Environment and Electrical Engineering (EEEIC), Venice, Italy, 18–25 May 2012; pp. 798–801.

10. Popelka, A.; Jurik, D.; Marvan, P. Actual line ampacity rating using PMU. In Proceedings of the 21st International Conference on Proceedings of the Electricity Distribution (CIRED), Frankfurt, Germany, 6–9 June 2011.

11. Puffer, R.; Schmale, M.; Rusek, B; Neumann, S.; Scheufen, M. Area-wide dynamic line ratings based on weather measurements. In Proceedings of the Conference on Cigre Session 44, Paris, France, 26–31 August 2012.

12. Abdelkader, S.; Morrow, D.J.; Fu, J; Abbott, S. Field measurement based PLS model for dynamic rating of overhead lines in wind intensive areas. In Proceedings of the International Conference on Renewable Energies and Power Quality, Bilbao, Spain, 20–22 March 2013.

13. International Council on Large Electric Systems, CIGRE. *Guide for Application of Direct Real-Time Monitoring Systems*; Technical Brochure 498; CIGRE: Paris, France, June 2012.

14. Michiorri, A.; Taylor, P.C.; Jupe, P.C.; Berry, C.J. Investigation into the influence of environmental conditions on power system ratings. *Proc. Inst. Mech. Eng. A J. Power Energy* **2009**, *223*, 743–757.

15. International Council on Large Electric Systems, CIGRE. *Guide for Thermal Rating Calculation of Overhead Lines*; Technical Brochure 601; CIGRE: Paris, France, December 2014.

16. *IEEE Std 738-2012: IEEE Standard for Calculation the Current-Temperature Relationship of Bare Overhead Conductors*; IEEE Standard Association: Washington, U.S.A. 23 December 2013.

17. Silva, A.A.P.; Bezerra, J.M.B. Applicability and limitations of ampacity models for HTLS conductors. *Electr. Power Syst. Res.* **2012**, *93*, 61–66.

18. Schmidt, N.P. Comparison between IEEE and CIGRE ampacity standards. *IEEE Trans. Power Deliv.* **1999**, *14*, 1555–1559.

19. International Council on Large Electric Systems, CIGRE. *Guide for Selection of Weather Parameters for Bare Overhead Conductor ratings*; Technical Brochure 299; CIGRE: Paris, France, 2006.

20. EN 50182:2001. Conductors for overhead lines. Round wire concentric lay stranded conductors. (ISBN 978 0 580 84034 0)

21. Abbott, S.; Abdelkader, S.; Bryans, L.; Flynn, D. Experimental validation and comparison of IEEE and CIGRE dynamic line models. In Proceedings of the 45th International Universities Power Engineering Conference (UPEC), Cardiff, UK, 31 August–3 September 2010; pp. 1–5.

energies

MDPI

Article

Response Based Emergency Control System for Power System Transient Stability

Huaiyuan Wang, Baohui Zhang and Zhiguo Hao *

State Key Laboratory of Electrical Insulation and Power Equipment, School of Electrical Engineering,
Xi'an Jiaotong University, No. 28, Xianning West Road, Xi'an 710049, Shaanxi, China;
wanghuaiy@stu.xjtu.edu.cn (H.W.); bhzhang@mail.xjtu.edu.cn (B.Z.)
*   Correspondence: Zhghao@mail.xjtu.edu.cn; Tel.: +86-029-82668598

Academic Editor: Ying-Yi Hong
Received: 8 August 2015; Accepted: 13 November 2015; Published: 30 November 2015

Abstract: A transient stability control system for the electric power system composed of a prediction method and a control method is proposed based on trajectory information. This system, which is independent of system parameters and models, can detect the transient stability of the electric power system quickly and provide the control law when the system is unstable. Firstly, system instability is detected by the characteristic concave or convex shape of the trajectory. Secondly, the control method is proposed based on the analysis of the slope of the state plane trajectory when the power system is unstable. Two control objectives are provided according to the methods of acquiring the far end point: one is the minimal cost to restore the system to a stable state; the other one is the minimal cost to limit the maximum swing angle. The simulation indicates that the mentioned transient stability control system is efficient.

Keywords: phase plane; transient instability prediction; transient stability control

---

1. Introduction

This is a follow-up paper of a series on closed loop control systems for power system transient stability. The previous work proposed a real-time approach to detect transient instability with high accuracy and wide applicability, while this paper mainly focuses on how and where to control the system after the instability is detected.

In China, transient stability control systems acquire some characteristic variables according to a large number of offline calculations and identify the stability by the combination of these characteristic variables. For unstable situations, the control quantity of different situations is obtained by repeated simulations to develop a cure table. When disturbances occur, protection devices operate on the basis of the cure table [1,2].

The samples consist of different situations, including power flow, grid topology and fault conditions making it hard to ensure an effective cure table. In order to decrease the number of samples and improve the utility of the cure table, a simulation which is helpful to detect transient instability and form control law adopts measurement results to update the online power flow with a refresh cycle of 3–5 min [3–6]. Corresponding approaches has been introduced in [5,6], including preplanned remedial action and system integrity protection schemes, software processes and hardware requirements. To speed up the simulation, the transient energy function method, the equal area criterion, topological energy function or quasi-real-time online transient analysis are employed in the time domain simulation, which can greatly improve the efficiency [7–12].

In any case these methods need repeated simulations to calculate control laws by taking into account the changes in operating conditions or parameter variations. The validity of the cure table depends on the similarity between the anticipated faults and actual faults and the accuracy of the model parameters. However, a practical operating model (especially a model of the load) or the system parameters are hard to acquire. Meanwhile, the rapid development and the operation variations lead to a large number of calculation samples.

With the development of computer science and communication technology, especially the application of phase measurement unit (PMU) based on global positioning system (GPS) in power systems, wider-area measurement systems (WAMS) offer an opportunity to develop real-time protection and control systems [13]. As a result, recently there has been a focus on online instability control, out-of-step protection and their corresponding theories.

It has been reported that the geometric characteristics of the system trajectory can be employed to detect the system instability [14]. Depending on the nature of the stability studied, Girgis has found that the characteristic shape (concave or convex) of a surface, on which a post-fault transient trajectory lies, can be used as an index for online instability detection [15]. The assumption is proved in a phase portrait, and an instability detection method is presented, which is independent of the network structure, system parameters and model because it only uses observation data. During the process of proof, a damping coefficient is considered in the system model to keep it more reasonable [16–18]. However, the discrete expression of the detection method in [18] requires differential calculations. The curve of the index looks ragged. For a stable situation after disturbances, it may cause some erroneous judgments.

This paper aims to introduce a real-time transient stability control system which employs the generator speed, power angle and unbalanced power to detect the system instability and develop a control strategy. In this paper, the system instability is detected by the characteristics concave or convex shape of the trajectory. Then, the control law is achieved based on the slope of the state plane trajectory, when the power system is unstable. Because the expected variables can be obtained from the trajectory information and the calculations are very simple, the control system can be realized in real time. The simulation indicates that the mentioned transient stability control system is fast, efficient, and realizable.

2. Instability Detection

2.1. Identification Method of Transient Instability for an Autonomous SMIB System

Previous research has found the geometric characteristics of the trajectory in a phase portrait can be employed to determine system instability through a great deal of simulation. It is found that stable trajectories are always concave with respect to the post-fault stable equilibrium point (SEP) and that unstable trajectories are convex with respect to the post-fault SEP immediately or a short time after the fault-clearing time as seen in Figure 1. The point at which the geometric characteristic of a system trajectory is convex, is defined as the no return point (NRP). All of these points (NRP) form of the interface that is defined as the no return point interface (NRPI).

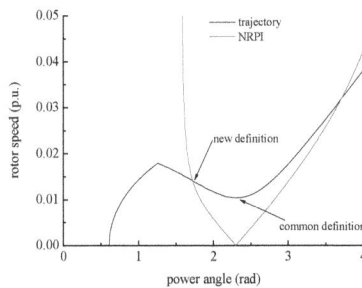

Figure 1. System trajectories of a SMIB.

It should be particularly pointed out that the definition in this paper is different from the usual definition. As a rule, the point where a trajectory shifts to accelerated progress ($\Delta P = 0$, $\delta = \delta_{uep}$, ΔP is the unbalanced power and δ_{uep} is the unstable equilibrium point) is defined as the NRP. The difference between both definitions is shown in Figure 1. Although Figure 1 is obtained for a specific set of system parameters, the characteristics of the system trajectory are typical of a single machine infinite bus (SMIB) system. Here, we use a SMIB system as an example to explain the foregoing concept. For the SMIB system, the system dynamics can be expressed as:

$$\dot{\delta} = w_0 \Delta w$$
$$\dot{\Delta w} = [P_m - P_{emi}\sin\delta - D\Delta w]/M \tag{1}$$

where δ and Δw are the generator angle and angular velocity with respect to a synchronous frame; D is the generator damping coefficient; M is the generator inertia; P_m is the generator mechanical power input; and P_{emi} is the generator maximum electric power output.

On the basis of the definition of NRP, the tangent slope at any point on the trajectory of system Equation (1) is given as follows:

$$\frac{d\Delta w/dt}{d\delta/dt} = \frac{d\Delta w}{d\delta} = k(\delta, \Delta w) = \Delta P/(Mw_0\Delta w) \tag{2}$$

Consequently, the one-order derivative of the tangent slope at any point of the trajectory with respect to the generator angle (that is, the two-order derivative of the angular velocity with respect to the generator angle) can be expressed as:

$$l = \frac{dk}{d\delta} = k'(\delta, \Delta w) = \frac{[-P_{emi}\cos(\delta) - D\frac{d\Delta w}{d\delta}]Mw_0\Delta w - [P_m - P_{emi}\sin\delta - D\Delta w]^2/\Delta w}{(Mw_0\Delta w)^2} \tag{3}$$

On the ground of the definition of NRP, the point of system trajectory at which $l = 0$ is actually NRP. The interface (NRPI) that all of NRP constitute can be described as:

$$-P_{emi}\cos\delta Mw_0\Delta w - D[P_m - P_{emi}\sin\delta - Dw] - [P_m - P_{emi}\sin\delta - D\Delta w]^2/\Delta w = 0 \tag{4}$$

Figure 2 shows the NRPI divides the whole plane into two regions, *i.e.*, a convex area and a concave area. In virtue of the mathematical definition, if $l > 0$, we consider the geometrical characteristics of the post-fault trajectory is convex, while it is considered as concave if $l < 0$.

We define the concavity and convexity of phase trajectory as follows:

(1) The phase trajectory is convex if $l \cdot \Delta w > 0$;
(2) The phase trajectory is concave if $l \cdot \Delta w < 0$;
(3) The trajectory is on the inflexion point if $l \cdot \Delta w = 0$.

In order to facilitate real-time computing, the unstable criterion can be expressed as:

$$\tau = l \cdot \Delta w = \frac{dk}{d\delta} \cdot \frac{d\delta}{dt} \cdot \frac{1}{w_0} = \frac{dk}{dt} \cdot \frac{1}{w_0} \geqslant 0 \tag{5}$$

The discrete form of the index is given by:

$$\tau(i) = \frac{\Delta P(i)}{M\Delta w(i)} - \frac{\Delta P(i-1)}{M\Delta w(i-1)} \tag{6}$$

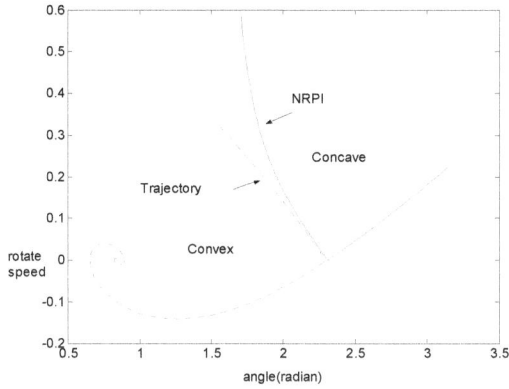

Figure 2. Relationship of NRPI and the critical post-fault trajectory.

If τ is always less than zero, the system will be stable, while the system will lose synchronization immediately or a short time after the fault-clearing time if τ is larger than zero. The proposition has proved that the proposed index is a sufficient and necessary condition for instability detection of SMIB system.

It should be pointed out that the moment when the trajectory passes through the NRI is just the detection time. As shown in Figure 1, it can be found that the unstable trajectory is sure to pass through the NRI before it arriving at dynamic saddle point (DSP). Thus, by employing the characteristic concave or convex shapes of the trajectory one can easily and quickly identify the transient instability.

2.2. Identification Theory of Transient Instability for a Non-Autonomous SMIB System

The equations of an electromechanical transient process for non-autonomous SMIB system can be described as follows:

$$\frac{d\delta}{dt} = \omega_0 \Delta\omega$$
$$M\frac{d\Delta\omega}{dt} = P_m(t) - P_e(t,\delta) = \Delta P(t,\delta) \tag{7}$$

where the right function of the motion equation involves the variable of time, and thus $\Delta P(t,\delta)$ doesn't represent a strict sine function any more. As a consequence, the index τ greater than zero in a certain time can only indicate the instability of the autonomous system under that parameter condition, but not the non-autonomous system:

$$P_e(t_i,\delta) = \lambda_0(t_i) - \lambda_1(t_i)\cos\delta - \lambda_2(t_i)\sin\delta$$
$$\Delta P(t_i,\delta) = P_c(t_i) - \lambda_1(t_i)\cos\delta - \lambda_2(t_i)\sin\delta \tag{8}$$

To fortify the accuracy of the detection, the index is established as follows. The feature index μ of trajectory in state-plane of unbalanced power and power angle is provided.

The feature index of the trajectory in state plane of unbalanced power and power angle is defined as:

$$\left.\frac{dl}{dt}\right|_{l=0} = \frac{d^2(\Delta P/M)}{d\delta^2} \tag{9}$$

The discrete form is proposed as:

$$\mu = \frac{\Delta P(i) - \Delta P(i-1)}{\delta(i) - \delta(i-1)} - \frac{\Delta P(i-1) - \Delta P(i-2)}{\delta(i-1) - \delta(i-2)} > 0 \tag{10}$$

The index means that the trajectory enters the convex region and does not return to the concave region in a short time. If the trajectory satisfies the index, the trajectory is unstable possibly.

225

Eventually, the instability criterion for a non-autonomous SMIB system can be described as:

$$\tau > 0 \ \& \ \mu > 0 \tag{11}$$

The detection method launches when the power system is disturbed. If the maximum swing angle is less than 50°, the detection method automatically ends.

3. Control Method

There are many transient stability control measures, including generator shedding and load shedding [19,20]. In North America, generator shedding has been proved to be one of the most effective discrete supplementary control means for maintaining stability [21]. In this paper, generator shedding is mainly discussed too.

3.1. The Slope of the State-Plane Trajectory

According to Equation (2), the expression of the slope is related to the unbalanced power which can be changed by generator shedding. It is supposed that the system returns to a stable state by generator shedding. If the value of the slope at control time is obtained, the corresponding generator shedding can be calculated by Equation (2). To obtain the value of the slope at control time, the characteristic of the slope is analyzed below. A single machine infinite bus system is shown in Figure 3.

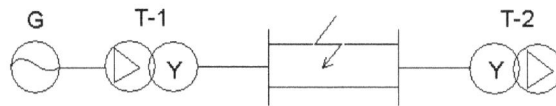

Figure 3. The topology network of the SMIB system.

Under normal conditions, the system operates at a stable equilibrium point (SEP). The initial state of the generator is 0.812 rad, 0 rad/s. A three-phase grounding fault occurs on one of the two transmission lines, and the fault is cleared by switching off the line. The clearing time is 0.18 s (critical clearing time) and 0.22 s. When the system is unstable, different control quantities are taken. The trajectory of the state-plane and the slope of the trajectory are shown in Figures 4 and 5.

For the black curve, the system is stable without control. The trajectory swings back at the far end point (FEP) which exists only in a stable system. The slope of the trajectory keeps decreasing with the mutation at the FEP.

For the red curve, the system is unstable without control. The slope begins to increase after the inflexion point and grows to zero at the dynamic saddle point (DSP).

For the green curve, the system is unstable with insufficient control. A sudden change of the slope is caused by the control at the control time. The larger the control quantity is, the greater the change of the slope at the control time is. However, the slope begins to increase at another inflexion point after control.

For the blue curve, the system is stable with enough control. The control causes a sudden reduction of the slope at the control time, and the slope keeps decreasing after the control with the mutation at the FEP. Apparently, the trajectory of the state-plane turns back at the unstable equilibrium point (UEP) by the minimum control quantity.

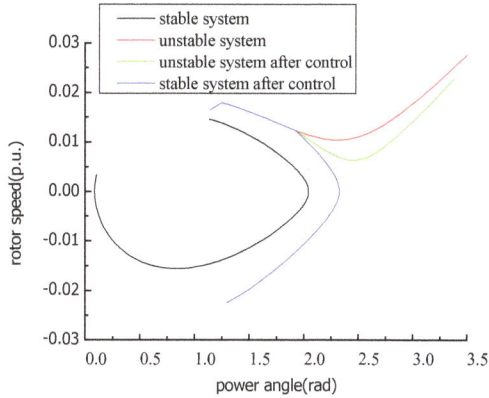

Figure 4. State-plane representation of speed and angle.

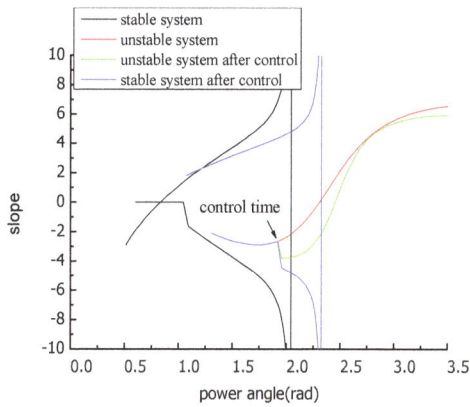

Figure 5. The slope over power angle.

The derivative of Equation (2) with respect to *t* is given by:

$$\frac{dk(\delta, \omega)}{dt} = \frac{dk(\delta, \omega)}{d\delta} \frac{d\delta}{dt} = l \cdot \Delta\omega = \tau \tag{12}$$

A similar equation can be found in the detection algorithm mentioned in the previous article. Obviously the decrease of the slope indicates that the trajectory runs in the concave area, and the trajectory runs in the convex area after the slope begins to increase. If the control is appropriate, the system will return to a stable state and the slope of the trajectory will keep decreasing until it reaches FEP.

3.2. Calculation of Control Quantity for SMIB

To find the relationship between the slope and angular speed, integrating both sides of Equation (2) gives:

$$\int_{\delta_a}^{\delta_b} k(\delta, \omega) d\delta = \int_{\delta_a}^{\delta_b} \frac{d\Delta\omega}{d\delta} d\delta = \Delta\omega_b - \Delta\omega_a \tag{13}$$

where δ_a and δ_b are the bounds of the integration; $\Delta\omega_a$ and $\Delta\omega_b$ are the angular speed corresponding to δ_a and δ_b. Let δ_b be the power angle at the control time t_a; and then $\Delta\omega_a$ is the angular speed at t_a. For stable state, $\Delta\omega_b$ is zero when δ_b is the FEP. For unstable state, if $\Delta\omega_b$ can be zero by the control, the system will return from δ_b to a stable state and δ_b will be the FEP. Therefore Equation (13) can be written as:

$$\int_{\delta_a}^{\delta_b} k(\delta, \omega)d\delta = -\Delta\omega_a \tag{14}$$

Therefore, the value of the slope at control time can be calculated by Equation (14), and the corresponding control quantity can be calculated by Equation (2). There are two unknowns δ_b and $k(\delta, \omega)$ in Equation (14). $k(\delta, \omega)$ is related to the unbalanced power on the basis of Equation (2). δ_b can be preset as the FEP as needed, but it must be no more than the UEP, so when δ_b is acquired, $k(\delta, \omega)$ can be calculated. If δ_b is just the UEP, the corresponding control quantity is the minimum.

3.3. Approximation Method to Calculate Control Quantity

As can be seen from the trajectories in Figure 3, $k(\delta, \omega)$ is nonlinear and changes corresponding to different control quantities at t_a. Because of the nonlinear nature of $k(\delta, \omega)$, the value of $k(\delta, \omega)$ at t_a could not be acquired exactly. Therefore, an approximation method is presented to acquire the value of $k(\delta, \omega)$ at t_a. Employing a constant k' instead of $k(\delta, \omega)$ in Equation (14):

$$\int_{\delta_a}^{\delta_b} k'd\delta = k'(\delta_b - \delta_a) = -\Delta\omega_a \tag{15}$$

$$k' = \frac{-\Delta\omega_a}{\delta_b - \delta_a} \tag{16}$$

Obviously, the angular speed and power angle at control time and the power angle at FEP are necessary to calculate k'. Above all δ_b must be preset less than or equal to UEP, then the system can be stable after control. Therefore, $k(\delta, \omega)$ will keep decreasing after the control and the trajectory will return at δ_b. Because that $k(\delta, \omega)$ will keep decreasing after the control, k' is surely less than the value of $k(\delta, \omega)$ at control time. Hence, the control quantity calculated by k' is a little bigger than actually needed.

For a SMIB system, suppose the ratio of generator shedding is λ, the relation of λ and k' after control is as follows:

$$k' = \frac{(1 - \lambda)P_m - P_{ea}}{(1 - \lambda)M\Delta\omega_a} \tag{17}$$

$$\lambda = 1 - \frac{P_{ea}}{P_m - M\Delta\omega_a k'} \tag{18}$$

where P_{ea} is the generator output electrical power at t_a. When k' is obtained by Equation (16), the generator shedding ratio of SMIB system can be calculated by Equation (18).

3.4. Seeking for FEP

The parameters needed in calculation can be collected by WAMS except for FEP. Two methods to acquire the FEP are provided in this paper:

Method 1: In order to obtain the minimum control quantity, it is intended to slow down the angular speed to zero at UEP by the control. It can be supposed that the power equilibrium point of the system after control is the UEP. Generator electrical power can be written as Equation (19), and mechanical power is considered as a constant in a short time period. It is assumed that generator shedding has a corresponding change on mechanical power:

$$P_e = P_c(t) + \lambda_1(t)\sin(\delta) + \lambda_2(t)\cos(\delta) \qquad (19)$$

where $P_c(t)$, $\lambda_1(t)$, $\lambda_2(t)$ are parameters to be identified at t instant. $P_c(t)$, $\lambda_1(t)$, $\lambda_2(t)$ are considered as constant in a short time when no other operations occur. At t_a instant, $P_c(t_a)$, $\lambda_1(t_a)$, $\lambda_2(t_a)$ are calculated by the least square method, and then the electrical power can be acquired.

In order to seek for the UEP after the control, the control should be known at first. Therefore an iterative method can be employed as follows:

(1) Obtain the prediction curve of the electrical power;
(2) Preset zero to the generator shedding ratio and UPE: $\lambda^{(0)} = 0$, $\delta_u^{(0)} = 0$;
(3) Mechanical power decreases at the same ratio:

$$P_m^{(k)} = (1 - \lambda^{(k-1)})P_m^{(k-1)}$$

(4) Search for the power equilibrium point as $\delta_b^{(n)}$;
(5) Acquire the $\lambda^{(k)}$ according to the $\delta_b^{(n)}$;
(6) If $\left|\delta_b^{(n)} - \delta_b^{(n-1)}\right| \leqslant \varepsilon$, complete the iterator; or return to step 3.

where ε is the convergence conditions; n is the number of the iterator.

Method 2: Preset a FEP as needed. On the basis of the operating requirement, the FEP, which must be less than or equal to the power equilibrium point, can be preset within the limits as needed.

Two control objectives are provided according to the way of acquiring the FEP: scheme one is the minimal cost to help restore the system to a stable state; scheme two is the minimal cost to limit the maximum angle.

3.5. Control Method for Multi-Machine System

Using the assumption made by [22] that the disturbed multi-machine system separates into two groups, leading group S and lagging group A, the partial center of angles, angular speed, mechanical power and electrical power of the two-machine power system are shown as follows:

$$\delta_s = \frac{\sum\limits_{i\in S} M_i\delta_i}{\sum\limits_{i\in S} M_i} \quad \delta_a = \frac{\sum\limits_{i\in A} M_i\delta_i}{\sum\limits_{i\in A} M_i} \qquad (20)$$

$$\Delta\omega_s = \frac{\sum\limits_{i\in S} M_i\Delta\omega_i}{\sum\limits_{i\in S} M_i} \quad \Delta\omega_a = \frac{\sum\limits_{i\in A} M_i\Delta\omega_i}{\sum\limits_{i\in A} M_i} \qquad (21)$$

$$P_{ms} = \sum_{i\in S} P_{mi} \quad P_{ma} = \sum_{i\in A} P_{mi}$$
$$P_{es} = \sum_{i\in S} P_{ei} \quad P_{ea} = \sum_{i\in A} P_{ei} \qquad (22)$$

Further, the two-machine power system can be equivalent to the SMIB system:

$$\delta = \delta_s - \delta_a \qquad (23)$$

$$\Delta\omega = \Delta\omega_s - \Delta\omega_a \qquad (24)$$

$$\Delta P = P_M - P_E \qquad (25)$$

where

$$M_T = \sum_{i=1}^{n} M_i, M = \frac{M_s M_a}{M_T},$$
$$P_M = \frac{M_a P_{ms} - M_s P_{ma}}{M_T},$$
$$P_E = \frac{M_a P_{es} - M_s P_{ea}}{M_T}.$$

According to the calculation method applied to SMIB equivalent system, the expected slope k' of SMIB equivalent system can be acquired. Transform the expected slope to control quantity:

$$\Delta P_{ms} = M_s(k - k')\omega_0(\Delta\omega_s - \Delta\omega_a) \tag{26}$$

where k is the original slope before control.

4. Simulation Result

4.1. SMIB System

The SMIB system as shown in Figure 3 is employed to illustrate the algorithm. Under normal conditions, the system operates at a stable equilibrium point. A three-phase grounding fault occurs on line L2, and the fault is cleared by switching off the line. The clear time differ from 0.17 s (critical clearing time) to 0.22 s. The system parameters are shown in Table 1.

Table 1. The parameters of the SMIB system.

Component	Parameters
The initial state	$\omega_0 = 2\pi f, f = 50Hz, \delta_0 = 34.49°$
The generator	$P_m = 120MW, T_j = 6s, x_d = 1.83$p.u., $x_q = 1.83$p.u., $x_d' = 0.3$p.u., $x_q' = 1.83$p.u., $x_q'' = 0.25$p.u., $x_d'' = 0.25$p.u.
The transmission line	$x_1 = x_2 = 0.486\Omega/km, x_0 = 4x_1$

It is needed to point out that the simulation just acquires the power angle, angular speed, mechanical power, electrical power, inertia constant of the generator without other information, which can be acquired on-line. When the sampling time interval is 10 ms, the detection results based on the instability criterion mentioned above are shown in Table 2.

Table 2. The transient stability detection results by 100 Hz sample frequency.

The Moment of Fault-Clearing	Simulation Results	Detection Results	The Moment of Instability Detected	The Angle of Instability Detected
0.17 s	Stable	Stable	\	\
0.18 s	Stable	Stable	\	\
0.187 s	Stable	Stable	\	\
0.188 s	Unstable	Unstable	0.56 s	121.9°
0.19 s	Unstable	Unstable	0.46 s	114.9°
0.20 s	Unstable	Unstable	0.37 s	106.2°
0.21 s	Unstable	Unstable	0.34 s	102.9°
0.22 s	Unstable	Unstable	0.33 s	102.8°

When the sampling time interval is 20 ms, the detection results based on the instability criterion mentioned above are shown in Table 3.

Table 3. The transient stability detection results by 50 Hz sample frequency.

The Moment of Fault-Clearing	Simulation Results	The Moment of Instability Detected	The Angle of Instability Detected	The Moment of Angle Reach the Threshold of 180°
0.17 s	Stable	Stable	-	-
0.18 s	Stable	Stable	-	-
0.187 s	Stable	Stable	-	-
0.188 s	Unstable	0.56 s	121.9°	1.44 s
0.19 s	Unstable	0.46 s	114.9°	1.14 s
0.20 s	Unstable	0.38 s	108.5°	0.86 s
0.21 s	Unstable	0.34 s	102.9°	0.76 s
0.22 s	Unstable	0.34 s	105.6°	0.68 s

The results in Tables 2 and 3 show that the instability criterion mentioned in this paper can distinguish unstable cases and stable cases correctly and rapidly. Even the critical cases of which the difference of fault-clearing time is only 1 ms can be rightly detected. The accuracy of the detection result is not influenced by the sample frequency. From the results in Table 3, the angles of instability detected vary from 102.9° to 121.9°, which are much less than 180°. Further, the detection moment of the method proposed in this paper is much earlier than it of the threshold of 180°, which can provide adequate time for implementation of control measures.

It is necessary to take some control actions when the system is detected as unstable. Two control objectives are provided by the way to acquire the FEP: scheme one is the minimal cost to help the system restore to stable state; scheme two is the minimal cost to limit the maximum angle.

When the system is unstable, the control quantity is calculated by scheme one. The results are shown in Table 4. The minimum control quantity calculated by scheme one is close to and a little bigger than the real minimum control quantity. The control quantity calculated by this paper is effective and conservative as expected. In addition, the conservative algorithm is helpful for the recovery of the power system.

The same condition introduced above, control quantity is calculated by scheme two, and the preset FEP is 130°. The result is shown in Table 5. The FEP after control is closed to and less than 130° as expected. It means that the scheme two intended to limit the FEP of the power system is effective. Transient stability margin is directly reflected in the angle swing range. The method can be applied to improve the transient stability margin when the power system suffers large disturbance.

Table 4. Calculated minimum control compare to the real minimum control.

Fault-Clearing Time	Distinction Time	Calculated Minimum Control Quantity (%)	Real Minimum Control Quantity (%)
0.18 s	None (stable)	None (stable)	None (stable)
0.19 s	0.46 s	5.61	4.55
0.20 s	0.37 s	14.1	13.5
0.21 s	0.34 s	22.4	20.6
0.22 s	0.33 s	30.1	28.4

Table 5. Calculated minimum control by limiting the maximum angle.

Fault-Clearing Time	Distinction Time	Calculated Controlled Quantity (%)	Real FEP after Control
0.18 s	None (stable)	None (stable)	None (stable)
0.19 s	0.46 s	8.91	129.1°
0.20 s	0.37 s	18.6	126.8°
0.21 s	0.34 s	22.5	128.1°
0.22 s	0.33 s	35.6	128.8°

4.2. IEEE 39-Bus System

The IEEE 39-bus system shown in Figure 6 is employed to illustrate the algorithm. Two control objectives are provided according to the way of acquiring the FEP: scheme one is the minimal cost to help the system restore to stable state; scheme two is the minimal cost to limit the maximum angle.

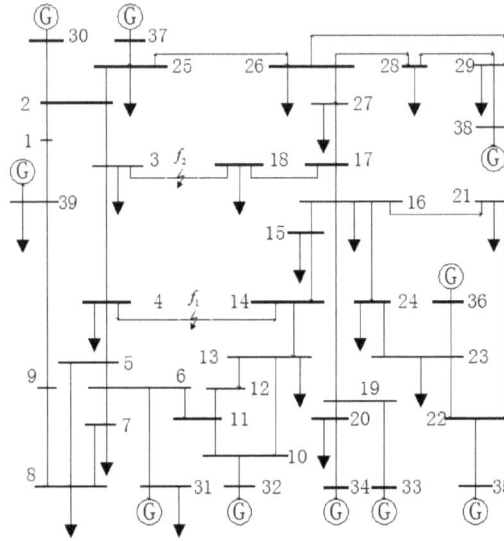

Figure 6. The diagram of IEEE 10-unit 39-bus power system.

A three-phase grounding fault occurs on the transmission line between bus 4 and bus 14 and the fault duration is 0.4 s. The power angle curves are shown in Figure 7.

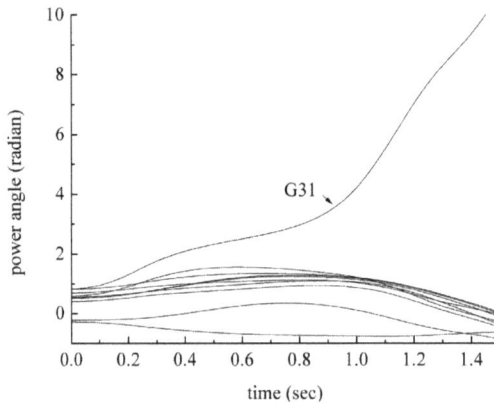

Figure 7. Multi-machine swing curves after fault.

The results of the closed loop control system are shown in Table 6.

Table 6. Results of the control system.

Fault	A Three-Phase Grounding Fault Occurs on the Line between Bus 4 And Bus 14	
Fault duration	0.23 s	
Detection time	0.41 s	
Detection angle	126.5904°	
Control objective	scheme one	scheme two (145°)
Control law (MW)	G31 (439)	G31 (521)
FEP after control	148.0°	144.8°

The power angle curves after scheme one are shown in Figure 8. It can be seen that both scheme one and scheme two quickly dampen the system oscillation and keep the system stable. The cost of scheme one is 439 MW, and the FEP after control is 148.0°. The cost of scheme two is 521 MW, and the FEP is 144.8° within the limit.

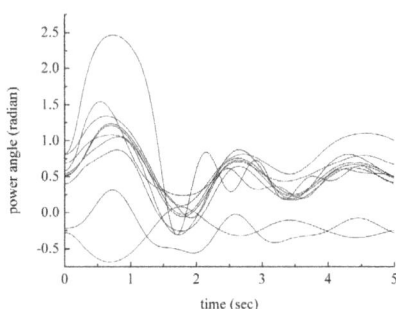

Figure 8. Power angle curves of scheme one.

5. Conclusions

In conclusion, the detection method and the control method deduced in this paper just use real-time information rather than the information of system models, parameters and disturbances to deal with the transient instability problem. As a result, the control system could be applicable in real time. The detection technique can identify instability accurately and quickly. Based on the tangent slope of the state plane trajectory, a new control method proposed in this paper forms a transient stability control system with the detection method. Two control objectives which are successfully tested in the simulations are provided based on the method of acquiring FEP in this paper. The simulation results in SMIB system and IEEE 39-bus system indicate the effectiveness of the proposed transient stability control system.

Author Contributions: Zhiguo Hao and Baohui Zhang checked and discussed the simulation results. Huaiyuan Wang confirmed the series of simulation parameters and arranged and organized the entire simulation process. Baohui Zhang and Zhiguo Hao made many useful comments and simulation suggestions. In addition, all authors reviewed the manuscript.

Conflicts of Interest: The authors declare no conflict of interest.

References

1. Anjan, B. Application of direct method to transient stability analysis of power system. *IEEE Trans. Power Appar. Syst.* **1984**, *103*, 1629–1635.
2. Ota, H.; Kitayama, Y.; Ito, H.; Fukushima, N.; Omata, K.; Morita, K.; Kokai, Y. Development of transient stability control system (TSC system) based on on-line stability calculation. *IEEE Trans. Power Syst.* **1996**, *11*, 1463–1472. [CrossRef]

3. Mahmud, M.A.; Pota, H.R.; Aldeen, M.; Hossain, M. Partial feedback linearizing excitation controller for multimachine power systems to improve transient stability. *IEEE Trans. Power Syst.* **2014**, *29*, 561–571. [CrossRef]
4. Beerten, J.; Cole, S.; Belmans, R. Modeling of multi-terminal VSC HVDC systems with distributed DC voltage control. *IEEE Trans. Power Syst.* **2014**, *29*, 34–42. [CrossRef]
5. Wang, L.; Morison, K. Implementation of online security assessment. *IEEE Power Energy Mag.* **2006**, *4*, 46–59. [CrossRef]
6. Madani, V.; Novosel, D.; Horowitz, S.; Adamiak, M.; Amantegui, J.; Karlsson, D.; Imai, S.; Apostolov, A. IEEE PSRC report on global industry experiences with system integrity protection schemes(SIPS). *IEEE Trans. Power Deliv.* **2010**, *25*, 2143–2155. [CrossRef]
7. Kim, S.; Overbye, T.J. Optimal subinterval selection approach for power system transient stability simulation. *Energies* **2015**, *8*, 11871–11882. [CrossRef]
8. Vega, R.; Glavic, M.; Ernst, D. Transient stability emergency control combining open-loop and closed-loop techniques. In Proceeding of the IEEE Power Engineering Society General Meeting, Toronto, ON, Canada, 13–17 July 2003; Volume 4, pp. 13–17.
9. Pai, A. *Energy Function Analysis for Power System Stability*; Springer Science & Business Media: Heidelberg, Germany, 2012.
10. Ruiz-Vega, D.; Pavella, M. A comprehensive approach to transient stability control. I. Near optimal preventive control. *IEEE Trans. Power Syst.* **2003**, *18*, 1446–1453. [CrossRef]
11. Xue, Y.; Wehenkel, L.; Belhomme, R.; Rousseaux, P.; Pavella, M.; Euxibie, E.; Heilbronn, B.; Lesigne, J.F. Extended equal area criterion revisited (EHV power systems). *IEEE Trans. Power Syst.* **1992**, *7*, 1012–1022. [CrossRef]
12. Pavella, M.; Ernst, D.; Ruiz-Vega, D. *Transient Stability of Power Systems: A Unified Approach to Assessment and Control*; Kluwer: Boston, MA, USA, 2000.
13. Jin, T.; Chu, F.; Ling, C.; Nzongo, D. A robust WLS power system state estimation method integrating a wide-area measurement system and SCADA technology. *Energies* **2015**, *8*, 2769–2787. [CrossRef]
14. Wang, L.; Girgis, A.A. A new method for power system transient instability detection. *IEEE Trans. Power Deliv.* **1997**, *12*, 1082–1088. [CrossRef]
15. Xie, H.; Zhang, B.; Yu, G.; Li, Y.; Li, P.; Zhou, D.; Yao, F. Power systems transient stability detection theory based on characteristic concave or convex of trajectory. *Proc. CSEE* **2006**, *26*, 38–42.
16. Xie, H.; Zhang, B. Power system transient stability detection based on characteristic concave or convex of trajectory. In Proceedings of the IEEE Transmission and Distribution Conference & Exhibition: Asia and Pacific, Dalian, China, 15–18 August 2005.
17. Zhang, B.; Yang, S.; Wang, H. Closed-loop control of power system transient stability (1): Transient instability detection principle of simple power system. *Electr. Power Autom. Equip.* **2014**, *8*, 1–6.
18. Zhang, B.; Yang, S.; Wang, H. Closed-loop control of power system transient stability (3): Initiation criterion of transient stability closed-loop control based on predicted response of power system. *Electr. Power Autom. Equip.* **2014**, *10*, 1–6.
19. Shao, H.; Lin, Z.; Norris, S.; Bialek, J. Application of emergency-single machine equivalent method for cascading outages. In Proceedings of the Power Systems Computation Conference (PSCC), Wrocław, Poland, 18–22 August 2014; pp. 1–6.
20. Jiang, Q.; Wang, Y.; Geng, G. A parallel reduced-space interior point method with orthogonal collocation for first-swing stability constrained emergency control. *IEEE Trans. Power Syst.* **2014**, *29*, 84–92. [CrossRef]
21. Fouad, A.A.; Ghafurian, A.; Nodehi, K.; Mansour, Y. Calculation of generation-shedding requirements of the BC hydro system using transient energy functions. *IEEE Trans. Power Syst.* **1986**, *1*, 17–23. [CrossRef]
22. Xue, Y.; van Custem, T.; Pavella, M. Extended equal area criterion justifications, generalizations, applications. *IEEE Trans. Power Syst.* **1989**, *4*, 44–52. [CrossRef]

![energies logo] *energies*

MDPI

Article

General Dynamic Equivalent Modeling of Microgrid Based on Physical Background

Changchun Cai [1,2,3,*], Bing Jiang [1,2] and Lihua Deng [1,2]

[1] Jiangsu Key Laboratory of Power Transmission & Distribution Equipment Technology, Hohai University, Changzhou 213022, Jiangsu, China; Jiangb@hhuc.edu.cn (B.J.); Denglh@hhuc.edu.cn (L.D.)
[2] College of IOT Engineering, Hohai University, Changzhou 213022, Jiangsu, China
[3] Changzhou Key Laboratory of Photovoltaic System Integration and Production Equipment, Hohai University, Changzhou 213022, Jiangsu, China
* Correspondence: caicc@hhu.edu.cn; Tel.: +86-519-8519-1711

Academic Editor: Ying-Yi Hong
Received: 17 September 2015 ; Accepted: 9 November 2015 ; Published: 17 November 2015

Abstract: Microgrid is a new power system concept consisting of small-scale distributed energy resources; storage devices and loads. It is necessary to employ a simplified model of microgrid in the simulation of a distribution network integrating large-scale microgrids. Based on the detailed model of the components, an equivalent model of microgrid is proposed in this paper. The equivalent model comprises two parts: namely, equivalent machine component and equivalent static component. Equivalent machine component describes the dynamics of synchronous generator, asynchronous wind turbine and induction motor, equivalent static component describes the dynamics of photovoltaic, storage and static load. The trajectory sensitivities of the equivalent model parameters with respect to the output variables are analyzed. The key parameters that play important roles in the dynamics of the output variables of the equivalent model are identified and included in further parameter estimation. Particle Swarm Optimization (PSO) is improved for the parameter estimation of the equivalent model. Simulations are performed in different microgrid operation conditions to evaluate the effectiveness of the equivalent model of microgrid.

Keywords: microgrid; equivalent modeling; trajectory sensitivity; parameter estimation

1. Introduction

In order to increase the reliability of the electricity supply to the sensitive load, microgrid concept was proposed and developed in recent years [1,2]. Microgrid normally consists of distributed energy resources (DER), energy storage devices and loads. Most of the time, microgrid can be regarded as a self-controlled system that separates and isolates itself from the utility when a severe disturbance nearby occurs, and reconnects itself to the grid automatically when the disturbance is cleared. Obviously, the operational characteristics of the microgrid are quite different from those of the traditional electrical equivalent. Hence, the increasing penetration of the microgrid will have significant impact on the dynamic performances of the distribution network.

To investigate the interactive effect between microgrid and distribution network, a suitable microgrid model is needed. The detailed model of the microgrid comprises dozens of differential equations of all dynamic and static components [2–4]. In a simple distribution network with a small number of microgrids, the detailed model of the microgrid is suitable for the dynamic simulation of the distribution network when the microgrid under connected operation mode [5]. However, with increasing penetration of the microgrid into the distribution network, the simulation of a large-scale distribution network becomes very difficult. Under this condition, if an equivalent model of the microgrid is employed, the simulation of the distribution network can be simplified.

Microgrid should operate under connected mode most time to take full advantages of distributed generator. Compared with the distribution network, microgrid can be seen as a controlled load or a controlled electric source under this operation mode. In connected mode, the interactions between loads and distributed generations can be ignored, and the microgrid synthesized dynamic characteristics will be considered in the simulation of distribution network.

Distributed generation is the basis of microgrid; if the equivalent model of the distributed generation is utilized, dynamic simulation of the microgrid can be simplified. An equivalent model compared with the detailed model of the photovoltaic was discussed in [6], and the equivalent model could well describe the dynamic characteristics of the photovoltaic under different faults in power grid. The authors of [7,8] proposed a photovoltaic source dynamic model, the parameters of which were identified based on a least-squares regression-based data processing algorithm. The singular perturbations theory was applied to reduce the model order of the wind farm in [9], and the dynamics of the reduced-order model matched well with those of the detailed model under different operational conditions. Aggregate modeling and detailed modeling for the transient interaction between a large wind farm and a power system were discussed in [10], and the aggregate modeling decreased the simulation time without significantly compromising the accuracy in different conditions. In [11,12], an equivalent method was proposed for integrating wind power generation system in power flow and transient simulation, the unit plants equivalent method and the multiply equivalent method were used for the power flow calculation and transient dynamics simulation, respectively. A probabilistic clustering concept for aggregate modeling of wind farms was proposed in [13], the support vector clustering technique was used to cluster wind turbines based on wind farm layout and incoming wind. Due to the short distances of the electric circuits in the microgrid, there is a strong electromagnetic coupling between the electrical components. These characteristics increase the difficulty in the microgrid analysis. A generalized homology equivalence theory based on differential geometry was used for the microgrid equivalent modeling in [14], and the mathematical analysis of its reduced-order nature was discussed.

Parameter estimation method is a very difficult and challenging task in system modeling. Recently, global optimization techniques such as genetic algorithm [15], evolutionary algorithm [16] and differential evolution [17] have been proposed to solve the parameter estimation problems. Though the genetic algorithm (GA) was employed successfully to solve complex non-linear optimization problems, some deficiencies of GA have been identified in recent research [18]. This degradation in efficiency is apparent when the parameters being optimized are highly correlated and the premature convergence of the GA degrades its performance in terms of reducing the search capability.

Particle swarm optimization (PSO) is an evolutionary computation technique in nature motivated by the simulation of social behaviors. In searching the optimal solution of a problem, information of the best position of each individual particle and the best position among the whole swarm are used to direct the searching. Due to the simple concept, easy implementation and quick convergence, nowadays PSO has gained much attention and wide applications in different fields. Authors of papers [19–24] showed that PSO is a feasible approach to parameter estimation of nonlinear systems. In [19], PSO was applied in harmonic estimation. A modified PSO was utilized in the maximum power point tracking for the photovoltaic system in [20]. In the field of parameter estimation, PSO-based parameter estimation technique of proton exchange membrane fuel cell models was proposed in [21], and PSO with quantum was introduced successfully in synchronous generator offline and online parameters estimation problem. Parameter estimation of an induction machine using PSO was shown in [22], and the dynamic PSO and chaos PSO were better than the standard PSO. PSO was used for jointly estimating both the parameters and states of the lateral flow immunoassay model in [23]. Diffusion particle swarm optimization was proposed to optimize the maximum likelihood function in [24], and the PSO technique has been shown to provide a good solution to bearing estimation as it alleviates the effects of multi-modality.

Research has been carried out in the fields of detailed modeling of microgrid and equivalent modeling of the distributed generation. With the increasing penetration of microgrids, the interaction between the microgrid and the distribution network should not be ignored in the power system real-time simulation. However, this will increase the complexity of the simulation with the detailed model of the microgrid components. Hence, a simplified equivalent model of the microgrid is extremely urgent for the simulation analysis of the distribution network. Based on the component detailed models and synthetically dynamic characteristics of the microgrid, an equivalent model of microgrid is proposed in this paper. The proposed equivalent model contains two parts: equivalent static component and equivalent machine component. In order to increase the accuracy of the parameters estimation, trajectory sensitivity is used to identify the key parameters for the further steps of parameter estimation. Particle Swarm Optimization (PSO) is improved and employed to estimate the parameters of the equivalent model. The presented equivalent model and modeling method are shown to be effective by the simulation study on a microgrid connected into distribution network.

2. Microgrid Equivalent Model

A microgrid is made up of a large number of distribution generations, electrical loads and storage devices. Typically, there are two types of components: static components and rotating machines [1]. Static components contain photovoltaic (PV) and static loads. It is common that PV connects to the microgrid through power electronics equipment. Maximum Point Power Tracking (MPPT) and constant power control strategy are applied to the power electronics equipment when the microgrid is connected in the grid-connected operation mode [25]. The output power of these distribution generations is controllable and the dynamic characteristics of them are similar with the static load. In principle, the static load is represented by an exponential of the voltage and frequency [26]. Hence, the output power of the static components can be described by an exponential of the voltage and frequency.

Rotating machine components contain induction motor load, asynchronous induction wind generator and synchronous generator. The structure and the mathematical equations of the asynchronous induction wind generator are similar with those of the induction motor load [27]. The synchronous machine generator and the asynchronous wind generator have similar dynamic characteristics during faults, and the only difference between them is the modeling reference frames. The synchronous machine rotor angular velocity is constant and the velocity voltage is zero in steady-state conditions [28]. However, the rotor angular velocity of synchronous will deviate slightly from the synchronous velocity during a fault since the synchronous machine capability is small in most microgrids. The rotor angular velocity is not equal to the system synchronous velocity and its electrical structure is similar with that of the asynchronous induction wind generator, so the synchronous machine generator can be regarded as an asynchronous machine generator during a fault. Furthermore, the synchronous machine generator, the asynchronous induction wind generator and the induction motor load can be described with a unified mathematical model in the transient dynamic analysis.

As shown in Figure 1, the equivalent model of the microgrid is comprised of an equivalent static component and an equivalent machine component. The equivalent static component is parallel to the equivalent machine component, and they are connected to the distribution network through the Point of Coupling Common (PCC).

2.1. Equivalent Machine Component

The stator and rotor circuits of the equivalent machine component are shown in Figure 2.

Rotor angular velocity ω_r is different from the stator angular velocity ω_s. Applying $dq0$ transformation [29], stator voltage equations can be written as

$$u_{ds} = p\psi_{ds} - \omega_s\psi_{qs} - r_s i_{ds}$$
$$u_{qs} = p\psi_{qs} + \omega_s\psi_{ds} - r_s i_{qs}$$

(1)

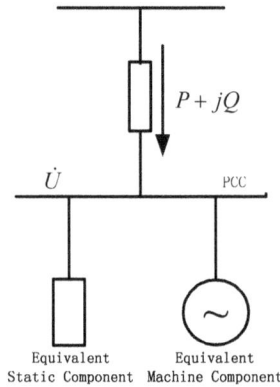

Figure 1. The equivalent model of microgrid.

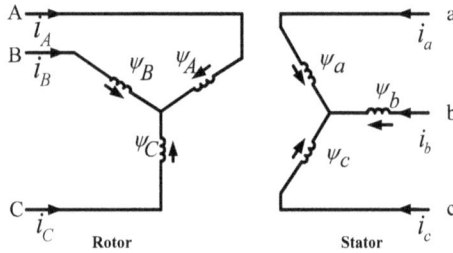

Figure 2. Equivalent circuit of the equivalent machine component.

Rotor voltage equations can be written as

$$u_{dr} = p\psi_{dr} - \omega_s s\psi_{qr} + r_{dr}i_{dr}$$
$$u_{qr} = p\psi_{qr} + \omega_s s\psi_{dr} + r_{qr}i_{qr}$$

(2)

where u_{ds} and u_{qs} are the stator voltages; ψ_{ds} and ψ_{qs} are the stator flux linkages; r_s is stator resistance; i_{ds} and i_{qs} are stator currents; u_{dr} and u_{qr} are rotor voltages; ψ_{dr} and ψ_{qr} are rotor flux linkages; r_{dr} and r_{qr} are rotor resistance; i_{dr} and i_{qr} are rotor currents; $s = \frac{\omega_s - \omega_r}{\omega_s}$ is the rotor slip; and $p = \frac{d}{dt}$ is the per time derivative.

The stator flux linkage is

$$\psi_{ds} = L_{ds}i_{ds} + L_{ad}i_{dr}$$
$$\psi_{qs} = L_{qs}i_{qs} + L_{aq}i_{qr}$$

(3)

The rotor flux linkage is

$$\psi_{dr} = L_{ad}i_{ds} + L_{dr}i_{dr}$$
$$\psi_{qr} = L_{aq}i_{qs} + L_{qr}i_{qr}$$

(4)

where L_{ds} and L_{qs} are the stator inductances; L_{dr} and L_{qr} are the rotor inductances; and L_{ad} and L_{aq} are mutual inductances.

Using the definitions below:

$$L'_d = L_{ds} - \frac{L^2_{ad}}{L_{dr}}, L'_q = L_{qs} - \frac{L^2_{aq}}{L_{qr}}$$

$$T'_{d0} = \frac{L_{dr}}{r_{dr}}, T'_{q0} = \frac{L_{qr}}{r_{qr}}$$

$$E'_d = -\frac{L_{aq}}{L_{dr}}\psi_{dr}, E'_q = \frac{L_{ad}}{L_{qr}}\psi_{qr}$$

$$E_{dr} = \frac{L_{ad}}{L_{dr}}u_{dr}, E_{qr} = -\frac{L_{aq}}{L_{qr}}u_{qr}$$

(5)

rotor voltage equations may be rewritten as follows:

$$\frac{dE'_d}{dt} = E_{dr} + w_s s\frac{L_{dr}}{L_{ad}}\frac{L_{aq}}{L_{qr}}E'_q - \frac{1}{T'_{q0}}(E'_d + (L_{qs} - L'_q)i_{qs})$$

$$\frac{dE'_q}{dt} = E_{qr} - w_s s\frac{L_{qr}}{L_{aq}}\frac{L_{ad}}{L_{dr}}E'_d - \frac{1}{T'_{d0}}(E'_q - (L_{ds} - L'_d)i_{ds})$$

(6)

When representing power system stability studies, $p\psi_{ds}$ and $p\psi_{qs}$ are neglected in the stator voltage relations. Their neglect corresponds to ignoring the dc component in the stator transient currents, permitting representation of only fundamental frequency components [28]. With the stator transients neglected, stator voltage equations may be rewritten as:

$$u_{ds} = -w_s[(L_{qs} - \frac{L^2_{aq}}{L_{qr}})i_{qs} + \frac{L_{aq}}{L_{qr}}\psi_{qr}] + r_s i_{ds}$$

$$u_{qs} = w_s[(L_{ds} - \frac{L^2_{ad}}{L_{dr}})i_{ds} + \frac{L_{ad}}{L_{dr}}\psi_{dr}] + r_s i_{qs}$$

(7)

From Equations (5) and (7), we have

$$u_{ds} = r_s i_{ds} - w_s L'_q i_{qs} + w_s E'_d$$

$$u_{qs} = r_s i_{qs} + w_s L'_d i_{ds} + w_s E'_q$$

(8)

The rotor acceleration equation, with time expressed in seconds, is

$$\frac{dw_r}{dt} = \frac{1}{T_j}(T_M - T_e)$$

(9)

where T_e is the electromagnetic torque, T_M is the mechanical torque, and T_j is the inertia constant of the rotor. Eliminating the rotor currents by expressing them in terms of the stator currents and rotor flux linkages, we find that the per unit electromagnetic torque is

$$T_e = \psi_{qr}i_{dr} - \psi_{dr}i_{qr} = -E'_q i_{qs} - E'_d i_{ds} - (w_s L'_d - w_s L'_q)i_{ds}i_{qs}$$

(10)

The system frequency of the microgrid is constant when the microgrid operation in connected mode. Thus, rotor acceleration equation with $f = w_s = 1$ pu can be written as:

$$\frac{dw_r}{dt} = \frac{1}{T_j}[T_M - (-E'_q i_{qs} - E'_d i_{ds} - (L'_d - L'_q)i_{ds}i_{qs})]$$

(11)

The output active power and reactive power of the equivalent machine component may be written as:

$$P_m = -u_{ds}i_{ds} - u_{qs}i_{qs}$$

$$Q_m = u_{qs}i_{ds} - u_{ds}i_{qs}$$

(12)

2.2. Equivalent Static Component

Static load model represents the load characteristics as an algebraic function of the bus voltage magnitude and frequency [26]. Equivalent static component, including static load and distribution generation such as PV in microgrid, is described using algebraic equations. The active and reactive power of the equivalent static component model are related to the system voltage and frequency in the following form:

$$P_s = P_{s0}(U/U_0)^{p_u}(f/f_0)^{p_f}$$
$$Q_s = Q_{s0}(U/U_0)^{q_u}(f/f_0)^{q_f} \tag{13}$$

where P_s and Q_s are the active and reactive power of the equivalent static component when the voltage magnitude is U and frequency is f, respectively. The subscript 0 identifies the values of the respective variables at the initial operating condition of PCC. The parameters of this model are the exponents p_u, q_u, p_f and q_f, where p_u is the coefficient of the active power and voltage, p_f is the coefficient of active power and frequency, q_u is the coefficient of reactive power and voltage, and q_f is the coefficient of reactive power and frequency.

The system frequency of the microgrid is constant when the microgrid operates in grid-connected mode. Thus, with $f = 1$ pu, the model of the equivalent static component can be written as:

$$P_s = P_{s0}(U/U_0)^{p_u}$$
$$Q_s = Q_{s0}(U/U_0)^{q_u} \tag{14}$$

2.3. Parameters of the Equivalent Model

From model Equations (6), (8), (11), (12) and (14), it can be seen that the equivalent model parameters include r_s, L_{ds}, L_{qs}, L'_d, L'_q, L_{dr}, L_{qr}, L_{ad}, L_{aq}, T'_{d0}, T'_{q0}, T_j, p_u, and q_u. In order to describe the physical characteristics of the equivalent model, the equivalent model parameters are initialized with corresponding fundamental parameters and will be further estimated in the microgrid modeling. Based on the definitions of the parameters in Equation (5), the fundamental parameters of the equivalent machine component are r_s, L_{sl}, L_{ad}, L_{aq}, r_{dr}, L_{drl}, r_{qr}, L_{qrl}, T_j, p_u, and q_u, where L_{sl} is the stator leakage inductance; L_{ad} and L_{aq} are mutual inductances of d and q axis; r_{dr} and r_{qr} are rotor resistances of d and q axis; and L_{drl} and L_{qrl} are rotor leakage inductances of d and q axis.

There are other two important parameters, namely s_0 and K_{mp}. Where s_0 is the initial slip of the equivalent machine component, and s_0 presents the type of equivalent machine. If $s_0 > 0$, the equivalent machine component absorbs power from the distribution network, and has the characteristics of induction motor load. Oppositely, $s_0 < 0$ means that the equivalent machine component injects power into the distribution network, and has the characteristics of asynchronous generator. K_{mp} is the fraction of the equivalent machine component active power with respect to the total initial active power P. The active power flow between the microgrid and the distribution network is bidirectional. Thus, $P > 0$ indicates that the microgrid absorbs power from the distribution network, and $P < 0$ indicates that the microgrid injects power into the distribution network. The similar definition is also applied to reactive power Q.

As a result, the 13 parameters, namely K_{mp}, s_0, r_s, L_{sl}, L_{ad}, L_{aq}, r_{dr}, L_{drl}, r_{qr}, L_{qrl}, T_j, p_u, and q_u, in the equivalent microgrid model need to be estimated in the microgrid equivalent dynamic modeling.

3. Equivalent Model Parameter Sensitivity Analysis

The number of the parameters to be estimated has significant impact on the accuracy of the parameters estimation. Parameter sensitivity analysis is an efficient method to determine the key parameters of the equivalent model. Parameter trajectory sensitivity is defined as:

$$\frac{\partial y(\theta, k)}{\partial \theta_j} = \lim_{\Delta\theta_j \to 0} \frac{y(\theta_1, \cdots, \theta_j + \Delta\theta_j, \cdots, \theta_m, k) - y(\theta_1, \cdots, \theta_j, \cdots, \theta_m, k)}{\Delta\theta_j} \tag{15}$$

where $y(\theta, k)$ is the time domain trajectory of the output variable, θ is the vector of the parameters of the equivalent model, θ_j is the jth parameter of the equivalent model, m is the number of the parameters, and k is the sampling sequence.

In order to improve the accuracy of parameter trajectory sensitivity, median method is used to calculate the trajectory sensitivity when $\Delta\theta_j$ is small enough, which is shown as:

$$\frac{\partial\,[y(\theta,k)/y_0]}{\partial\,[\theta_j/\theta_{j0}]} = \frac{\left[y(\theta_1,\cdots,\theta_j+\Delta\theta_j,\cdots,\theta_m,k) - y(\theta_1,\cdots,\theta_j-\Delta\theta_j,\cdots,\theta_m,k)\right]/y_0}{2\Delta\theta_j/\theta_{j0}} \tag{16}$$

where θ_{j0} is the initial value of the parameter θ_j, $\Delta\theta_j$ is the variation of θ_j, and y_0 is the initial value of the output variable in steady-state.

3.1. Trajectory Sensitivity

Time domain parameter trajectory sensitivity curve can describe the behaviors of the output variable. For convenience of comparison, the average sensitivity can be calculated as

$$A_j = \frac{1}{N}\sum_{k=1}^{N}\left|\frac{\partial\,[y(\theta,k)/y_0]}{\partial\,[\theta_j/\theta_{j0}]}\right| \tag{17}$$

where A_j is the average of the j-th parameter with respect to the trajectory of the output variable N is the number of sample points, y_0 is the initial value of the output variable.

Trajectory sensitivity demonstrates the impact of the variation of the parameter on that of the output variable's trajectory. If the trajectory sensitivity of a parameter is larger than that of the other parameters, the parameter plays a more important role on the dynamics of the output variable; in other words, the parameter can be estimated easily by using the dynamics of the output variable. In contrast, if the trajectory sensitivity of a parameter is very small, e.g., even close to zero, it is difficult to estimate the parameter using the dynamics of the output variable.

3.2. Trajectory Sensitivity Phase

If a couple of parameters have an unknown relationship between each other, they are dependent on each other and unidentifiable as well. However, these unidentifiable parameters can be identified using trajectory sensitivity analysis [29].

Assuming that the parameters θ_i and θ_{i+1} are coupling with each other, the output of the power system can be written as

$$y = f[\theta_1,\theta_2,\cdots,\varphi(\theta_i,\theta_{i+1}),\cdots,\theta_n] \tag{18}$$

The sensitivities of the parameters θ_i and θ_{i+1} can be analyzed as [27]

$$\begin{cases} \dfrac{\partial y}{\partial\theta_i} = \dfrac{dy}{d\varphi}\dfrac{\partial\varphi}{\partial\theta_i} \\[2mm] \dfrac{\partial y}{\partial\theta_{i+1}} = \dfrac{dy}{d\varphi}\dfrac{\partial\varphi}{\partial\theta_{i+1}} \end{cases} \tag{19}$$

Then

$$\frac{\partial y}{\partial\theta_{i+1}} = \frac{dy}{d\theta_i}\left[\frac{\partial\varphi/\partial\theta_{i+1}}{\partial\varphi/\partial\theta_i}\right] \tag{20}$$

It should be pointed out that $\partial y/\partial\theta_{i+1}$ and $\partial y/\partial\theta_i$ vary with time, while $\partial\varphi/\partial\theta_{i+1}$ and $\partial\varphi/\partial\theta_i$ are constant [29,30]. Hence, $\partial y/\partial\theta_{i+1}$ and $\partial y/\partial\theta_i$ reach zero at the same time. In other words, the trajectory sensitivities of these two couple unidentifiable parameters are either in phase or in reverse with each other. The trajectory sensitivity of these two parameters will pass zero at the same time. Oppositely, if the output variable curves of the two parameters do not pass zero at the same time approximately, they are independent and can be identified using the dynamics of output variables of the system [31].

4. Microgrid Parameter Estimation

4.1. Rotor Voltage Equivalence

Rotor voltage is an important operational variable in synchronous machine, and the parameters of the excitation system are estimated individually [32]. In the microgrid equivalent model, the excitation system is a virtual polymerization system, which maintains the terminal voltage. In [33], a synchronous machine with an excitation system is described by an asynchronous machine and a constant current load in load modeling. The dynamic of the synchronous machine is similar with the dynamic of the asynchronous machine and the constant current load. Hence, the rotor voltage of the equivalent microgrid model is replaced by a constant current load, which can be seen as a part of the equivalent static component in the modeling of microgrid.

4.2. Parameter Estimation Based on PSO

4.2.1. Review of Particle Swarm Optimization

Particle Swarm Optimization (PSO) is a heuristic optimization algorithm. It was first introduced by Kennedy and Eberhart, based on the observations of social behaviors of animals, such as bird flocking, fish schooling, and swarm theory [34]. This algorithm implements a global method that performs a search of parameters over a specified problem space. Like other evolutionary algorithms, PSO performs using a population of individuals that are updated iteratively. Swarm members communicate good position with each other and dynamically adjust their own position and velocity. Velocity adjustment is based upon the historical behaviors of the particles themselves as well as their neighbors. In each iteration, the velocity and position of each particle are updated according to the following equations.

$$v_i(k) = wv_i(k-1) + c_1 rand_1[pbest_i - x_i(k-1)] + c_2 rand_2[gbest - x_i(k-1)] \tag{21}$$

$$x_i(k) = v_i(k) + x_i(k-1) \tag{22}$$

where x_i is the position of the ith particle, v_i is the velocity of the i-th particle, w is the inertia weight which decreases linearly determined by Equation (23), $pbest_i$ is the best position of the i-th particle, $gbest$ is the global optimal position of the current swarm with the best objective value, c_1 and c_2 are acceleration constants, and $rand_1$ and $rand_2$ are two independent random numbers uniformly distributed over [0 1].

$$w = w_{min} + \frac{w_{max} - w_{min}}{k_{max}} \times k \tag{23}$$

The algorithm achieves the optimal solution by two types of search memory, "cognitive" component and "social" component, which are shown as the second and third parts of Equation (21). The cognitive component makes the particle move toward its own best positions. The social component makes the particle toward the best position found by its neighbors, which means the collaborative behavior of particles.

4.2.2. Particle Swarm Optimization (PSO) with Chaos Neighborhood Searching

As optimization problem becomes more and more complex, the random characteristics of PSO will reduce the convergence velocity. In this paper, chaotic mutation theory is used to improve the global convergence velocity of the standard PSO. Chaos is a common phenomenon in non-linear systems that include infinite unstable period motions. Chaos-based neighborhood searching method with multiple different neighborhoods is designed and incorporated to enrich the searching behaviors, so as to avoid premature convergence [35]. In addition, an effective nonlinear adaptive inertia weight is employed to further enhance the exploitation ability after the chaos-based neighborhood searching.

Chaotic mapping is a discrete-time dynamical system, in which the chaotic sequences are considered as sources of random sequences, which can avoid getting stuck in a local optimum during the search process and overcomes the premature convergence phenomenon. Logistic mapping is a common chaos mapping, which is employed in the global optimal position neighborhood searching with the following steps.

(1) Generating the initial variable u_{0j} randomly based on the global optimal position of the particles.

(2) Getting the chaotic sequences u_{1j} using the logistic mapping $u_{1j} = 4u_{0j}(1 - u_{0j})$.

(3) Generating a local neighborhood mutation variable Δx_j: $\Delta x_j = -\beta + 2\beta u_{1j}$, where β is the radius of the local neighborhood and is updated by Equation (24).

$$\beta = (x_{j\max} - x_{j\min})\cos(\frac{\pi(t-1)}{2(t_{\max}-1)}) \tag{24}$$

where, t and $t_{\max}$ are the round of the iteration and the maximal iteration times of the neighborhood searching, respectively. $x_{j\max}$ and $x_{j\min}$ are the upper and lower bounds of the variable x_j, respectively.

(4) Local neighborhood searching: A temporary global optimal position is defined as $gbest' = gbest + \Delta X$, where ΔX is the chaos mutation variables vector $[\Delta x_1 \, \Delta x_j \cdots \Delta x_N]$ and $gbest = [x_{g1} \cdots x_{gN}]$ is the current global optimal position. Comparing $gbest'$ with $gbest$, the current global optimal position is updated by the larger one.

4.2.3. Improved Particle Swarm Optimization(IPSO) Based Parameter Estimation

IPSO algorithm is employed to search the optimal parameters of the microgrid equivalent model to achieve an optimal matching between the detailed model and the equivalent model of the microgrid. The steps of the IPSO-based microgrid equivalent model parameter estimation are shown in Figure 3.

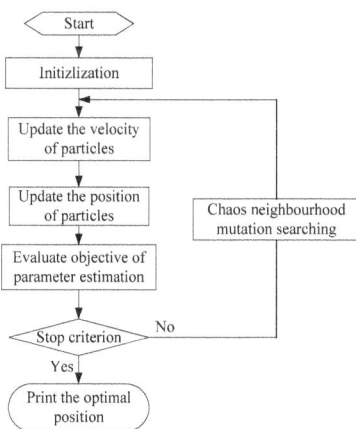

Figure 3. Chart of Improved Particle Swarm Optimization (IPSO)-based parameter estimation.

(1) *Initialization*: From Section 2, there are 13 parameters, K_{mp}, s_0, r_s, L_{sl}, L_{ad}, L_{aq}, r_{dr}, L_{drl}, r_{qr}, L_{qrl}, T_j, p_u and q_u for the optimal parameter of the microgrid model. The initial particles may be generated randomly with a specified upper and lower bounds, $X_{\max}$ and $X_{\min}$, respectively. The dimension of each particle is equal to the number of the parameters. Measured data include voltages $(u_x + ju_y)$ and powers $(P + jQ)$ of PCC.

The initial velocity of particles are generated randomly between the upper and lower bounds, $v_{i,max}$ and $v_{i,min}$. Where the upper and lower bounds is defined by

$$v_{i,max} = \frac{x_{i,max} - x_{i,min}}{N}$$

$$v_{i,min} = -v_{i,max}$$

(25)

where $x_{i,max}$ and $x_{i,min}$ are the upper and lower bounds of particles; and $v_{i,max}$ and $v_{i,min}$ are the maximum and minimum velocities of particles. N is the interval of dimension, which is normally chosen to be between 5 and 10.

(2) *Evaluation*: The Objective function is employed to evaluate each particle, and it is usually defined as the error between the measured power and the output power of the microgrid equivalent model as follows:

$$E(\theta) = \min \sum_{k=1}^{N} \{[P(k) - P_M(k, \theta)]^2 + [Q(k) - Q_M(k, \theta)]^2\}$$

(26)

where θ is the parameter vector of the equivalent model, $P(k)$ and $Q(k)$ are the measured power of sampling time k, and $P_M(k, \theta)$ and $Q_M(k, \theta)$ are the calculated power of the microgrid equivalent model. The aim of the objective function is that the output power of equivalent model can match the measured power.

(3) *Updating movement velocities of the particles:* Movement velocity updating is an important step in the process of evolution. The movement velocity of each particle is updated by Equation (21), and the velocity is limited in the upper and lower bounds as follows:

$$if\ v_{ij}(t+1) > v_{j,max}\ then\ v_{ij}(t+1) = v_{j,max}$$
$$if\ v_{ij}(t+1) < v_{j,min}\ then\ v_{ij}(t+1) = v_{j,min}$$

(27)

(4) *Updating positions of particles:* The position of each particle is updated by Equation (22). The position of each particle is limited as follows

$$if\ x_{ij}(t+1) > x_{j,max}\ then\ x_{ij}(t+1) = x_{j,max}$$
$$if\ x_{ij}(t+1) < x_{j,min}\ then\ x_{ij}(t+1) = x_{j,min}$$

(28)

(5) *Chaos neighborhood searching:* Following the steps described in the previous Section 4.2.2, the global optimal particle position is updated.

Firstly, an initial variable u_{0j} is generated randomly, and the chaotic sequences u_{1j} is generated by using the formula $u_{1j} = 4u_{0j}(1 - u_{0j})$. Secondly, a local neighborhood mutation variable Δx_j is generated by using the formula $\Delta x_j = -\beta + 2\beta u_{1j}$. Lastly, a new temporary global optimal position is generated which is the neighbor of the global optimal position. The fitness of the temporary global optimal position is compared with that of the current global optimal position, and the current global optimal position is replaced by the better one.

(6) *Termination:* The parameter estimation algorithm will be terminated if the iteration times exceeds the specified maximal iteration times or the fitness value of the global best is smaller than a given value.

5. Microgrid Equivalent Modeling and Discussion

5.1. Microgrid System

In this work, a microgrid system is built in DIgSILENT PowerFactory to test the effectiveness of the microgrid equivalent model and the feasibility of the modeling method. The structure of the microgrid is shown in Figure 4, and the detailed parameters and the operation mode can be found in [36]. The microgrid consists of wind generation, PV generation, micro-gas turbine, asynchronous induction motor and static load. The detailed parameters of microgrid are shown in Appendix A. The microgrid is radial and is connected to the distribution network through a 10 kv/20 kv transformer. The distribution network is a standard Benchmark medium voltage distribution network model in [37].

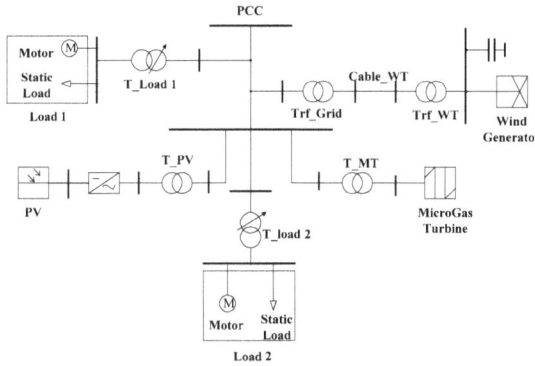

Figure 4. Microgrid system.

A single-phase short circuit is applied in the distribution network, and is cleared in 0.06 s. Using the dynamics of the exchange power between the microgrid and the distribution network, all the parameters of the equivalent model are estimated by the proposed PSO. The values of parameters used in PSO are depicted in Table 1. The algorithm has been implemented in C#.NET. The program runs on a 1.8 GHz, Intel Core2 Duo, with 2 GB RAM PC. The detail of the key parameters selected are: $w_{max} = 0.9$, $w_{min} = 0.4$, $c_1 = c_2 = 2.0$, $t_{max} = 10$, $k_{max} = 500$. The bound of estimated parameters are shown in Appendix B. The estimated parameters are listed in Table 1, and the dynamics of the output power of the detailed model and the equivalent model are illustrated in Figure 5. It can be seen that the dynamics of the output power of the equivalent model is approximate to that of the detailed model.

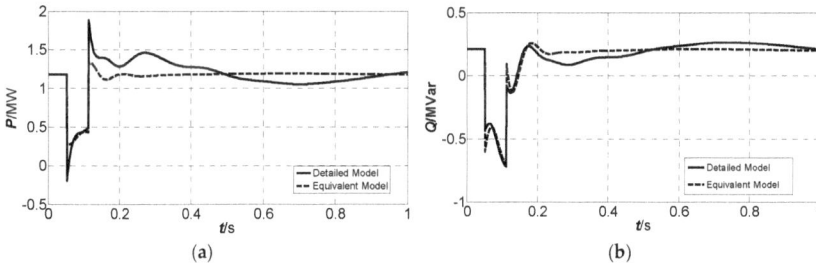

Figure 5. (**a**) The active power dynamics of the detailed model and the equivalent model of the Microgrid; (**b**) The reactive power dynamics of the detailed model and the equivalent model of the Microgrid.

Table 1. A group of feasible parameters for the equivalent model.

Parameter	K_{mp}	s_0	r_s	L_{sl}	L_{ad}	L_{ap}	r_{dr}
Value	−8.70	−0.206	0.031	0.340	0.848	3.14	0.00023
Parameter	L_{drl}	r_{qr}	L_{qrl}	T_j	p_u	q_u	Error
Value	2.18	0.171	0.756	9.66	7.06	0.731	2.401

The evolution of function objective is presented along the iterations in order to observe the behavior of improved PSO, and to verify if the PSO works as expected. Figure 6 shows the evolution of function objective along with the iterations. It can be seen from the graph that the objective function has converged to a constant value of 2.401 after 57 iterations. It is quite apparent that the improved PSO has the faster convergence than the standard PSO algorithm. Moreover, the proposed IPSO and standard PSO achieved in the same machine converged their corresponding optimal solutions in 25.3 s and 37.5 s, respectively.

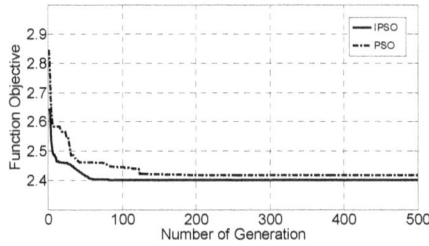

Figure 6. Convergence of the improved particle swarm optimization (IPSO) and PSO Algorithms.

5.2. Parameter Sensitivity Analysis and Error Analysis

The number of parameters to be estimated has a significant impact on the accuracy of the parameter estimation. If only the key parameters are estimated, the convergence of the PSO can be improved and the probability finding the true value of the parameters will be higher.

The sensitivities of the equivalent microgrid model parameters, $\theta = [K_{mp}, s_0, r_s, L_{sl}, L_{ad}, L_{aq}, r_{dr}, L_{drl}, r_{qr}, L_{qrl}, T_j, p_u, q_u]$, with respect to the trajectory of the apparent power S, are calculated. S is defined as:

$$S = \sqrt{P_M^2 + Q_M^2} \tag{29}$$

where P_M and Q_M are the calculated active and reactive power based on the dynamic equivalent model of equal Equation (26).

The average trajectory sensitivities of the parameters are listed in Table 2, and the dynamics of the trajectory sensitivities are shown in Figure 7. It can be seen in Table 2 that the parameters $K_{mp}, s_0, L_{ad}, L_{aq}, r_{dr}, L_{drl}, L_{qrl}$ and p_u have more significant impacts on the trajectory than the other parameters. Figure 7 shows that the phases of the sensitivity of K_{mp}, is almost reverse to that of r_{dr}, which means that K_{mp} and r_{dr} are related with each other and are unidentifiable by using the dynamics of S. As a result, only the seven parameters, *i.e.*, $K_{mp}, s_0, L_{ad}, L_{aq}, L_{drl}, L_{qrl}$ and p_u, are included in the further steps of the parameter estimation, while the other parameters are set as default values.

Table 2. Parameter sensitivity absolute value.

Parameter	K_{mp}	s_0	r_s	L_{sl}	L_{ad}	L_{ap}	r_{dr}
Sensitivity	0.0160	0.0352	0.0016	0.0040	0.1367	0.0377	0.0149
Parameter	L_{drl}	r_{qr}	L_{qrl}	T_j	p_u	q_u	
Sensitivity	0.1543	0.00043	0.0583	0.0015	0.0120	0.0011	

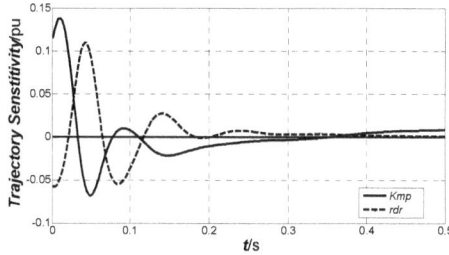

Figure 7. The trajectory sensitivity of parameter K_{mp} and r_{dr}.

5.3. Key Parameters Estimation and Error Analysis

By using the dynamics of the output power of the detailed model of microgrid, the key parameters are estimated and listed in Table 3. The dynamics of the output active power and reactive power are also illustrated in Figure 8a,b, respectively. As can be seen in Figure 8, the amplified windows show the differences of the two estimated method, and it is clear that the error between the dynamics of the output power of the equivalent model and the detailed model is smaller than that when all the parameters are estimated simultaneously, as shown in Table 1.

The simulation results reveal that the proposed dynamic equivalent model is accurate enough to describe the dynamic characteristics of the microgrid detailed model quickly during the simulation period. The equivalent model can track the fluctuation of active power and reactive power quickly and flexibly, especially in the moment of fault cutting. The estimated results of parameters are suitable according to the electrical parameter standards.

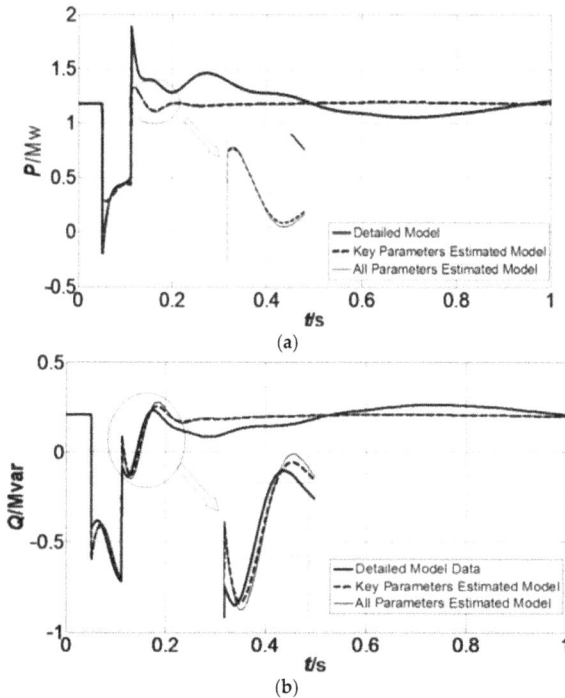

Figure 8. (a) The active power dynamics of microgrid; (b) The reactive power dynamics of microgrid.

247

Table 3. Parameter estimated value.

Parameter	K_{mp}	s_0	L_{ad}	L_{ap}	L_{drl}	L_{qrl}	p_u	Error
Value	−8.637	−0.2161	0.8042	3.417	2.120	0.7008	7.039	2.389

Because the proposed model is an equivalent model in electromechanical transient, the stator transient is neglected during the process of equivalent modeling, which will bring some error both in parameters and dynamic of the microgrid. In Figure 8a, it can be seen that at the time of the single-phase short fault occurs and cleared, the error between equivalent model and the detailed model is larger than other time relatively. At this moment, the active power of equivalent model cannot track the dynamic of microgrid completely. The components of microgrid are various and the dynamic of the output power of microgrid contains different components. In order to improve the representational capacity of the equivalent model, some elements are ignored, such as the stator transient of wind turbine and micro-gas turbine, the transient of PV inverter.

5.4. Comparing to Other Models

The proposed microgrid equivalent model is also compared with the equivalent machine component model and black-box model. The equivalent microgrid model contains two parts: equivalent machine component and equivalent static component. The equivalent machine component is used to describe the rotor characteristics of electrical equipment, and the equivalent static component is used to describe the characteristics of steady state of microgrid.

The equivalent machine model is compared with the comprehensive model in this paper. This means that the values of the parameters are $K_{mp} = 1$, $p_u = 0$ and $q_u = 0$. Figure 9 shows the dynamics between different models of the microgrid. It can be seen that the error between the proposed equivalent microgrid model and detailed model is smaller than that of the equivalent machine model and detailed model. As can be seen in Figure 9b, the output reactive power of the proposed equivalent microgrid model can track the detailed model more accurately than that of the equivalent machine component model during the fault. It means that the equivalent static component is important in the proposed equivalent microgrid model. The equivalent static component can describe the dynamic characteristics of static component in the microgrid, such as: photovoltaic system, static load, and other static equipment.

The proposed microgrid equivalent model is based on the physics characteristics of the electrical components in microgrid, and can describe the electrical physical characteristics of microgrid components. Non-mechanism model is also used in the dynamic equivalent of power system. In the equivalent modeling of microgrid, a dynamic equivalent black-box model based on prony analysis is presented in [38], the proposed equivalent model is compared with the black-box equivalent model. Figure 10 shows the comparison the proposed model and the black-box model; it can be seen that the two models are in good agreement with the detailed model, and the proposed model is more accurate.

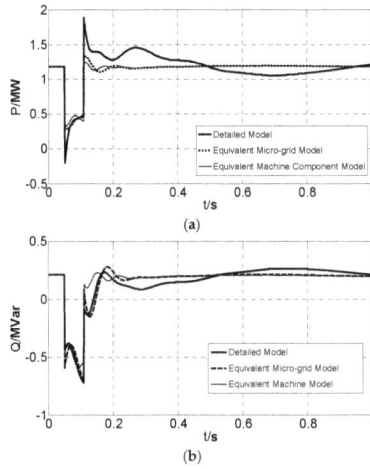

Figure 9. (**a**) Dynamics of Active Power of different models; (**b**) Dynamics of Reactive Power of different models.

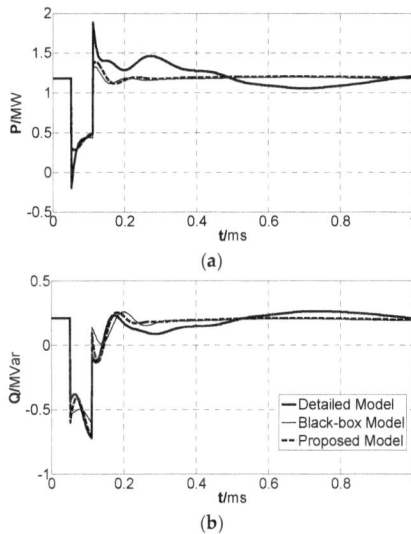

Figure 10. (**a**) Dynamics of Active Power of different models; (**b**) Dynamics of Reactive Power of different models.

5.5. Parameters Estimation under Different Operational Conditions

The operational condition of the microgrid changes with the fluctuation of the distributed generation output power and the variation of loads, which leads to the power exchange between the microgrid and the distribution network changes accordingly. Therefore, the parameters of the microgrid equivalent model should be adjusted to adapt the changes of the microgrid operational condition. Three different typical operational conditions are studied in the paper, the detailed operational parameters are shown in Table 4 and the parameters estimated results are shown in Table 5.

Table 4. Describe of operational conditions.

Operational Condition	Description
Condition A	The output power of gas turbine reduces 40 percent with the output active power is 3 MW.
Condition B	The output power of gas turbine reduces 40 percent with the output active power is 3 MW, 40 percent of Static Load 2, which is 2 MW active power and 0.4 Mvar reactive power, is removed.
Condition C	The output power of gas turbine reduces 40 percent with the active power is 3 MW. The output active power of Wind Generation reduces a half, which become 0.75 MW. PV is removed. 40 percent of Static Load 1, which is 0.4 MW and 0.2 Mvar, is removed.

As can be seen in Table 5, the equivalent machine component outputs active power as a generator with $K_{mp} < 0$ and $s_0 < 0$, while the microgrid absorbs power from the distribution network in condition A. This means that the consume power of equivalent static component comes from distribution network and equivalent machine component. In conditions B and C, the equivalent machine component absorbs power as an asynchronous induction motor with $K_{mp} > 0$ and $s_0 > 0$, while the microgrid absorbs power from the distribution network.

Table 5. Parameter value of different operational conditions.

Parameter	K_{mp}	s_0	r_s	L_{sl}	L_{ad}	L_{aq}	r_{dr}	L_{drl}	r_{qr}	L_{qrl}	T_j	p_u	q_u
Condition A	−4.69	−0.17	0.0590	0.36	1.94	2.50	0.0003	3.04	0.2560	2.24	5.62	4.87	0.94
Condition B	1.64	0.07	0.0610	0.10	4.05	7.54	0.1186	1.49	0.0024	4.00	5.18	8.95	0.61
Condition C	0.99	0.12	0.0492	0.31	3.70	4.76	0.1450	0.65	0.0057	4.00	6.55	3.58	0.70

In condition A, with the power reduction of gas turbine, microgrid may absorb more power from distribution network. Figure 11 shows microgrid response to the single-phase short fault in distribution network. The exchange power between microgrid and distribution network fluctuates violently, but the results show the great agreement between proposed equivalent model and the detailed model, and the reactive power of equivalent model can track the detailed model with a small time delay.

Figures 12 and 13 show the response of dynamic equivalent model during the fault of condition B and condition C. The microgrid construction of condition B and C is similar, and it can be seen that the dynamics of the equivalent model coincide with that of the detailed model well. Especially during the fault, the errors between equivalent model and detailed model are very small.

Furthermore, Table 5 shows the stability of estimated parameters of equivalent model in different microgrid operational conditions, and the equivalent model can descript the dynamic of microgrid under different operational condition, which means that the equivalent model is robust.

(a)

(b)

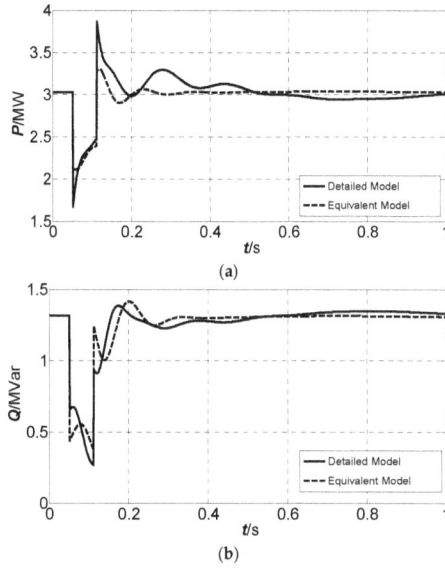

Figure 11. (**a**) The active power dynamics of the detailed model and the equivalent model under condition A; (**b**) The reactive power dynamics of the detailed model and the equivalent model under condition A.

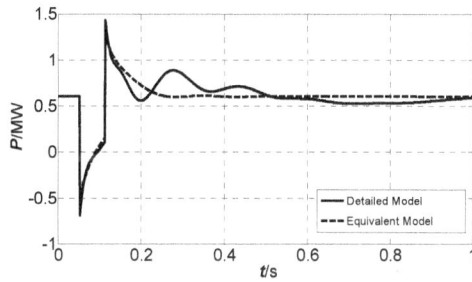

Figure 12. The active power dynamics of the detailed model and the equivalent model under condition B.

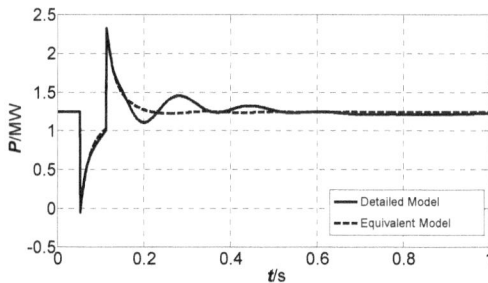

Figure 13. The active power dynamics of the detailed model and the equivalent model under condition C.

6. Conclusions

The equivalent model of the microgrid, based on electrical equipment physical characteristics, is proposed in this paper. The equivalent model consists of an equivalent machine component and an equivalent static component. Equivalent machine component is used to describe the dynamic of rotor machine characteristics and equivalent static component is used to describe the comprehensive characteristics of static component. In order to clear up the importance of each parameter toward the dynamics of the model, time-domain sensitivity of the parameters of the equivalent model with respect to the output power of the microgrid has been analyzed. The key parameters with important impact on the dynamics of the equivalent model have been detected. An improved particle with variable neighborhood searching is proposed where the radius of neighborhood is changed based on chaos iteration. The improved particle swarm optimization (IPSO) algorithm is used for parameters estimation, which improves the accuracy of the parameter estimation. The simulation results of microgrid under different operation conditions show that the simplified equivalent model is in good agreement with the detailed model under fault. The proposed model can be combined with other electrical component model in the power system simulation platform. With the proposed equivalent model, we can simplify the detailed model of microgrid obviously, and improved the simulation speed of power system.

Acknowledgments: This paper was sponsored by the Open Fund of Jiangsu Key Laboratory of Power Transmission & Distribution Equipment Technology (2011JSSPD11) and Science and technology project of Changzhou (CE20130043). The authors are also grateful to College of the Internet of Things Engineering and Jiangsu Key Laboratory of Power Transmission & Distribution Equipment Technology, at Hohai University, China.

Author Contributions: Changchun Cai presented the equivalent model of microgrid and designed the experiments, performed the experiments, written the article. Bing Jiang and Lihua Deng analyzed the data and modified the article.

Conflicts of Interest: The authors declare no conflicts of interest.

Appendix A. Detail Data of Microgrid

Table A1. Wind Generator Parameters.

U_N/kV	S_N/MVA	$\cos\varphi$	R_s/pu	X_s/pu	R_r/pu	X_r/pu	X_μ/pu	T_j/s
0.96	2.4	0.8756	0.0100	0.1000	0.0100	0.1000	3.0000	1.188

Table A2. Gas Turbine Parameters.

U_N/kV	S_N/MVA	r_s/pu	X_d/pu	X_q/pu	X'_d/pu	X''_d/pu	X'_q/pu	X''_q/pu	T'_{d0}/s	T'_{q0}/s	T''_{d0}/s	T''_{q0}/s
0.44	8	0	1.5	1.5	0.256	0.07	0.3	0.07	0.0171	0	0.0125	0.0057

Table A3. Rotor Load Parameters.

Load	U_N/kV	P_N/kVA	s	r_s/pu	X_s/pu	R_r/pu	X_r/pu	X_μ/pu	T_j/s
Load 1	0.4150	315.0 × 2	0.02756	0	0.0200	0.0347	0.2022	2.390	0.6230
Load 2	0.4150	315.0 × 5	0.02756	0	0.0200	0.0347	0.2022	2.390	0.6230

Table A4. Static Load Parameters.

Load	$P + Q$
Load 1	1 MW + 0.5 MVar
Load 2	5 MW + 1 MVar

Appendix B. Parameters Used in the PSO

Table B1. The bounds of the parameters.

x_{min}	-10	-0.3	0	0	0	0	0	0	0	0	0	-1	-1
x_{max}	10	0.3	0.2	0.5	5	10	0.2	5	0.5	5	10	10	5

References

1. Lasseter, B. Microgrids. In Proceeding of the Power Engineering Society Winter Meeting, New York, NY, USA, 28 January–1 February 2001; pp. 146–149.
2. Delghavi, M.B.; Yazdani, A. Islanded-Mode control of electronically coupled distributed-resource units under unbalanced and nonlinear load conditions. *IEEE Trans. Power Deliv.* **2011**, *26*, 661–673. [CrossRef]
3. Zhang, M.R.; Chen, J. The Energy Management and Optimized Operation of Electric Vehicles Based on Microgrid. *IEEE Trans. Power Deliv.* **2014**, *29*, 1427–1435. [CrossRef]
4. Xiong, L.; Peng, W.; Chiang, L.P. A hybrid AC/DC microgrid and its coordination control. *IEEE Trans. Smart Grid* **2011**, *2*, 278–286. [CrossRef]
5. Logenthiran, T.; Srinivasan, D.; Khambadkone, A.M.; Aung, H.N. Multiagent system for real-time operation of a Microgrid in real-time digital simulator. *IEEE Trans. Smart Grid* **2012**, *3*, 925–933. [CrossRef]
6. Yang, B.; Li, W.H.; Zhao, Y.; He, X.N. Design and analysis of a Grid-Connected photovoltaic power system. *IEEE Trans. Power Electron.* **2010**, *25*, 992–1000. [CrossRef]
7. Piazza, M.C.D.; Luna, M.; Vitale, G. Dynamic PV model parameter identification by least-squares regression. *IEEE J. Photovolt.* **2013**, *3*, 799–806. [CrossRef]
8. Keyhani, A. Modeling of photovoltaic microgrids for bulk power grid studies. In Proceeding of the Power and Energy Society General Meeting, Detroit, MI, USA, 24–28 July 2011; pp. 1–6.
9. Castro, R.M.G.; de Jesus, J.M.F. A wind park reduced-order model using singular perturbations theory. *IEEE Trans. Energy Convers.* **1996**, *11*, 735–741. [CrossRef]
10. Conroy, J.; Watson, R. Aggregate modelling of wind farms containing full-converter wind turbine generators with permanent magnet synchronous machines: Transient stability studies. *IET Renew. Power Gener.* **2009**, *3*, 39–52. [CrossRef]
11. Wang, H.H.; Tang, Y.; Hou, J.X.; Zhou, J.F.; Liang, S.; Su, F. Equivalent method of integrated power generation system of wind, photovoltaic and energy storage in power flow calculation and transient simulation. *Proc. CSEE* **2011**, *32*, 1–8.
12. Rubio, J.J.; Figueroa, M.; Pacheco, J.; Jimenez-Lizarraga, M. Observer design based in the mathematical model of a wind turbine. *Int. J. Innov. Comput. Inf. Control* **2011**, *7*, 6711–6725.
13. Ali, M.; Ilie, I.S.; Milanovic, J.V.; Chicco, G. Wind farm model aggregating using probabilistic clustering. *IEEE Trans. Power Syst.* **2013**, *28*, 309–316. [CrossRef]
14. Zha, X.M.; Zhang, Y.; Cheng, Y.; Fan, Y.P. New method of extended coherency for Microgrid based on homology in differential geometry. *Trans. China Electrotech. Soc.* **2012**, *27*, 24–31.
15. Ju, P.; Handschin, E.; Karlsson, D. Nonlinear dynamic load modelling model and parameter estimation. *IEEE Trans. Power Syst.* **1996**, *11*, 1689–1697. [CrossRef]
16. Abido, M.A. Multi-objective evolutionary algorithms for electric power dispatch problem. *IEEE Trans. Evolut. Comput.* **2006**, *10*, 315–329. [CrossRef]
17. Wang, S.K.; Chiou, J.P.; Liu, C.W. Parameters tuning of power system stabilizers using improved ant direction hybrid differential evolution. *Int. J. Electr. Power Energy Syst.* **2009**, *31*, 34–42. [CrossRef]
18. Fogel, D.B. *Evolutionary Computation: Toward a New Philosophy of Machine Intelligence*, 2nd ed.; IEEE Press: New York, NY, USA, 2000.
19. Lu, Z.; Ji, T.Y.; Tang, W.H.; Wu, Q.H. Optimal harmonic estimation using a particle swarm optimizer. *IEEE Trans. Power Deliv.* **2008**, *23*, 1166–1174. [CrossRef]
20. Iashif, K.; Salam, Z.; Amjad, M.; Mekhilef, S. An improved particle swarm optimization (PSO)-Based MPPT for PV with reduced steady-state oscillation. *IEEE Trans. Power Electron.* **2012**, *27*, 3627–3638.
21. Ye, M.Y.; Wang, X.D.; Xu, Y.S. Parameter identification for proton exchange membrane fuel cell model using particle swarm optimization. *Hydrog. Energy* **2009**, *34*, 981–989. [CrossRef]

22. Huynh, D.C.; Dunnigan, M.W. Parameter estimation of an induction machine using advanced particle swarm optimization algorithms. *IET Electr. Power Appl.* **2010**, *49*, 748–760. [CrossRef]

23. Zeng, N.Y.; Wang, Z.D.; Li, Y.R.; Du, M.; Liu, X.H. A hybrid EKF and Switching PSO algorithm for Joint State and Parameter Estimation of Lateral Dlow Immunoassay Models. *IEEE Trans. Comput. Biol. Bioinf.* **2012**, *9*, 321–329. [CrossRef] [PubMed]

24. Panigrahi, T.; Panda, G.; Mulgrew, B. Distributed bearing estimation techinque using diffusion particle swarm optimation algorithm. *IET Wirel. Sens. Syst.* **2012**, *2*, 385–393. [CrossRef]

25. Kanchev, H.; Lu, D.; Colas, F.; Lazarov, V.; Francois, B. Energy management and operational planning of a Microgrid with a PV-Based active generator for smart grid applications. *IEEE Trans. Ind. Electron.* **2011**, *58*, 4583–4592. [CrossRef]

26. IEEE. Task Force on Load Representation for Dynamic Performance Load representation for dynamic performance analysis. *IEEE Trans. Power Syst.* **1993**, *8*, 472–482.

27. Wu, F.; Zhang, X.P.; Godfrey, K.; Ju, P. Small signal stability analysis and optimal control of a wind turbine with doubly fed induction generator. *IET Gener. Transm. Distrib.* **2007**, *1*, 751–760. [CrossRef]

28. Kunder, P. *Power System Stability and Control*, 1st ed.; China Electric Press: Beijing, China, 2001.

29. Ju, P.; Qin, C.; Wu, F.; Xie, H.L.; Ning, Y. Load modeling for wide area power system. *Int. J. Electr. Power Energy Syst.* **2011**, *33*, 909–917. [CrossRef]

30. Rubio, J.J. Modified optimal control with a back propagation network for robotic arms. *IET Control Theory Appl.* **2012**, *6*, 2216–2225. [CrossRef]

31. Pérez-Cruz, J.H.; Rubio, J.J.; Ruiz-Velázquez, E.; Solís-Perales, G. Tracking control based on recurrent neural networks for nonlinear systems with multiple inputs and unknown dead-zone. *Abstr. Appl. Anal.* **2012**, *2012*. [CrossRef]

32. Puma, J.Q.; Colome, D.G. Parameters identification of excitation system models using genetic algorithms. *IET Gener. Transm. Distrib.* **2008**, *2*, 456–467. [CrossRef]

33. Wang, J.L. Study on Special Load Modeling in Power System. Ph.D. Thesis, North China Electric Power University, Beijing, China, 2010.

34. Giaquinto, A.; Fornarelli, G. PSO-Based cloning template design for CNN associative memories. *IEEE Trans. Neural Netw.* **2009**, *20*, 1837–1841. [CrossRef] [PubMed]

35. Jiang, C.W.; Bompard, E. A hybrid method of chaotic particle swarm optimization and linear interior for reactive power optimization. *Math. Comput. Simul.* **2005**, *68*, 57–65.

36. Cao, X.Q.; Ju, P.; Cai, C.C. Simulative analysis equivalent reduction for Microgrid. *Electr. Power Autom. Equiv.* **2011**, *31*, 94–98.

37. Rudion, K.; Styczynski, Z.A.; Hatziargyriou, N.; Papathanassiou, S. Development of benchmarks for low and medium voltage distribution networks with high penetration of dispersed generation. In Proceedings of the 3rd International Symposium on Modern Electric Power Systems, Wroclaw, Poland, 6–8 September 2006.

38. Papadopoulos, P.N.; Papadopoulos, T.A.; Crolla, P.; Roscoe, A.J.; Papagiannis, G.K.; Burt, G.M. Black-box dynamic equivalent model for microgrid using measurement data. *IET Gener. Transm. Distrib.* **2014**, *8*, 851–861. [CrossRef]

energies

MDPI

Article

An Analysis of Decentralized Demand Response as Frequency Control Support under Critical Wind Power Oscillations

Jorge Villena [1,*], **Antonio Vigueras-Rodríguez** [2,†], **Emilio Gómez-Lázaro** [3,†], **Juan Álvaro Fuentes-Moreno** [4,†], **Irene Muñoz-Benavente** [4,†] and **Ángel Molina-García** [4,†]

[1]   CF Power Ltd., Calgary, AB T2M 3Y7, Canada
[2]   Department of Civil Engineering, University of Cartagena, Cartagena 30203, Spain;
      avigueras.rodriguez@upct.es
[3]   Renewable Energy Research Institute and DIEEAC/EDII-AB, Universidad de Castilla-La Mancha,
      Albacete 02071, Spain; emilio.gomez@uclm.es
[4]   Department of Electrical Eng., Technical University of Cartagena, Cartagena 30202, Spain;
      juanalvaro.fuentes@upct.es (J.A.F.-M.); irene.munoz@upct.es (I.M.-B.); angel.molina@upct.es (A.M.-G.)
*   Correspondence: jvillena@cfpowerltd.com; Tel.: +1-587-719-3947

Academic Editor: Ying-Yi Hong
Received: 4 September 2015 ; Accepted: 4 November 2015; Published: 13 November 2015

Abstract: In power systems with high wind energy penetration, the conjunction of wind power fluctuations and power system inertia reduction can lead to large frequency excursions, where the operating reserves of conventional power generation may be insufficient to restore the power balance. With the aim of evaluating the demand-side contribution to frequency control, a complete process to determine critical wind oscillations in power systems with high wind penetration is discussed and described in this paper. This process implies thousands of wind power series simulations, which have been carried out through a validated offshore wind farm model. A large number of different conditions have been taken into account, such as frequency dead bands, the percentages of controllable demand and seasonal factor influence on controllable loads. Relevant results and statistics are also included in the paper.

Keywords: wind power generation; frequency control; load management; demand response

---

1. Introduction

The integration of intermittent renewable energy sources into power systems can be limited due to their disruptive effects on power quality and reliability. In the case of wind energy, the increasing penetration of this type of power generation may involve changes in power system design and management, such as grid reinforcements [1] and the necessity of studying its impact on grid frequency control [2,3]. In a power system, keeping a close balance between the generated and demanded power is an important operational requirement to maintain the grid frequency within a narrow interval around its nominal value [4]. Nowadays, grid frequency is controlled by conventional power plants driven by conventional generation sources. The main goal of this control is to keep the frequency within specified limits according to each country's grid code, addressing power imbalance exclusively by modifying the generated power, since demand is generally considered as not controllable. In this way, conventional generators are usually equipped with so-called primary and secondary control, which are part of this grid frequency control. Primary frequency control involves all actions performed locally at the generator to stabilize the system frequency after a power disturbance. These actions achieve a stable grid frequency, but different from its nominal value. The goal of the secondary frequency control is then to maintain the power balance within a bigger area, not just locally, as well as to recover the system frequency to its nominal value. These frequency control mechanisms are mainly performed by conventional power plants. In recent years, the rapid development of wind turbine technology and increasing wind power penetration in the generation mix have resulted in a continuous reformulation of wind power requirements to be integrated with traditional generation sources. Some transmission system operators have unified requirements and connection rules for all production units, whether they are driven by conventional energy sources or not, which are very difficult for wind turbine producers and wind farm developers to fulfill [5]. Under such conditions, in regions where there is a high wind energy penetration and whose interconnections are weak, difficulties in maintaining the nominal frequency could arise if sufficiently large wind power fluctuations occur.

System inertia also plays an important role in the grid frequency control, limiting the rate of frequency change under power imbalances. The lower the system inertia, the higher the rate of frequency change when demand-side or supply-side variations appear. System inertia is directly related to the amount of synchronous generators in the power system. This inherent relation is not as obvious when dealing with wind turbine generators due to the electromechanical characteristics of the currently prevailing variable speed technologies, whose turbine speed is decoupled from the grid frequency [6]. The inertia contribution of wind turbines is much less than that of conventional power plants [6–8]. Actually, some variable speed wind turbines use back-to-back power electronic converters, which create an electrical decoupling between the machine and the grid, leading to an even lower participation of wind generation to the system stored kinetic energy.

Some authors suggest that this drawback can be compensated by an adequate implementation of the machine control. In [9,10], a power reserve is obtained following a power reference value lower than the maximum power, which can be extracted from the wind, thus decreasing the turbine power efficiency. A method to let variable-speed wind turbines emulate inertia and support primary frequency control using the kinetic energy stored in the rotating mass of the turbine blades is proposed in [11]. In [12], a power reserve is obtained with the help of pitch control when the wind generator works close to the rated power.

With the aim of reducing the impact of wind power fluctuations and wind turbines lower inertia contribution to grid frequency control, demand-side actions can also be considered: switching-off some loads has similar effects on a grid power imbalance as increasing in the supply-side, reducing the need for ramp up/down services provided by conventional generators [13]. However, demand-side actions have been usually contemplated only in emergency situations to save the power system, such as load shedding [14,15] actions or load curtailment [16] and considering a minimum level of aggregated load power.

In this context, different load shedding schemes that have been proposed take into account a certain frequency threshold, as well as a certain rate of change of frequency (ROCOF) [17–20]. This demand-side complementary control could significantly help in maintaining grid frequency, as some authors consider that 40% of residential appliances are compatible with the proposed load control strategies [21]. Due to the high penetration of cooling and heating loads, about 20 percent of the load in the U.S. comes from consumer appliances that cycle on and off and which could make a contribution to frequency control during the normal operation state [22]. In [23], the authors present a centralized management system focused on electric water heaters for areas with a high penetration of renewable energy sources, in order to smoothen the imbalances between generation and demand within the controlled area. In [24], the authors assess what real-time operation could be like with a significant amount of active frequency-sensitive fridge/freezer load for the national grid system in Great Britain. The advantages of a higher proportion of these types of loads when wind penetration increases are also discussed. In [25], a decentralized approach for using thermal controllable loads (TCL) for providing primary frequency response is shown. The authors argue that a two-way communication between these loads and the control center is not essential. They thus propose a frequency-responsive load controller, allowing the loads to respond under frequency changes in a similar way as conventional generators do. They also show that, using this approach, the demand side can make a significant and reliable contribution to primary frequency control without affecting the customers' comfort. Recently, the authors have also discussed the effectiveness of demand-side participation in primary frequency control together with the action of auxiliary frequency control carried out by variable-speed wind turbines, focused on evaluating the potential of additional controls and the compatibility between those controls [26]. In [27], following a similar decentralized approach, TCLs are grouped according to their essentiality for the customer, and in the event of a frequency drop, each of them is switched off for a predefined time, which depends on the frequency deviation.

Considering previous contributions, this paper discusses and describes a complete process to determine realistic wind speed oscillations. This process is able to evaluate the contribution of the demand side to primary frequency control in power systems with high wind penetration, providing a wide range of realistic wind speed variations and allowing one to analyze the demand response as frequency-controlled reserves under critical circumstances. The rest of the paper is structured as follows: In Section 2, the different components of the power system model used in this study are described. The proposed methodology for critical wind power fluctuations is described in Section 3. Simulations and results are given in Section 4. Finally, conclusions are discussed in Section 5.

2. Power System Model

2.1. General Description

For frequency control study purposes, power systems are usually modeled according to the general scheme shown in Figure 1. As can be seen, all turbine generators are lumped into a single equivalent rotating mass (M), and similarly, all individual system loads are lumped into an equivalent load with an equivalent damping coefficient (D) [28]. Additionally, frequency deviations are used as feedback signals for primary and secondary frequency control.

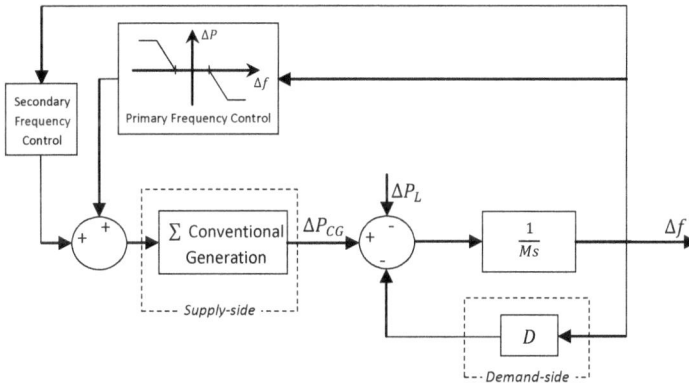

Figure 1. General scheme of a power system.

According to [28], the mathematical model of the power system can be expressed as:

$$\Delta P_{mec} - \Delta P_L = D'\Delta f + 2H\frac{d\Delta f}{dt} \tag{1}$$

In this equation, $\Delta P_{mec} - \Delta P_L$ represents the imbalance between power supply and demand, Δf is the consequent variation of the grid frequency; D' is the damping coefficient expressed in pu and H is the inertia constant, in seconds; with $M = 2H$. In this study; ΔP_L is not considered, so frequency excursions are caused by typical fluctuations in wind power generation. Figure 2 shows the schematic block diagram of the power system model used in this work.

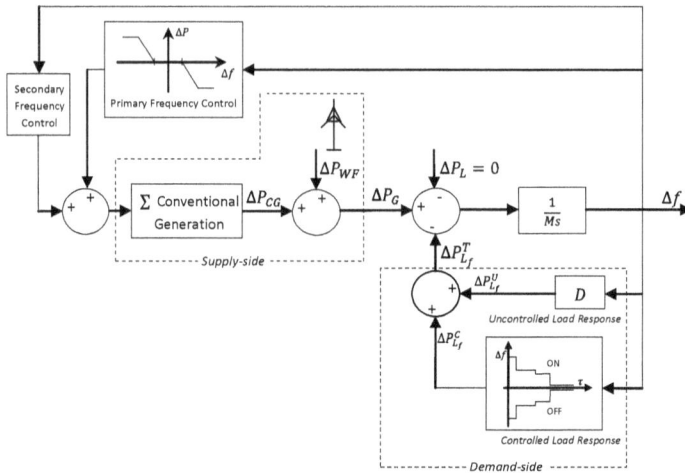

Figure 2. General scheme of the power system with the demand-side contribution to primary frequency control and wind power generation.

2.2. Supply-Side Model

The generation part of the power system model shown in Figure 2 consists of a thermal power plant and an offshore wind farm, which allows for simulating different energy mixes.

2.2.1. Conventional Generation

The conventional generation block shown in Figure 2 consists of a thermal power plant with a reheating system. The power plant model includes the transfer functions for the two main elements of the control loop: the primary energy-mechanical torque converter (governor) and the mechanical torque-electrical power converter (turbine), as detailed in [28,29]. The block diagram corresponding to this system is shown in Figure 3. The power plant is modeled to provide load-frequency control. The aim of the primary frequency control or speed governor is to change the primary energy input in order to maintain the generator's rotating speed as close as possible to the rated speed. The generator's rotating speed may change as a consequence of any modification in the power demanded by the customer side or by a rise or fall in the power produced by other generators. The speed governor response takes place within a few seconds, according to the speed-droop characteristic shown in Figure 3, in order to restore the active power balance. However, after the governor response, the grid frequency stabilizes, but differs from its nominal value. Secondary frequency control or automatic generation control (AGC) then takes place within one to several minutes to restore the power balance in a bigger area, not just locally, and to take the system frequency back to its nominal value. In this work, as only one generator models the controllable supply side, AGC is implemented through the loop shown in Figure 2, which consists of an integral control that aims to reduce the frequency error to zero. Limitations on slope and maximum power output variations are included for the generator, as well as a dead band (DB) to model the sensors sensibility and the precision of frequency measuring [30,31]. Changes in power from the supply side come solely from the wind farm power fluctuations, whilst the average demanded power remains constant during the simulations.

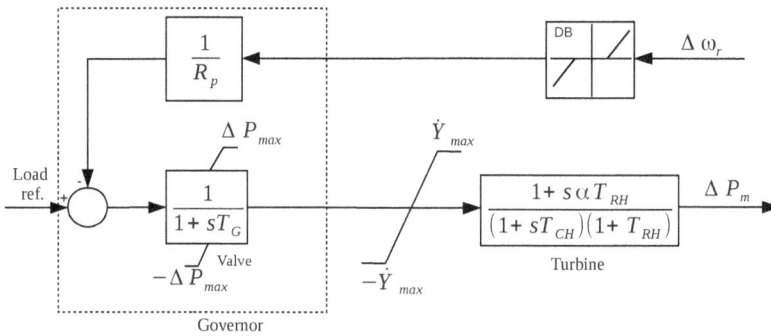

Figure 3. Model of the thermal plant.

2.2.2. Wind Power Generation

The wind power generation input in the supply side of Figure 2 consists of a wind power time series ($\Delta P_{WF}(t)$) obtained from an aggregated model of power fluctuations in an offshore wind farm (WF). The WF model was previously validated by comparing its results with real power fluctuations measured at the Nysted offshore wind farm. The WF model can be divided into two main blocks: a generator of spatially-averaged wind speed time series and an aggregated power curve that relates WF wind speed with WF power generation; see Figure 4. The first block takes the input to the simulator, which is the average wind speed upstream from the wind farm ($\overline{V}_\infty$), *i.e.*, far enough not to be affected by the wind farm, and outputs a spatially-averaged time series of the wind speed within the wind farm. The second block takes this wind speed and passes it through the wind farm model, which consists of an aggregated wind farm power curve, yielding the simulator output: a second 2-h series of wind farm power generation ($P_{WF}(t)$).

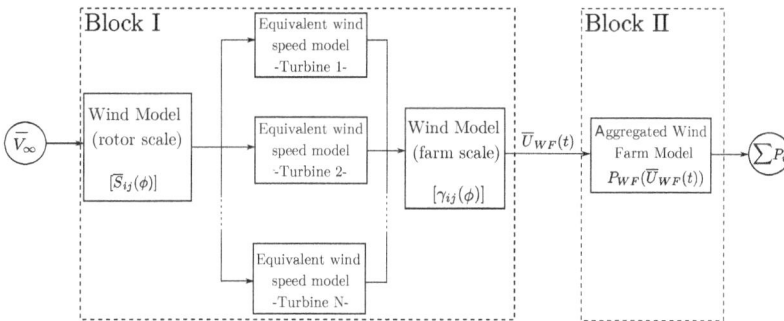

Figure 4. Wind farm power simulator scheme.

As mentioned in the previous paragraph, wind speed fluctuations within a wind farm can be modeled through spectral tools, like power spectral density (PSD) and spectral coherence. Such functions are defined in the frequency domain. In this section, frequency is represented as ϕ in order to avoid confusions with the electrical frequency, f.

Particularly, wind speed fluctuations in a single point are described by the following PSD ($S(\phi)$) function:

$$S(\phi) = \beta_{LF}^2 \frac{\frac{z}{\overline{V}}}{\left(\frac{z \cdot \phi}{\overline{V}}\right)^{\frac{5}{3}} \cdot \left(1 + 100 \frac{z \cdot \phi}{\overline{V}}\right)} + \sigma_V^2 \frac{2 \frac{L_1}{\overline{V}}}{1 + 6 \frac{L_1}{\overline{V}} \cdot \phi} \tag{2}$$

where ϕ is the frequency; z is the turbine hub height above sea level; $\overline{V}$ is the average wind speed within the wind farm and σ_V is its standard deviation. Finally, β_{LF} is an empirical parameter suggested by [9]. A numerical value of $\beta_{LF} = 0.04$ ms^{-1} is proposed by [32], based on experimental data from the Nysted and Horns Rev offshore wind farms.

Some quick variations of wind speed are directly smoothed at the rotor disk. Actually, wind turbines obtain their power from the wind speed within an area where part of these quick oscillations are not correlated. This issue is taken into account through the equivalent wind speed rotor model ($F_{EWS}(\phi)$) suggested by [33]; see Figure 4. The equivalent wind speed, as shown in Figure 5, is the wind speed that, when applied uniformly on the entire rotor surface, produces the same aerodynamic torque as with the actual wind speed. Moreover, the spatial aggregation of wind turbines can contribute to smoothening part of the wind oscillations affecting individual WTs. This is taken into account by means of the spectral coherence model $\gamma_{i,j}(\phi)$, which measures the relation of wind speed between two wind turbines (e.g., turbine numbers i and j). The coherence model implemented for this simulator is the one presented in [32].

Combining $S(\phi)$ and $\gamma_{i,j}(\phi)$ as described in [34], the WF average wind speed PSD ($S_{av}(\phi)$) can be obtained by:

$$S_{av}(\phi) = \frac{1}{N^2} \left(\sum_{i=1}^{N} \sum_{j=1}^{i} 2\text{Re}\left(\gamma_{i,j}(\phi)\right) \right) \left(F_{EWS}(\phi) \cdot S(\phi)\right) \tag{3}$$

being N the number of wind turbines in the wind farm and Re$\left(\gamma_{i,j}(\phi)\right)$ the real part of $\gamma_{i,j}(\phi)$.

From the above PSD, $S_{av}(\phi)$, time series of WF average wind speed are generated following the algorithm suggested in [32,35], whose scheme is shown in Figure 4. These wind speed series are converted into WF generated power series through an aggregated power curve, such as the one shown in Figure 6.

The power curve shown in Figure 6 in particular is an empirical curve measured at the Nysted offshore wind farm [36]. The curve is calculated by monitoring the total wind farm power output and the average wind speed within the wind farm, measured at the nacelle of each wind turbine. Usually, in order to calculate the power curve of a single wind turbine, the wind speed is measured ahead of the nacelle. The actual wind speed at the nacelle is slightly smaller. The wind speed shown in the figure is the average within the wind farm. Therefore, when the wind farm average wind speed is around 23 m/s, some wind turbines are already at their cut-out point (25 m/s) and automatically stop producing power. As wind speed rises, more wind turbines cut out, up to an average wind speed of around 27 m/s, at which all wind turbines are disconnected, hence the slope of the right part of the curve, which differs from the vertical shape of a single wind turbine power curve. Besides, it can be seen in the figure that the maximum power is below 1 pu. This is due to the effect of the wakes within the wind farm, especially when considering wind directions that are not optimal [37]. Thus, when the wind speed is around 15 m/s, the first row extracts the maximum power from the wind; the wind flowing to the second row, due to the turbulence, has less available energy, and so on. Overall, the total wind farm power is lower than 1 pu, although the wind speed is above the nominal speed. As wind speed increases, more rows will produce the maximum power, but the available power at the last rows will still drop below the nominal value. As wind speed goes above 23 m/s, the cut-out speed, the first rows (the higher the wind speed, the more rows will cut out) will automatically stop producing power, hence the slope of the right part of the curve, which differs from the vertical shape of a single wind turbine power curve.

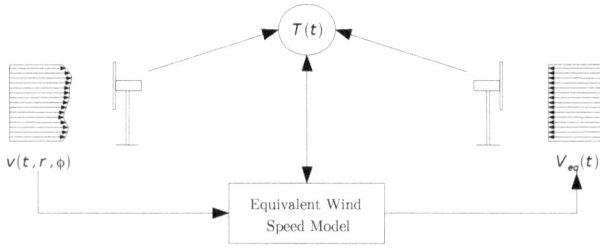

Figure 5. Scheme of the equivalent wind speed.

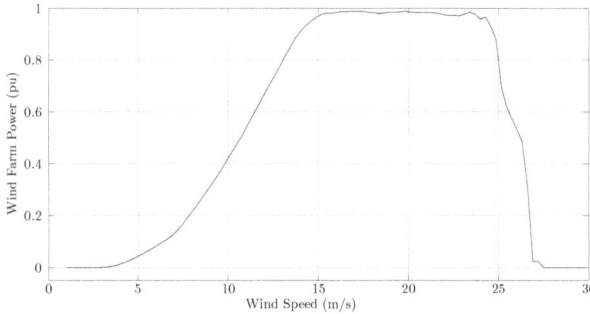

Figure 6. Aggregated wind farm power curve.

2.3. Demand-Side Model

The power demand is divided into two types of loads: uncontrollable and controllable loads. Uncontrollable loads are lumped into a single equivalent load, where the power consumption depends only on the system frequency changes through parameter D; see Figure 2. Controllable loads are modeled individually and involve those appliances with high thermal inertia mainly due to the following reasons: they comprise a significant portion of electricity consumption in the residential sector and their operation can be shifted in time without noticeable effects on consumers. Over 30% of the total electricity demand is consumed in households [38,39], of which more than 40% is compatible with the proposed forced connection/disconnection actions [21]. The number of controllable loads belonging to the residential sector is higher than the commercial sector. Moreover, their rated power is usually lower in comparison with the commercial sector. Thus, for the same amount of controlled power, a greater number of individual loads results in a smoother global response.

The load controller previously proposed in [25] is used to turn off (on) the controllable demand, which is assumed dependent on the grid frequency through parameter D. Actually, the influence of grid frequency excursions on the controllable power demand can be considered negligible in comparison with the forced disconnection (connection) commands set by the load controllers. These controllers consider not only the frequency deviation Δf, but also its evolution over time τ. In this way, each load controller has an associated $\Delta f - \tau$ profile that determines when the load starts contributing to the frequency control. As long as the frequency deviation does not exceed a certain threshold for a certain time, the load controller remains inactive, and the load maintains its normal operation demand. If the frequency deviation enters into the control region, the controller will switch off (on) the load, reducing (increasing) the power demand. Figure 7 shows the $\Delta f - \tau$ characteristics for different types of loads. Controllable loads have been grouped into three main categories according to the specific operating characteristics of each type of load and the patterns of use: fridges/freezers (Load Group I), air-conditioners/heat pumps (Load Group II) and electric water heaters (Load Group III); see Table 2. These operating characteristics must be set in the load controller, and they define its behavior, *i.e.*, contingency response speed (given by t_{delay}), maximum off time and the minimum recovery time. A suitable configuration of these parameters allows us to preserve the minimum standards of customers' comfort and to consider mechanical and electrical load requirements. For example, randomly setting t_{delay} (assigning different values to each individual controlled load) avoids instantaneous and massive disconnections and, consequently, undesired frequency oscillations, therefore ensuring a smoother demand response. Moreover, the recovery time is required after forced disconnections (or connections) for the thermal variables to recover their ordinary values. Further information and a detailed description of the load controller algorithm can be found in [25].

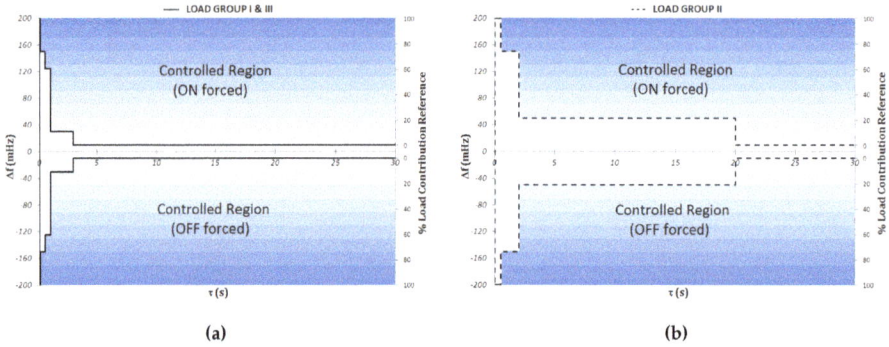

Figure 7. Frequency responsive load controller: Δf–τ characteristics. (**a**) Load Groups I and III; (**b**) Load Group II.

With the aim of providing a demand response that is proportional to the frequency excursion, controllable loads are called up to participate in the frequency control progressively depending on the depth of such an excursion; see Figure 7, the percentage of load contribution reference. Under a certain frequency event, the actual percentage of load contribution will depend on the connection status of each individual load according to its duty-cycle.

3. Simulation of Critical Wind Power Fluctuations

Realistic wind power series with high fluctuations are estimated with the aim of evaluating the wind power fluctuations' impact on the grid frequency. These series are determined as follows; see Figure 8.

- A set of 10,000 2-h series, with a one-second sample rate, of WF wind speed ($V_{WF}(t,j)$) is firstly estimated. The inputs to the wind farm model are an upstream 2-h average wind speed ($\overline{V}_j$) according to a Weibull probability distribution, as well as a spectral wind farm model [32,35,40].

  In this case, the wind is simulated considering a 506-MW offshore wind farm, with 10 rows with 22 wind turbines in each row.
- Realistic wind power data series ($P_{WF}(t,j)$) are obtained from $V_{WF}(t,j)$ through an aggregated wind farm power curve [36]. These series correspond to a global period of time of around 2.5 years, which is large enough for obtaining significant wind fluctuations.
- In order to characterize the power oscillations within the series, ramp power rates each of 2-min intervals are calculated ($P_{ramp}(n,j)$). This interval length is between the characteristic times of frequency control and wind power oscillations.
- Calculated ramp rates ($\forall j$ & n) are then sorted in descending order, obtaining the duration curve.
- From the stability point of view, the most critical cases are those where both wind power drops are steep, as well as the wind power share in the current mix of generation is high. Indeed, the

ramp rate around the 99th-percentile with the highest wind power share is selected (n_{P99}), and the corresponding 2-h series where this drop happens is identified (j_{P99}) within the set of WF power series ($P_{WF}(t, j)$).

- A 10-min time interval around the n_{P99} event is selected to provide suitable frequency oscillations in the modeled power system $P_{WF_s}(t)$. Such a 10-min interval is highlighted in red color in Figure 9.
- Finally, wind power deviation (ΔP_{WF}) shown in Figure 2 is determined as the difference between $P_{WF}(t, j_{P99})$ and the expected wind power within this time interval (P_{WF_0}),

$$\Delta P_{WF}(t) = P_{WF}(t, j_{P99}) - P_{WF_0}; \forall t \in [t_{\min}, t_{\max}] \tag{4}$$

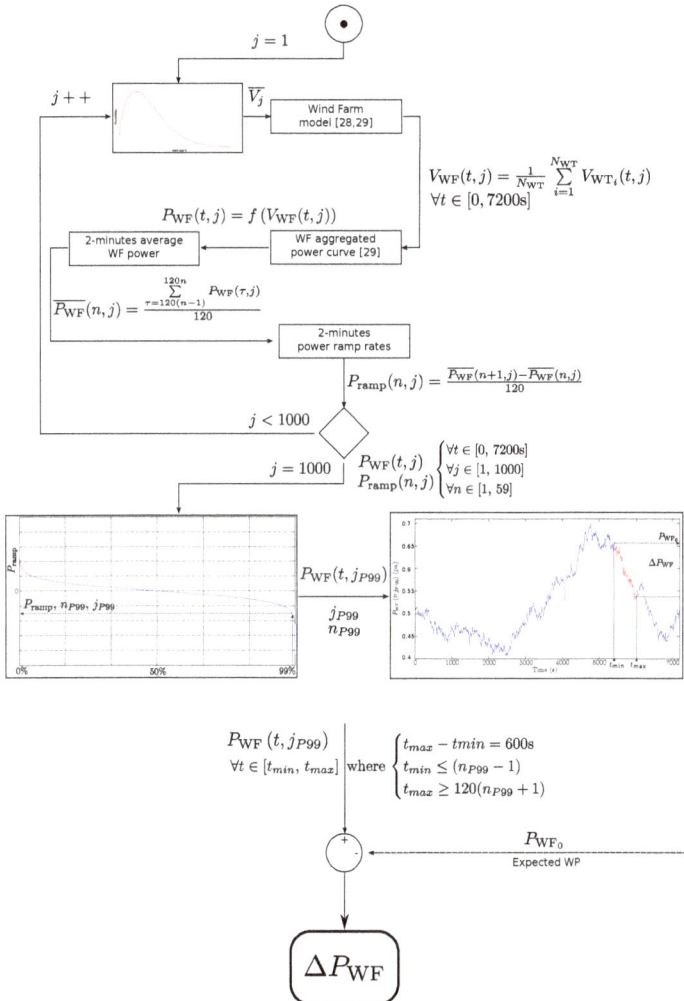

Figure 8. Sketch of the algorithm used for the simulation of critical wind power fluctuations.

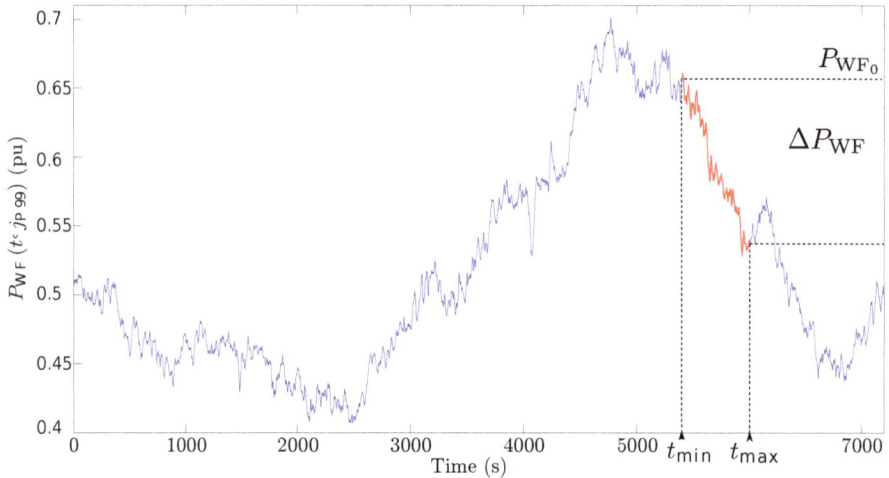

Figure 9. Wind farm power in 2-h time interval, expressed as per unit of installed wind power. In red, the 10-min interval used in the simulations.

4. Simulation and Results

4.1. Preliminaries

The simulated power system includes a thermal power plant and the previous wind farm. The parameters of the thermal power plant are shown in Table 1 based on [28]; see Figure 3.

Table 1. Parameters for the conventional generator model.

R_P	ΔP_{max}	$\dot{Y}_{max}$	$\dot{Y}_{min}$	α	T_{RH}	T_{CH}	H
5 %	0.05 pu	0.05 pu/s	−0.1 pu/s	0.3 pu	7 s	0.3 s	4 s

In order to reflect a low inertia in the modeled power system, the following assumptions are made: first, the contribution of the wind turbines to the system inertia is considered as negligible, since they are electrically decoupled from the grid; second, the inertia of the motor loads is also not considered. Therefore, it is assumed that the only inertial support comes from the thermal synchronous generator. An equivalent inertia constant (H_{eq}) can then be calculated by dividing the generator's kinetic energy at the rated speed by the system base power (S_b) as:

$$H_{eq} = \frac{\frac{1}{2}J_{th}\omega_0^2}{S_b} = \frac{\frac{1}{2}J_{th}\omega_0^2}{S_{th}}\frac{S_{th}}{S_b} = H\frac{S_{th}}{S_b} \tag{5}$$

J_{th} is the moment of inertia of the thermal generator, and ω_0 is its rated speed, so $\frac{1}{2}J_{th}\omega_0^2$ is the generator's kinetic energy at the rated speed. If the thermal plant relative size is chosen such that $\frac{S_{th}}{S_b} = \frac{3}{4}$, the equivalent inertia constant used for the simulations is $H_{eq} = 3$ s. For the damping coefficient (D' in Equation (1)) a typical value used in dynamic studies for isolated power systems, $D' = 1$, is chosen [28].

4.2. Implemented Scenarios

The worldwide residential sector accounts for about 30% of total electric energy consumption (TEEC) [21,38,39,41]. Considering two representative cases (winter and summer), the participation of total residential electricity consumption (TREC) in the TEEC presents a maximum share of residential controllable loads of around 13% (30% · 43.0%) for the winter case and 10% (30% · 33.3%) for the summer case. Table 2 [21] shows the share of TREC by major end-use. For the sake of simplicity, a maximum share of residential controllable loads, in both the winter and summer cases, of 10% will be considered. For the proposed 1-GW power system model, 10% of the demand means 100,000 individual controllable loads.

Table 2. Share of residential electricity consumption by major end-use.

Group	Type of load	Share percentage (%)	
		Winter	*Summer*
I	Refrigeration and freezing	13.4	13.4
II	Space cooling	–	6.4
II	Space heating	16.1	–
III	Water heating	13.5	13.5
	Total	43.0	33.3

To evaluate the impact of demand-side participation on frequency control, a set of different scenarios are defined modifying the share of controllable loads. Specifically, 10-min time interval simulations are carried out for 2.5%, 5%, 7.5% and 10% of residential load share. In addition, and regarding the governor speed control for the supply side, different dead band values (20, 50, 80 and 100 mHz) are considered to give the demand side a more active role in frequency control. Subsequently, simulations for each governor's dead band are then compared to a reference case, in which demand-side response is not considered.

Finally, due to the presence of a certain degree of randomness in the load controller, simulations are repeated five times for each case study to include such variability.

4.3. Analysis of a Case Study: Winter Scenario

A detailed analysis of a specific case study is discussed in this subsection. A winter scenario is simulated considering 10% of controllable loads and a ±20-mHz governor's dead band.

Figure 10 depicts the corresponding frequency excursions and power deviations. To evaluate the demand response effect on frequency control, this figure also includes the $\Delta f(t)$ profile when demand side participation is not considered (blue line). As can be seen, there is a relevant reduction of oscillations in comparison with the reference case. Indeed, some of the peaks that surpass the upper bound are cut out, the frequency deviations within the thermal plant governor dead band being only dependent on the controllable load participation. Due to the very low system inertia, there are significant oscillations of the frequency around the limits of the thermal plant governor's dead band (20 mHz and −20 mHz). This effect of the inertia deficit on the grid stability was analyzed in a previous work [42].

Figure 10c shows the aggregated behavior of the controlled loads ($\Delta P^c_{L_f}$). It represents the forced load connection ($\Delta P^c_{L_f} > 0$) or disconnection ($\Delta P^c_{L_f} < 0$) with respect to their expected demand power profile. In this case study, the maximum positive power deviation is about 0.8% of the total demand, whereas the maximum negative power deviation is around 0.6%. The greater negative power deviations correspond to those time intervals where wind power drops steeply. Such time intervals are highlighted in Figure 10a. During these critical time intervals, the demand response contributes significantly to reducing the corresponding under-frequency excursions. In fact, the greater the ratio of controllable loads, the smoother the frequency deviations. To justify this assessment, Figure 11 compares frequency deviations for different controllable load ratios during the first critical time interval: $\forall t \in [5630, 5635]s$. In addition, Table 3 summarizes the reductions of the maximum under-frequency deviations for the different percentages of controllable loads.

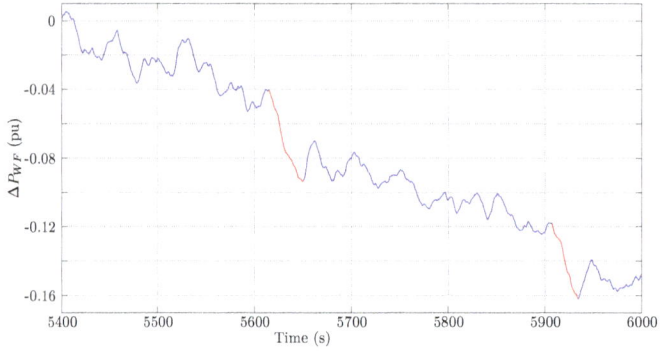

(a)

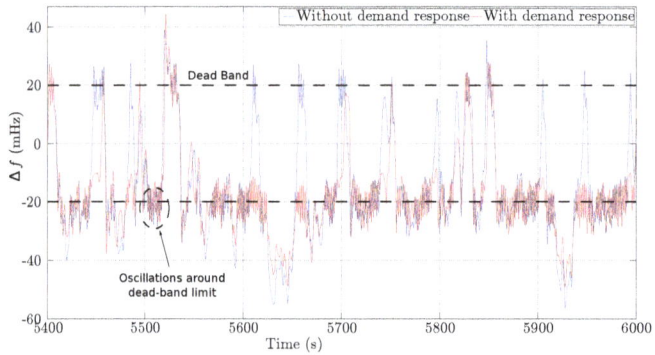

(b)

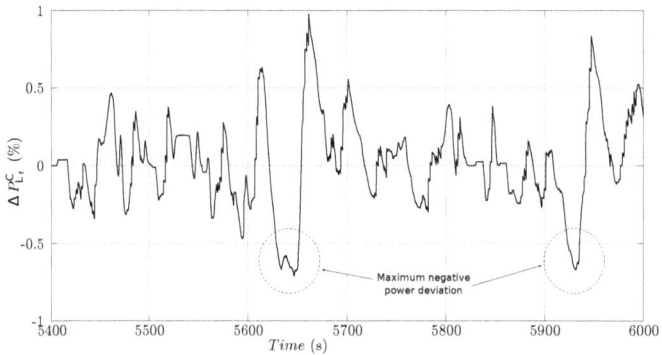

(c)

Figure 10. Results for a specific case study. (**a**) ΔP_{WF} for the most critical interval considered in the simulations; (**b**) example of the temporal evolution of frequency deviations when demand response is or not applied; (**c**) aggregated forced demand change.

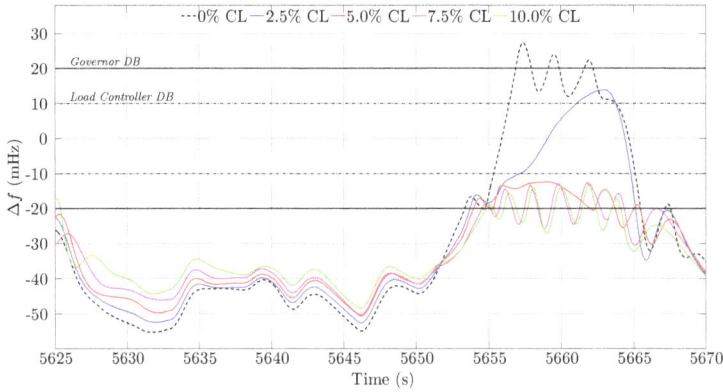

Figure 11. Δf for different percentages of controllable load, including the reference case.

Table 3. Δf_{min} variation (%) for $t \in [5630, 5635]$ s.

CL (%)	0	2.5	5	7.5	10
Winter	-	5.08	9.36	16.48	19.97
Summer	-	4.05	8.14	14.67	18.93

4.4. Summary of Case Studies

This subsection is devoted to providing an extensive analysis of the demand-response behavior for the set of winter and summer case studies.

In line with previously-observed results for the winter case, Figure 12 shows the increment in time during which frequency deviations are within the governor's dead band as a consequence of the controllable load actions. Such an increment is represented in percentages with respect to the uncontrolled reference case, according to Equation (6):

$$Improvement(\%) = \frac{T_{OK}^C - T_{OK}^{UC}}{T_{OK}^{UC}} \times 100 \tag{6}$$

being T_{OK}^C and T_{OK}^{UC} the total time during which $|\Delta f| < 20$ mHz with controlled and uncontrolled loads, respectively. For a given dead band, the growth of the time interval within the band is in general higher as the amount of controllable loads increases, *i.e.*, the higher the number of controllable loads, the lower the time in which the thermal generator's primary frequency controller remains active. For a ± 20-mHz dead band, both winter and summer cases, and also the winter case ± 100-mHz dead band, an increment of the controllable load rate from 7.5% to 10% does not lead to a significant improvement of the time interval within the dead band for the supply side. In fact, a certain saturation effect is detected in these cases.

In Figure 13, extreme deviations on power demand due to the individual load controllers are represented through both the 1- and 99-percentile of the power deviation with respect to the uncontrolled demand. Such deviations are given as per mille (‰) of the global demand, with relatively small demand modifications. Indeed, low values of Δf imply a reduced number of active controlled loads, enough remaining available controllable loads to provide a response under critical contingencies, as discussed by the authors in [25]. Controllable load thus significantly reduces the frequency excursion under large wind power fluctuations, keeping the capacity for the critical contingency response.

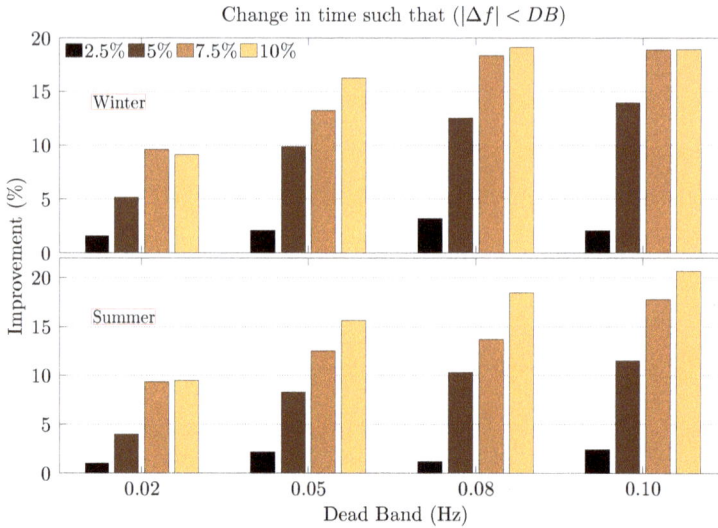

Figure 12. Increase of the time during which $|\Delta f| <$ dead band (DB) for different governor dead bands and controllable loads ; see Equation (6).

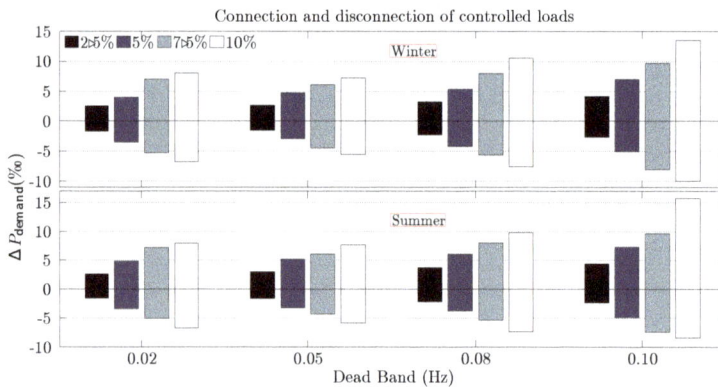

Figure 13. First and 99-percentiles (expressed in ‰) of connected or disconnected controlled loads for different governor dead bands and controllable loads.

5. Conclusions

A complete process to determine realistic wind oscillations is proposed to evaluate the contribution of the demand-side to primary frequency control in power systems with high wind penetration. These wind oscillations are determined based on a model for offshore wind power fluctuations, previously assessed through real data from the Nysted offshore wind farm. In this paper, extreme wind oscillations have been selected from 10,000 2-h series.

A steam turbine with conventional primary energy input control and an offshore wind farm are considered to model the supply side of a simplified power system. A highly fluctuating wind power data series is then used to emulate power system conditions. In particular, 10-min periods under the presence of negative and persistent ramps are selected as case studies. Primary and secondary frequency control actions are only provided by the thermal power plant, evaluating different governor dead bands and controllable load levels.

The demand-side contribution to primary frequency control is considered using a decentralized approach not requiring any explicit communication. Thus, simple hardware can thermostatically control individual residential loads in response to deviations between the frequency and its nominal value over time. The behavior of different amounts of such controllers is simulated to study their effect at the system level when power fluctuations due to high wind energy penetration are considered. This high wind power penetration, when considering variable-speed wind turbines, implies that the system inertia can be critically low, resulting in high system instabilities.

Results show that it is possible to decrease primary frequency reserves with a relatively low demand-side participation in frequency control. Indeed, the time interval in which frequency deviations are within the governor's dead band is increased up to 20% with 10% of demand-side participation. Consequently, controllable loads significantly reduce frequency excursions under large power fluctuations, providing additional capacity for critical contingencies.

Acknowledgments: This work was supported by "Ministerio de Economía y Competitividad" and the European Union (ENE2012-34603), as well as by *"Fundación Séneca-Agencia de Tecnología de la Región de Murcia"* PCTIRM2011-14 (08747/PI/08 and 19379/PI/14).

Author Contributions: Jorge Villena developed and implemented the overall power system model, performed, analyzed and compared the simulations and prepared the manuscript. Antonio Vigueras-Rodríguez developed the original wind power generation model, proposed some simulation algorithms to evaluate the scenarios and provided comments on the manuscript. Emilio Gómez-Lázaro suggested several improvements on the analysis and comparisons of the simulation results and provided comments on the manuscript. Juan Álvaro Fuentes-Moreno prepared the conventional generation model and provided comments and suggestions on the manuscript. Irene Muñoz-Benavente implemented the supply-side model and helped with preparing the manuscript. Ángel Molina-García developed the supply-side model and provided comments on the results' comparison, as well as on the manuscript. All of the authors read and approved the final manuscript.

Conflicts of Interest: The authors declare no conflict of interest.

References

1. Weisser, D.; Garcia, R.S. Instantaneous wind energy penetration in isolated electricity grids: Concepts and review. *Renew. Energy* **2005**, *30*, 1299–1308.
2. Doherty, R.; Mullane, A.; Nolan, G.; Burke, D.; Bryson, A.; O'Malley, M. An assessment of the impact of wind generation on system frequency control. *IEEE Trans. Power Syst.* **2010**, *25*, 452–460.
3. Klempke, H.; McCulloch, C.; Wong, A.; Piekutowski, M.; Negnevitsky, M. Impact of high wind generation penetration on frequency control. In Proceedings of the 20th Australasian Universities Power Engineering Conference (AUPEC), Christchurch, New Zealand, 5–8 December 2010; pp. 1–6.
4. Gómez-Expósito, A.; Conejo, A.; Cañizares, C. *Electric Energy Systems: Analisys and Operation*; CRC Press: Boca Raton, FL, USA, 2009.
5. Fagan, E.; Grimes, S.; McArdle, J.; Smith, P.; Stronge, M. Grid code provisions for wind generators in Ireland. In Proceedings of the IEEE Power Engineering Society General Meeting, San Francisco, CA, USA, 12–16 June 2005; Volume 2, pp. 1241–1247.
6. Mullane, A.; O'Malley, M. The inertial response of induction-machine-based wind turbines. *IEEE Trans. Power Syst.* **2005**, *20*, 1496–1503.
7. Lalor, G.; Ritchie, J.; Rourke, S.; Flynn, D.; O'Malley, M.J. Dynamic frequency control with increasing wind generation. In Proceedings of the IEEE Power Engineering Society General Meeting, Denver, CO, USA, 6–10 June 2004; pp. 1715–1720.
8. Lalor, G.; Mullane, A.; O'Malley, M. Frequency control and wind turbine technologies. *IEEE Trans. Power Syst.* **2005**, *20*, 1905–1913.
9. Soerensen, P.; Hansen, A.D.; Thomsen, K.; Madsen, H.; Nielsen, H.A.; Poulsen, N.K.; Iov, F.; Blaabjerg, F.; Donovan, M.H. Wind farm controllers with grid support. In Proceedings of the 5th International Workshop on Large-Scale Integration of Wind Power and Transmission Networks for Offshore Wind Farms, Glasgow, UK, 7–8 April 2005.

10. Mokadem, M.E.; Courtecuisse, V.; Saudemont, C.; Robyns, B.; Deuse, J. Experimental study of variable speed wind generator contribution to primary frequency control. *Renew. Energy* **2009**, *34*, 833–844.

11. Morren, J.; de Haan, S.W.H.; Kling, W.L.; Ferreira, J.A. Wind turbines emulating inertia and supporting primary frequency control. *IEEE Trans. Power Syst.* **2006**, *21*, 433–434.

12. Bousseau, P.; Belhomme, R.; Monnot, E.; Laverdure, N.; Boeda, D.; Roye, D.; Bacha, S. Contribution of wind farms to ancillary services. In Proceedings of the Council on Large Electric Systems (CIGRE) General Meeting, Paris, France, 27 August–1 September 2006.

13. Yousefi, A.; Iu, H.C.; Fernando, T.; Trinh, H. An approach for wind power integration using demand side resources. *IEEE Trans. Sustain. Energy* **2013**, *4*, 917–924.

14. Concordia, C.; Fink, L.; Poullikkas, G. Load shedding on an isolated system. *IEEE Trans. Power Syst.* **1995**, *10*, 1467–1472.

15. Chuvychin, V.; Gurov, N.; Venkata, S.; Brown, R. An adaptive approach to load shedding and spinning reserve control during underfrequency conditions. *IEEE Trans. Power Syst.* **1996**, *11*, 1805–1810.

16. Cirio, D.; Demartini, G.; Massucco, S.; Morim, A.; Scalera, P.; Silvestro, F.; Vimercati, G. Load control for improving system security and economics. In Proceedings of the 2003 IEEE Bologna Power Tech Conference Proceedings, Bolgna, Italy, 23–26 June 2003; Volume 4, p. 8.

17. Delfino, B.; Massucco, S.; Morini, A.; Scalera, P.; Silvestro, F. Implementation and comparison of different under frequency load–shedding schemes. In Proceedings of the IEEE Power Engineering Society Summer Meeting, Vancouver, BC, Canada, 15–19 July 2001; Volume 1, pp. 307–312.

18. Zhao, Q.; Chen, C. Study on a system frequency response model for a large industrial area load shedding. *Int. J. Electr. Power Energy Syst.* **2005**, *27*, 233–237.

19. Vieira, J.; Freitas, W.; Wilsun, X.; Morelato, A. Efficient coordination of ROCOF and frequency relays for distributed generation protection by using the application region. *IEEE Trans. Power Deliv.* **2006**, *21*, 1878–1884.

20. Gu, W.; Liu, W.; Zhu, J.; Zhao, B.; Wu, Z.; Luo, Z.; Yu, J. Adaptive decentralized under-frequency load shedding for Islanded smart distribution networks. *IEEE Trans. Sustain. Energy* **2014**, *5*, 886–895.

21. International Energy Agency. *Cool Appliances: Policy Strategies for Energy-Efficient Homes*; Organization for Economic Co-operation and Development (OECD): Paris, France, 2003.

22. Bertoldi, P.; Atanasiu, B. *Electricity Consumption and Efficiency Trends in the Enlarged European Union*; Technical Report; European Commission – Institute for Environment Sustainability, 2007. Available online: http://ies.jrc.ec.europa.eu (accessed on 9 April 2011).

23. Malik, O.; Havel, P. Active demand-side management system to facilitate integration of RES in low-voltage distribution networks. *IEEE Trans. Sustain. Energy* **2014**, *5*, 673–681.

24. Short, J.A.; Infield, D.G.; Freris, L.L. Stabilization of grid frequency through dynamic demand control. *IEEE Trans. Power Syst.* **2007**, *22*, 1284–1293.

25. Molina-García, A.; Bouffard, F.; Kirschen, D. Decentralized demand-side contribution to primary frequency control. *IEEE Trans. Power Syst.* **2011**, *26*, 411–419.

26. Molina-García, A.; Muñoz Benavente, I.; Hansen, A.; Gómez-Lázaro, E. Demand-side contribution to primary frequency control with wind farm auxiliary control. *IEEE Trans. Power Syst.* **2014**, *29*, 2391–2399.

27. Samarakoon, K.; Ekanayake, J. Demand side primary frequency response support through smart meter control. In Proceedings of the 44th International Universities Power Engineering Conference (UPEC), Glasgow, UK, 1–4 September 2009; pp. 1–5.

28. Kundur, P. *Power System Stability and Control*; McGraw-Hill: New York, NY, USA, 1994.

29. Ullah, N.R.; Thiringer, T.; Karlsson, D. Temporary primary frequency control support by variable speed wind turbines—Potential and applications. *IEEE Trans. Power Syst.* **2008**, *23*, 601–612.

30. REE. P.O. 7.1. *Servicio Complementario de Regulación Primaria*; Technical Report; Red Eléctrica de España: Madrid, Spain, 1998. Available online: http://www.ree.es (accessed on 20 May 2010).

31. UCTE. Operation Handbook – ver. 2.5. Technical Report; European Network of Transmission Network, Brussels, Belgium, 2004. Available online: https://www.entsoe.eu/resources/publications/system-operations/operation-handbook/ (accessed on 8 April, 2011).

32. Vigueras-Rodríguez, A.; Sørensen, P.; Cutululis, N.; Viedma, A.; Donovan, M. Wind model for low frequency power fluctuations in offshore wind farms. *Wind Energy* **2010**, *13*, 471–482.

33. Sørensen, P.; Hansen, A.D.; Carvalho-Rosas, P.E. Wind models for simulation of power fluctuations from wind farms. *J. Wind Eng. Ind. Aerodyn.* **2002**, *90*, 1381–1402.

34. Vigueras-Rodríguez, A. Modelling of the Power Fluctuations in Large Offshore Wind Farms. Ph.D. Thesis, Universidad Politécnica de Cartagena, Cartagena, Spain, 2008.

35. Sørensen, P.; Cutululis, N.; Vigueras-Rodríguez, A.; Madsen, H.; Pinson, P.; Jensen, L.; Hjerrild, J.; Donovan, M. Modelling of power fluctuations from large offshore wind farms. *Wind Energy* **2008**, *11*, 29–43.

36. Norgaard, P.; Holttinen, H. A multi-turbine power curve approach. In Proceedings of the Nordic Wind Power Conference (NWPC'04), Gothenburg, Sweden, 1–4 March 2004; Volume 1.

37. Wan, Y.; Ela, E.; Orwig, K. Development of an equivalent wind plant power curve. Proc. Wind Power, 2010, pp. 1–20.

38. Iain MacLeay, K.H.; Annut, A. *Digest of United Kingdom Energy Statistics 2014*; National Statistics Publication: London, UK, 2014.

39. International Energy Agency (IEA). *World Energy Outlook* 2014. Technical Report; IEA: Paris, France, 2014. Available online: http://dx.doi.org/10.1787/weo-2014-en (accessed on 14 July 2015).

40. Vigueras-Rodríguez, A.; Sørensen, P.; Viedma, A.; Donovan, M.H.; Gómez-Lázaro, E. Spectral coherence model for power fluctuations in a wind farm. *J. Wind Eng. Ind. Aerodyn.* **2012**, *102*, 14–21.

41. Saidur, R.; Masjuki, H.H.; Jamaluddin, M.Y. An application of energy and exergy analysis in residential sector of Malaysia. *Energy Policy* **2007**, *35*, 1050–1063.

42. Villena-Lapaz, J.; Vigueras-Rodríguez, A.; Gómez-Lázaro, E.; Molina-García, A.; Fuentes-Moreno, J.A. Stability assessment of isolated power systems with high wind power penetration. In Proceedings of the European Wind Energy Asociation Conference (EWEA), Copenhagen, Denmark, 16–19 April 2012.

MDPI

Article

Modeling and Optimization of the Medium-Term Units Commitment of Thermal Power

Shengli Liao, Zhifu Li, Gang Li *, Jiayang Wang and Xinyu Wu

Institute of Hydropower System & Hydroinformatics, Dalian University of Technology, Dalian 116024, China; shengliliao@dlut.edu.cn (S.L.); lizhifu@mail.dlut.edu.cn (Z.L.); aroundtheworld@mail.dlut.edu.cn (J.W.); wuxinyu@dlut.edu.cn (X.W.)

* Correspondence: glee@dlut.edu.cn; Tel.: +86-411-8470-8468; Fax: +86-411-8470-8768

Academic Editor: Ying-Yi Hong
Received: 25 September 2015 ; Accepted: 4 November 2015 ; Published: 12 November 2015

Abstract: Coal-fired thermal power plants, which represent the largest proportion of China's electric power system, are very sluggish in responding to power system load demands. Thus, a reasonable and feasible scheme for the medium-term optimal commitment of thermal units (MOCTU) can ensure that the generation process runs smoothly and minimizes the start-up and shut-down times of thermal units. In this paper, based on the real-world and practical demands of power dispatch centers in China, a flexible mathematical model for MOCTU that uses equal utilization hours for the installed capacity of all thermal power plants as the optimization goal and that considers the award hours for MOCTU is developed. MOCTU is a unit commitment (UC) problem with characteristics of large-scale, high dimensions and nonlinearity. For optimization, an improved progressive optimality algorithm (IPOA) offering the advantages of POA is adopted to overcome the drawback of POA of easily falling into the local optima. In the optimization process, strategies of system operating capacity equalization and single station operating peak combination are introduced to move the target solution from the boundary constraints along the target isopleths into the feasible solution's interior to guarantee the global optima. The results of a case study consisting of nine thermal power plants with 27 units show that the presented algorithm can obtain an optimal solution and is competent in solving the MOCTU with high efficiency and accuracy as well as that the developed simulation model can be applied to practical engineering needs.

Keywords: medium-term unit commitment (UC); progressive optimality algorithm (POA); target isopleths; utilization hours of installed capacity

---

1. Introduction

In the past several decades, China's power system has experienced rapid development along with rapid economic growth. Among all electric power sources, thermal power, which mainly consists of coal-fired plants, accounts for the largest proportion of China's power system. By the end of 2012, China's total installed generation capacity reached 1146.76 GW, of which 819.68 GW (approximately 71.5%) was from thermal power, and the total electric energy production reached 4986.5 TWh, with 3925.5 TWh (approximately 78.7%) coming from thermal power. Simultaneously, the utilization hours of the installed capacity for thermal power plants are larger than those of hydro power or wind power. The statistical results of 2012 showed that the annual utilization hours of power plants with an installed capacity over 6 MW are 4982 h for thermal power, 3591 h for hydropower and 1929 h for wind power [1–5]. However, because the preheating and cooling of the stream-turbine requires more time than other resource units, the coal-fired thermal unit is inflexible between start-up and shut-down resulting in a slow response to system load demands. Moreover, the start-up and shut-down of coal-fired thermal units cause high fuel costs [6–9]. Generally, thermal plants are usually scheduled to meet base load demands to ensure that the generation process runs smoothly and reduce the start-up and shut-down times of thermal units. Furthermore, the different values of power system demand become larger among adjacent days due to various factors, including heavy weather, minor vacation, and transition periods between dry and wet season for hydro power [10,11]. This situation causes remarkable difficulties in preparing the generation schedule for power systems, especially for ones with numerous thermal plants. Consequently, the power dispatch centers of China are in charge of the medium-term optimal commitment of thermal units (MOCTU), which usually span from one month to half a year and take one day as the time interval in calculations to obtain the numbers and times of the start-up and shut-down units for every thermal plant each day before actual short-term and daily-ahead scheduling.

The objective of MOCTU is to find the optimal set of thermal generating units or the boot capacity (sum capacity of units that putting into operation in the first operation of a plant) scheme of a power system to satisfy the system load demand, operational restrictions, reliability constraints, and security requirements in each time period. Thus the MOCTU is a form of unit commitment (UC) problem [8,9,12]. In contrast with short-term UC in power systems, which involves determining a start-up and shut-down schedule of units to meet the required demand during a short-term period such as one day or one week [13,14], MOCTU is responsible for dispatching load to each plant for maintenance scheduling and hydrothermal coordination. The main difference between medium-term and short-term thermal UC optimization is that medium-term optimization takes each plant's operating capacity or unit numbers as decision variables instead of scheduling a single unit by short-term operation [15]. It is well-known that the utilization rate of power generation equipment reflects the generation ability of a thermal plant, and the standby condition of generation equipment reflects the ability of a thermal plant to deal with sudden accidents, and utilization hours of the installed capacity for thermal power plants can reasonably reflect the two abovementioned factors, so that operators can operate and manage plants better. Hence, many power grids take equal accumulated operating hours of installed capacity for all thermal power plants as the primary objective of MOCTU during the selected period. However, due to thermal power's main characteristics, including slow start-up and shut-down, complex constraints, large-scale as well as large complicated demands by power grid, developing and establishing computer simulation and optimization models for thermal UC optimization have been most challenging and complex issues [16–18]. Furthermore, to achieve the best possible optimal solution and the largest benefit for thermal power plants and power grid, the model should be simulated as close to reality as possible to suit the practical needs of power dispatch centers, especially in China [19–21]. However, according to our retrieval results, there are few studies on MOCTU, especially on the development of simulation models for MOCTU that meet the demands of practical engineering [22,23]. In this paper, a medium-term thermal UC optimization model with equal capacity utilization hours for all thermal power plants is established. The model not only respects the basic principles of electric power dispatch (equality, impartiality and transparency), but also adopts a reward principle that gives extra operation hours to the thermal plants with low energy consumption and dust discharge and high efficiency, namely reward hours.

The work described in this paper is originated from the real, practical demands for power dispatch centers of China to develop a medium-term optimal commitment of thermal units. The aim of this paper is to develop a simulation model for MOCTU to satisfy the needs of practical engineering and present a feasible and effective algorithm to optimize the model. MOCTU is a multi-stage decision problem that involves a highly nonlinear and computationally expensive objective function with a large number of constraints. The progressive optimality algorithm (POA) has been shown to be an effective method for solving multi-stage optimization problems by decomposing a multi-stage decision problem into a series of non-linear programming two-stage problems [24–28], and it is suitable for solving MOCTU. However, it is a difficult task to find feasible solutions for a large-scale MOCTU problem using POA due to its drawback, namely the easily encountered local optimum for complex problems. Therefore, in this paper, an improved progressive optimality algorithm (IPOA) is proposed for a large-scale MOCTU problem with nine thermal plants and 27 units in Yunnan Province of China to validate the effectiveness and practicality of the developed simulation model as well as to improve the quality of optimal solutions of POA. The original optimization problem is first optimized by using a heuristic method and an initial feasible solution is obtained, then POA is adopted to search the optimal solution, and finally two strategies are utilized to adjust the solution from the constraint boundaries into the feasible zone's interior to continue to search for the global solution. Actually, the mid-term boot scheduling plan of thermal plants has been generated by operators in Yunnan power grid (YNPG) using this proposed simulation model and method since 20 May 2013. Over a two-year implementation period, it has given operators more confidence to use it.

This paper is organized as follows: Section 2 gives the formulation of the mathematical model of MOCTU; Section 3 analyzes the characteristics of this problem and presents the IPOA used to solve it; Section 4 applies the model to a real-world scenario with nine thermal plants and 27 units in Yunnan province of China; Finally, Section 5 presents the conclusions.

2. Mathematical Model

2.1. Objective Function

MOCTU is an important and complex task for medium-term scheduling of power grids. On the one hand, reasonable MOCTU results should not only satisfy the complicated operation requirements of thermal plants, but respond to system load changes rapidly. Furthermore, they should meet the requirements of the national energy conservation policy as well as balance the interests between different thermal plants. In actual management and operation of thermal power in China, utilization hours of installed capacity for thermal plants can reflect the utilization rate of power equipment and standby application of thermal plants. Thus, controlling the utilization hours of installed capacity for thermal plants has always been the goal that needs to be considered in MOCTU for many power grids. Based on the abovementioned information, taking the equal utilization hours of installed capacity for thermal plants as the optimization goal, and at the same time, the concept of reward-hours is introduced in the model, though which power grids can reward stations that with small coal consumption flexibly according to energy saving principles. Assuming the total number of plants participating in the calculation is m, the objective function can be expressed as follows:

$$h_i = h_j \ (i \neq j) \tag{1}$$

where h_i and h_j are the accumulated operating hours of installed capacity for plants i, j, respectively, in the selected period.

Although the above objective function (Equation (1)) is conceptually clear, it is very difficult to solve directly because the objective formulation is related to all plants and the target result cannot be obtained by computing a single mathematical expression. Considering the fact that variance can be used to describe the dispersion degree of data series, the original objective function is replaced by minimizing the variance of vector $\mathbf{h} = [h_1, h_2, \cdots, h_i, \cdots, h_m]$, which is composed of plants' capacity utilization hours. The replacement function can be represented as:

$$\min f = \frac{1}{m} \sum_{i=1}^{m} (h_i - \bar{h})^2 \tag{2}$$

$$\bar{h} = \frac{1}{m} \sum_{i=1}^{m} h_i \tag{3}$$

$$h_i = h_i^P + h_i^T - h_i^G \tag{4}$$

$$h_i^T = \frac{1}{N_i} \times \overline{C_i} \times T \times 24 \times r_i \qquad (5)$$

$$\overline{C_i} = \frac{1}{T} \sum_{t=1}^{T} C_i^t \qquad (6)$$

where the following notations are used:

i, j plant i and plant j, $i, j = 1, 2, \cdots m$;

T Number of time steps. $t = 1, 2, \cdots, T$;

h_i Accumulated operating hours of installed capacity of plant i in the selected period;

h_i^P Operating hours of installed capacity of plant i accumulated in pre-calculation period;

h_i^T Operating hours of installed capacity of plant i accumulated in the calculation period;

h_i^G Extra award hours assigned to plant i;

N_i Installed capacity of plant i;

$\overline{C_i}$ Average operating capacity of plant i in the calculation period;

r_i Given load factor of plant i;

C_i^t Operating capacity of plant i in period t. It is also the decision variable to be solved.

In this model, the values of h_i^P and h_i^G are given and do not require special treatment while solving the problem. It should be noted that, under the condition of h_i^P being unequal to h_i^G, to equalize h_i, h_i^T has to be made unequal. Thus, it realizes the differentiation of capacity utilization hours of all thermal power plants in the calculation and ensures the flexibility of the model.

2.2. Constraints

For MOCTU, each individual thermal plant is subjected to its own set of constraints, while the power system is subjected to system power balance constraints. Specifically, we consider the following constraints:

(1) Unit number constraints:

$$\underline{n_i^t} \leqslant n_i^t \leqslant \overline{n_i^t} \qquad (7)$$

where $\underline{n_i^t}$ and $\overline{n_i^t}$ are the minimum and maximum unit number of plant i in period t, respectively; n_i^t is the active unit number of plant i in period t.

(2) System power balance constraints:

$$C_S^t \times \underline{r} \times \Delta h_t \leqslant P_D^t \times \Delta h_t \leqslant C_S^t \times \bar{r} \times \Delta h_t \qquad (8)$$

where C_S^t is the operating capacity of the system in period t in MW; Δh_t is the number of hours in period t in h, and here $\Delta h_t = 24$; P_D^t is the load demand of the system in period t in MW; $\underline{r}$ and $\bar{r}$ are, respectively, the minimum and maximum load factors of thermal power, which are calculated from the long-term real-world operation of power grid, $\underline{r}, \bar{r} \in [0, 1]$, and $r_m = (\underline{r} + \bar{r})/2$ is recorded as the mean-value of the load factor of thermal power.

(3) Peak and valley duration constraints:

The medium-term thermal UC optimization only dispatches system load to plants, and there is no need to distribute load among units. Therefore, we transform minimum up- and down-time constraints into minimum peak and valley time constraints:

$$\begin{cases} T_i^{\text{up}} \geqslant \underline{T_i^{\text{up}}} \\ T_i^{\text{down}} \geqslant \underline{T_i^{\text{down}}} \end{cases} \qquad (9)$$

where T_i^{up} is the minimum duration time during the peak in the operating capacity process of plant i in calculation period; $\underline{T}_i^{up}$ is the limit value of the minimum duration time in the peak in the operating capacity process of plant i in the calculation period; T_i^{down} is the minimum duration time in the valley period in the operating capacity process of plant i; $\underline{T}_i^{down}$ is the limit value of the minimum duration time in the valley period in the operating capacity process of plant i.

(4) Boundary condition constraints:

The plants' generating scheme is not only related to the calculation periods, but also relevant to the boot capacity scheme in previous periods, namely pre-calculation periods, where the start-up mode has been determined, because the boot capacity scheme in the pre-calculation period has an impact on the generating scheme in current periods. Therefore, the target start-up mode in calculation periods should link up with existing p days start-up mode in calculation periods. The relationship among all periods is shown in Figure 1.

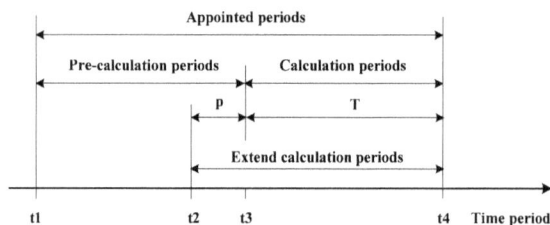

Figure 1. Sketch map of the time relationship.

Note that the actual unit number and operating capacity of a plant are variables to be calculated, and the others are given input data.

3. Model Solution

3.1. Solution Approach

MOCTU is a form of the UC problem with a large number of constraints, including unit number constraints (Constraint Group 1), system power balance constraints (Constraint Group 2), peak and valley duration constraints (Constraint Group 3), and boundary condition constraints (Constraint Group 4). It is obvious that Constraint Groups 1 and 2 are single-period constraints that can be treated to reduce the search space. Although Constraint Groups 3 and 4 are multi-period constraints, Constraint Group 4 can be satisfied when Constraint Group 3 is satisfied in extended calculation periods. On the other hand, MOCTU is a high dimensional problem to be optimized. For a power grid with two thermal plants, each having three units. For one plant with three units, it has $(3 + 1)$ kinds of combination in one period, and for two plants, it reaches $(3 + 1)^2$, so the total number of unit combinations will reach $[(3 + 1)^2]^{30} \approx 1.329 \times 10^{36}$ for a horizon of one month with 30 periods, and the overhead of optimization will be computationally expensive. Considering the complex constraints, especially Constraint Group 3, the solving process can be divided into two procedures:

Procedure I: Obtaining the feasible solution space $S = [S_1, S_2, \cdots, S_t, \cdots, S_T]$, where S_t represents all of the possible unit-commit combinations that satisfy Constraint Groups 1 and 2 in period t. First, combine all units of all plants to obtain all possible unit-commit combinations in each period. Second, obtain the solution space S by filtering out the combinations that cannot satisfy Constraint Groups 1 and 2 period-by-period. The solution for the UC scheme in the calculation period can be acquired from the unit-commit combinations in S. The above process is relatively simple.

Procedure II: Search for the global optimal solution that can meet Constraint Group 3. Because the solution in S from *Procedure I* already satisfies Constraint Groups 1 and 2, and as well as Constraint

Group 4 being automatically satisfied after satisfying Constraint Group 3, the aim of this search process is to find the solution that satisfies Constraint Group 3. The following discussion will introduce how to find the optimal scheduling that meets Constraint Group 3 in the solution space *S*. First, generate an initial feasible solution in *S* that satisfies Constraint Group 3 by using heuristic method. Second, produce the global optimal solution by the POA. However, POA is sensitive to initial trajectories and sometimes cannot guarantee a global optima, which is the reason that restricts its application. Thus, POA easily converges to a local optimum while being directly used for MOCTU, which is a strong constrained optimization problem. For the medium-term thermal UC optimization, a local optimum means that the system power reaches the upper or lower limit boundary constraints or that the duration periods of peak and valley in the operating capacity process are equal to the limit value of the minimum duration periods. In this case, it is very difficult to further optimize the result. Therefore, an improved POA combining POA with two adjustment strategies is presented to overcome the mentioned demerits.

3.2. Initial Feasible Solution Generation

As *Procedure I* has achieved the solution space *S* that meets Constraint Groups 1 and 2, the initial feasible solution can be acquired by searching the solution space *S* by heuristic method only when the operating load rate approximates to the given target load rate that obtained from long-term schedule of power dispatch centers in each period. The feasible region in the latter period will be sharply contracted when the generating scheme in the previous period is determined and the duration period constraints (Constraint Group 3) are considered. To avoid the infeasible region, this paper suggests r_m as the target load rate. Before searching, sort the generating schemes in S_t in every period according to the ascending order of the index value $|r_k - r_m|$. Where r_k represents the operating load rate of the *k*th generating scheme in S_t, and its formula is as follows:

$$r_k = \sum_{i=1}^{m} C_i^k / \sum_{i=1}^{m} N_i \tag{10}$$

The generating scheme S_t', which is sorted according to the ascending order of the index value $|r_k - r_m|$ in every period, can be obtained by using Equation (10). Then, the feasible solution can be obtained from the sorted solution space $\mathbf{S'} = [S_1', S_2', \cdots, S_t', \cdots, S_T']$ by the heuristic method. The detailed procedure is as follows:

Step 1: Construct an integer array *ks* of length *T*. Set *ks*[1] = 1 and *t* = 1. Make sure that the combination of boot capacity in period 1 is the first element of the solution space S_t'.

Step 2: Set *t* = *t* + 1. If *t* > *T*, go to Step 6. Otherwise, set *ks*[1] = 1 and set the combination of boot capacity in period *t* as the first element of the solution space S_t'.

Step 3: Verify whether the start-up mode lying in the interval of [0,1] satisfies the constraints of Equation (3). If it does satisfy them, go to Step 2; otherwise, go to Step 4.

Step 4: Set *ks*[*t*] = *ks*[*t*] + 1. If *ks*[*t*] is greater than the number of elements in solution space S_t', go to Step 5. If not, replace the combination of boot capacity in period *t* with the *ks*[*t*]th element of the solution space S_t' and go to Step 3.

Step 5: Set *t* = *t* + 1, and go to Step 1.

Step 6: Output the result.

3.3. Optimization Process of Progressive Optimality Algorithm (POA)

Considering the complex requirements by power grids and the constraints mentioned above, the present optimization problem exhibits multi-stage, large-scale and high dimension characteristics. A suitable solution algorithm is required for solving the MOCTU. Based on Bellman's Principle, the POA, which is proposed by Howson and Sancho for reducing dimensionality difficulties by decomposing a multi-stage decision problem into a series of non-linear programming two-stage problems [29], has been shown to have great advantages over classical optimization methods as one of the most widely used techniques for hydroelectric generator scheduling and water resources problems [30,31]. The advantages of POA over other optimization techniques are that it can decompose a multi-state decision problem into several nonlinear programming sub-problems to reduce the dimensionality. More specifically, the merits of POA have clearly been elaborated in [32], including no need to discretize the state variables, no resolution or linearization of nonlinear objective functions and constraints and minimal storage requirements.

To understand the main principle of POA, a general multi-stage optimization example was given. In this example, a set of state variables $x_0^*, x_1^*, x_2^*, \cdots, x_{T-1}^*, x_T^*$ is given to determine the optimal solution by minimizing objective functio $f(x_0, x_1, x_2, \cdots x_T)$, where x_0 and x_T are given as the initial state values. The algorithm starts with an initial trajectory $x_0, x_1^0, x_2^0, \cdots, x_{T-1}^0, x_T$ which is gained in a certain way. Then, this multi-stage decision problem would be decomposed into two problems, each of which minimizes $f(x_{i-1}^j, x_i^j, x_{i+1}^j)$ to obtain optimal $x_i^{j,*}$ at stage i during the jth iteration by fixing values of x_{i-1}^j and x_{i+1}^j. In other words, x_0^j and x_2^j are fixed to optimize x_1^j to yield $x_1^{j,*}$, as shown in Figure 2a. Then $x_1^j = x_1^{j,*}$ and x_3^j are fixed to optimize x_2^j for $x_2^{j,*}$, as shown in Figure 2b, in turn, the jth iteration isn't finished until $x_{T-1}^{j,*}$ is obtained, as shown in Figure 2c,d. Based on the optimized results from the last iteration, another iteration is restarted again. The process isn't over until the difference between the last two iterations meets the predefined precision limit. Figure 2 illustrates the optimization process during the jth iteration.

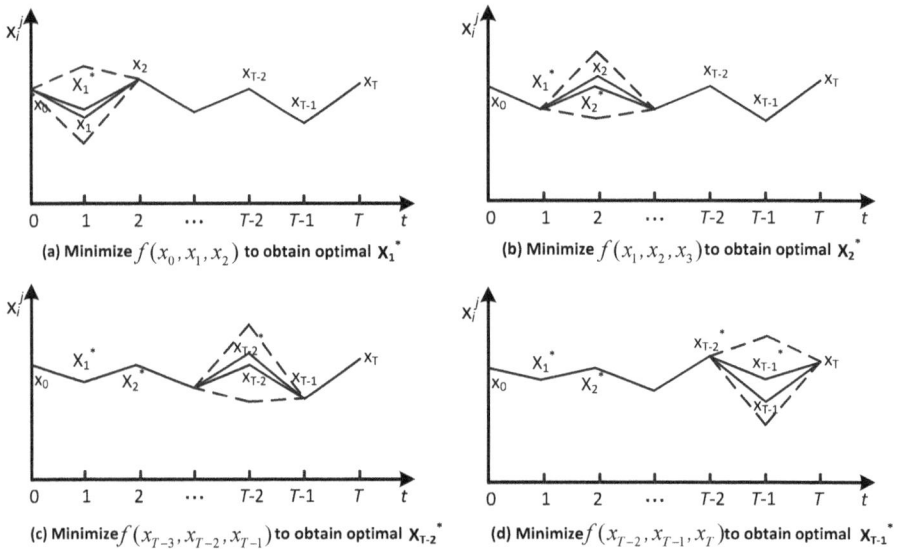

Figure 2. Optimization process of progressive optimality algorithm (POA).

The initial feasible solution obtained in Section 3.3 is used for POA to search for optimal solution. The general process can be mainly described in the following steps:

Step 1: Set the generating scheme to an initial feasible solution generated by the heuristic search and an initial objective function value *value*. Set $t = 1$ and $k = 1$.

Step 2: Obtain a generating scheme by replacing the generating scheme in period t with the kth element of the solution space S_t. If the new generating scheme meets Constraint Group 3, go to Step 3. If not, go to Step 4.

278

Step 3: Calculate the objective function value *value'* according to Equation (5). If *value'* < *value*, adjust the results by using the equalization operation and combination operation and set *value'* = *value*.

Step 4: Set $k = k + 1$. If k is greater than the number of elements in solution space S_t, go to Step 5. If not, go to Step 2.

Step 5: Set $t = t + 1$ and $k = 1$. If $t < T$, go to Step 2. If not, it means one iteration has been performed, then check whether the objective value has been improved over the previous iteration. If so, set the initial solution as the acquired result and go to Step 2. If not, go to Step 6.

Step 6: Output the result.

3.4. Equalization Operation and Combination Operation

Although possessing advantages of quick and strong convergence, POA easily falls into sub-optimal results because each iterative calculation is merely associated with a single period. Unable to make full use of the multi-stage information of the original problem, the sub-optimal results are easily attained at the bounds of constraints. To enhance the searching space and obtain the global optimal result, two strategies are presented including: equalization operation of the system operating capacity and single station boot peak combination operation.

The equalization operation of the system operating capacity is for making the start-up units' capacity equal to the shut-down units' capacity as much as possible by means of adjusting the ratio of the system load demand for each time and operating capacity to approach r_m. The equalization operation has two purposes: (1) to make the system generating scheme move from the power constraint boundary into the feasible solution's interior; and (2) to make it more convenient to move and merge the peak of the operating capacity. The equalization process is equivalent to solving the following programming problem:

$$\min f = \sum_{t=1}^{T} \left| C_s^t r_m - P_D^t \right| \tag{11}$$

The constraints are the same as those we have previously mentioned. The approach needs to traverse the generating scheme of each plant in turn and to find whether it satisfies all constraints when performing start-up (shut-down) of one unit in period t and shut-down (start-up) of one unit in period $t + j$ ($j > 0$). If it satisfies the constraints, the problem can be optimized with Equation (11) as the target. In the search process, the system's operating capacity remains unchanged, and it can guarantee that the capacity utilization hours of every plant are unchanged. The search process is shown in Figure 3.

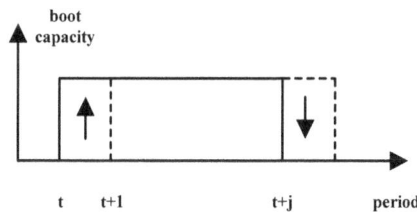

Figure 3. Sketch map of the system's boot capacity equalization.

As the equalization operation makes the start-up unit's capacity equal to the shut-down unit's capacity as much as possible, it provides the peak moving of a single plant's operating capacity some space for adjustment. The single station boot peak combination process searches the single station boot capacity in turn for the peak with the minimum number of duration periods and finds whether it should move and merge it with other peaks. If so, it executes the combination operation.

Without changing the objective function value, this method can make an adjustment to the operating capacity process, which requires that the operating durations of peak and valley periods equalize to the demanded minimum duration, and forces it to leave the local optimum. The peak combination operation is shown in Figure 4.

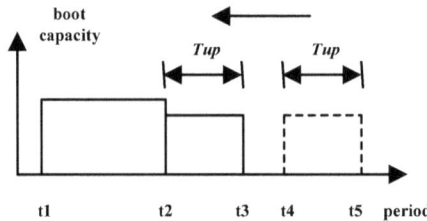

Figure 4. Sketch map of the single station boot peak combination.

From Equations (2)–(6), it can be seen that the installed capacity utilization hours of a thermal power plant is relative to the accumulative total but not the process of the scheme. Utilizing the two operations mentioned above, the results from POA can be converted by making the generating scheme move from the constraint boundary to the feasible region internal area under the condition of the unchanged objective function. As a result, POA acquires its new search space and will not stop iterating until convergence is obtained. The search principle of POA with the two strategies is shown in Figure 5. The lines of $f = F_0$, $f = F_1$ and $f = F_2$ represent the isopleths of the objective function value. According to Equation (5), the objective function value is zero for the best solution x^*, namely $f(x^*) = 0$. Two main steps are defined in the optimization process. Step α is defined as the optimization process of POA and β presents the adjustment process of equalization operation and merge operation. First, the initial feasible solution x_0 converges to the local optimal solution x_1 through Step α which cannot be further optimized because x_1 lies on the constraint boundary. Secondly, through Step β the local optimal solution x_1 can be moved along the target isopleth $f = F_1$ to x_2 which is still in the feasible solution internal area, then step α is reutilized and solution x_3 is obtained. If x_3 still lies on the constraint boundary, then step β is utilized again and the local optimal solution x_3 is moved along the target isopleth $f = F_2$ to x_4. Based on the optimized results from the adjustment process, another iteration is restarted again until the global optimal solution x^* is obtained.

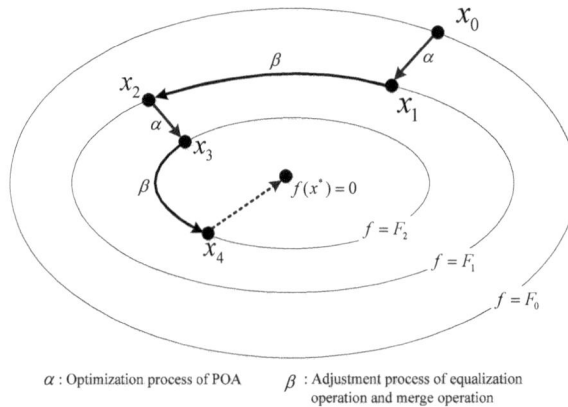

α : Optimization process of POA β : Adjustment process of equalization operation and merge operation

Figure 5. Sketch map of the search principle.

3.5. Solution Architecture

With a large amount of constraints and complex requirements, MOCTU is a challenging optimization problem that involves a highly nonlinear and computationally expensive objective functions. The optimization process is divided into two procedures: generation of an initial feasible solution and optimization of POA. However, POA is sensitive to initial trajectories and easily converges to local optima. Thus any changes from the decision or state variables may have a slight improvement on the solutions. Therefore, IPOA is adopted with the equalization operation and the combination operation to adjust the results acquired from POA. By doing this, it moves the solutions from constraint boundaries into the feasible zone's interior and then continues to search for the optimal solution with POA. The solution flowchart of MOCTU is shown in Figure 6 and the general process for the IPOA procedure can be mainly described in the following steps:

Step 1: Obtain the initial feasible solution by heuristic search and calculate the objective function *value*;

Step 2: Obtain the optimal results by using POA and calculate the objective function *value'* (the detailed process is described in Section 3.3);

Step 3: Judge whether the current isopleths of the objective function value are equal to zero, if not, use the two strategies to move the optimal results from the constraint boundary to the feasible region internal area to search again. Otherwise, go to Step 4.

Step 4: Output optimal scheduling.

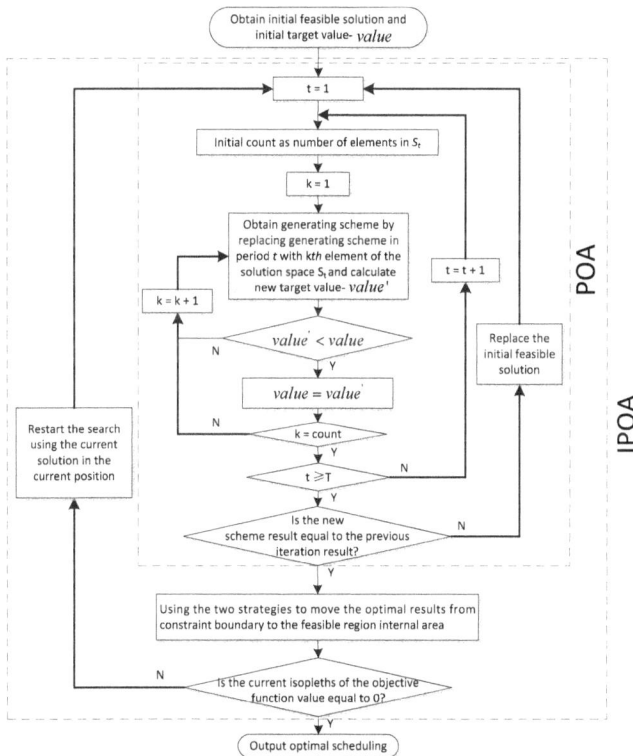

Figure 6. Solution flowchart of medium-term optimal commitment of thermal units (MOCTU) by improved progressive optimality algorithm (IPOA).

4. Simulation Results

The proposed simulation model and method have been applied to the YNPG in China. Currently, it is used as the primary tool to determine the mid-term boot scheduling of thermal plants by the operators of YNPG. By the end of 2013, the total installed capacity of YNPG had reached 26.1 GW among which hydro power was responsible for 16.7 GW and thermal power was responsible for 8.935 GW. The reasonable scheduling of thermal power is beneficial for optimizing hydropower systems and system security.

The thermal power system in the YNPG consists of nine thermal power plants and 27 units. As mentioned above in Section 3.1, it is a highly dimensional and complex problem to optimize. Therefore, it is a substantial challenge to determine the medium-term operating policies for these thermal plants. The dispatch center in the YNPG is in charge of these thermal power systems. Table 1 lists the basic data of these plants. IPOA has been implemented in Java on a PC with an Intel® Core™2 Duo CPU, operating at 2.93 GHz, with 2 GB of memory. A real scheduling for 2013 in the YNPG is used to test the validity and computational efficiency of the proposed method.

October, a typical month and the beginning of the dry season, was selected to demonstrate the actual availability of the simulation model as well as the practicality and efficiency of the method. In October, the hydro-power system comes to its lowest output and the demand on the thermal system increases. As a result, the operating scheduling of thermal power plants will obviously change. The simulation results and rationality of the algorithm will be presented in the paper. According to the real-world operating experience and users' actual demands, the parameters are set as follows:

1) Given load rate of plant i, $r_i = 0.8$;
2) Minimum load rate of thermal power, $\underline{r} = 0.7$;
3) Maximum load rate of thermal power, $\bar{r} = 0.9$;
4) Minimum duration periods of peak in the operating capacity process of plant i during the calculation period, $T_i^{up} = 7d$;
5) Minimum duration periods of valley in the operating capacity process of plant i during the calculation period, $T_i^{down} = 3d$;
6) Maximum unit number of plant k in period t, set $\overline{n_i^t}$ as the number of installed units
7) Minimum unit number of plant i in period t, $\underline{n_i^t} = 1(i = A \sim H)$, except $\underline{n_i^t} = 0$ while $i = I$;
8) Actual capacity utilization hours of plant i in the pre-calculation period, $h_i^P = 0$;
9) Extra award hours of plant i, $h_i^G = 0$;
10) System load demands in each period and the first ten days' boot process ($p = 10d$) are all given.

Table 1. Thermal power system in Yunnan power grid (YNPG).

Thermal plant	Units (number × capacity, MW)	Capacity (MW)
A	4 × 600	2400
B	6 × 300	1800
C	4 × 300	1200
D	2 × 200 + 2 × 300	1000
E	2 × 300	600
F	2 × 300	600
G	2 × 300	600
H	2 × 300	600
I	1 × 135	135

Comparison of each power plant's capacity utilization hours calculated by IPOA in October is shown in Table 2.

Table 2. Comparison of each power plant's capacity utilization hours in October (h).

Items	Heuristic search	POA	First adjustment	IPOA
Plant A	547.2	422.4	422.4	422.4
Plant B	297.6	416.0	416.0	422.4
Plant C	369.6	422.4	422.4	422.4
Plant D	240.0	422.4	422.4	422.4
Plant E	489.6	422.4	422.4	422.4
Plant F	489.6	432.0	432.0	422.4
Plant G	499.2	422.4	422.4	422.4
Plant H	528.0	432.0	432.0	422.4
Plant I	422.4	422.4	422.4	422.4
Average value	431.5	423.8	423.8	422.4
Max-min difference	307.2	16.0	16.0	0
Objective value (h^2)	10283	23.0	23.0	0

First adjustment means the results from the first equalization operation and combination operation.

Figures 7 and 8 respectively, show system energy balance maps and boot modes of plant *F* by different methods (the horizontal axis represents the time period). As can be seen from Figure 7, the results of the simulation model match the actual characteristics of thermal power system in October that the load demand on the thermal system is increasing. It demonstrates that the proposed simulation model for MOCTU is very practical and can satisfy the actual project requirements.

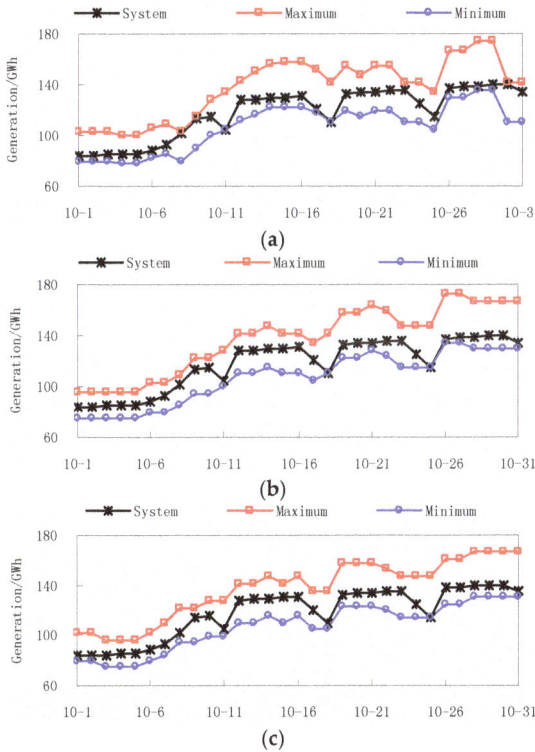

Figure 7. Energy balance maps by three methods: (**a**) POA; (**b**) first adjustment; (**c**) IPOA.

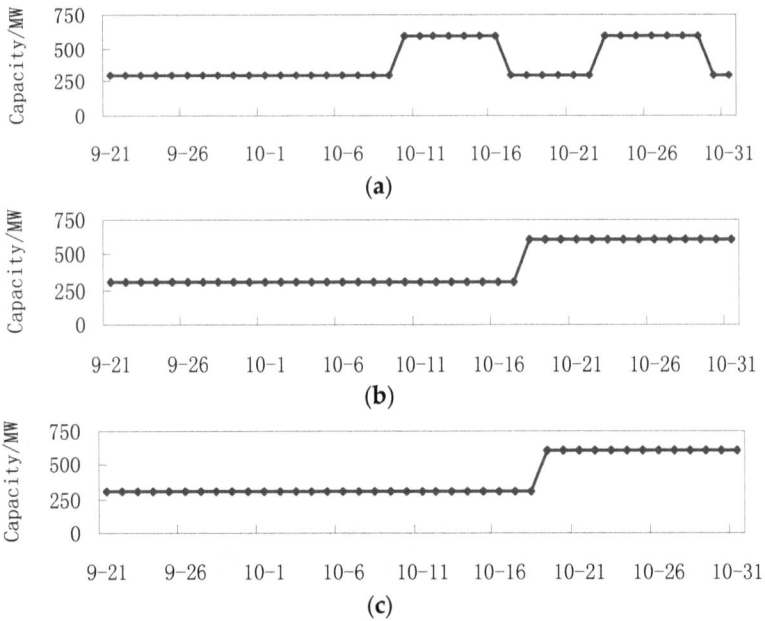

Figure 8. Boot capacity of Plant F for three methods: (**a**) POA; (**b**) first adjustment; (**c**) IPOA.

The calculation results on the YNPG are shown in Table 2. IPOA is compared with heuristics and POA. Max-min difference and objective value are taken as the measurement index. When the initial solution is obtained by heuristics, most plants' capacity utilization hours are different. As shown in Table 2, calculated by heuristic and POA in turn, the max-min differences are 307.2 h and 16 h, and the objective values are 10,283.0 h^2 and 23.0 h^2. Although the results have been significantly improved, the optimal solution is still not found. The reason is that the results obtained from POA run into local optima. The results by POA in Figure 7a illustrate that it has reached the system's power constraint boundary in many periods (10-08, 10-09, 10-30) and Figure 8a shows that two peak values of Plant F's boot mode reach the duration periods of peak in the operating capacity constraint, and it cannot be optimized any further by POA. Table 2, Figures 7b and 8b show that the equalization operation and combination operation can change the structure of solutions and make the solution leave the constraint boundary under the condition of the unchanged objective function value (23.0 h^2). After the first combination operation, the boot process of Plant F combines two peaks (Figure 8a) into one peak (Figure 8b) and makes the duration periods of the peak change from 7 d ($T^{up} = 7$) to 14 d, leaving the constraint boundary. After applying IPOA, the second unit's boot time of Plant F is adjusted from 18 October (Figure 8b) to 19 October (Figure 8c), and the capacity utilization hours change from 432 h to 422.4 h.

At the same time, Table 2 shows that it has found the optimal results of the problem, which indicates that the strategies of equalization operation and combination operation can provide a new round of the search for the optimization space and finding the global optimum. At this moment, the optimal solution is obtained, as shown in Table 2, which manifests that the equalization operation and combination operation provide the search space for a new iteration to find the optimal solution. The generation requirements for the nine plants in October are shown in Figure 9 and the details of the results are listed in Table 3.

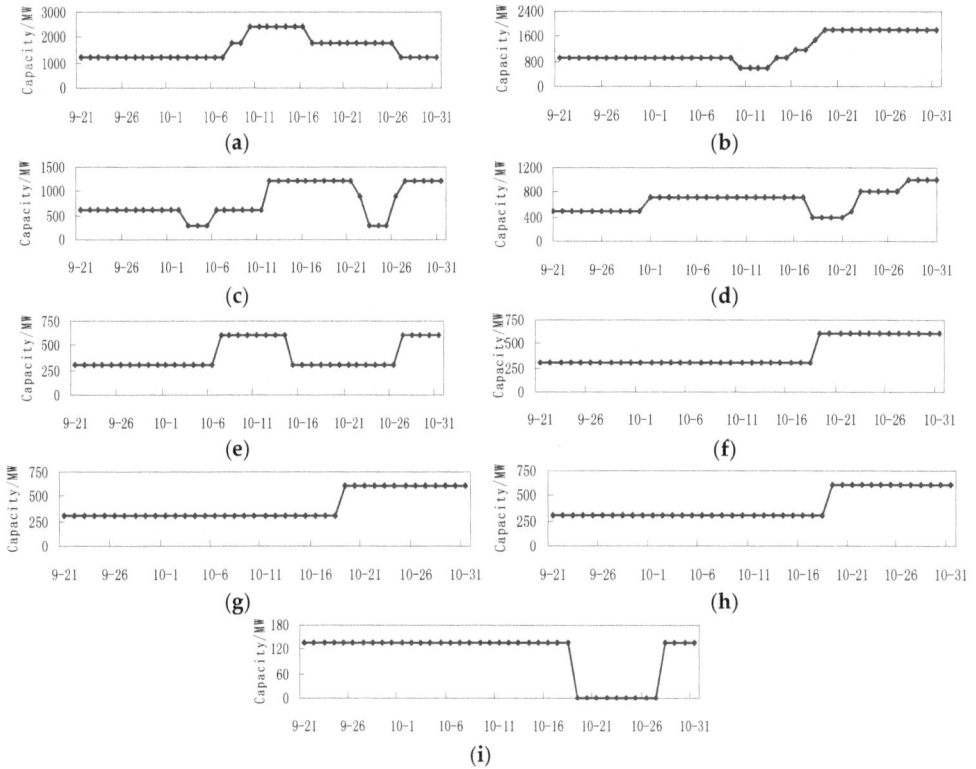

Figure 9. Results of thermal plants: (**a**) Plant A; (**b**) Plant B; (**c**) Plant C; (**d**) Plant D; (**e**) Plant E; (**f**) Plant F; (**g**) Plant G; (**h**) Plant H; (**i**) Plant I.

From Table 3, it can be seen that the boot capacity result of each plant can meet all constraints. In Plant B, for example, the days of the first valley in plant output are 4 (from 10th to 13th) which is larger than the minimum duration of valley in plant output $T_i^{down} = 3d$, while the days of first peak in plant output are 13 (from 19th to 31st) which is larger than minimum duration of peak in plant output $T_i^{up} = 7d$. Figure 9 lists the boot processes for each plant achieved by IPOA and all of the boot processes are satisfied Constraint Groups 3 and 4, which demonstrate the effectiveness of the proposed algorithm.

For comparison and illustration the impact of extra award hours, namely h_i^G, attach extra award hours for Plant C, D and F, and the award hours are 30, 20 and 10, respectively. The other conditions are unchanged. The optimization results are shown in Table 4. Because values of installed capacity utilization hours of each plant is discrete in calculation periods, the objective function value corresponding to optimal solutions is not always zero, and as a result, the objective value becomes 0.057 h^2. Moreover, installed capacity utilization hours of Plants C, D and F are more 29.52, 19.36 and 9.68 h than other plants, respectively, and the calculation precision meets the demand of practical engineering.

Table 3. Results of generation requirements for plants in October (MW).

Date	Plant A	Plant B	Plant C	Plant D	Plant E	Plant F	Plant G	Plant H	Plant I
1st	1200	900	600	700	300	300	300	300	135
2nd	1200	900	600	700	300	300	300	300	135
3rd	1200	900	**300**	700	300	300	300	300	135
4th	1200	900	300	700	300	300	300	300	135
5th	1200	900	300	700	300	300	300	300	135
6th	1200	900	**600**	700	300	300	300	300	135
7th	1200	900	600	700	**600**	300	300	300	135
8th	**1800**	900	600	700	600	300	300	300	135
9th	1800	900	600	700	600	300	300	300	135
10th	**2400**	**600**	600	700	600	300	300	300	135
11th	2400	600	600	700	600	300	300	300	135
12th	2400	600	**1200**	700	600	300	300	300	135
13th	2400	600	1200	700	600	300	300	300	135
14th	2400	**900**	1200	700	600	300	300	300	135
15th	2400	900	1200	700	**300**	300	300	300	135
16th	2400	**1200**	1200	700	300	300	300	300	135
17th	**1800**	1200	1200	700	300	300	300	300	135
18th	1800	**1500**	1200	**400**	300	300	300	300	135
19th	1800	**1800**	1200	400	300	**600**	**600**	**600**	**0**
20th	1800	1800	1200	400	300	600	600	600	0
21st	1800	1800	1200	400	300	600	600	600	0
22nd	1800	1800	**900**	**500**	300	600	600	600	0
23rd	1800	1800	**300**	**800**	300	600	600	600	0
24th	1800	1800	300	800	300	600	600	600	0
25th	1800	1800	300	800	300	600	600	600	0
26th	1800	1800	**900**	800	300	600	600	600	0
27th	**1200**	1800	**1200**	800	**600**	600	600	600	0
28th	1200	1800	1200	**1000**	600	600	600	600	**135**
29th	1200	1800	1200	1000	600	600	600	600	135
30th	1200	1800	1200	1000	600	600	600	600	135
31st	1200	1800	1200	1000	600	600	600	600	135

Table 4. Comparison of capacity utilization hours of each plant (h).

Items	h_i^G	Heuristic Search		IPOA	
		$h_i^p + h_i^T$	h_i	$h_i^p + h_i^T$	h_i
Plant A	0	240.00	240.00	422.40	422.40
Plant B	0	297.60	297.60	422.40	422.40
Plant C	30	399.12	369.12	451.92	421.92
Plant D	20	508.96	488.96	441.76	421.76
Plant E	0	422.40	422.40	422.40	422.40
Plant F	10	499.28	489.28	432.08	422.08
Plant G	0	547.20	547.20	422.40	422.40
Plant H	0	499.20	499.20	422.40	422.40
Plant I	0	528.00	528.00	422.40	422.40
Average value	-	-	431.50	-	422.24
Max-min difference	-	-	307.20	-	0.64
Objective value (h²)	-	-	10,277.489	-	0.057

5. Conclusions

According to the real and practical demands of power dispatch centers of China, a simulation model for medium-term thermal UC is presented and solved by using an efficient solution named IPOA. As a new contribution to thermal power UC for power grids, the simulation model gives plenty of considerations to load demands, plant characteristics, practical engineering needs and

Energies **2015**, *8*, 12848–12864

other important factors. IPOA, which is developed to improve the performance of POA, is another contribution. The major advantage of IPOA is that strategies of operating capacity equalization and single station operating peak combination are performed to move the feasible solution away from constraint boundaries, along target isopleths, into the feasible zone's interior, in order to provide the search space for a new iteration to overcome the drawback of easily converging to local optima of POA. Basically, the mid-term boot scheduling plan of thermal plants has been generated by operators in YNPG using the proposed simulation model and method since 20 May 2013. Two-year implementation and the results of thermal power plants with 27 units in YNPG demonstrate that the proposed model has made a closer step from theoretical developments to real-world implementations, and that IPOA is an effective method for solving the MOCTU problem.

Acknowledgments: This study is supported by the National Natural Science Foundation of China (No. 51209029), the Fundamental Research Funds for the Central Universities (20120041120002) and the Fundamental Research Funds for the Central Universities (No. DUT14QY15).

Author Contributions: Shengli Liao and Gang Li devised the experimental strategy and carried out this experiment. Shengli Liao and Zhifu Li wrote the manuscript and contributed to the revisions. Jiayang Wang and Xinyu Wu contributed to the experiment and analysis of the data.

References

1. China Electric Power Yearbook Editorial Board. *China Electric Power Yearbook*; China Electric Power Press: Beijing, China, 2011.
2. China Electric Power Yearbook Editorial Board. *China Electric Power Yearbook*; China Electric Power Press: Beijing, China, 2012.
3. China Electric Power Yearbook Editorial Board. *China Electric Power Yearbook*; China Electric Power Press: Beijing, China, 2013.
4. Ministry of Planning and Statistics. *National Power Industry Statistics Express*; China Electricity Council: Beijing, China, 2013.
5. Liu, L.; Zong, H.; Zhao, E.; Chen, C.; Wang, J. Can China realize its carbon emission reduction goal in 2020: From the perspective of thermal power development. *Appl. Energy* **2014**, *124*, 199–212. [CrossRef]
6. Mario, C.; Antonio, M.; Giuseppina, N. A Dynamic Fuzzy Controller to Meet Thermal Comfort by Using Neural Network Forecasted Parameters as the Input. *Energies* **2014**, *7*, 4727–4756.
7. Padhy, N.P. Unit commitment—A bibliographical survey. *IEEE Trans. Power Syst.* **2004**, *19*, 1196–1205. [CrossRef]
8. Niknam, T.; Khodaei, A.; Fallahi, F. A new decomposition approach for the thermal unit commitment problem. *Appl. Energy* **2009**, *86*, 1667–1674. [CrossRef]
9. Senjyu, T.; Shimabukuro, K.; Uezato, K.; Funabashi, T. A fast technique for unit commitment problem by extended priority list. *IEEE Trans. Power Syst.* **2003**, *18*, 882–888. [CrossRef]
10. Baldwin, C.J.; Dale, K.M.; Dittrich, R.F. A study of the economic shutdown of generating units in daily dispatch. *Trans. Am. Inst. Electr. Eng.* **1959**, *78*, 1272–1284. [CrossRef]
11. Iguchi, M.; Yamashiro, S. An efficient scheduling method for weekly hydro-thermal unit commitment. In Proceedings of the 2002 IEEE Region 10 Conference on Computers, Communications, Control and Power Engineering, Beijing, China, 28–31 October 2002; Volume 3, pp. 1772–1777.
12. Dudek, G. Genetic algorithm with integer representation of unit start-up and shut-down times for the unit commitment problem. *Eur. Trans. Electr. Power* **2007**, *17*, 500–511. [CrossRef]
13. Tong, S.K.; Shahidehpour, S.M.; Ouyang, Z. A heuristic short-term unit commitment. *IEEE Trans. Power Syst.* **1991**, *6*, 1210–1216. [CrossRef]
14. Hyeon, G.P.; Jae, K.L.; Kang, Y.C.; Jong, K.P. Unit commitment considering interruptible load for power system operation with wind power. *Energies* **2014**, *7*, 4281–4299.
15. Liao, S.L.; Cheng, C.T.; Wang, J.; Feng, Z.K. A hybrid search algorithm for midterm optimal scheduling of thermal power plants. *Math. Probl. Eng.* **2015**, *2015*. [CrossRef]

16. Venkata, S.P.; Istvan, E.; Kurt, R.; Jan, D. A stochastic model for the optimal operation of a wind-thermal power system. *IEEE Trans. Power Syst.* **2009**, *24*, 940–949.

17. Saadawi, M.M.; Tantawi, M.A.; Tawfik, E. A fuzzy optimization-based approach to large scale thermal unit commitment. *Electr. Power Syst. Res.* **2004**, *72*, 245–252. [CrossRef]

18. Dang, C.Y.; Li, M.Q. A floating point genetic algorithm for solving the unit commitment problem. *Eur. J. Oper. Res.* **2007**, *181*, 1670–1395. [CrossRef]

19. Simonovic, S.P. Reservoir systems analysis: Closing gap between theory and practice. *J. Water Resour. Plan. Manag.* **1992**, *118*, 262–280. [CrossRef]

20. Yang, G.; James, M.C.; Ming, N.; Rui, B. Economic modeling of compressed air energy storage. *Energies* **2013**, *6*, 2221–2241.

21. Rogelio, P.M.; Juan, A.M.; Miguel, J.P.; Lourdes, A.B.; Juan, M.M.S. A novel modeling of molten-salt heat storage systems in thermal solar power plants. *Energies* **2014**, *7*, 6721–6740.

22. Wei, S.Y.; Xu, F.; Min, Y. Study and modeling on maintenance strategy for a thermal power plant in the new market environment. In Proceedings of the IEEE International Conference on Electric Utility Deregulation, Restructuring and Power Technologies, Hong Kong, China, 5–8 April 2004; pp. 200–204.

23. Janusz, B.; Andrzej, O. Modelling of thermal power plants reliability. In Proceedings of the International Conference on Power Engineering, Energy and Electrical Drives, Istanbul, Turkey, 13–17 May 2013; pp. 1038–1043.

24. Turgeon, A. Optimal short-term hydro scheduling from the principle of progressive optimality. *Water Resour. Res.* **1981**, *17*, 481–486. [CrossRef]

25. Nanda, J.; Bijwe, P.R.; Kothari, D.P. Application of progressive optimality algorithm to optimal hydrothermal scheduling considering deterministic and stochastic data. *Int. J. Electr. Power Energy Syst.* **1986**, *8*, 61–64. [CrossRef]

26. Nanda, J.; Bijwe, P.R. Optimal hydrothermal scheduling with cascaded plants using progressive optimality algorithm. *IEEE Trans. Power Syst.* **1981**, *100*, 2093–2099. [CrossRef]

27. Lucas, N.J.D.; Perera, P.J. Short-term hydroelectric scheduling using the progressive optimality algorithm. *Water Resour. Res.* **1985**, *21*, 1456–1458. [CrossRef]

28. Cheng, C.T.; Shen, J.J.; Wu, X.Y.; Chau, K.W. Short-term hydro scheduling with discrepant objectives using multi-step progressive optimality algorithm. *J. Am. Water Resour. Assoc.* **2012**, *48*, 464–479. [CrossRef]

29. Labadie, J.W. Optimal operation of multireservoir systems: State-of-the-art review. *J. Water Resour. Plan. Manag.* **2004**, *130*, 93–111. [CrossRef]

30. Howson, H.R.; Sancho, N.G. A new algorithm for the solution of multi-stage dynamic programming problems. *Math. Program.* **1975**, *8*, 104–116. [CrossRef]

31. Marino, M.A.; Loaiciga, H.A. Dynamic model for multireservoir operation. *Water Resour. Res.* **1985**, *21*, 619–630. [CrossRef]

32. Marino, M.A.; Loaiciga, H.A. Quadratic model for reservoir management: Application to the Central Valley Project. *Water Resour. Res.* **1985**, *21*, 631–641. [CrossRef]

energies

MDPI

Article

Optimal Subinterval Selection Approach for Power System Transient Stability Simulation

Soobae Kim [1] and Thomas J. Overbye [2,*]

[1] Korea Electric Power Research Institute (KEPRI), Korea Electric Power Corporation (KEPCO), 105 Munji-Ro, Yuseong-Gu, Daejeon 305-760, Korea; soobkim@kepco.co.kr

[2] Department of Electrical and Computer Engineering, University of Illinois at Urbana-Champaign, 306 N. Wright St., Urbana, IL 61801, USA

* Author to whom correspondence should be addressed; overbye@illinois.edu; Tel.: +1-217-333-4463; Fax: +1-217-333-1162.

Academic Editor: Ying-Yi Hong
Received: 2 September 2015; Accepted: 16 October 2015; Published: 21 October 2015

Abstract: Power system transient stability analysis requires an appropriate integration time step to avoid numerical instability as well as to reduce computational demands. For fast system dynamics, which vary more rapidly than what the time step covers, a fraction of the time step, called a subinterval, is used. However, the optimal value of this subinterval is not easily determined because the analysis of the system dynamics might be required. This selection is usually made from engineering experiences, and perhaps trial and error. This paper proposes an optimal subinterval selection approach for power system transient stability analysis, which is based on modal analysis using a single machine infinite bus (SMIB) system. Fast system dynamics are identified with the modal analysis and the SMIB system is used focusing on fast local modes. An appropriate subinterval time step from the proposed approach can reduce computational burden and achieve accurate simulation responses as well. The performance of the proposed method is demonstrated with the GSO 37-bus system.

Keywords: transient stability simulation; numerical integration; time step; multi-rate method; subinterval; computational efficiency

1. Introduction

Modern power systems operate closer to their operation and stability limits. This is because of load demand growth, the open access of transmission network, and economic operation [1]. Therefore, dynamic security assessment has become increasingly important. Power system transient stability analysis is a fundamental tool for this dynamic security assessment. It determines whether a contingency a power system will reach a new operating point after, and it examines how system values change during transient deviations from an equilibrium following a contingency [2,3]. Based on the simulation results from potentially many contingency cases, system operators can take corrective and preventive actions to maintain stable operation of power systems.

A transient stability simulator solves a set of nonlinear differential and algebraic equations (DAEs) representing machines and controllers placed in the power systems, and the power system network, respectively. It utilizes numerical integration methods to estimate dynamic states at the next time step and uses iterative techniques to solve the nonlinear algebraic equations. Numerical integration methods are divided into two main categories: implicit and explicit methods [4]. Each method has its own advantages and disadvantages in regards to stability and reliability calculations [5]. Due in part from their ease of implementation, explicit numerical integration methods are widely used in commercial transient stability packages [6–8]. However, for a stiff power system in which a wide variety of different time frames exist, explicit numerical integration methods might require very small time steps to avoid numerical instability. Entire simulation with very small time steps would require substantial computation.

The multi-rate method provides an efficient integration technique for systems exhibiting a wide variety of time responses. The multi-rate method integrates different variables with different time steps [9]. It uses small time steps, called subintervals, for fast varying variables and larger time steps for more slowly varying ones. The subinterval integration approach, coupled with a longer main time step, allows for the network algebraic equations and slow dynamic equations to be evaluated much less frequently and thus computational benefits can be achieved, in particular for a large power system with very few fast variables. The method was applied to power system transient stability problem for the first time in [10] and advanced in [11,12]. Some widely used transient stability packages employ the multi-rate method [6–8].

As modern systems have become increasingly complex such as with the introduction of load dynamics, other power electronic models, and renewable energy sources, the system has become potentially more susceptible to numerical instability issues. Appropriate time steps should be carefully determined to prevent the numerical instability. However, commercial transient stability packages that utilize the multi-rate method often do not provide guidelines or a built-in function to select an appropriate subinterval time step. Instead, they use either a fixed value for all models of a type, or in the case of PowerWorld Simulator, heuristics based on model parameters. The tradeoff is between excessive computation if the interval is too short, and potential numeric instability if it is too long.

In this paper, a new subinterval selection approach is presented to determine the optimal numerical integration time step for the use with the multi-rate method. An appropriate subinterval can be determined by identifying how fast dynamic states are varying in the considered power system. The proposed approach uses modal analysis for single machine infinite bus (SMIB) models [3]. It analyzes the SMIB model eigenvalues, and thus fast dynamic states can be recognized. A primary source of fast dynamics in power systems are generators and their controllers. Instead of considering the whole system, the SMIB approach focuses on each generator with its controllers to identify the fast dynamics. Thus, the required computation for modal analysis with the SMIB approach can be quite modest, even for large systems. Depending on the SMIB approach eigenvalues, appropriate subinterval values can be determined to avoid numerical instability issues and to minimize the required computations.

This paper is organized as follows: Section 2 presents background information including the explicit numerical integration method, the multi-rate method and eigenvalue analysis with a SMIB system. In Section 3, a problem is defined and a new optimal time step selection method is then proposed. Section 4 illustrates simulation results with the GSO 37-bus case. The conclusion is made in Section 5.

2. Preliminary

2.1. Explicit Numerical Integration Method

Many commercial power system transient stability packages use explicit numerical integration methods, which estimate values for the next time step explicitly using present values. The second-order Runge-Kutta (RK2) method is one example of the explicit methods [2]. With an Ordinary Differential Equation (ODE), the RK2 method approximates the next time step value as:

$$x_n = x_{n-1} + h \cdot \left(\frac{1}{2}k_1 + \frac{1}{2}k_2\right) \tag{1}$$

where, $k_1 = f(x_{n-1}, t_{n-1})$, $k_2 = f(x_{n-1} + k_1, t_{n-1} + h)$, h: numerical integration time step.

When the simplest test equation $\dot{x} = \lambda x$ is given, the region of stability is defined with Equation (2) and the region is shown in Figure 1. More detailed information about Equation (2) can be found in [12]:

$$\left| 1 + h\lambda + \frac{1}{2}(h\lambda)^2 \right| < 1 \tag{2}$$

As shown in Figure 1, for a real valued λ, stable region can be defined with Equation (3):

$$-2 < h\lambda < 0 \tag{3}$$

Based on Equation (3), minimum eigenvalues that the RK2 method can cover without numerical instability can be determined depending on the time step as shown in Table 1.

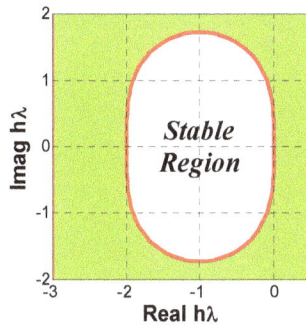

Figure 1. Region of stability of the second-order Runge-Kutta (RK) method.

Table 1. Range of eigenvalues depending on the time step using the RK2 method.

Time Step (Cycle)	Range of Eigenvalue
1	$-120 < \lambda < 0$
0.25	$-480 < \lambda < 0$
0.1	$-1200 < \lambda < 0$
0.05	$-2400 < \lambda < 0$

2.2. Multi-Rate Method

In practical power systems, only a small portion of the dynamic states are associated with fast dynamics and thus the use of very small time step for the entire simulation is not efficient in terms of computational time and storage. To avoid this computational burden, the multi-rate method that uses different time steps in a numerical integration scheme, has been commonly used [9]. The approach is shown in Figure 2. For the fast changing variables, a small time step (h) is used, while for the relatively slower changing variables, an integer multiple of the small time step (H) is used. As one can see from Figure 2, the ratio of the fast time step to the slow time step for this case is four. The equations for the fast variables must be solved at points where the slow ones have not been solved. This can be done by using a linear solution value interpolated from the slow variables [12].

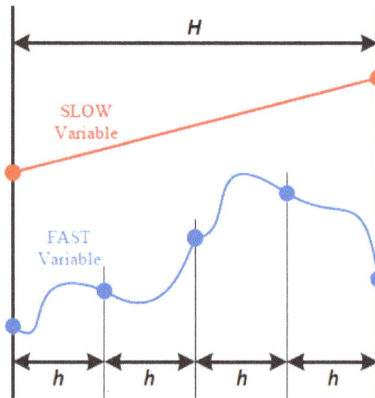

Figure 2. Multi-rate method.

2.3. SMIB System and Eigenvalue Analysis

The SMIB system models the machine of interest in detail, while the rest of the system is represented with a Thevenin equivalent circuit of a voltage behind an impedance. The Thevenin equivalent impedance at bus k is equal to the kth diagonal element of the Z matrix representing the entire power system [13,14]. The infinite bus voltage is set such that the total machine complex power output is the same in between the SMIB model and the entire system. In the SMIB system depicted in Figure 3, the multi-machine power system is reduced to a single machine that is connected to an infinite bus system in order to simplify the analysis and focus on one machine. The SMIB system does not capture the inter-area modes attributed to interaction among generators, but the local modes related to the generator including controllers can be identified.

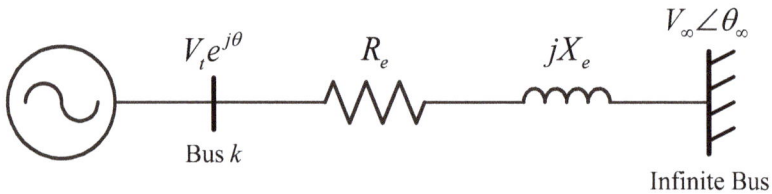

Figure 3. Single machine infinite bus (SMIB) system.

For small perturbations, it is sufficient to analyze the linearized power system model. Linear models are simpler to understand and have many useful tools for analysis. One such tool is eigenvalue analysis. Eigenvalues indicate the system stability and how close the system is to becoming unstable. It also shows what frequencies and modes exist in the system, as well as how the system states interact with these modes.

The SMIB system can be modeled with Equations (4) and (5) which represent the power system dynamics and the stator and network algebraic equations, respectively [3]. The x variables show the dynamic state variables such as generator rotor angle and speed. The y variables show the algebraic variables such as the network bus voltage and angle:

$$\dot{x}_{SMIB} = f(x_{SMIB}, y_{SMIB}) \tag{4}$$

$$0 = g(x_{SMIB}, y_{SMIB}) \tag{5}$$

where:

f: a vector of dynamic equations;

g: a vector of algebraic equations;

x_{SMB}: a vector of dynamic state variables of SMIB system;

y_{SMIB}: a vector of algebraic state variables of SMIB system.

The power system equations shown in Equations (4) and (5) are nonlinear. Hence, they need to be linearized around the operating point in order to find out the eigenvalues. The equations of the linearized power system are given by Equations (6) and (7):

$$\Delta \dot{x}_{SMIB} = A \Delta x_{SMIB} + B \Delta y_{SMIB} \tag{6}$$

$$0 = C \Delta x_{SMIB} + D \Delta y_{SMIB} \tag{7}$$

where:

$A = \left. \dfrac{\partial f}{\partial x_{SMIB}} \right|_{(x^o_{SMIB}, y^o_{SMIB})}; B = \left. \dfrac{\partial f}{\partial y_{SMIB}} \right|_{(x^o_{SMIB}, y^o_{SMIB})}; C = \left. \dfrac{\partial g}{\partial x_{SMIB}} \right|_{(x^o_{SMIB}, y^o_{SMIB})}; D = \left. \dfrac{\partial g}{\partial y_{SMIB}} \right|_{(x^o_{SMIB}, y^o_{SMIB})}; x^o_{SMIB}, y^o_{SMIB}$: SMIB system operating points.

Then, the system modal matrix is obtained by incorporating Equation (7) into (6) as following:

$$\Delta \dot{x} = A_{sys} \Delta x = (A - BD^{-1}C)\Delta x \tag{8}$$

With the system modal matrix (A_{sys}), eigenvalues corresponding to fast local modes can be identified. Real components of eigenvalues provide information about how fast the corresponding modes are varying. The required time step to prevent the numerical instability can then be determined.

3. Problem Definition and Proposed Approach

3.1. Problem Definition

As described in Section 2.2, the multi-rate method requires the ratio of the fastest variables to the main time step (H in Figure 2) in order to determine the subinterval step size. However, it is not easy to identify the ratio with higher order, nonlinear power system models. Commercial transient stability packages usually use either a fixed value based on the model type, or heuristics that depend on model parameters. In designing such heuristics, in case of numerical instability for particular parameters, the subinterval time step is changed to a smaller value. On the other hand, choosing too small an interval can cause increased computation. Therefore, an optimal subinterval time step for use with the multi-rate method could remove unnecessary computations while avoiding numerical instability. Such an approach is presented in the next section.

3.2. Proposed Approach

In the approach presented here, the optimal subinterval time step is determined by analyzing the fast dynamics in each SMIB system. This works because the fast varying states are related to the local oscillations rather than inter-area oscillations. In the proposed method, the modal matrix is constructed by linearizing the SMIB system equations around an operating point. The SMIB eigenvalues can then be found, of which their magnitudes describe how rapidly the dynamics vary. By the use of the participation factor in linear system theory [2,3], it can be determined which dynamic states are associated with the fast modes. However, we note that in commercial packages, the subinterval is usually applied to all the differential equations for a particular model, such as an exciter. Hence, the participation factors only need to determine the particular model (e.g., the machine model, the exciter model, *etc.*) not the individual differential equations.

Finally, the subinterval step size for the fast dynamic states is determined by considering the region of stability equation of numerical integration method in use. For the RK2 method, explained in Section 2.1, Equation (3) is used to determine the appropriate step size. For the multi-rate method, the slow time step is obtained from the user specified value and the ratio between fast and slow time steps are then identified with Equation (9):

$$\text{Minimum subinterval ratio} = \frac{\text{Step size for slow variables (User specified)}}{\text{Required step size for fast variables}} \qquad (9)$$

Figure 4 shows the overall procedure of the proposed approach. Note that the SMIB systems can be determined and analyzed quite quickly, so this approach could be efficiently used for a large system.

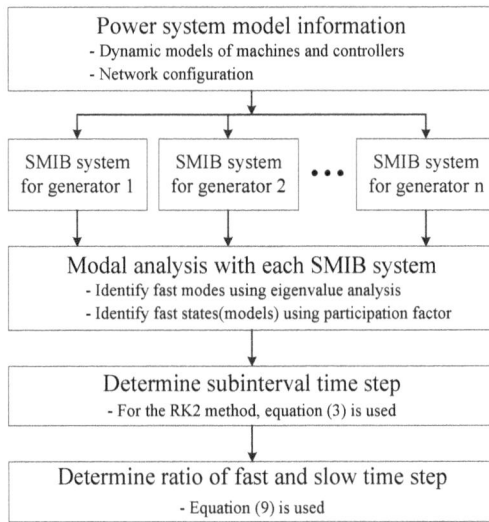

Figure 4. Flowchart of the proposed approach.

4. Case Study

The proposed approach was tested with the PowerWorld Simulator, which provides SMIB eigenvalue analysis and uses a multi-rate numerical integration method [5]. The GSO 37-bus system shown in Figure 5 is used to demonstrate the proposed method. The case has nine generators, 25 loads and 57 branches [15]. One common exciter, the EXST1 (IEEE Type ST1 excitation system model) shown in Figure 6, quite commonly introduces very fast modes to the system because fast electronic controllers are incorporated. For test purposes, parameters for each of the EXST1 exciters for the two generators at bus 28 were changed as shown in Table 2.

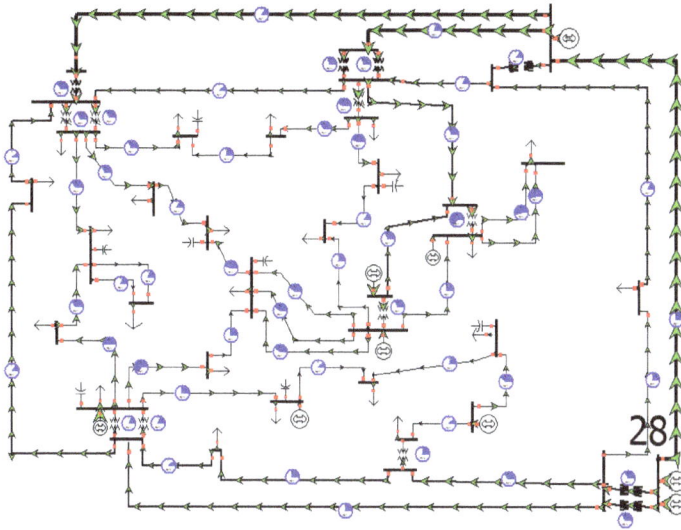

Figure 5. GSO 37-bus system.

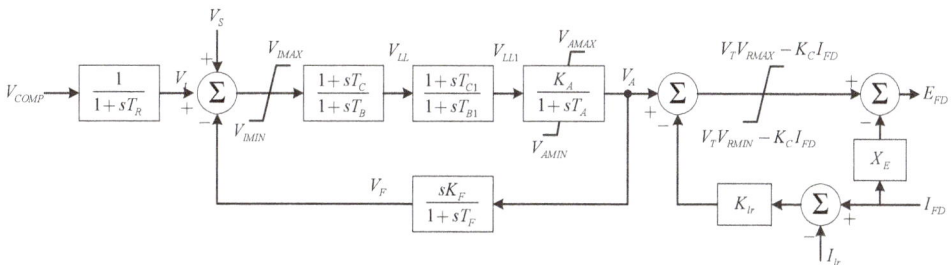

Figure 6. Block diagram of EXST1 exciter.

Table 2. EXST1 Exciter model parameters.

$T_r = 0$	$V_{i\max} = 10$	$V_{i\min} = -10$	$T_c = 1$
$T_b = 1$	$K_a = 150$	$T_a = 0.01$	$V_{r\max} = 3.6$
$V_{r\min} = 0$	$K_c = 0$	$K_f = 0.04$	$T_f = 0.4$
$T_{c1} = 1$	$T_{b1} = 1$	$V_{a\max} = 99$	$V_{a\min} = -99$
$X_e = 0$	$I_{lr} = 0$	$K_{lr} = 0$	

4.1. SMIB Eigenvalue Analysis

With the test case considered, SMIB systems for all generators were created and modal analysis of each SMIB system was then performed. The results in Table 3 indicate that two big negative eigenvalues (-1602) are originated from the two generators at bus 28. The participation factors in Table 4 identify that these modes are solely contributed by exciter dynamic state V_A. Therefore, the required time step for the state V_A to avoid numerical instability can be determined with the real part of eigenvalue information, which represents how much the dynamic state varies.

Table 3. Single machine infinite bus (SMIB) eigenvalue analysis with the GSO 37-bus case.

Bus Number	Generator ID	Max Eigenvalues	Bus Number	Generator ID	Max Eigenvalues
28	1	−1602	54	1	−44
28	2	−1602	53	1	−42
31	1	−49	44	1	−42
14	1	−45	50	1	−38
48	1	−44	–	–	–

Table 4. Participation factor of a generator at bus 28.

Real Part of Eigenvalues	Machine Angle	Machine Speed	Machine Eqp	Machine PsiDp	Machine PsiQpp	Exciter V_A	Exciter V_F
−1602	0	0	0.0001	0	0	1	0.0015
−45	0.0178	0.0183	0	0	0.9997	0	0
−32	0.0008	0.0007	0.0511	0.9987	0.0001	0.0001	0.0002
−22	0	0.0017	0	0	0.0002	0	0
−0.6	0.709	0.6983	0.0329	0.0056	0.0915	0	0.0007

4.2. Subinterval Step Size

Due to the extremely fast modes (−1602), if a single rate approach was used, the time step for the entire system would need to be just 0.05 cycles to avoid numerical instability; this can be determined with Equation (3) and Table 1. On the other hand, the use of the multi-rate method can increase the step size by using the subinterval time step only for the dynamic states associated with the fastest modes (or as previously mentioned more commonly for all the states associated with the fast exciter model). When a quarter-cycle is selected as the step size for the entire simulation, the subinterval ratio should be greater than five based on Equation (9). The following equation shows how the ratio can be determined:

$$\text{Minimum subinterval ratio}$$
$$= \frac{\text{Step size for slow variables (User speficied)}}{\text{Required step size for fast variables}}$$
$$= \frac{0.25 \text{ cycles}}{0.05 \text{ cycles}} = 5$$

4.3. Simulation Comparisons

PowerWorld Simulator has options that allow the user to override the built-in heuristics and directly specify the desired number of subintervals (in powers of two from two up to 128) for specific models. Hence, eight subintervals were chosen here to meet the minimum. In order to validate the performance of the proposed method, simulation comparisons were made by changing the subinterval step ratio. A three-phase bus to ground fault was applied at bus 28, which has the two EXST1 exciter generators. The fault was applied at 1.0 s and cleared at 1.1 s.

Figure 7 shows the bus voltage magnitude at bus 28 with two different subintervals (eight and four). When the subinterval was set to four, the simulation results show numerical instability even before the fault. This is because the subinterval step size for the fast variables is still bigger than the requirements. However, when the number of subinterval is increased to eight, it shows a stable simulation response. The appropriate step size for the fast ones is correctly estimated with the proposed method.

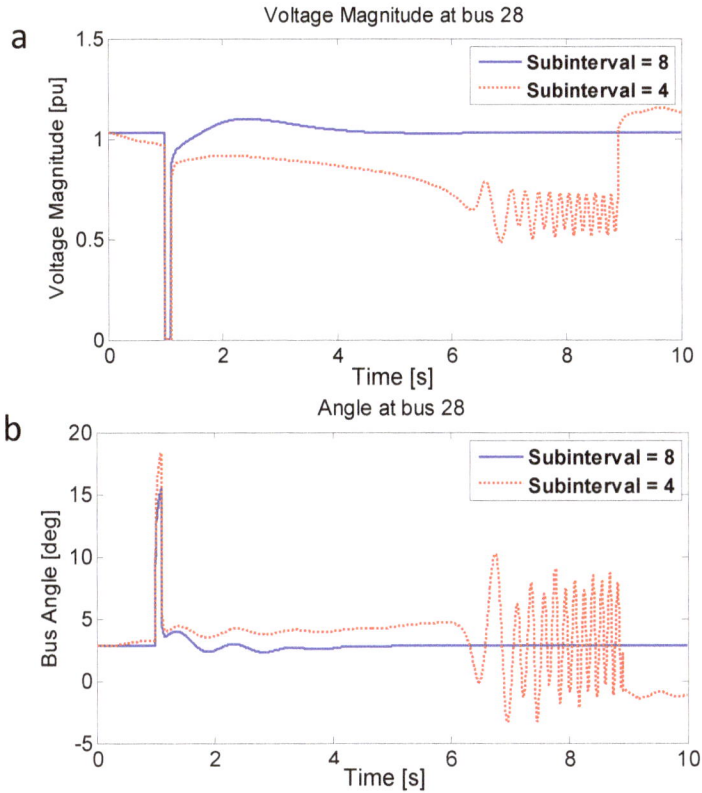

Figure 7. Simulation comparison between four and eight subintervals: (**a**) Voltage magnitude; (**b**) Angle.

Next, a comparison was made between single rate and multi-rate methods. With the single rate method, 0.05 cycles was used for the time step, which is determined based on Table 1 to prevent numerical instability. For the multi-rate method, a quarter-cycle (0.25 cycles) for the slow dynamics and an eight subinterval option was used, which showed a stable response in Figure 7. As shown in Figure 8, the two simulation results are essentially identical. Table 5 shows the computation time with two different simulation approaches. The computation time is an average of running the simulation ten times and it does not include the required computation for SMIB eigenvalue analysis. Because the SMIBs are only associated with the local machines, they can be computed quite quickly. For example, in benchmarking, the PowerWorld simulator can perform complete SMIB eigenvalue analysis for a 3500 generator system within two seconds; for this nine generator system, the time would be less than a millisecond. The multi-rate method shows excellent computational performance by reducing the simulation time by about 80% compared to the single rate method.

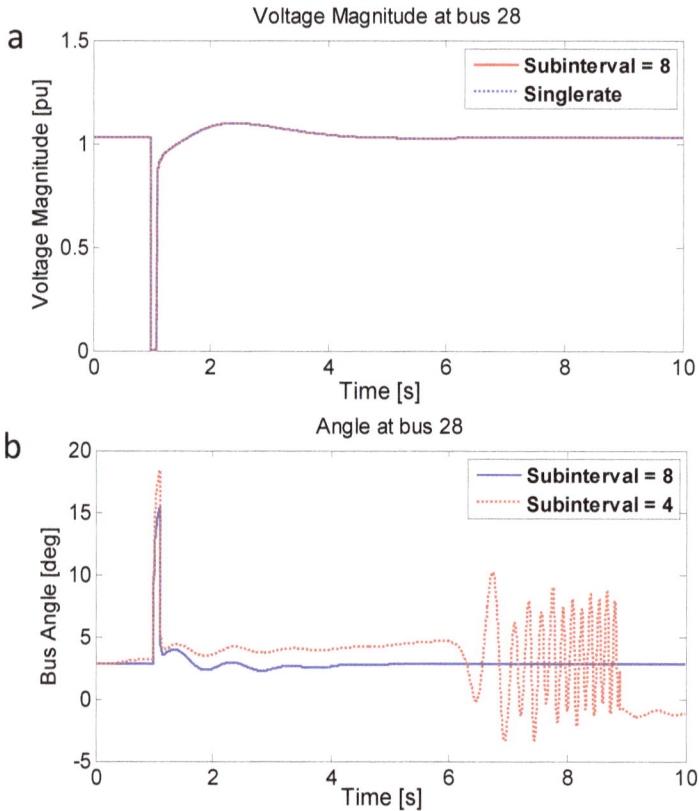

Figure 8. Simulation comparison with single-rate and multi-rate methods: (**a**) Voltage magnitude;
(**b**) Angle.

Table 5. Computational time comparison.

Method Used	Time Step (Cycle)	Subinterval for Fast States	Computation Time (s)	Ratio of the Computation Time
Singlerate	0.05	–	51.3	1
Multirate	0.25	8	10.6	0.21

5. Conclusions

When extremely large negative eigenvalues exist with power system dynamics being integrated using explicit methods, special considerations should be required to avoid numerical instability during power system transient stability analysis. The multi-rate numerical integration method is a common approach to reduce this computational issue. For the use of the multi-rate approach, the ratio between the user-specified time step and the subinterval for fast variables should be determined. This task is not easily identified in commercial transient stability packages. In this paper, an optimal subinterval selection method has been proposed. The method utilizes SMIB eigenvalue analysis, which identifies fast modes and dynamic states related to those modes. Based on the region of stability of numerical integration method in use, the appropriate step size can then be determined by comparing a user-defined time step for slow variables with very small time steps for fast variables. Optimum step size can reduce computational demands as well as avoid numerical instability. Test simulations with

Energies **2015**, *8*, 11871–11882

the GSO 37-bus case confirm that the proposed method provides quite a good performance in terms of accuracy and computational speed.

Acknowledgments: The authors gratefully acknowledge the support for this work provided by Korea Electric Power Corporation (KEPCO), Seoul, Korea and the U.S. Department of Energy through award number DE-OE0000097.

Author Contributions: Soobae Kim proposed the approach and performed simulation and wrote the vast majority of paper. Thomas J. Overbye overviewed the approach, made some additions, and proofread the manuscript.

Conflicts of Interest: The first author declares no conflict of interest. The second author is the original developer of PowerWorld Simulator, helped to develop the PowerWorld transient stability code, and is a co-owner of PowerWorld Corporation.

References

1. Fouad, A.A.; Aboytes, F.; Carvalho, V.F.; Corey, S.L.; Dhir, K.J.; Vierra, R. Dynamic security assessment practices in North America. *IEEE Trans. Power Syst.* **1998**, *3*, 1310–1321. [CrossRef]
2. Kundur, P. *Power System Stability and Control*; McGraw-Hill: New York, NY, USA, 1994.
3. Sauer, P.W.; Pai, M.A. *Power System Dynamics and Stability*; Prentice Hall: Upper Saddle River, NJ, USA, 1998.
4. Griffths, D.F.; Higham, D.J. *Numerical Methods for Ordinary Differential Equations: Initial Value Problems*; Springer: London, UK, 2010.
5. Stott, B. Power system dynamic response calculation. *Proc. IEEE* **1979**, *67*, 219–241. [CrossRef]
6. PowerWorld Corporation. Available online: http://www.powerworld.com/ (accessed on 1 September 2015).
7. Simens PSS/E. Available online: http://www.siemens.com/ (accessed on 1 September 2015).
8. General Electric Company. *PSLF Version 18.0 User Manual*; General Electric Company: Schenectady, NY, USA, 2011.
9. Gear, C. *Multirate Methods for Ordinary Differential Equations*; Technical Report; University of Illinois at Urbana-Champaign: Champaign, IL, USA, 1974.
10. Crow, M.L.; Chen, J.G. The multirate method for simulation of power system dynamics. *IEEE Trans. Power Syst.* **1994**, *9*, 1684–1690. [CrossRef]
11. Crow, M.L.; Chen, J.G. The multirate simulation of FACTS devices in power system dynamics. *IEEE Trans. Power Syst.* **1996**, *11*, 376–382. [CrossRef]
12. Chen, J.G.; Crow, M.L. A Variable Partitioning Strategy for the Multirate Method in Power Systems. *IEEE Trans. Power Syst.* **2008**, *23*, 259–266. [CrossRef]
13. Haque, M.H. A fast method for determining the voltage stability limit of a power system. *Electr. Power Syst. Res.* **1995**, *32*, 35–43. [CrossRef]
14. Haque, M.H. Novel method of assessing voltage stability of a power system using stability boundary in P–Q plane. *Electr. Power Syst. Res.* **2003**, *64*, 35–40. [CrossRef]
15. Glover, J.D.; Sarma, M.S.; Overbye, T.J. *Power System Analysis and Design*; Cengage Learning: Stamford, CT, USA, 2012.

energies

MDPI

Article

Designing a Profit-Maximizing Critical Peak Pricing Scheme Considering the Payback Phenomenon

Sung Chan Park [1], Young Gyu Jin [2,*] and Yong Tae Yoon [1]

[1] Department of Electrical Engineering and Computer Science, Seoul National University, Daehak-dong, Gwanak-gu, Seoul 151-742, Korea; sweetpe4@snu.ac.kr (S.C.P.); ytyoon@snu.ac.kr (Y.T.Y.)

[2] Center for Advanced Power & Environmental Technology (APET), the University of Tokyo, 7-3-1 Hongo, Bunkyo-ku, Tokyo 113-8656, Japan

* Correspondence: ygjin93@snu.ac.kr; Tel.: +82-2-880-9143; Fax: +82-2-885-4958.

Academic Editor: Ying-Yi Hong

Received: 15 July 2015; Accepted: 8 October 2015; Published: 13 October 2015

Abstract: Critical peak pricing (CPP) is a demand response program that can be used to maximize profits for a load serving entity in a deregulated market environment. Like other such programs, however, CPP is not free from the payback phenomenon: a rise in consumption after a critical event. This payback has a negative effect on profits and thus must be appropriately considered when designing a CPP scheme. However, few studies have examined CPP scheme design considering payback. This study thus characterizes payback using three parameters (duration, amount, and pattern) and examines payback effects on the optimal schedule of critical events and on the optimal peak rate for two specific payback patterns. This analysis is verified through numerical simulations. The results demonstrate the need to properly consider payback parameters when designing a profit-maximizing CPP scheme.

Keywords: critical peak pricing; critical event scheduling; optimal peak rate; payback phenomenon; load recovery; load serving entity

---

1. Introduction

Demand response (DR) programs give customers a more active role in the operation of the power system, allowing them to change their consumption patterns. They have been implemented to ensure secure power system operation when the system suffers from severe supply-demand imbalances [1]. The main operators of DR programs are load serving entities (LSEs) that supply electricity to contracted customers (*i.e.*, utilities). Recent deregulation of the power industry has made it possible for DR programs to be implemented in a market environment [2]. Accordingly, an LSE could become a demand-side participant in the market, aiming to maximize its profit through the DR program [3].

In a market context, the LSE's profit depends on the difference between the purchase and resale prices. The purchase prices are determined for a specified time interval (e.g., every five min) based on the supply and demand for electricity [4]. They are inherently time-varying, and will be denoted here as "real-time market clearing prices" (RTMCPs). Since the resale (or retail) rates are relatively fixed compared to the RTMCPs, a sudden increase in the RTMCP leads to a corresponding reduction in the LSE's profits [5]. As a result, dynamic pricing schemes are typically designed and included in DR programs to take variations in the RTMCPs into consideration [6].

Such dynamic pricing schemes include three main approaches: real-time pricing (RTP), time-of-use (TOU), and peak pricing (CPP) [6]. This study focuses on CPP, which has certain advantages over RTP and TOU. In RTP, customers are exposed to continuously changing prices; they must thus repeatedly decide whether to respond to price changes in order to reduce their electricity bills. By contrast, in a CPP scheme, customers must make decisions about whether to reduce consumption only when critical events occur. Thus, CPP is simpler to implement than RTP, particularly for residential customers [7]. TOU is also easy to implement, as it consists only of a few block rates. However, these rates must be announced to customers in advance, making TOU unable to manage sudden increases in the RTMCP. CPP can thus complement TOU by dynamically applying the peak rate during critical situations of high RTMCP [7].

In terms of the design of a CPP scheme, two main problems have been addressed in the literature. One, the event scheduling problem, seeks to determine when critical events should be triggered to maximize or minimize a certain target outcome, such as profits. Past research has formulated and solved the event scheduling problem so as to maximize profit via dynamic programming based on the forecasted price and demand [8]. In [9], the problem is solved through a stochastic approach considering the uncertainties inherent in price and temperature, also with the goal of profit maximization. Integer programming is used to solve the problem in [10]. The second problem in CPP design involves selection of the peak rate. Recent research [11] presents guidelines for determining the optimal peak rate (along with other CPP parameters) to achieve maximum profits for an LSE. In [12], a methodology is proposed to determine the peak rate as well as the optimal events schedule considering variable wind power generation.

In various DR programs, the interrupted or curtailed demand later appears as delayed consumption after the restriction is lifted [13–18]. This phenomenon is referred to differently in the literature—as payback in [13–15] and load recovery in [16,17]—but here we refer to it as payback. Further, in [18], the payback is represented concisely as the cross-elasticity in a mathematical form of the elasticity matrix. In DR programs, the payback phenomenon occurs because demand is shifted in time, but the reduction in overall consumption is very small [19]. Moreover, the paid-back amount may exceed the amount of curtailed demand because of losses from the energy conversion processes of customers' appliances [20]. Because of the costs incurred in serving this delayed demand, the payback lessens the market value of demand-side resources [21]. Therefore, some studies model the payback phenomenon as part of their economic analyses. In [13], the effect of payback on generation costs and the amount of peak reduction are evaluated for air-conditioning loads. In [16], a mathematical model of payback is developed, and its effects on each market participant are analyzed from an economic perspective.

The payback phenomenon also occurs in CPP following a critical event [22]. Accordingly, it affects the profits enjoyed by the LSE and thus must be appropriately considered when designing a profit-maximizing CPP scheme. The optimal peak rate in a CPP scheme is determined in [11] based on the assumption that customers' demand is not recovered but rather purely reduced when critical events take place. However, if the payback phenomenon occurs, the optimal peak rate given in [11] no longer ensures optimality. In addition, the main concerns of an LSE operating a CPP scheme are the optimal schedule of critical events and the resulting profit. However, if payback is present, the optimal event schedule and the LSE's profit, which are determined without considering payback as in [11], would change in a manner differing from the characteristics of the payback phenomenon. Nonetheless, few studies have examined the payback within CPP schemes; even fewer have presented how CPP parameters, such as the peak rate, should be chosen to maximize LSE profits considering the payback.

This study thus extends the work presented in [11] and presents several analyses to fill these gaps in knowledge. First, the payback phenomenon is characterized using three parameters: duration, amount, and pattern. Further, a payback ratio and a payback function are introduced and defined for the payback amount and pattern, respectively. Second, the payback effects on the critical event scheduling problem are demonstrated based on the characteristics of customers' responses to a critical event. Finally, an analytical expression for the optimal peak rate considering payback is derived for a general payback pattern. Then, the payback effects on the optimal peak rate are further analyzed for two specific payback models: exponentially decreasing payback (EDP), to model an intense demand recovery a short time after a critical event, and uniformly distributed payback (UDP), to represent a redistribution of demand over a longer time period. In all these analyses, we adopt the customer price response model used in [23] to quantify the reduction in electricity consumption in response to a critical event.

The remainder of this paper is organized as follows. As background information, Section 2 briefly describes the customer price response model [23] and CPP design in the absence of payback [11]. Section 3 characterizes the payback phenomenon. The effects of payback on the design of a CPP scheme are analyzed in Section 4, the results of which are verified through numerical simulations in Section 5. Finally, Section 6 offers some concluding remarks.

2. Backgrounds

2.1. Price Responsiveness Model

The response of customers to a price change is described in [23] as given by:

$$q_k = q_{0,k}\left\{1 + \frac{\beta_k(\rho_k - \rho_{0,k})}{\rho_{0,k}}\right\} \tag{1}$$

where $q_{0,k}$ and $\rho_{0,k}$ are the nominal demand and price, respectively; q_k and ρ_k are the modified demand and price, respectively; and β_k is the customers' demand elasticity, defined as:

$$\beta_k = \frac{\rho_k}{q_k}\frac{dq_k}{d\rho_k} \tag{2}$$

The variable β_k is negative because a price increase leads to a demand reduction. For example, when β_k is equal to -0.01, demand decreases by 1% following a 100% increase in price.

CPP consists of two price levels: the off-peak rate, ρ_{BASE}, and the peak rate, ρ_{PEAK}. The off-peak rate is applied in most periods while the peak rate is applied only rarely, when critical events are triggered. When a critical event is triggered, the price changes from ρ_{BASE} to ρ_{PEAK}, and demand changes accordingly. Assuming that elasticity of demand is constant at β, the modified consumption, $q_{CR,k}$, during the critical event triggered in period k can be determined by replacing $\rho_{0,k}$ and ρ_k in Equation (1) with ρ_{BASE} and ρ_{PEAK}, respectively, as [11]:

$$q_{CR,k} = q_{0,k}\left\{1 + \beta\left(\frac{\rho_{PEAK}}{\rho_{BASE}} - 1\right)\right\} \tag{3}$$

In reality, $q_{0,k}$ does not occur when consumption has already changed to q_k in response to ρ_{PEAK}. Thus, $q_{0,k}$ should be interpreted as forecast demand based on ρ_{BASE}. The cumulative curtailed demand $Q_{CUR,k}$ for a critical event starting in period k can then be represented as:

$$Q_{CUR,k} = \sum_{i=k}^{k+D_{CPP}-1}(q_{0,i} - q_{CR,i}) = \beta\left(1 - \frac{\rho_{PEAK}}{\rho_{BASE}}\right) \cdot \sum_{i=k}^{k+D_{CPP}-1}q_{0,i} \tag{4}$$

where D_{CPP} is the duration of the critical event. It should be noted that $Q_{CUR,k}$ takes a positive value because both β and $(1 - \rho_{PEAK}/\rho_{BASE})$ are normally negative.

2.2. Designing a Critical Peak Pricing (CPP) Scheme without Payback

The event scheduling problem of CPP to maximize LSE profits can be formulated as:

$$\max_{u_k}\sum_{k=1}^{H}\{R_k - C_k\} \tag{5}$$

where u_k is a binary event decision variable that takes a value of one if a critical event is triggered and zero otherwise, H is the scheduling time horizon of the problem, and R_k and C_k are the LSE's revenues and costs, respectively, in period k. When payback does not occur, R_k and C_k are defined as:

$$R_k = u_k \rho_{PEAK} q_{CR,k} + (1 - u_k) \rho_{BASE} q_{0,k} \tag{6}$$

$$C_k = u_k \rho_{RTMCP,k} q_{CR,k} + (1 - u_k) \rho_{RTMCP,k} q_{0,k} \tag{7}$$

where $\rho_{RTMCP,k}$ is the forecasted RTMCP in period k. The constraints consist of the conditions on the maximum number of events, the maximum event duration, and the minimum interval between successive events [8,11]. The specific forms of these constraints are as follows:

Maximum Number of Events (N_{CPP}):

$$\sum_{k=1}^{H} u_k(1 - u_{k-1}) \leq N_{CPP} \tag{8}$$

Maximum Event Duration (D_{CPP}):

$$\sum_{i=k}^{k+D_{CPP}} u_i \leq D_{CPP}, \quad \forall k \in \{1, 2, \cdots, H - D_{CPP}\} \tag{9}$$

Minimum Interval between Successive Events (Δk):

$$u_{k-1}(1 - u_k) \sum_{i=k}^{k+\Delta k-1} |u_i - u_{i+1}| = 0, \quad \forall k \in \{1, 2, \cdots, H - \Delta k + 1\} \tag{10}$$

In Equations (8)–(10), $u_k = 0$ is assumed for $k \leq 0$. These constraints are imposed in order to avoid inconveniencing customers by interrupting consumption through critical events. For example, Equation (8) prevents triggering an excessive number of critical events, and Equation (10) allows customers to return to normal consumption within a reasonable time. The existing techniques for solving such an optimization problem include dynamic programming [8] and integer programming [10]. In this study, we use the former methodology, as in [8], to solve the event scheduling problem.

In terms of the optimal peak rate, we first need to define the profit index, which means an additional profit that the LSE will receive from triggering a critical event. Suppose that $\mathbf{K}|\rho_{PEAK} = \{k_1, k_2, \cdots, k_{N_{CPP}}|\rho_{PEAK}\}$ denotes the optimal solution of the event scheduling problem for a given ρ_{PEAK}. Then, the profit index, $PI_{N,k}(\rho_{PEAK})$, for a critical event in period $k \in \mathbf{K}|\rho_{PEAK}$ without payback can be expressed as [11]:

$$PI_{N,k}(\rho_{PEAK}) = \sum_{i=k}^{k+D_{CPP}-1} \left\{ q_{CR,i}(\rho_{PEAK} - \rho_{RTMCP,i}) - q_{0,i}(\rho_{BASE} - \rho_{RTMCP,i}) \right\} \tag{11}$$

where the event duration is equal to the maximum event duration, D_{CPP}, because the LSE's profit is always maximized when the maximum event duration applies. Substituting Equations (3) into (11) makes $PI_{N,k}(\rho_{PEAK})$ a quadratic function of ρ_{PEAK}. Then, the critical point of the function is determined as the optimal peak rate without payback, $\rho_{N,PEAK}^*$, which has a form as [11]:

$$\rho_{N,PEAK}^* = \frac{\rho_{BASE}}{2}\left(1 - \frac{1}{\beta}\right) + \frac{\displaystyle\sum_{k\in\mathbf{K}^*}\sum_{i=k}^{k+D_{CPP}-1} q_{0,i}\rho_{RTMCP,i}}{2\displaystyle\sum_{k\in\mathbf{K}^*} Q_{0,k}} \tag{12}$$

where $\mathbf{K}^* = \mathbf{K}|\rho^*_{N,PEAK}$ is the optimal event schedule for optimal peak rate and $Q_{0,k}$ is cumulative consumption during all critical event periods starting from period k, which is expressed as:

$$Q_{0,k} = \sum_{i=k}^{k+D_{CPP}-1} q_{0,i} \tag{13}$$

3. Characterization of the Payback Phenomenon

There are two key aspects that characterize the payback phenomenon. The first is the amount of paid-back demand. The curtailed demand, $Q_{CUR,k}$, may be under-, equally, or over-recovered following a critical event [13–15]. Thus, a payback ratio, denoted as $\alpha_{PB,k}$, is introduced as the ratio of recovered consumption to curtailed demand. The specific value of $\alpha_{PB,k}$ depends on the composition of customer demand. For example, one does not compensate for turning off lights during a critical event period by greater light use later on; such demand thus tends to be connected with under-payback, or $\alpha_{PB,k} > 1$. In contrast, a heating or air-conditioning system may require more post-event energy for transition to the target value from the decreased or increased room temperature arising during the critical event period; this tends to result in over-payback, or $\alpha_{PB,k} > 1$. Let $Q_{PB,k}$ be defined as the amount of paid-back demand for a critical event in period k. Then $Q_{PB,k}$ can be represented as:

$$Q_{PB,k} = \sum_{i=k+D_{CPP}}^{k+D_{CPP}+D_{PB,k}-1} q_{PB,i} = \sum_{n=1}^{D_{PB,k}} q_{PB,n+k+D_{CPP}-1} = \alpha_{PB,k} Q_{CUR,k} \tag{14}$$

where $q_{PB,i}$ is the recovered demand in period i and $\alpha_{PB,k}$ and $D_{PB,k}$ are the payback ratio and duration, respectively, for the critical event in period k.

The other key aspect of the payback phenomenon is its pattern. Let the payback function, $f_{PB,k}(n)$ for $n \in \{1, 2, \cdots, D_{PB,k}\}$, be defined as the ratio of paid-back demand to $Q_{CUR,k}$ in the n-th time period from the end of the critical event. Then, $q_{PB,n+k+D_{CPP}-1}$ in Equation (14) is expressed as:

$$q_{PB,n+k+D_{CPP}-1} = Q_{CUR,k} f_{PB,k}(n) \tag{15}$$

Comparing Equations (14) and (15), $f_{PB,k}(n)$ should satisfy the condition:

$$\sum_{n=1}^{D_{PB,k}} f_{PB,k}(n) = \alpha_{PB,k} \tag{16}$$

Let the normalized unit payback function, $f^U_{PB,k}(n)$, be defined as $f^U_{PB,k}(n) = f_{PB,k}(n)/\alpha_{PB,k}$, which satisfies the condition:

$$\sum_{n=1}^{D_{PB,k}} f^U_{PB,k}(n) = 1 \tag{17}$$

Then, $f_{PB,k}(n)$ is represented as $f_{PB,k}(n) = \alpha_{PB,k} f^U_{PB,k}(n)$ such that the payback function can be separately expressed by the payback ratio, $\alpha_{PB,k}$, and the payback pattern, $f^U_{PB,k}(n)$.

In real-world situations, it is difficult to specify a particular form for $f^U_{PB,k}(n)$. Nonetheless, some studies suggest a payback function model for analytic purposes particularly for water heating and air conditioning loads. Empirical results in [14] show that the payback pattern can be represented with an exponentially decreasing function for water heating loads. In [13] and [15], sets of decreasing values are specified as the payback function values for water heating loads and both water heating and air conditioning loads, respectively. This study thus adopts an EDP function as a specific payback pattern to model an intensive recovery of demand over a short time period immediately after a critical event. This takes the form of:

$$f^U_{PB,k}(n) = e^{-n\lambda}, \quad n \in \{1, 2, \cdots, D_{PB,k}\} \tag{18}$$

where λ is a constant, which is determined by solving the equation $\sum_{n=1}^{D_{PB,k}} e^{-n\lambda} = 1$ derived from Equation (17). For example, $\lambda = \ln((1 + \sqrt{5})/2)$ can be obtained when $D_{PB,k} = 2$. Figure 1 shows the shape of the EDP function for various values of $D_{PB,k}$ when $\alpha_{PB,k} = 1$. It should be noted from Figure 1 that the EDP functions are almost identical for $D_{PB,k}$ greater than five.

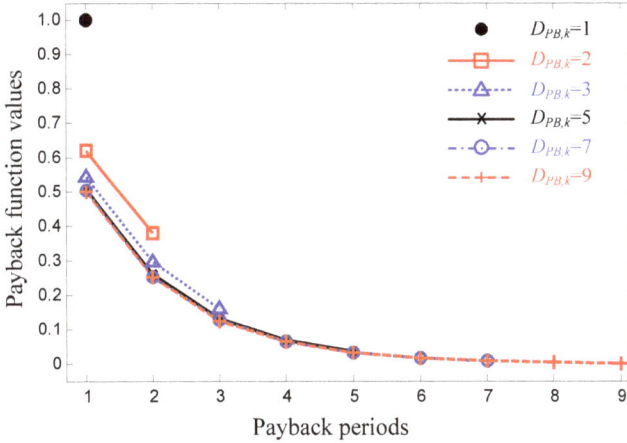

Figure 1. Exponentially decreasing payback functions for several payback duration values of 1, 2, 3, 5, 7, and 9.

Despite past studies [13–15] mentioning the EDP pattern, we cannot rule out the possibility that the curtailed demand is recovered fairly evenly during the payback period. Consequently, a constant function is used to model UDP and is analyzed as an additional specific payback pattern for comparison purposes. The UDP function is expressed as:

$$f_{PB,k}^{U}(n) = c, \quad n \in \{1, 2, \cdots, D_{PB,k}\} \tag{19}$$

where c is a constant that takes the value $c = 1/D_{PB,k}$ from Equation (17).

4. Payback Effects on CPP Design

4.1. Payback Effects on the Event Scheduling Problem

The optimal schedule of critical events that was determined without considering payback may no longer be optimal once the payback phenomenon is taken into account. Further, the LSE's profits may decrease if the additional costs arising from payback exceed the cost savings reaped through the critical event. A simple example in Figure 2 demonstrates such a scenario. Suppose that $H = 4$, $N_{CPP} = 1$, $D_{PB,2} = D_{PB,3} = 1$, $\alpha_{PB,2} = \alpha_{PB,3} = 1$, $\rho_{BASE} = 4$ cents/kWh, and $\rho_{PEAK} = 44$ cents/kWh. The nominal demand and RTMCPs are given in Figure 2a. The customers' price responsiveness is assumed as $\beta = -0.05$.

Under these parameter settings, customers eliminate half of their nominal demand when a critical event is triggered according to Equation (3). Figure 2b and Table 1 present the modified consumption and profit levels under four different scenarios. Without payback, the profit is largest when a critical event is triggered in period $k = 2$; when payback is considered, however, the profit-maximizing critical event period shifts to $k = 3$. In other words, the optimal event schedule changes due to payback. In addition, profit decreases from $30 in the case without payback to $28 in the case of payback with optimal scheduling. This clearly indicates that payback may have a negative effect on an LSE's profits, suggesting that the event scheduling problem in the presence of payback must be solved as a separate optimization problem.

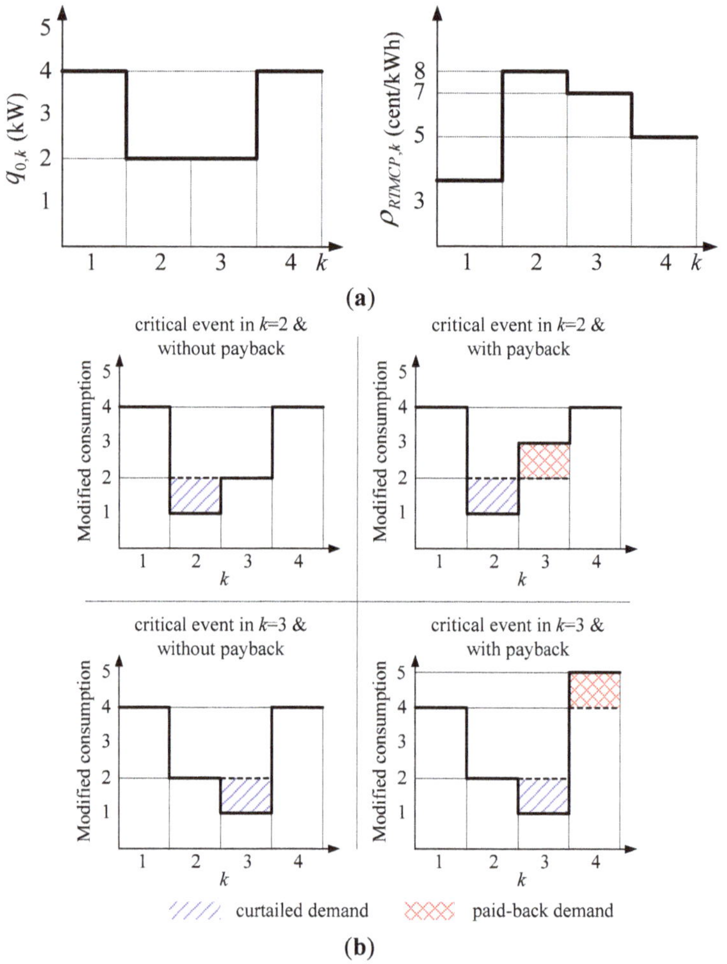

Figure 2. (a) Demand and real-time market clearing prices (RTMCPs); (b) A simple example to show how payback affects the optimal schedule of critical events.

Table 1. Profits of the load serving entity (LSE) with and without payback for two event schedules in the example.

Critical Event Period		Without Payback ($)	With Payback ($)
	revenue	84	88
$k = 2$	cost	54	61
	profit	30	27
	revenue	84	88
$k = 3$	cost	55	60
	profit	29	28

4.2. Payback Effects on the Optimal Peak Rate

As in Equation (11), the profit index including the payback arising from a critical event in period k, $PI_{PB,k}(\rho_{PEAK})$, is represented as:

$$
\begin{aligned}
PI_{PB,k}(\rho_{PEAK}) = {} & \sum_{i=k}^{k+D_{CPP}-1} \left\{ q_{CR,i}(\rho_{PEAK} - \rho_{RTMCP,i}) - q_{0,i}(\rho_{BASE} - \rho_{RTMCP,i}) \right\} \\
& + \sum_{n=1}^{D_{PB,k}} \left\{ \alpha_{PB,k} Q_{CUR,k} f_{PB,k}^{U}(n)(\rho_{BASE} - \rho_{RTMCP,n+k+D_{CPP}-1}) \right\}
\end{aligned}
\tag{20}
$$

As with the procedures for $\rho_{N,PEAK}^{*}$ in Equation (12), the optimal peak rate considering payback, $\rho_{PB,PEAK}^{*}$, can be obtained by substituting Equations (4) into (20), differentiating with respect to ρ_{PEAK}, and solving the resulting equation for ρ_{PEAK}. After rearranging the terms, a specific form of $\rho_{PB,PEAK}^{*}$ is obtained as:

$$
\begin{aligned}
\rho_{PB,PEAK}^{*} = {} & \left[\frac{\rho_{BASE}}{2}\left(1 - \frac{1}{\beta}\right) + \frac{\sum_{k \in \mathbf{K}^*} \sum_{i=k}^{k+D_{CPP}-1} q_{0,i}\rho_{RTMCP,i}}{2 \sum_{k \in \mathbf{K}^*} Q_{0,k}} \right] \\
& + \left[\frac{\rho_{BASE} \sum_{k \in \mathbf{K}^*} \alpha_{PB,k} Q_{0,k}}{2 \sum_{k \in \mathbf{K}^*} Q_{0,k}} - \frac{\sum_{k \in \mathbf{K}^*} \left\{ \alpha_{PB,k} Q_{0,k} \sum_{n=1}^{D_{PB,k}} f_{PB,k}^{U}(n)\rho_{RTMCP,n+k+D_{CPP}-1} \right\}}{2 \sum_{k \in \mathbf{K}^*} Q_{0,k}} \right]
\end{aligned}
\tag{21}
$$

The terms in the first square bracket in Equation (21) are equal to $\rho_{N,PEAK}^{*}$ in Equation (12). Thus, $\rho_{PB,PEAK}^{*}$ can be represented as:

$$
\rho_{PB,PEAK}^{*} = \rho_{N,PEAK}^{*} + \Delta\rho_{PB,PEAK}
\tag{22}
$$

where $\Delta\rho_{PB,PEAK}$ indicates the payback effect on the optimal peak rate, the expression of which is as given in the second bracket in Equation (21). The optimal event schedule, $\mathbf{K}^*$, depends on α_{PB}. Thus, α_{PB} should be included within the summation sign in Equation (21). However, if $\mathbf{K}^*$ does not change and the payback parameters are the same for $k \in \mathbf{K}^*$, that is, $\alpha_{PB,k} = \alpha_{PB}$, $D_{PB,k} = D_{PB}$, and $f_{PB,k}^{U}(n) = f_{PB}^{U}(n)$ for $k \in \mathbf{K}^*$, then α_{PB} can be pulled out of the sum and $\Delta\rho_{PB,PEAK}$ can be represented as:

$$
\Delta\rho_{PB,PEAK} = \frac{\alpha_{PB}}{2} \left\{ \rho_{BASE} - \frac{\sum_{k \in \mathbf{K}^*} \left\{ Q_{0,k} \sum_{n=1}^{D_{PB}} f_{PB}^{U}(n)\rho_{RTMCP,n+k+D_{CPP}-1} \right\}}{\sum_{k \in \mathbf{K}^*} Q_{0,k}} \right\}
\tag{23}
$$

Equation (23) shows that, while maintaining the optimal schedule of critical events, the payback ratio, α_{PB}, has a linear relationship with the amount of change in the optimal peak rate. α_{PB} is not, however, related to whether the payback causes an increase or decrease in the optimal peak rate. On the other hand, the payback pattern, $f_{PB}^{U}(n)$, and duration, D_{PB}, affect both the amount and sign of the change in the optimal peak rate.

Despite the relationships between the payback parameters and the optimal peak rate, it is not evident whether payback causes an increase or decrease in the optimal peak rate based only on Equation (23). This is because $\Delta\rho_{PB,PEAK}$ depends on the specific RTMCPs and nominal demand as well as the payback parameters; the interrelation among these factors is difficult to define conclusively, particularly in cases where $N_{CPP} \geq 2$. As a result, the payback effects on the optimal peak rate will first be examined analytically for the simplest case ($N_{CPP} = 1$). These results will then be extrapolated to the general cases with $N_{CPP} \geq 2$.

Suppose that $N_{CPP} = 1$ and the optimal event schedule is determined as $\mathbf{K}^* = \{k^*\}$. Then, $Q_{0,k}$ terms in the numerator and denominator of Equation (23) cancel one another out and $\Delta\rho_{PB,PEAK}$ can be simplified to:

$$\Delta\rho_{PB,PEAK} = \frac{\alpha_{PB}}{2}\left\{\rho_{BASE} - \sum_{n=1}^{D_{PB}} f_{PB}^{U}(n)\rho_{RTMCP,n+k^*+D_{CPP}-1}\right\} \qquad (24)$$

For EDP, most of the paid-back demand is concentrated in the initial time periods following the critical event. In addition, a critical event is usually triggered when the real-time market clearing price (RTMCP) is high, such that the RTMCPs in the early time periods of the payback phase are likely to be higher than ρ_{BASE}. This implies that the second term inside the bracket in Equation (24), which refers to the average RTMCP weighed by $f_{PB}^{U}(n)$ during the payback periods, is also likely to exceed ρ_{BASE}; $\Delta\rho_{PB,PEAK}$ is thus negative. Consequently, for EDP, where the curtailed demand is recovered quickly, $\rho_{PB,PEAK}^*$ tends to be lower than $\rho_{N,PEAK}^*$. Additionally, as described in Section 3, the EDP functions hardly change for $D_{PB} \geq 5$. Therefore, for EDP, as D_{PB} increases, $\rho_{PB,PEAK}^*$ converges to a value, which is likely less than $\rho_{N,PEAK}^*$.

For UDP, $\Delta\rho_{PB,PEAK}$ in Equation (24) can be simplified as:

$$\Delta\rho_{PB,PEAK} = \frac{\alpha_{PB}}{2}\left\{\rho_{BASE} - \frac{1}{D_{PB}}\sum_{n=1}^{D_{PB}} \rho_{RTMCP,n+k^*+D_{CPP}-1}\right\} \qquad (25)$$

In Equation (25), all relevant RTMCPs are equally weighted in the calculation of $\Delta\rho_{PB,PEAK}$. In addition, the RTMCPs are likely to be small, as the times in question are far from the critical event period. As a result, the absolute value of $\Delta\rho_{PB,PEAK}$ for UDP might be smaller than for EDP, suggesting that $\rho_{PB,PEAK}^*$ is greater for UDP than for EDP. In the extreme situation when the RTMCPs below ρ_{BASE} are dominant over a long payback duration, it is possible that $\Delta\rho_{PB,PEAK}$ becomes positive and thus $\rho_{PB,PEAK}^*$ exceeds $\rho_{N,PEAK}^*$. Moreover, assuming that ρ_{BASE} is set close to the average of the RTMCPs over all periods, $\rho_{PB,PEAK}^*$ for UDP approaches $\rho_{N,PEAK}^*$ as the payback duration increases. Nonetheless, in real situations, the payback duration is usually limited to a few time periods, and the RTMCPs around the critical event periods are likely to exceed ρ_{BASE}. Therefore, $\rho_{PB,PEAK}^*$ for UDP is still likely to be below $\rho_{N,PEAK}^*$, even though it increases and approaches $\rho_{N,PEAK}^*$ as the payback duration increases.

Until now, the analysis has considered payback effects on the optimal peak rate for the simplest case, $N_{CPP} = 1$. As indicated in Equation (23), $\Delta\rho_{PB,PEAK}$ in the general case $N_{CPP} \geq 2$ can be interpreted as a superposition of the effects for the $N_{CPP} = 1$ case, weighted by the $Q_{0,k}$ terms. In other words, $\Delta\rho_{PB,PEAK}$ for $N_{CPP} \geq 2$ can be determined as the weighted sum of N_{CPP} terms of $\Delta\rho_{PB,PEAK}$ for $N_{CPP} = 1$. Therefore, the above-presented analysis of payback effects on the optimal peak rate remains valid in cases with $N_{CPP} \geq 2$ unless $Q_{0,k}$ takes a very abnormal value for a certain critical event. Nonetheless, the payback effects on the optimal peak rate still depend strongly on the specific conditions of the RTMCPs and demand levels. As a result, the following section will perform numerical simulations for $N_{CPP} = 3$ given specific values of the RTMCP and demand; this will allow verification of the payback effects for $N_{CPP} = 1$ and validate their application to cases with $N_{CPP} \geq 2$.

5. Simulations and Verification

5.1. Simulation Methods

Actual data for future RTMCPs and demand is unavailable when an LSE designs a CPP scheme. Thus, these quantities must be forecasted for all periods within the scheduling time horizon. In this study's simulations, the autoregressive moving average (ARMA) method in [24] is used for the forecasting. Historical data on RTMCPs and demand levels, as announced by the Pennsylvania-New Jersey-Maryland Interconnection for 31 days in January 2014 [25], are used as input data for the ARMA method. The resulting forecasted data are shown in Figure 3. The time period length is assumed to be one hour, making the simulations' scheduling time horizon, H, equal to 744. The simulations are performed using the CPP parameters of $N_{CPP} = 3$, $\Delta k = 48$ hours, and $\rho_{BASE} = 12$ cents/kWh and a customer price responsiveness of $\beta = -0.05$. It is assumed that the payback parameters are equal for all critical event time period, that is, $\alpha_{PB,k} = \alpha_{PB}$, $D_{PB,k} = D_{PB}$, and $f^U_{PB,k}(n) = f^U_{PB}(n)$.

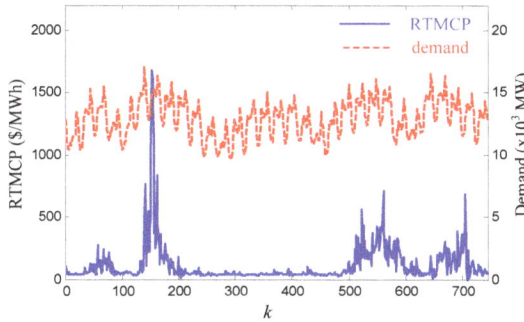

Figure 3. Forecasted data for real-time market clearing price (RTMCP) and demand.

5.2. Results: Payback Effects on the Optimal Event Schedule

The effects of payback on the event scheduling problem are examined under the conditions that $\alpha_{PB} = 1$ and $D_{PB} = 3$ h. For set-ups both with and without payback, the optimal peak rate is arbitrarily selected as $\rho_{N,PEAK} = \rho_{PB,PEAK} = 120$ cents/kWh. Simulations of three different scenarios were undertaken, as listed in Table 2, including without payback, with EDP, and with UDP. The specific values of the payback functions are also given in Table 2 (first column).

The results for the simulated optimal schedule and corresponding profits are also listed in Table 2. The optimal schedules differ from one another according to not only whether or not payback occurs but also the payback pattern. This suggests that the payback phenomenon must be considered to properly solve the event scheduling problem. Furthermore, the profit is larger in the non-payback case than in either case with payback, verifying that the payback phenomenon has a negative effect on the LSE's profits due to the additional cost of the paid-back demand.

Table 2. Simulation results for the optimal schedule and the corresponding profit of the LSE.

$f_{PB}(n) = \{f_{PB}(1), f_{PB}(2), f_{PB}(3)\}$	Optimal Schedule of Critical Events	Profit (Million Dollars)
Without payback	{154, 561, 668}	46.476
Exponentially decreasing payback {0.54, 0.30, 0.16}	{157, 561, 644}	40.265
Uniformly distributed payback {1/3, 1/3, 1/3}	{157, 561, 704}	40.843

5.3. Result: Payback Effects on the Optimal Peak Rate

The effects of payback on the optimal peak rate are simulated by changing the payback duration and ratio. The payback duration is set to change from one to ten ($D_{PB} \in \{1, 2, \cdots, 10\}$), and the range of the payback ratio is taken from [13] as $0.80 \leq \alpha_{PB} \leq 1.06$, with 271 equidistant values of α_{PB} selected within this range. For each combination of D_{PB} and α_{PB}, the optimal peak rate is determined from Equation (21) and the corresponding profit for the LSE is calculated. For the case without payback, we find that $\rho^*_{N,PEAK} = 177.68$ (cents/kWh), which yields a profit of \$52.355 million.

The simulation results for the optimal peak rate, $\rho^*_{PB,PEAK}$, for three values of α_{PB} are presented with respect to α_{PB} in Figure 4, where $D_{PB} = 0$ indicates the case without payback. As described in Section 4.2, the values of $\rho^*_{PB,PEAK}$ are below $\rho^*_{N,PEAK}$ in all simulations because the RTMCP values around critical event periods are larger than ρ_{BASE}. Therefore, $\rho^*_{PB,PEAK}$ should be set below the usual level if the payback phenomenon is expected. In addition, Figure 4 shows that $\rho^*_{PB,PEAK}$ tends to be smaller for a short payback duration than for a long one, regardless of the payback pattern. This suggests that the LSE should select a lower optimal peak rate if the payback period is expected to be short.

Figure 4 also demonstrates how the payback pattern affects the optimal peak rate. The value of $\rho^*_{PB,PEAK}$ is smaller for EDP than for UDP, particularly for a long payback duration. This is because the RTMCPs in the late time periods of the payback duration are smaller than those in the early time periods, but they are all equally weighted when determining $\rho^*_{PB,PEAK}$ for UDP. Moreover, in contrast with the fact that $\rho^*_{PB,PEAK}$ for UDP increases as D_{PB} increases, $\rho^*_{PB,PEAK}$ for EDP hardly changes for $D_{PB} \geq 5$.

As shown in Figure 4, the effect of α_{PB} on the changing shape of $\rho^*_{PB,PEAK}$ is insignificant when compared with the effect of D_{PB}. Nevertheless, to examine the effects of α_{PB} on $\rho^*_{PB,PEAK}$ in greater detail, the change in $\rho^*_{PB,PEAK}$ with respect to α_{PB} is represented in Figure 5 for a few values of D_{PB}. As Equation (24) and the associated analyses indicate, $\rho^*_{PB,PEAK}$ decreases linearly as α_{PB} increases, irrespective of the payback duration and pattern. This linear relationship between α_{PB} and $\rho^*_{PB,PEAK}$ holds only as long as the optimal event schedule, $\mathbf{K}^*$, does not change. When the optimal schedule changes, a step change of $\rho^*_{PB,PEAK}$ occurs, as can be clearly observed from the cases $D_{PB} = 1$ for EDP and UDP and $D_{PB} = 5$ for UDP in Figure 5. However, Figure 5 shows that $\rho^*_{PB,PEAK}$ decreases linearly after the step change as long as the modified optimal schedule is maintained.

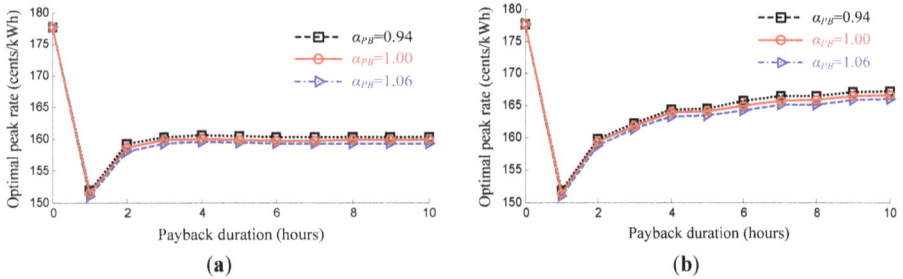

Figure 4. Simulation results of the optimal peak rate with respect to the payback duration for (a) Exponentially decreasing payback (EDP); (b) Uniformly distributed payback (UDP).

Finally, it is necessary to check whether the values of $\rho^*_{PB,PEAK}$ determined in the simulations are optimal. Figure 6 illustrates the profits with payback for $D_{PB} \in \{1, 3, 5\}$ and $\alpha_{PB} = 1$ with respect to the peak rate for the two payback patterns. The profit without payback is also shown in Figure 6. The values for $\rho^*_{N,PEAK}$ and $\rho^*_{PB,PEAK}$ determined via the simulation are clearly the extreme ones, yielding maximum profits. The existence of payback decreases profits in all cases. To emphasize the significance of this analysis, the profits resulting when $\rho^*_{N,PEAK} = 177.68$ cents/kWh and $\rho^*_{PB,PEAK} = 151.41$ cents/kWh for $D_{PB} = 1$ and $\alpha_{PB} = 1$ are indicated in Figure 6. (Note that the functions for the two payback patterns are the same for $D_{PB} = 1$). Comparing the two profits reveals that using $\rho^*_{PB,PEAK}$ increases profits by 2.83% (from $40.307 million for $\rho^*_{N,PEAK}$ to $41.448 million for $\rho^*_{PB,PEAK}$). In practical terms, properly designing a CPP scheme by considering payback effects could lead to a significant, if not dramatic, increase in profits for the LSE.

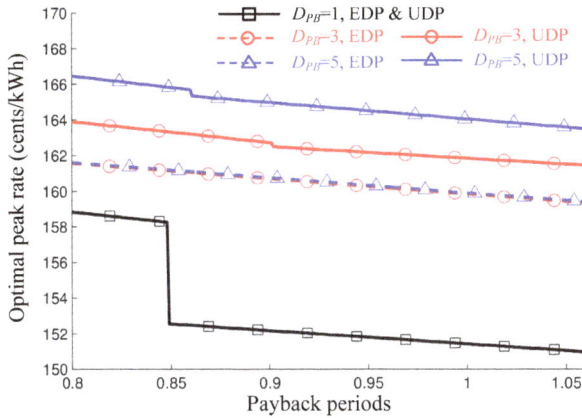

Figure 5. Simulation results of the optimal peak rate with respect to the payback ratio for selected values of D_{PB}.

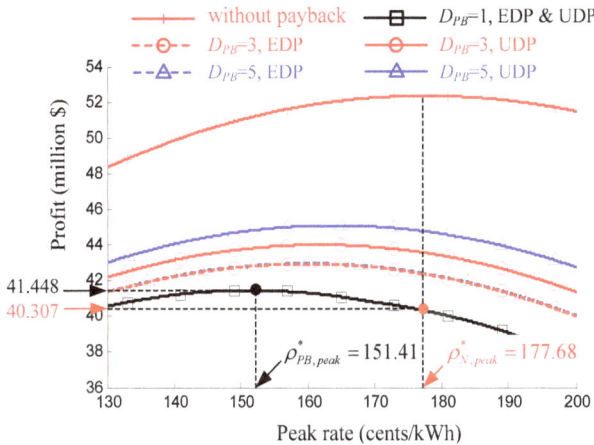

Figure 6. Profit of the load serving entities (LSE) with respect to the peak rate for the case without payback and several cases with payback.

6. Conclusions

A CPP scheme is a useful demand response program that enables an LSE to increase profits by controlling customers' demand at key moments. However, these profits are later reduced by the payback phenomenon. This study considered optimal strategies for designing a profit-maximizing CPP scheme taking payback into consideration. After characterizing payback through the appropriate parameters, the resulting change in optimal event scheduling was demonstrated, and the optimal peak rate under payback was analytically derived. The validity of this analysis was then verified through numerical simulations.

The results yield certain practical suggestions for designing a CPP scheme in a payback scenario. When payback occurs, it is better to set the peak rate to a lower value than would be optimal without payback. Moreover, if the paid-back demand is expected to be concentrated in the time periods soon after a critical event, the peak rate should be set at an even lower value. As long as the optimal event schedule does not change, payback results in a slight linear decrease in the optimal peak rate. However, if the schedule changes, there is a step change in the optimal peak rate. Consequently, the LSE should jointly optimize the event schedule and the peak rate.

Although the results of the proposed method are helpful for designing a CPP scheme, there remain open questions regarding their practical applications. In particular, the availability of payback parameters, which are not constant but depend on the levels of demand and price, can be challenging in the implementation of a CPP scheme. This is one reason why payback has never been considered in the operation of a real-world CPP scheme; as such, it is not obvious that the extended characterization of the payback concept with other unknown and arbitrary parameters will have meaningful implications for actual operations. Additionally, the effects of nonlinear and unpredictable behavior resulting from different compositions of customer loads need to be examined. Therefore, further empirical research will be necessary to demonstrate the practical implications and real-world effectiveness of the CPP design strategy presented here.

Author Contributions: All the authors contributed to this work. Sung Chan Park designed the study, performed the analysis, and wrote the first draft of the paper. Young Gyu Jin contributed to developing the mathematical model and thoroughly revised the paper. Yong Tae Yoon contributed to the conceptual approach and provided important comments on the modeling and analysis.

Conflicts of Interest: The authors declare no conflicts of interest.

Nomenclature

ρ_{BASE}	off-peak rate of a critical peak pricing scheme
ρ_{PEAK}	peak rate of a critical peak pricing scheme
$\rho_{RTMCP,k}$	real-time market clearing price in period k
$\rho^*_{N,PEAK}$	optimal peak rate for a normal situation without payback
$\rho^*_{PB,PEAK}$	optimal peak rate considering payback effects
$\Delta\rho_{PB,PEAK}$	difference of $\rho^*_{PB,PEAK}$ from $\rho^*_{N,PEAK}$
$q_{0,k}$	nominal consumption of customers in period k
$q_{CR,k}$	consumption of customers in period k when a critical event is triggered
$q_{PB,k}$	recovered demand due to payback in period k
$Q_{0,k}$	cumulative consumption during the critical event periods starting from the period k
$Q_{CUR,k}$	cumulative curtailed demand for a critical event starting in period k
$Q_{PB,k}$	paid-back demand for the critical event in period k
R_k	revenue of a load serving entity in period k
C_k	cost of a load serving entity in period k
PI_k	profit index in period k
$PI_{N,k}$	profit index in a normal situation without payback in period k
$PI_{PB,k}$	profit index considering payback effects in period k
u_k	binary event decision variable in period k
N_{CPP}	maximum number of critical events
D_{CPP}	duration of the critical event
D_{PB}	payback duration
$f_{PB}(n)$	payback function
$f^U_{PB}(n)$	normalized payback function
α_{PB}	payback ratio
β	price elasticity of customers
H	scheduling time horizon of the event scheduling problem
Δk	minimum interval between successive events
λ	constant for the exponentially decreasing payback function
c	constant for the uniformly distributed payback function
$\mathbf{K}\vert\rho_{PEAK}$	solution of the events scheduling problem for a given peak rate ρ_{PEAK}
$\mathbf{K}^*$	solution of the events scheduling problem for the optimal peak rate

Abbreviations

ARMA	autoregressive moving average
CPP	critical peak pricing
DR	demand response
EDP	exponentially decreasing payback
LSE	load serving entity
RTMCP	real-time market clearing price
RTP	real-time pricing
TOU	time-of-use
UDP	uniformly distributed payback

References

1. U.S. Department of Energy. *Benefits of Demand Response in Electricity Markets and Recommendations for Achieving Them: A Report to the United States Congress*; U.S. Department of Energy: Washington, DC, USA, 2006.
2. Kirschen, D.S. Demand-side view of electricity markets. *IEEE Trans. Power Syst.* **2003**, *18*, 520–527. [CrossRef]
3. Doostizadeh, M.; Ghasemi, H. A day-ahead electricity pricing model based on smart metering and demand-side management. *Energy* **2012**, *46*, 221–230. [CrossRef]
4. Ferrero, R.W.; Rivera, J.F.; Shahidehpour, S.M. Application of games with incomplete information for pricing electricity in deregulated power pools. *IEEE Trans. Power Syst.* **1998**, *13*, 184–189. [CrossRef]
5. Kirschen, D.S.; Strbac, G. *Fundamentals of Power System Economics*; John Wiley & Sons: Chichester, UK, 2004; pp. 75–79.
6. Albadi, M.H.; El-Saadany, E.F. A summary of demand response in electricity markets. *Elect. Power Syst. Res.* **2008**, *78*, 1989–1996. [CrossRef]
7. Herter, K. Residential implementation of critical peak pricing of electricity. *Energy Policy* **2007**, *35*, 2121–2130. [CrossRef]
8. Joo, J.-Y.; Ahn, S.H.; Yoon, Y.T.; Choi, J.W. Option valuation applied to implementing demand response via critical peak pricing. In Proceedings of the IEEE PES General Meeting, Tampa, FL, USA, 24–28 June 2007; pp. 1–7.
9. Chen, W.; Wang, X.; Petersen, J.; Tyagi, R.; Black, J. Optimal scheduling of demand response events for electric utilities. *IEEE Trans. Smart Grid* **2013**, *4*, 2309–2319. [CrossRef]
10. Zhang, Q.; Wang, X.; Fu, M. Optimal implementation strategies for critical peak pricing. In Proceedings of the 6th International Conference on the European Energy Market (EEM), Leuven, Belgium, 27–29 May 2009; pp. 1–6.
11. Park, S.C.; Jin, Y.G.; Song, H.Y.; Yoon, Y.T. Designing a critical peak pricing scheme for the profit maximization objective considering price responsiveness of customers. *Energy* **2015**, *83*, 521–531. [CrossRef]
12. Zhang, X. Optimal scheduling of critical peak pricing considering wind commitment. *IEEE Trans. Sustain. Energy* **2014**, *5*, 637–645. [CrossRef]
13. Wei, D.; Chen, N. Air conditioner direct load control by multi-pass dynamic programming. *IEEE Trans. Power Syst.* **1995**, *10*, 307–313.
14. Ericson, T. Direct load control of residential water heaters. *Energy Policy* **2009**, *37*, 3502–3512. [CrossRef]
15. Cohen, A.I.; Patmore, J.W.; Oglevee, D.H.; Berman, R.W.; Ayers, L.H.; Howard, J.F. An integrated system for residential load control. *IEEE Trans. Power App. Syst.* **1987**, *2*, 645–651. [CrossRef]
16. Nguyen, D.T.; Negnevitsky, M.; de Groot, M. Modeling load recovery impact for demand response applications. *IEEE Trans. Power Syst.* **2013**, *28*, 1216–1225. [CrossRef]
17. Karangelos, E.; Bouffard, F. Towards full integration of demand-side resources in joint forward energy/reserve electricity markets. *IEEE Trans. Power Syst.* **2012**, *27*, 280–289. [CrossRef]
18. Kirschen, D.S.; Strbac, G.; Cumperayot, P.; de Paiva Mendes, D. Factoring the elasticity of demand in electricity. *IEEE Trans. Power Syst.* **2000**, *15*, 612–617. [CrossRef]
19. Strbac, G. Demand side management: Benefits and challenges. *Energy Policy* **2008**, *36*, 4419–4426. [CrossRef]
20. Shaw, R.; Attree, M.; Jackson, T.; Kay, M. The value of reducing distribution losses by domestic load-shifting: A network perspective. *Energy Policy* **2009**, *37*, 3159–3167. [CrossRef]

21. Strbac, G.; Kirschen, D.S. Assessing the competitiveness of demand-side bidding. *IEEE Trans. Power Syst.* **1999**, *14*, 120–125. [CrossRef]

22. Piette, M.A.; Watson, D.; Motegi, N.; Kiliccote, S. *Automated Critical Peak Pricing Field Tests: 2006 Pilot Program Description and Results*; Lawrence Berkeley National Laboratory: Berkeley, CA, USA, 2007.

23. Schweppe, F.C.; Caramanis, M.C.; Tabors, R.D.; Bohn, R.E. *Spot Pricing of Electricity*; Springer: New York, NY, USA, 1988; pp. 327–333.

24. Weron, R. *Modeling and Forecasting Electricity Loads and Prices: A Statistical Approach*; John Wiley & Sons: Chichester, UK, 2007; pp. 84–85; 109–110.

25. PJM Market & Operations: Real-time Energy Market Data. Available online: http://www.pjm.com/markets-and-operations/energy/real-time.aspx (accessed on 14 February 2015).

energies

MDPI

Article

Active Participation of Air Conditioners in Power System Frequency Control Considering Users' Thermal Comfort

Rongxiang Zhang, Xiaodong Chu *, Wen Zhang and Yutian Liu

Key Laboratory of Power System Intelligent Dispatch and Control of Ministry of Education, Shandong University, 17923 Jingshi Road, Jinan 250061, China; zrx900105@163.com (R.Z.); zhangwen@sdu.edu.cn (W.Z.); liuyt@sdu.edu.cn (Y.L.)

* Correspondence: chuxd@sdu.edu.cn; Tel./Fax: +86-531-8169-6127.

Academic Editor: Ying-Yi Hong
Received: 21 August 2015; Accepted: 21 September 2015; Published: 28 September 2015

Abstract: Air conditioners have great potential to participate in power system frequency control. This paper proposes a control strategy to facilitate the active participation of air conditioners. For each air conditioner, a decentralized control law is designed to adjust its temperature set point in response to the system frequency deviation. The decentralized control law accounts for the user's thermal comfort that is evaluated by a fuzzy algorithm. The aggregation of air conditioners' response is conducted by using the Monte Carlo simulation method. A structure preserving model is applied to the multi-bus power system, in which air conditioners are aggregated at certain load buses. An inner-outer iteration scheme is adopted to solve power system dynamics. An experiment is conducted on a test air conditioner to examine the performance of the proposed decentralized control law. Simulation results on a test power system verify the effectiveness of the proposed strategy for air conditioners participating in frequency control.

Keywords: power system; frequency control; air conditioner; thermal comfort; fuzzy mathematics

---

1. Introduction

In a power system, real-time balance between generation and load is a fundamental requirement for its stable operations. However, some disturbances, such as a sudden drop of generation or short-circuit faults, will cause the mismatch between the two sides bringing damage to the facilities and instability to the power system [1]. Power system frequency is an accurate indicator for generation-load balance. The primary frequency control operates at a time-scale of seconds and uses a speed governor to adjust the mechanical power responding to the frequency deviation. The secondary frequency control operates at a time-scale of minutes and regulates the frequency according to the area control error (ACE) [2]. Traditionally, these two types of control are implemented on the generation side.

Relying solely on the generation side is not sufficient for frequency control. Conventional generating units are of high wear-and-tear cost and low thermal efficiency when responding to control signals in intervals of seconds [3,4]. With the environmental cost to be accounted for, reserve provision from generation, especially fossil-fueled generators, will be much more expensive. The large-scale integration of intermittent energy sources [5–7], e.g., wind and solar energy, will cause high variety and uncertainty in power outputs because of the stochastic nature of weather conditions [8–10]. Supplementary to the generation side, resources from the load side have the potential to provide low cost and fast responsive regulation.

The feasibility and efficiency of load control have been justified [11–14]. Direct load control (DLC) is a centralized scheme that is exerted directly by the control center for large-scale industrial loads [15–17]. As opposed to DLC, load control through small-scale residential and commercial loads is implemented on disperse end-users [18,19]. Thermostatically controlled loads (TCLs), e.g., air conditioners, heat pumps and refrigerators, are good options for primary frequency control because of their inherent heat capacity [20]. Compared with DLC, decentralized control through TCLs is much more flexible. Moreover, TCLs, especially air conditioners take a large proportion in load portfolio. For instance, in China, the proportion of air conditioners in peak power is almost 30%–40% during summer and grows year by year [21]. Consequently, air conditioners can take an active role for frequency control [22].

The fast advancement of smart grid technologies provides a gradually open platform for electricity users. Almost simultaneously, the smart home domain attracts building designers, telecommunication suppliers and appliance manufactures to join in to make home appliances connected and controlled flexibly [23]. Active participation of residential air conditioners in power system control appropriately interlinks the two smart domains, which can be implemented in an economical way.

The feasibility and effectiveness of air conditioners participating in power system operation have been demonstrated by recent research and field programs [24–29]. Studies in [24,25] verified the technical feasibility of air conditioners to participate in frequency control, which requires fast response. More and more utility companies carry out demand response programs involving residential air conditioners, e.g., the SmartAC program of Pacific Gas & Electric Company (PG&E) [26,27], the energy smart thermostat program by Southern California Edison [28] and the smart thermostat program at San Diego Gas & Electric Company [29].

Two mechanisms for air conditioners participating in frequency control are available: (1) directly switch ON/OFF and (2) modulation of temperature setting. In [14], a primary frequency control strategy by switching ON/OFF of demand-side devices was proposed. A frequency control scheme was demonstrated in [16] by adjusting temperature set points of air conditioners in high-density residential buildings. Two types of logics were respectively developed in [19] by directly switching ON/OFF or modulating temperature set points. Compared with the first mechanism, the second one is much more flexible and causes less negative impacts on users' comfort. Many of the current control designs focus on manipulation of loads in a centralized way by using a general dynamic equation to describe the collective thermodynamics of loads [19,20]. The decentralized load control strategies were also studied. A decentralized primary-dual algorithm was proposed for optimal load control in [4,30].

For air conditioners, the end-use utility is evaluated mainly by users' thermal comfort. Various factors may influence human's thermal feelings, in which temperature and relative humidity take the largest part for indoor users [31]. In general, humans feel more comfortable when the temperature is between 24 °C and 27 °C, while relative humidity is between 55% and 70% in summer [32–36]. Accounting for the user's thermal comfort, the accurate description of conditioned space is required. First-order, second-order and third-order models were built, respectively, in [37–39] to depict thermodynamics of the conditioned space. In [30], the control strategy accounts for the utility of loads, but it focuses on the group's characteristic rather than on each individual's.

Air conditioners can actively respond to power system dynamics. Efforts have been made in the field of dynamics modeling. A single-generator power system model was used in [4] to test the feasibility and efficiency of load control, which is too simplified and less accurate. A classical multi-generator model was introduced in [3], which simplifies the network by grouping each generator bus with its nearby load buses. Since load buses are eliminated, the model is not able to accurately depict load characteristics. The structure preserving model is required with load buses retained [30]. Most of previous studies adopted DC power flow to describe the connectivity between buses, ignoring the influence of voltage characteristics on system dynamics [3,4,18,19].

In this paper, we propose a strategy to facilitate the active participation of air conditioners. To evaluate users' thermal comfort, fuzzy mathematics is employed to establish membership functions for various temperature and relative humidity values. The states of air conditioners are described by a third-order model, with its parameters sampled by using the Monte Carlo simulation method. When a disturbance occurs, the temperature set points of air conditioners will be adjusted in response to the degree of frequency deviation while keeping users' thermal comfort within the appropriate range. A structure preserving model is applied to multi-bus power system, with both generator and load buses retained. Full AC power flow is adopted instead of DC power flow.

The remainder of this paper is organized as follows. Section 2 proposes a decentralized control law for air conditioners accounting for individual user's thermal comfort. Section 3 models power system dynamics with the aggregation of air conditioners integrated. Section 4 presents results of case studies. Conclusions are drawn in Section 5.

2. Fuzzy Rule Based Decentralized Control for Frequency Regulation

2.1. Fuzzy Evaluation of Human Thermal Comfort

Thermal comfort is the subjective evaluation of occupants' satisfaction with the thermal environment. It is influenced by multiple factors such as ambient temperature, relative humidity, wind velocity and solar radiation. For indoor occupants, temperature and relative humidity are two major factors. It is difficult to rank thermal comfort precisely with crisp boundaries. Fuzzy mathematics is an appropriate tool for assessment of thermal comfort. A comprehensive fuzzy algorithm involving indoor temperature and relative humidity is proposed to evaluate occupants' thermal comfort.

Let χ = {very low, low, slightly low, moderate, slightly high, high, very high} stand for the domain of temperature ranks. Each rank is depicted by its own membership function shown in Figure 1. Triangular and trapezoid functions are used. The membership degree of a point x with respect to a triangular function defined in the interval $[\alpha, \gamma]$ and middle value in β is obtained as:

$$\mu_{Triangular}(x) = \begin{cases} \frac{x-\alpha}{\beta-\alpha}, & \text{if } \alpha \leq x \leq \beta \\ 0, & \text{if } \alpha > x \text{ or } \gamma < x \\ \frac{\gamma-x}{\gamma-\beta}, & \text{if } \beta \leq x \leq \gamma \end{cases} \tag{1}$$

The membership degree of a point x with respect to a trapezoid function defined in the interval $[\alpha, \gamma]$ and middle values in β and φ ($\beta < \varphi$) is obtained as:

$$\mu_{Trapezoid}(x) = \begin{cases} \frac{x-\alpha}{\beta-\alpha}, & \text{if } \alpha \leq x \leq \beta \\ 1, & \text{if } \beta < x < \varphi \\ \frac{\gamma-x}{\gamma-\varphi}, & \text{if } \varphi \leq x \leq \gamma \\ 0, & \text{otherwise} \end{cases} \tag{2}$$

Let σ = {very dry, dry, slightly dry, temperate, slightly wet, wet, very wet} stand for the domain of relative humidity ranks. Each rank is depicted by its own membership functions of triangular or trapezoid form as shown in Figure 2. The unit of humidity is g/m^3 and the relative humidity is expressed in the form of percentage as the ratio of the water vapor density to the saturation water vapor density, which is used in this paper.

The mutual influence of temperature and relative humidity on thermal comfort is within a fuzzy domain. Human's sensitivity to relative humidity is associated directly with temperature. When temperature is low, relative humidity has small influence on thermal comfort. However, when temperature rises, the influence of relative humidity will increase. Let ρ = {small, medium, large} stand for the domain of relative humidity influence rank. As shown in Figure 3, trapezoid membership functions are adopted for various relative humidity influence ranks. Once the temperature value is known, the influence rank of relative humidity will be obtained.

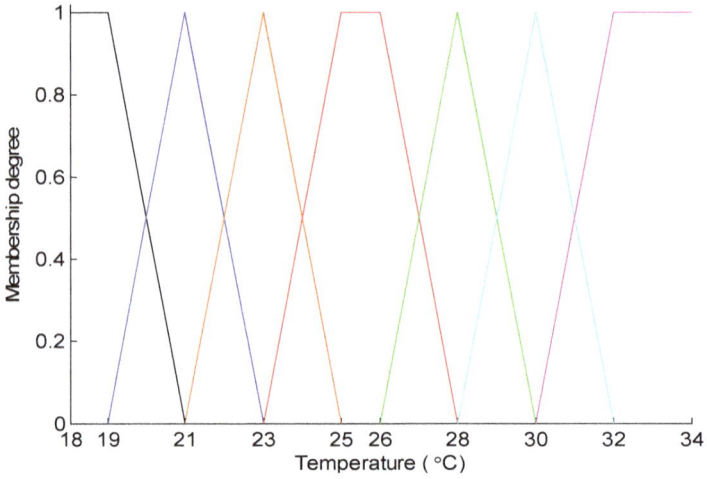

Figure 1. Temperature membership functions in summer.

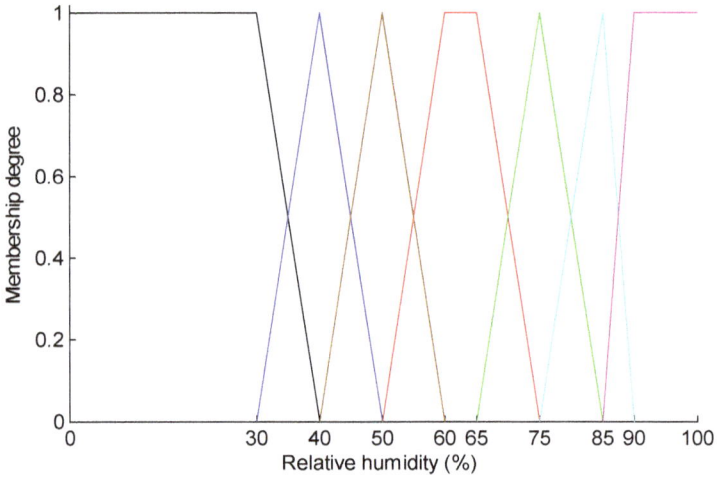

Figure 2. Relative humidity membership functions.

Let $\lambda = [\lambda_1, \lambda_2]$ be the weight set for temperature and relative humidity, respectively, which satisfies

$$\lambda_1 + \lambda_2 = 1 \tag{3}$$

where λ_1 weighs the influence of temperature on thermal comfort and λ_2 weighs the influence of relative humidity. Note that λ_2 is different for three influence ranks shown in Figure 3. When the influence rank of relative humidity is obtained, the weight set λ can be determined.

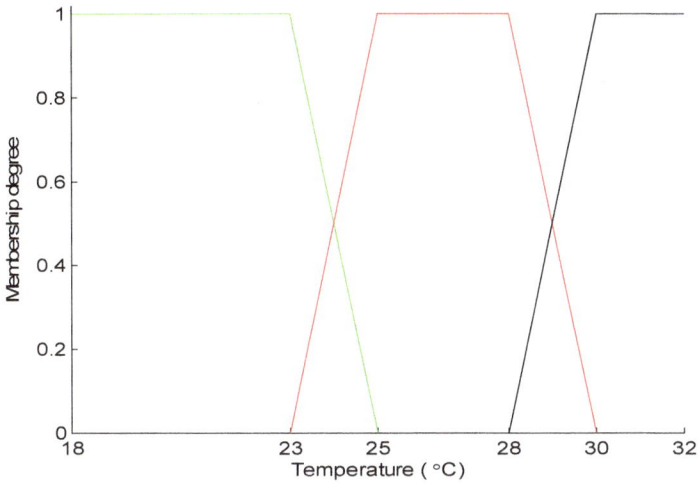

Figure 3. Relative humidity influence membership functions in summer.

When temperature and relative humidity values are provided, the thermal comfort can be evaluated as

$$H = [\lambda_1, \lambda_2] \begin{bmatrix} \mu_{11} & \mu_{12} & \mu_{13} & \mu_{14} & \mu_{15} & \mu_{16} & \mu_{17} \\ \mu_{21} & \mu_{22} & \mu_{23} & \mu_{24} & \mu_{25} & \mu_{26} & \mu_{27} \end{bmatrix} \tag{4}$$
$$= \begin{bmatrix} h_1 & h_2 & h_3 & h_4 & h_5 & h_6 & h_7 \end{bmatrix}$$

where μ_{1j} is the jth temperature membership degree and μ_{2j} is the jth relative humidity membership degree ($j = 1, 2, \ldots ,7$). Ranks of the comprehensive thermal comfort are {very cold, cold, slightly cold, comfortable, slightly hot, hot, very hot}. The thermal comfort rank is defuzzified by the maximum operator:

$$h_{\max} = \max\{h_1, h_2, h_3, h_4, h_5, h_6, h_7\} \tag{5}$$

For instance, if $h_{\max} = h_4$, the thermal comfort rank is "comfortable".

2.2. Decentralized Control Design for Frequency Regulation

Local measurements enable each individual air conditioner to sense frequency deviation from the nominal value. When the frequency reaches the limit value of normal range, controller of the air conditioner will adjust its temperature set point to regulate power demand. The adjustment of the temperature set point accounts for the user's thermal comfort evaluated by the fuzzy algorithm in Section 2.1.

In practice, under-frequency load shedding (UFLS) is a special control scheme for power system stability, which will cause a great cost once triggered. A frequency value is set for UFLS under which load shedding should be activated [40,41]. Compared with UFLS, the control scheme proposed in this paper is a "soft" one with much less cost and it will respond to the abnormal frequency earlier than the activation of UFLS. As shown in Figure 4, the activation frequency range of the proposed scheme is

$$\Delta f_{range} = f_u - f_l \tag{6}$$

where f_u is the activation frequency value of UFLS and f_l is the lowest value of normal frequency.

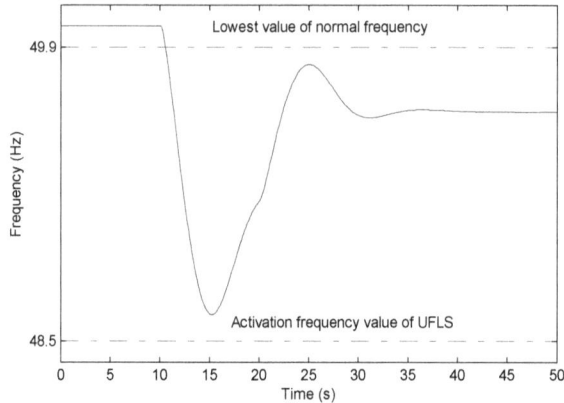

Figure 4. Activation frequency range.

Thermal comfort rank can be evaluated at a certain temperature and relative humidity. In summer, if the thermal comfort rank is "very cold", "cold" or "slightly cold", the temperature will be raised to a higher value at which the rank is "comfortable". Otherwise, if the rank is among "comfortable", "slightly hot", "hot" or "very hot", the temperature should not be adjusted. In winter, the adjustment is in the reverse direction. Relative humidity is treated as constant during a control span, which is reasonable since the span is short. The span is the time duration of an air conditioner from OFF state to ON state after it responds to the control signal, which is determined by the time constant of thermal mass in the conditioned room. The maximum temperature adjustment is

$$\Delta T_{max} = T_{new} - T_{old} \tag{7}$$

where T_{old} is the initial temperature; and T_{new} is the higher temperature.

For each air conditioner, the actual temperature adjustment responding to frequency deviation is

$$\Delta T_{set} = \frac{\Delta T_{max}}{\Delta f_{range}} \Delta f \tag{8}$$

where Δf is actual frequency deviation.

Within the activation frequency range, the proposed control scheme will regulate air conditioners in response to the frequency deviation while maintaining users' thermal comfort. However, when the frequency value is lower than f_u, the air conditioners will be switched off directly.

2.3. Aggregation of Air Conditioners' Response

Air conditioners are of periodic on/off working characteristics. The temperature set point is the boundary for an air conditioner to switch from on to off or *vice versa*. By increasing the temperature set point, the air conditioner tends to be off. Power demand will decrease significantly if a large number of air conditioners switch off within a short time range.

Assume that the power system is working at an operating point where generation and load are in balance. Then, a sudden generation drop occurs and consequently the system frequency decreases. Each air conditioner responds to the frequency deviation through temperature set point adjustment according to the control law in Equation (8).

Changing the temperature set point will influence states of an air conditioner and its conditioned space, which can be depicted by a dynamical model [39]

$$\dot{x} = Ax + Bu$$
$$y = Cx$$

(9)

where x, u, y, A, B and C are detailed in Appendix A.

The Monte Carlo simulation method is applied to the aggregation of air conditioners' response. Each air conditioner in the control group is treated as a sample with its parameters randomized. The aggregated power demand of the group of air conditioners is

$$P_{ac}(t) = \sum_{j \in \psi} p_j(t)$$

(10)

where $\psi = \{1, 2, \dots, N\}$ represents individual air conditioners; and $p_j(t)$ is the power demand of air conditioner j at time t.

Note that the following three aspects of using air conditioners for frequency control are considered but not directly treated in this paper:

(1) Energy payback phenomenon is caused by the synchronization of air conditioners during the recovery process. A careful recovery strategy can be used to avoid the phenomenon [42,43].
(2) To attract more electricity users to participate in active load control, a power grid can design various incentive tools. Incentive-based demand response programs have been carried out in recent years [44–46]. Users can contract with the power grid on the compensation and award packages for their comfort loss and contributions for the power grid.
(3) Conventionally, generating units supply spinning reserves and non-spinning reserves for the power system. At demand peak periods, the cost of maintaining high reserve level is very large. Air conditioners are capable of providing reserves with much lower cost than generating units. The economic saving through the proposed strategy can be quantified by the value of reduced reserve capacity procured from the generation side.

3. Power System Dynamics Model

3.1. Dynamical Model of Generating Units

A governor-turbine-generator model is used to represent a generating unit as shown in Figure 5, which accounts for primary frequency regulation of the unit.

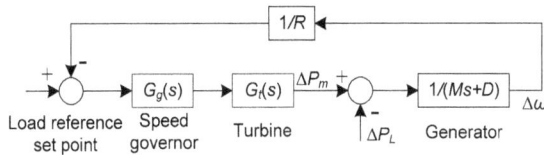

Figure 5. Generating unit model.

Dynamics of the generator can be described by the swing equations as

$$\begin{cases} \frac{d\delta}{dt} = \omega_s(\omega - 1) \\ M\frac{d\omega}{dt} = P_m - P_e \end{cases}$$

(11)

where δ is generator rotor angle; ω is rotor angle speed; ω_s is nominal angle speed; P_m and P_e are, respectively, mechanical power and electrical power; and M is angular momentum of the generator. Electrical power can be derived by

$$P_e = P_L + D\omega \tag{12}$$

where D is damping coefficient; and P_L is power demand of load.

The turbine block has the transfer function as

$$G_t(s) = \frac{1 + sF_{HP}T_{RH}}{(1 + sT_{CH})(1 + sT_{RH})} \tag{13}$$

where F_{HP} is intermediate pressure cylinder mechanical ratio; and T_{CH} and T_{RH} are volume effect time constants.

The transfer function of the speed governor is

$$G_g(s) = \frac{1}{1 + sT_G} \tag{14}$$

where T_G is governor time constant. As shown in Figure 5, R is the gain of the feedback loop for the governor, which determines the speed droop characteristic.

3.2. Network Model

A power network is modeled by a graph $\zeta = \{v, \varepsilon\}$, where $v = \{1, 2, \ldots, n\}$ is the set of buses and ε is the set of transmission lines connecting the buses. Bus voltage equations of the power network can be expressed as

$$YV = I \tag{15}$$

where I and V are injection current vector and bus voltage vector, respectively; and Y is nodal admittance matrix.

A power network has three types of buses including generator buses, load buses and contact buses. The injection currents at a generator bus i can be described as

$$\begin{bmatrix} I_{xi} \\ I_{yi} \end{bmatrix} = \begin{bmatrix} \frac{\sin \delta_i}{X'_{di}} \\ -\frac{\cos \delta_i}{X'_{di}} \end{bmatrix} E'_i - \begin{bmatrix} 0 & \frac{1}{X'_{di}} \\ -\frac{1}{X'_{di}} & 0 \end{bmatrix} \begin{bmatrix} V_{xi} \\ V_{yi} \end{bmatrix} \tag{16}$$

where I_{xi} and I_{yi} are, respectively, real and imaginary part of injection current; V_{xi} and V_{yi} are, respectively, real and imaginary part of voltage; X'_{di} is direct-axis transient reactance; and E'_i is electromotive force of the generator.

At contact buses, the net injection currents are zero:

$$\begin{bmatrix} I_{xi} \\ I_{yi} \end{bmatrix} = 0 \tag{17}$$

A load bus may consist of conventional loads and aggregation of air conditioners. The injection currents at load bus i are presented by

$$\begin{bmatrix} I_{xi} \\ I_{yi} \end{bmatrix} = \begin{bmatrix} I_{xi\_c} \\ I_{yi\_c} \end{bmatrix} + \begin{bmatrix} I_{xi\_ac} \\ I_{yi\_ac} \end{bmatrix} \tag{18}$$

where I_{xi_c} and I_{yi_c} are, respectively, real and imaginary part of injection current of conventional loads; and I_{xi_ac} and I_{yi_ac} are corresponding values of air conditioners, respectively. The conventional loads can be described by their static characteristics and dynamic characteristics, respectively.

Quadratic polynomial is usually used to describe the static characteristics of loads

$$\begin{cases} P_L = P_{L(0)}[a_P(\frac{V_L}{V_{L(0)}})^2 + b_P(\frac{V_L}{V_{L(0)}}) + c_P] \\ Q_L = Q_{L(0)}[a_Q(\frac{V_L}{V_{L(0)}})^2 + b_Q(\frac{V_L}{V_{L(0)}}) + c_Q] \end{cases} \tag{19}$$

where $P_{L(0)}$ and $Q_{L(0)}$ are pre-disturbance value of real power and reactive power of a load, respectively; and $V_{L(0)}$ is pre-disturbance value of voltage at the load bus. The load comprises constant impedance, constant current and constant power proportions and a, b, c are corresponding coefficients satisfying

$$\begin{cases} a_P + b_P + c_P = 1 \\ a_Q + b_Q + c_Q = 1 \end{cases} \tag{20}$$

The injection currents of the constant impedance proportion are

$$\begin{bmatrix} I_{xi\_c} \\ I_{yi\_c} \end{bmatrix} = \begin{bmatrix} \frac{-P_{i\_c(0)}}{V_{i(0)}^2} & \frac{-Q_{i\_c(0)}}{V_{i(0)}^2} \\ \frac{Q_{i\_c(0)}}{V_{i(0)}^2} & \frac{-P_{i\_c(0)}}{V_{i(0)}^2} \end{bmatrix} \begin{bmatrix} V_{xi} \\ V_{yi} \end{bmatrix} \tag{21}$$

where $P_{i\_c(0)}$ and $Q_{i\_c(0)}$ are, respectively, pre-disturbance value of real power and reactive power; and $V_{i(0)}$ is pre-disturbance value of voltage. Constant current and constant power proportions are also readily to be included in the power network model.

Dynamic characteristics of conventional loads are usually associated with induction motors, which can be described as

$$\begin{bmatrix} I_{xi\_m} \\ I_{yi\_m} \end{bmatrix} = -\begin{bmatrix} \frac{R_{i\_m}}{R_{i\_m}^2 + X_{i\_m}^2} \\ \frac{-X_{i\_m}}{R_{i\_m}^2 + X_{i\_m}^2} \end{bmatrix} E'_{i\_m} \tag{22}$$

where E'_{i_m} is the transient electromotive force of the induction motor; and R_{i_m} and X_{i_m} are equivalent resistance and reactance, respectively.

Real power and reactive power of the aggregation of air conditioners will change with time in response to frequency dynamics. Consequently, the relationship between the injection current and bus voltage is nonlinear. The injection currents of the aggregation of air conditioners can be described as

$$\begin{cases} I_{xi\_ac} = -\frac{P_{i\_ac}V_{xi} + Q_{i\_ac}V_{yi}}{V_{xi}^2 + V_{yi}^2} \\ I_{yi\_ac} = -\frac{P_{i\_ac}V_{yi} - Q_{i\_ac}V_{xi}}{V_{xi}^2 + V_{yi}^2} \end{cases} \tag{23}$$

where P_{i_ac} and Q_{i_ac} are real power and reactive power of the aggregation of air conditioners, respectively. When I_{xi}, I_{yi}, V_{xi} and V_{yi} are determined, the electrical power of generator i can be computed as

$$P_{ei} = I_{xi}V_{xi} + I_{yi}V_{yi} \tag{24}$$

3.3. Solving Dynamics

The dynamical model for the power system is constructed by integrating dynamical equations of air conditioners in Equations (9) and (10) and generating units in Equations (11)–(14) with the network model in Equation (15). The numerical integration method is applied to solving the differential-algebraic equations.

Since the relationship between injection current and voltage is nonlinear in Equation (23), an iteration process should be used to solve it. In each iteration k, the correction formula of voltage variables is

$$\begin{bmatrix} \Delta I_x^{(k)} \\ \Delta I_y^{(k)} \end{bmatrix} = \begin{bmatrix} H^{(k)} & N^{(k)} \\ M^{(k)} & L^{(k)} \end{bmatrix} \begin{bmatrix} \Delta V_x^{(k)} \\ \Delta V_y^{(k)} \end{bmatrix} \tag{25}$$

where

$$\begin{cases} H_{ij}^{(k)} = \partial I_{xi}^{(k)} / \partial V_{xj}^{(k)} - G_{ij} \\ N_{ij}^{(k)} = \partial I_{xi}^{(k)} / \partial V_{yj}^{(k)} + B_{ij} \\ M_{ij}^{(k)} = \partial I_{yi}^{(k)} / \partial V_{xj}^{(k)} - B_{ij} \\ L_{ij}^{(k)} = \partial I_{yi}^{(k)} / \partial V_{yj}^{(k)} - G_{ij} \end{cases} \tag{26}$$

where G_{ij} and B_{ij} are the real and imaginary part of nodal admittance, respectively.

At bus i, the error items for injection currents are

$$\begin{cases} \Delta I_{xi}^{(k)} = I_{xi}^{(k)} - \sum_{m \in i} (G_{im} V_{xm}^{(k)} - B_{im} V_{ym}^{(k)}) \\ \Delta I_{yi}^{(k)} = I_{yi}^{(k)} - \sum_{m \in i} (G_{im} V_{ym}^{(k)} + B_{im} V_{xm}^{(k)}) \end{cases} \tag{27}$$

where $m \in i$ represents each bus that connects with bus i by a transmission line.

A two stage iterative scheme is designed as shown in Figure 6. The inner iteration process is to solve voltage variables at the load bus with the aggregation of air conditioners. The outer iteration process is for solving dynamics of generating units and air conditioners. An alternating method is adopted for the outer iteration process involving both differential and algebraic equations. The modified Euler algorithm is used for numerical integration.

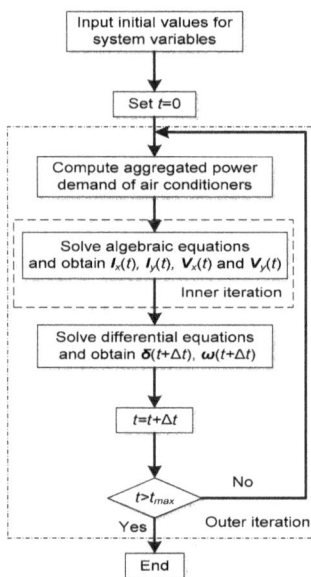

Figure 6. Solving process for power system dynamics.

4. Case Studies

4.1. Experimental Results of Decentralized Control Design

We design an experiment to verify the effectiveness of the fuzzy rule based decentralized control for an air conditioner. The experiment was conducted in a winter season, when the air conditioner was operated in the heat pump mode. Except for the setting parameters, the control law is the same with that in summer. As shown in Figure A1 of Appendix B, the temperature rank membership functions in winter are different from those in summer. However, the relative humidity rank membership functions are the same with those in summer. The relative humidity influence rank membership functions are not same with those in summer, which are displayed in Figure A2 of Appendix B. Weights of temperature and relative humidity for three influence ranks are shown in Table 1.

Table 1. Weight proportions.

Weights	Relative humidity influence ranks		
	Small	Medium	Large
λ_1	0.9	0.8	0.6
λ_2	0.1	0.2	0.4

The test rig consists of an air conditioner, a power system state monitoring module, temperature and humidity sensors, data communication units, an infrared controller and a control desk, which is displayed in Figures 7 and 8. In the experiment, the power system state monitoring module is responsible for measuring frequency, voltage, and current at the node where the air conditioner plugs into the power system. Two sensors are used to sense indoor temperature and relative humidity, respectively. The measurement data are transmitted to the control desk through data communication units. The infrared controller is the external agent for the air conditioner to regulate its temperature set point. A supervisory system runs in the control desk to make control decision and coordinate various modules, which was developed with LabVIEW. When the frequency deviates from the normal range, the control desk will determine the new temperature set point according to the control law in Section 2.2. The control signal then will be sent to the air conditioner via the infrared controller.

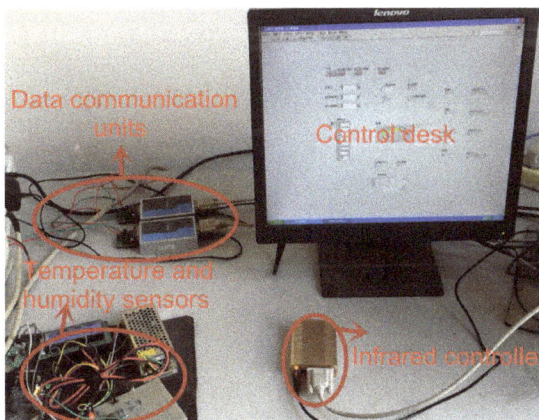

Figure 7. Measurement, communication and control modules.

In this experimental setup, the cost of temperature-humidity sensors and power system frequency meter are about 20 RMB Yuan (3.2 dollars) and 10 RMB Yuan (1.6 dollars), respectively. It is feasible to expect that the expense will be affordable for such a module to be equipped in an air conditioner or installed separately.

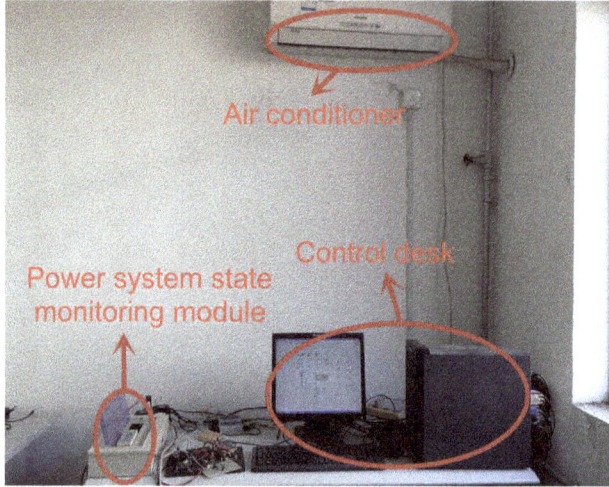

Figure 8. Test rig.

As shown in Figure 9, a frequency signal is generated to emulate the typical characteristic curve of frequency dynamics caused by a power system disturbance. Assume that frequency initially keeps at its nominal value of 50 Hz and then a generation drop occurs at the 10-s instant. Driven by the imbalance between generation and load, frequency decreases from 50 Hz. After dropping to the lowest value, frequency recovers to a new steady state value. The frequency signal is fed into the test rig and sampled every 0.1 s to trigger the control flow. The experiment was conducted for 240 runs starting from various initial values of the indoor temperature that follow the uniform distribution between 13 °C and 29 °C. The assumption of uniform distribution intends to generally cover more indoor conditions, and is not limited to certain specific situations.

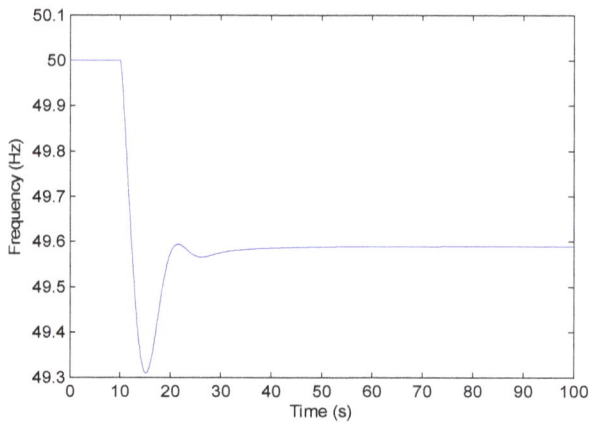

Figure 9. Frequency signal for experiment.

In each experimental run, the air conditioner will be controlled in response to the frequency signal in accordance with the proposed control law. If the thermal comfort rank at the initial temperature value is among "very hot", "hot" or "slightly hot", the temperature set point will be adjusted to decrease. Consequently, the indoor temperature will decrease since the air conditioner intends to switch off. However, if the thermal comfort rank at the initial temperature value is among "very cold", "cold" or "slightly cold", the air conditioner will not respond to the frequency deviation. From results of the 240 experimental runs, the distribution of various thermal comfort ranks is derived. As shown in Figure 10, the percentages of "very hot" and "hot" ranks decrease significantly, whereas the percentages of "comfortable" and "slightly hot" ranks increase most after control. However, the percentages of "very cold", "cold" and "slightly cold" ranks change less. Table 2 lists the detailed percentage numbers. To summarize, the indoor temperature has a large probability to be moved to the rank of "comfortable" after control. The control design for the air conditioner has positive impact on the user's thermal comfort.

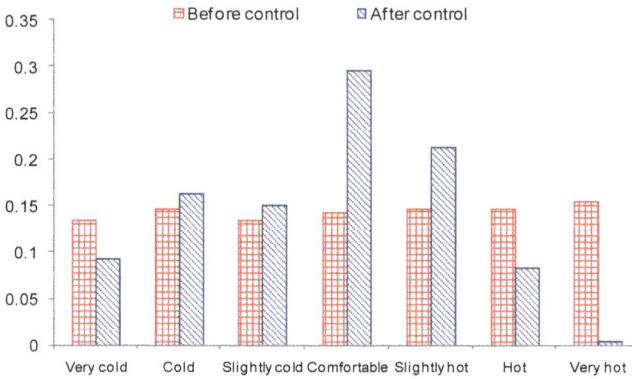

Figure 10. Thermal comfort rank distribution.

Table 2. Thermal comfort rank distribution.

Thermal comfort ranks	Very cold	Cold	Slightly cold	Comfortable	Slightly hot	Hot	Very hot
Before control	0.133	0.146	0.133	0.142	0.146	0.146	0.154
After control	0.092	0.1625	0.150	0.296	0.2125	0.083	0.004

4.2. Simulation Results of Test Power System

To examine the control performance on multi-machine, multi-bus power system, simulations are conducted on the IEEE 9-bus test system. The test system consists of nine buses, three generating units and three loads. The single line diagram of the test system is displayed in Figure 11. Parameters of generating units are given in Table 3. Simulations are programmed and run in MATLAB.

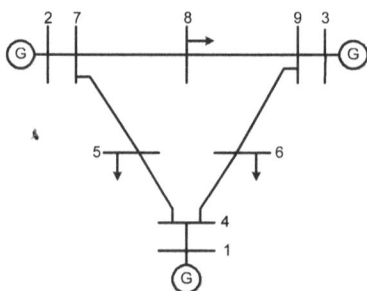

Figure 11. Single line diagram of the IEEE 9-bus test system.

Table 3. Parameters of generating units.

Parameters	Values (s)	Parameters	Values(p. u.)
T_G	0.2	D	1
T_{RH}	7	R	0.05
T_{CH}	0.3	F_{HP}	0.3
M	10		

Assume that a total of 10,000 air conditioners are aggregated at bus 8. Each air conditioner responds in accordance with the decentralized control law proposed in Section 2. The aggregation of air conditioners' response is conducted on the basis of random sampling. The typical values of thermal parameters characterizing air-conditioned indoor space are given in Tables A1 and A2 of Appendix A. The parameters are randomly sampled, which follows the normal distribution (the parameters shown in Table A1) or uniform distribution (the parameters shown in Table A2).

The power system is stable before the disturbance, with its frequency fluctuating very slightly from the nominal value of 50 Hz. At the 10-s instant, a generation drop occurs at bus 2, which is simulated by regulating the steam valve opening U:

$$U(t) = \begin{cases} 1\,\text{p. u.} & 0 \leq t < 10\,\text{s} \\ 0.5\,\text{p. u.} & t \geq 10\,\text{s} \end{cases} \tag{28}$$

Then, the system frequency drops down from its nominal value. To mitigate the frequency drop, reserve resources are activated from both the generation side and load side. In this study, air conditioners are the major reserve resource from the load side. When the power balance is rebuilt between generation and load around the 40-s instant, the system frequency reaches a new steady state.

The participation of air conditioners improves the system frequency characteristics by lessening drop depth and lifting steady state, which is displayed in Table 4 and Figure 12. Six cases are studied representing various participation ratios of air conditioners, from zero ($r = 0$) to 50% ($r = 0.5$). The larger the ratio is, the more significant the improvement is. For instance, the lowest point of frequency drop, the nadir, is 49.49 Hz for the case of $r = 0.3$ while it is 49.31 Hz for the case of $r = 0$, *i.e.*, no air conditioners participating. With the new steady state built, the frequency recovers to 49.71 Hz for the case of $r = 0.3$ compared with 49.59 Hz for the case of $r = 0$.

Table 4. Frequency characteristics.

Frequency (Hz)	Ratio of air conditioning demand in load at bus 8					
	r = 0	*r* = 0.1	*r* = 0.2	*r* = 0.3	*r* = 0.4	*r* = 0.5
Lowest point	49.31	49.36	49.45	49.49	49.56	49.64
Steady state	49.59	49.62	49.68	49.71	49.73	49.81

As displayed in Figure 13, aggregated power demand of air conditioners decreases since it is fairly sensitive to the system frequency, which services to the power rebalance between generation and load after disturbance. For the case of $r = 0.3$, the aggregated power demand is around 30 MW before disturbance. After the disturbance occurs at the 10-s instant, it decreases sharply in response to the frequency drop. At the 40-s instant, the aggregated power demand levels off to around 18 MW. The contribution of air conditioners is estimated to be 12 MW (= 30 MW−18 MW), which is about 40% of its baseline demand before disturbance.

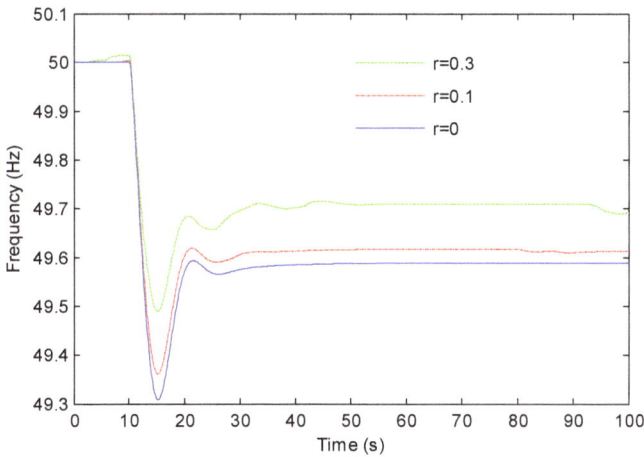

Figure 12. Frequency dynamics.

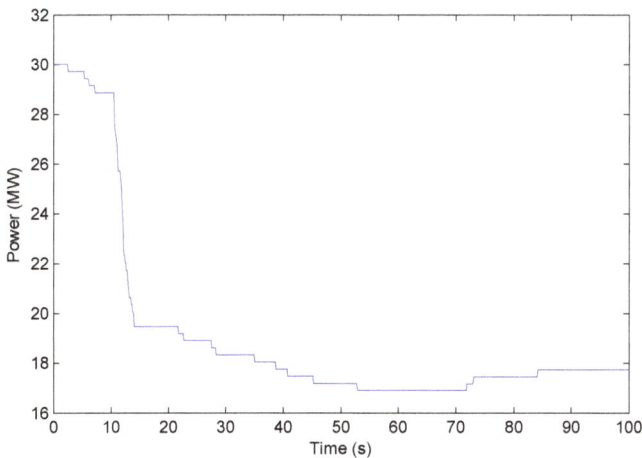

Figure 13. The demand of air conditioners after the disturbance.

As shown in Figure 14, generating units respond to the system frequency dynamics by regulating their mechanical power along with electrical power. Each unit is controlled locally to stabilize the rotor angle speed, with the regulation of its mechanical power contributing to compensate for the power imbalance caused by the disturbance. Mechanical power of the generating unit at bus 2 decreases because of the generation drop. Mechanical power of the unit at bus 1 (or bus 3) increases to compensate for the power deficit. The rebalance between mechanical power and electrical power is achieved at each generator bus around the 40-s instant when the frequency stabilizes to a steady state as shown in Figure 12.

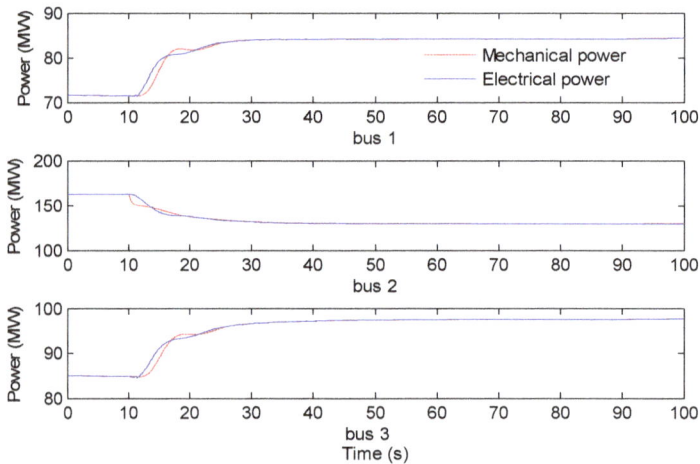

Figure 14. Mechanical power and electrical power at generator buses 1, 2 and 3.

5. Conclusions

This paper proposes a strategy for air conditioners participating in power system frequency regulation. According to the decentralized control law proposed, each air conditioner will respond to the frequency deviation by regulating its temperature set point while maintaining the user's thermal comfort. A fuzzy algorithm is used to evaluate user's thermal comfort. The population of air conditioners is aggregated through the Monte Carlo simulation method. A structure preserving model for power system is adopted with characteristics of aggregated air conditioners being depicted correctly, which is solved by an inner-outer iteration algorithm. Experimental and simulation results verify the effectiveness and performance of the proposed strategy for air conditioners to actively participate in power system frequency control.

Acknowledgments: This work was funded by the National Natural Science Foundation of China (51207082).

Author Contributions: Rongxiang Zhang and Xiaodong Chu proposed the models; Rongxiang Zhang conducted the experiments and simulations; Rongxiang Zhang and Xiaodong Chu analyzed the data; Rongxiang Zhang and Xiaodong Chu wrote the manuscript; and Wen Zhang and Yutian Liu helped revising the manuscript.

Conflicts of Interest: The authors declare no conflict of interest.

Appendix A Thermodynamics Model for Air-Conditioned Space

The thermodynamics model for an air conditioner and its conditioned space is described by Equation (9) in Section 2 as

$$\dot{x} = Ax + Bu$$
$$y = Cx$$

The state variables are

$$x = \left[X_{ew},\ X_{in},\ X_{iw}\right]^{\mathrm{T}}$$

where X_{ew} is external wall temperature; X_{in} is indoor temperature; and X_{iw} is internal wall temperature. The input variables are

$$u = \left[X_{ext},\ P_{e-w},\ P_{eq},\ P_{HVAC} \cdot m_{on-off},\ X_{adj-r}\right]^{\mathrm{T}}$$

where X_{ext} is external temperature evolution; P_{e-w} is solar radiation on external wall faces; P_{eq} is solar radiation which introduces through glazed surfaces plus internal load generation; P_{HVAC} is a value associated with power supply; m_{on-off} is a discrete variable which represents the operating state of the device (1 for ON and 0 for OFF); and X_{adj-r} is adjoining room temperature evolution.

The output variable is

$$y = X_{in}$$

A, B and C are as follows:

$$A = \begin{bmatrix} \frac{1}{C_{ew}}\left[\frac{R_{cew}}{R_{ew}(R_{cew}+R_{ew})} - \frac{2}{R_{ew}}\right] & \frac{1}{R_{ew}C_{ew}} & 0 \\ \frac{1}{C_{in}R_{ew}} & \frac{-1}{C_{in}}\left(\frac{1}{R_{ew}} + \frac{1}{R_{iw}} + \frac{1}{R_{gs}}\right) & \frac{1}{C_{in}R_{iw}} \\ 0 & \frac{1}{C_{iw}R_{iw}} & \frac{-2}{C_{iw}R_{iw}} \end{bmatrix}$$

$$B = \begin{bmatrix} \frac{1}{C_{ew}}\frac{1}{R_{cew}+R_{ew}} & \frac{R_{cew}}{C_{ew}(R_{cew}+R_{ew})} & 0 & 0 & 0 \\ \frac{1}{C_{in}R_{gs}} & 0 & \frac{1}{C_{in}} & \frac{-1}{C_{in}} & 0 \\ 0 & 0 & 0 & 0 & \frac{1}{C_{iw}R_{iw}} \end{bmatrix}$$

$$C = [0,\ 1,\ 0]$$

where C_{ew} and C_{iw} are, respectively, thermal capacity of external walls and internal walls; C_{in} is indoor thermal capacity; R_{cew} is external convection resistance between external environment and external wall faces; R_{ew} is half equivalent thermal resistance of external walls; R_{iw} is half equivalent thermal resistance of the internal walls; and R_{gs} is equivalent thermal resistance of external glazed surfaces.

Table A1. Thermal parameters of air-conditioned indoor space.

Parameters	Expectation	Variance
C_{ew}	1.875 MJ/°C	0.03 MJ/°C
C_{in}	1.2 MJ/°C	0.03 MJ/°C
C_{iw}	3.1 MJ/°C	0.05 MJ/°C
R_{gs}	0.08 °C/W	0.0002 °C/W
R_{iw}	0.01 °C/W	0.00002 °C/W
R_{ew}	0.028 °C/W	0.00002 °C/W
R_{cew}	0.005 °C/W	0.00002 °C/W

Table A2. Thermal parameters of air-conditioned indoor space.

Parameters	Range (°C)	Parameters	Range (W)
X_{ext}	[25,28]	P_{eq}	[490,510]
X_{in}	[18,33]	P_{e-w}	[190,210]
X_{iw}	[20,30]	P_{HVAC}	[2500,3000]
X_{ew}	[24,27]		
X_{adj-r}	[20,30]		

Appendix B   Thermal Membership Functions in Winter

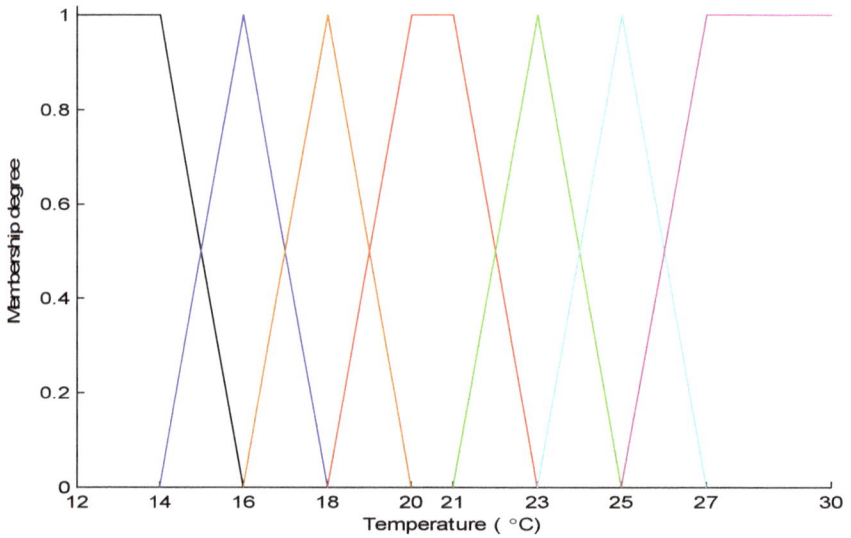

Figure A1. Temperature membership functions in winter.

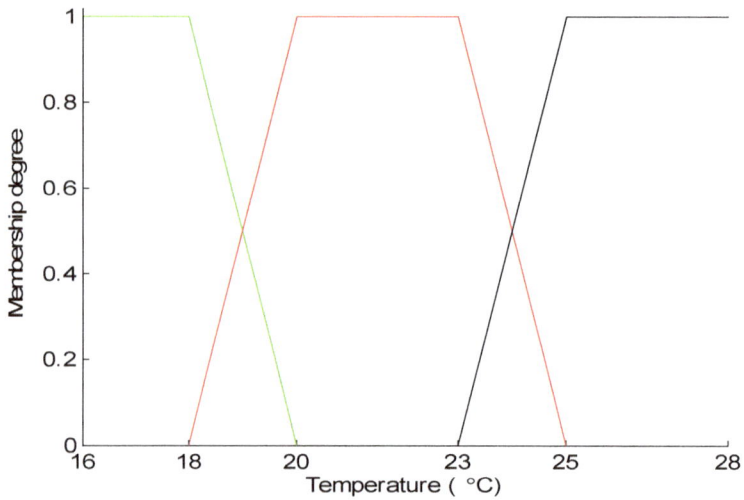

Figure A2. Relative humidity influence membership functions in winter.

References

1. Wang, X.; Song, Y.; Irving, M. *Modern Power System Analysis*, 1st ed.; Springer U.S.: New York, NY, USA, 2008.
2. Wood, A.J.; Wollenberg, B.F. *Power Generation, Operation and Control*, 2nd ed.; Wiley & Sons: New York, NY, USA, 1996.
3. Zhao, C.; Ufuk, T.; Low, S. Swing dynamics as primary-dual algorithm for optimal load control. In Proceedings of the IEEE 3rd International Conference on Smart Grid Communications, Tainan, Taiwan, 5–8 November 2012; pp. 570–575.

4. Zhao, C.; Ufuk, T.; Low, S. Optimal load control via frequency measurement and neighborhood area communication. *IEEE Trans. Power Syst.* **2013**, *28*, 3576–3587.

5. Ummels, B.C.; Gibescu, M.; Pelgrum, E.; Kling, W.L.; Brand, A.J. Impacts of wind power on thermal generation unit commitment and dispatch. *IEEE Trans. Energy Convers.* **2007**, *22*, 44–51. [CrossRef]

6. Doherty, R.; Mullane, A.; Nolan, G.; Burke, D.J.; Bryson, A.; O'Malley, M. An assessment of the impact of wind generation on system frequency control. *IEEE Trans. Power Syst.* **2010**, *25*, 452–460. [CrossRef]

7. Kayikci, M.; Milanovic, J.V. Dynamic contribution of DFIG-based wind plants to system frequency disturbances. *IEEE Trans. Power Syst.* **2009**, *24*, 859–867. [CrossRef]

8. Kaspirek, M.; Mezera, D.; Jiricka, J. Problems of voltage stabilization in MV and LV distribution grids due to the operation of renewable energy sources. In Proceedings of the 22nd International Conference and Exhibition on Electricity Distribution (CIRED 2013), Stockholm, Sweden, 10–13 June 2013; pp. 1–4.

9. Lalor, G.; Mullane, A.; O'Malley, M. Frequency control and wind turbine technologies. *IEEE Trans. Power Syst.* **2005**, *20*, 1905–1913. [CrossRef]

10. Lee, D.J.; Wang, L. Small-signal stability analysis of an autonomous hybrid renewable energy power generation/energy storage system part I: Time-domain simulations. *IEEE Trans. Energy Convers.* **2008**, *23*, 311–320. [CrossRef]

11. Faria, P.; Vale, Z.; Baptista, J. Demand response programs design and use considering intensive penetration of distributed generation. *Energies* **2015**, *8*, 6230–6246. [CrossRef]

12. Douglass, P.J.; Garcia-Valle, R.; Nyeng, P.; Ostergaard, J.; Togeby, M. Smart demand for frequency regulation: Experimental results. *IEEE Trans. Smart Grid* **2013**, *4*, 1713–1720. [CrossRef]

13. Tan, Z.; Li, H.; Ju, L.; Song, Y. An optimization model for large-scale wind power grid connection considering demand response and energy storage systems. *Energies* **2014**, *7*, 7282–7304. [CrossRef]

14. Biegel, B.; Hansen, L.H.; Andersen, P.; Stoustrup, J. Primary control by ON/OFF demand-side devices. *IEEE Trans. Smart Grid* **2013**, *4*, 2061–2071. [CrossRef]

15. Chu, C.M.; Jong, T.L. A novel direct air-conditioning load control method. *IEEE Trans. Power Syst.* **2008**, *23*, 1356–1363.

16. Pedrasa, M.A.A.; Oro, M.M.; Reyes, N.C.R.; Pedrasa, J.R.I. Demonstration of direct load control of air conditioners in high density residential buildings. In Proceedings of the 2014 IEEE Innovative Smart Grid Technologies-Asia (ISGT Asia), Kuala Lumpur, Malaysia, 20–23 May 2014; pp. 400–405.

17. Salehfar, H.; Patton, A.D. Modeling and evaluation of the system reliability effects of direct load control. *IEEE Trans. Power Syst.* **1989**, *4*, 1024–1030. [CrossRef]

18. Xu, Z.; Ostergaard, J.; Togeby, M.; Marcus-Moller, C. Design and modeling of thermostatically controlled loads as frequency controlled reserve. In Proceedings of the 2007 IEEE Power Engineering Society General Meeting, Tampa, FL, USA, 24–28 June 2007; pp. 1–6.

19. Xu, Z.; Ostergaard, J.; Togeby, M. Demand as frequency controlled reserve. *IEEE Trans. Power Syst.* **2011**, *26*, 1062–1071. [CrossRef]

20. Short, J.A.; Infield, D.G.; Freris, L.L. Stabilization of grid frequency through dynamic demand control. *IEEE Trans. Power Syst.* **2007**, *22*, 1284–1293. [CrossRef]

21. Yang, H.; Tang, S.; Zeng, Z.; He, X.; Zhao, R.; Kuroda, M. Demand response of inverter air conditioners and applications in distribution system voltage stability enhancement. In Proceedings of the 2013 International Conference on Electrical Machines and Systems, Busan, Korea, 26–29 October 2013; pp. 954–959.

22. Zhang, Y.; Lu, N. Demand-side management of air conditioning cooling loads for intra-hour load balancing. In Proceedings of the 2013 Power & Energy Society on Innovative Smart Grid Technologies (ISGT), Washington, DC, USA, 24–27 February 2013; pp. 1–6.

23. Son, J.Y.; Park, J.H.; Moon, K.D.; Lee, Y.H. Resource-aware smart home management system by constructing resource relation graph. *IEEE Trans. Consum. Electron.* **2011**, *57*, 1112–1119. [CrossRef]

24. Dehghanpour, K.; Afsharnia, S. Electrical demand side contribution to frequency control in power systems: A review on technical aspects. *Renew. Sust. Energ. Rev.* **2015**, *41*, 1267–1276. [CrossRef]

25. Molina-Garcia, A.; Munoz-Benavente, I.; Hansen, A.D.; Gomez-Lazaro, E. Demand-side contribution to primary frequency control with wind farm auxiliary control. *IEEE Trans. Power Syst.* **2014**, *29*, 2391–2399. [CrossRef]

26. Sullivan, M.; Bode, J.; Kellow, B.; Woehleke, S.; Eto, J. Using residential AC load control in grid operations: PG&E's ancillary service pilot. *IEEE Trans. Smart Grid* **2013**, *4*, 1162–1170.

27. Albert, A.; Rajagopal, R. Thermal profiling of residential energy use. *IEEE Trans. Power Syst.* **2015**, *30*, 602–611. [CrossRef]

28. Southern California Edison (SCE). *Program Impact Evaluation of 2004 SCE Energy Smart Thermostat Program Final Report*; SCE: Rosemead, CA, USA, 2005.

29. Keuring Van Elektrotechnische Materialen (KEMA). *2005 Smart Thermostat Program Impact Evaluation*; KEMA: San Diego, CA, USA, 2006.

30. Zhao, C.; Ufuk, T.; Li, N.; Low, S. Design and stability of load-side primary frequency control in power systems. *IEEE Trans. Autom. Control* **2014**, *59*, 1177–1189. [CrossRef]

31. Alajmi, A.F.; Baddar, F.A.; Bourisli, R.I. Thermal comfort assessment of an office building served by under-floor air distribution (UFAD) system-a case study. *Build. Environ.* **2015**, *85*, 153–159. [CrossRef]

32. Nikolaou, T.G.; Kolokotsa, D.S.; Stavrakakis, G.S.; Skias, I.D. On the application of clustering techniques for office buildings' energy and thermal comfort classification. *IEEE Trans. Smart Grid* **2012**, *3*, 2196–2210. [CrossRef]

33. Hamdi, M.; Lachiver, G. A fuzzy control system based on the human sensation of thermal comfort. In Proceedings of the 1998 IEEE International Conference on IEEE World Congress on Computational Intelligence, Anchorage, AK, USA, 4–9 May 1998; pp. 487–492.

34. Bermejo, P.; Redondo, L.; de la Ossa, L.; Rodriguez, D.; Flores, J.; Urea, C.; Gamez, J.A.; Puerta, J.M. Design and simulation of a thermal comfort adaptive system based on fuzzy logic and on-line learning. *Energy Build.* **2012**, *49*, 367–379. [CrossRef]

35. Ku, K.L.; Liaw, J.S.; Tsai, M.Y.; Liu, T. Automatic control system for thermal comfort based on predicted mean vote and energy saving. *IEEE Trans. Autom. Sci. Eng.* **2015**, *12*, 378–383. [CrossRef]

36. Yan, F.; Li, H.; Wang, L.; Zhou, H. Simulation of indoor dynamic thermal comfort based on CFD. In Proceedings of the 2011 4th International Conference on Intelligent Computation Technology and Automation, Shenzhen, Guangdong, China, 28–29 March 2011; pp. 860–865.

37. Malhame, R.; Chong, C.Y. Electric load model synthesis by diffusion approximation of a high-order hybrid-state stochastic system. *IEEE Trans. Autom. Control* **1985**, *30*, 854–860. [CrossRef]

38. Balan, R.; Cooper, J.; Chao, K.M.; Stan, S.; Donca, R. Parameter identification and model based predictive control of temperature inside a house. *Energy Build.* **2011**, *43*, 748–758. [CrossRef]

39. Molina, A.; Gabaldon, A.; Fuentes, J.A.; Alvarez, C. Implementation and assessment of physically based electrical load models: Application to direct load control residential programmes. *IET Gener. Transm. Distrib.* **2003**, *150*, 61–66. [CrossRef]

40. Turner, A.; Chan, T.N.; Gibbs, A.N. A fast reacting power system load shedding management system. In Proceedings of the 9th Conference on the Electric Supply Industry (CEPSI), Hong Kong, China; 1992.

41. Tang, J.; Liu, J.; Ponci, F.; Monti, A. Adaptive load shedding based on combined frequency and voltage stability assessment using synchrophasor measurements. *IEEE Trans. Power Syst.* **2013**, *28*, 2035–2047. [CrossRef]

42. Molina-Garcia, A.; Bouffard, F.; Kirschen, D.S. Decentralized demand side contribution to primary frequency control. *IEEE Trans. Power Syst.* **2011**, *26*, 411–419. [CrossRef]

43. Angeli, D.; Kountouriotis, P. A stochastic approach to dynamic demand refrigerator control. *IEEE Trans. Control Syst. Technol.* **2012**, *20*, 581–592. [CrossRef]

44. Palensky, P.; Dietrich, D. Demand side management: Demand response, intelligent energy systems, and smart loads. *IEEE Trans. Ind. Inform.* **2011**, *7*, 381–388. [CrossRef]

45. Zhong, H.; Xie, L.; Xia, Q.; Kang, C.; Rahman, S. Multi-stage coupon incentive-based demand response in two-settlement electricity markets. In Proceedings of the 2015 IEEE Power & Energy Society on Innovative Smart Grid Technologies Conference (ISGT), Washington, DC, USA, 18–20 February 2015; pp. 1–5.

46. Zhong, H.; Xie, L.; Xia, Q. Coupon incentive-based demand response: Theory and case study. *IEEE Trans. Power Syst.* **2013**, *28*, 1266–1276. [CrossRef]

energies

MDPI

Article

A Two-Stage Algorithm to Estimate the Fundamental Frequency of Asynchronously Sampled Signals in Power Systems

Joon-Hyuck Moon [1], Sang-Hee Kang [1], Dong-Hun Ryu [2], Jae-Lim Chang [2] and Soon-Ryul Nam [1,*]

[1] Department of Electrical Engineering, Myongji University, Yongin 449-728, Korea; mjhmjh03@naver.com (J.-H.M.); shkang@mju.ac.kr (S.-H.K.)
[2] Industrial Physical Instrument Center, Korea Testing Laboratory, Ansan 426-901, Korea; dhryu@ktl.re.kr (D.-H.R.); cjl@ktl.re.kr (J.-L.C.)
* Author to whom correspondence should be addressed; ptsouth@mju.ac.kr; Tel.: +82-31-330-6361; Fax: +82-31-330-6816.

Academic Editor: Ying-Yi Hong
Received: 9 July 2015; Accepted: 21 August 2015; Published: 28 August 2015

Abstract: A two-stage algorithm is proposed for the estimation of the fundamental frequency of asynchronously sampled signals in power systems. In the first stage, time-domain interpolation reconstructs the power system signal at a new sampling time and the reconstructed signal passes through a tuned sine filter to eliminate harmonics. In the second stage, the fundamental frequency is estimated using a modified curve fitting, which is robust to noise. The evaluation results confirm the efficiency and validity of the two-stage algorithm for accurate estimation of the fundamental frequency even for asynchronously sampled signals contaminated with noise, harmonics, and an inter-harmonic component.

Keywords: fundamental frequency estimation; two-stage algorithm; time-domain interpolation; tuned sine filter; modified curve fitting

1. Introduction

Sampling-based power measurements are typically carried out using a clock signal that is synchronized with the signal from the power system under analysis. However, often there may be no means of synchronizing the clock signal to the power system signal, and power measurements are instead based on asynchronous sampling of power system signals. These circumstances include industrial measurements and high-precision calibrations in metrology laboratories. Under such circumstances, accurate estimation of the fundamental frequency of power system signals is essential to minimize the errors caused by asynchronous sampling.

Numerous studies have estimated the fundamental frequency of asynchronously sampled signals, and several interesting results have been published in recent decades. Spectral techniques [1–5] are among the most popular approaches because of their low computational burden and feasibility for real-time applications. When asynchronously sampled data are processed using a discrete Fourier transform (DFT) to extract frequency information, the spectral leakage error is appended to the calculated result. Spectral techniques compensate for this leakage error to estimate the fundamental frequency. Orthogonal techniques [6–14] are another of the most popular approaches for real-time applications. In this approach, filters generate orthogonal signals and complex vectors, which are used to estimate the fundamental frequency. However, the calculation accuracy of spectral and orthogonal techniques is limited, and does not satisfy the precision required for certain applications.

In contrast, time-domain techniques, such as curve fitting [15–19] and parametric interpolations [20–23], are capable of highly accurate frequency estimation, at the cost of a higher computational burden. Curve fitting algorithms are commonly used in time-domain techniques because of their robustness to noise. In particular, the four-parameter sine fit (4PSF) algorithm [16] performs close to the Cramer Rao lower bound (CRLB), which is the theoretical limit of variance for an unbiased estimator, for almost all noise levels [24,25]. The 4PSF algorithm is more suited to less distorted steady-state waveforms, since its performance significantly degrades as total harmonic distortion (THD) increases. Parametric interpolations, such as cubic spline interpolations [20,21] and Newton interpolations [22,23], offer an alternative time-domain approach and are used to modify the sampling rate of an analog-to-digital conversion in software. In particular, the time-domain interpolation and scanning (TDIS) algorithm [20] reconstructs a waveform by cubic spline interpolation to find a new sampling rate that is an integer multiple of the fundamental frequency using a scanning procedure. The phase drift in the reconstructed waveform is used to estimate the fundamental frequency and iteratively correct the waveform. The TDIS algorithm gives greater frequency errors in the presence of inter-harmonics. In the opposite sense to the TDIS algorithm, which modifies the sampled signal to synchronize with the estimated frequency, the phase sensitive frequency estimation with interpolated phase (PSFEi) algorithm [26,27] modifies the frequency of the sine waves used in curve fitting to synchronize with the sampled data. The PSFEi algorithm calculates the phase increase of the fundamental component between two groups within the sampled data using the three-parameter sine fit (3PSF) algorithm [28] and then updates the fundamental frequency to be consistent with the phase increase. To minimize the effect of harmonic components, the phase increase in the final iteration is interpolated to an exact integer number of fundamental cycles. The PSFEi gives greater frequency errors in the presence of fluctuating harmonics.

In this paper, a two-stage algorithm is proposed for accurate estimation of the fundamental frequency of asynchronously sampled signals regardless of noise and harmonics. The remainder of the paper is organized as follows: frequency estimation using the two-stage algorithm is formulated in Section 2, and the performance of the algorithm is evaluated in Section 3 using both computer simulations and a hardware implementation. Conclusions are presented in Section 4.

2. Two-Stage Algorithm for Estimating a Fundamental Frequency

Assuming that a power system signal has a purely sinusoidal waveform with a fundamental frequency ω_f (rad/s), amplitude A_f (V or A), and phase θ_f (rad), the N_M-point data uniformly sampled with a sampling time Δt can be described in discrete time steps as follows:

$$y(n) = A_f \cdot \cos\left(\omega_f \cdot n \frac{2\pi}{\omega_0 N_0} + \theta_f\right) = A_f \cdot \cos(\omega_f \cdot n\Delta t + \theta_f) \quad n = 1, \cdots, N_M \qquad (1)$$

where ω_0 (rad/s) is a nominal frequency and N_0 is the number of samples per cycle at ω_0.

2.1. Tuned Sine Filtering Followed by Time-Domain Interpolation

Harmonics within a power system signal make accurate estimation of the fundamental frequency difficult. To eliminate these harmonics from the power system signal, a sine filter is usually used and its reference frequency is typically set to the nominal frequency [11]. When the fundamental frequency is equal to the reference frequency, the sine filter can eliminate harmonics from the power system signal perfectly. However, when the fundamental frequency deviates from the reference frequency, harmonics may not be eliminated effectively. To overcome this drawback, this paper proposes a tuned sine filter that adjusts its reference frequency to synchronize with an estimate of the fundamental frequency. The filter length should be matched to the number of samples per cycle at the reference frequency to keep its ability to eliminate harmonics. To meet this constraint, time-domain interpolation is used to reconstruct the power system signal at a new sampling time corresponding to the reference frequency, prior to applying the tuned sine filter.

Assuming that $\hat{\omega}_f^{(i-1)}$ is an estimate of the fundamental frequency at the previous iteration $(i-1)$, which will be used as the reference frequency for the tuned sine filter, the new sampling time $\Delta\hat{t}^{(i)}$ is given as:

$$\Delta\hat{t}^{(i)} = \frac{2\pi}{\hat{\omega}_f^{(i-1)} N_0} \tag{2}$$

To reconstruct the power system signal at the new sampling time, application of cubic spline interpolation to Equation (1) yields:

$$\hat{y}^{(i)}(n) = A_f \cdot \cos(\omega_f \cdot n\Delta\hat{t}^{(i)} + \theta_f) \quad n = 1, \cdots, \hat{N}_M^{(i)} \tag{3}$$

where $\hat{N}_M^{(i)}$ is the number of samples in the reconstructed signal and is given as $\left\lfloor \frac{N_M \Delta t}{\Delta\hat{t}^{(i)}} \right\rfloor$. The algorithm used to compute the splines and interpolation is taken from [29]. Using the estimate of the fundamental frequency as the reference frequency for the tuned sine filter, the coefficients of the filter are given by:

$$H_S(n) = \frac{2}{N_0} \sin(\hat{\omega}_f^{(i-1)} \cdot n\Delta\hat{t}^{(i)}) = \frac{2}{N_0} \sin(2\pi\frac{n}{N_0}) \quad n = 1, \cdots, N_0 \tag{4}$$

It is noted that the coefficients of the filter are independent of the estimate of the fundamental frequency since this has already been considered in the calculation of the new sampling time, given in Equation (2). Applying Equation (4) to $\hat{y}^{(i)}(n)$ in Equation (3) yields a sine-filtered signal:

$$\hat{y}_S^{(i)}(n) = A_f \left| H_S(\omega_f) \right| \cdot \cos(\omega_f \cdot n\Delta\hat{t}^{(i)} + \theta_f + \angle H_S(\omega_f)) \quad n = N_0, \cdots, \hat{N}_M^{(i)} \tag{5}$$

where $\left| H_S(\omega_f) \right|$ and $\angle H_S(\omega_f)$ are the amplitude and phase response of the tuned sine filter at ω_f, respectively. Some minor calculations yield:

$$\hat{y}_S^{(i)}(n) = C_f \cdot \cos(\omega_f \cdot n\Delta\hat{t}^{(i)}) + S_f \cdot \sin(\omega_f \cdot n\Delta\hat{t}^{(i)}) \quad n = N_0, \cdots, \hat{N}_M^{(i)} \tag{6}$$

where:

$$C_f = A_f \left| H_S(\omega_f) \right| \cdot \cos(\theta_f + \angle H_S(\omega_f))$$

$$S_P = -A_f \left| H_S(\omega_f) \right| \cdot \sin(\theta_f + \angle H_S(\omega_f))$$

2.2. Modified Curve Fitting with an Unknown Frequency

The 3PSF algorithm performs a least-squares curve fitting on the sampled data to find the DC component, and the amplitude and phase of the component at a known frequency [28]. In the 4PSF algorithm, the frequency itself is treated as an unknown parameter to be found. Although the 4PSF algorithm is more computationally complex, this algorithm is recommended for sampled data containing five or more cycles of the power system signal [16,30], because the 4PSF algorithm usually determines the frequency to greater accuracy even if it is accurately known *a priori*. Since the tuned sine filter, which is used prior to applying the modified curve fitting, can eliminate the DC component, it can be assumed that there is no DC component in the sine-filtered signal given in Equation (6). Therefore, the number of modeling parameters in the 4PSF algorithm can be reduced from four to three: the fundamental frequency, amplitude, and phase. Assuming that $\hat{C}_f^{(i)}$, $\hat{S}_f^{(i)}$, and $\hat{\omega}_f^{(i)}$ are estimates of C_f, S_f, and ω_f in Equation (6) respectively, the sine-filtered signal of Equation (6) can be expressed as:

$$\hat{y}_S^{(i)}(n) = \hat{C}_f^{(i)} \cdot \cos(\hat{\omega}_f^{(i)} \cdot n\Delta\hat{t}^{(i)}) + \hat{S}_f^{(i)} \cdot \sin(\hat{\omega}_f^{(i)} \cdot n\Delta\hat{t}^{(i)}) \quad n = N_0, \cdots, \hat{N}_M^{(i)} \tag{7}$$

To find an iterative form of Equation (7), a Taylor series expansion of the cosine function in Equation (7) is taken around the previous frequency estimate $\hat{\omega}_f^{(i-1)}$:

$$
\begin{aligned}
\cos(\hat{\omega}_f^{(i)} \cdot n\Delta\hat{t}^{(i)}) &= \cos((\hat{\omega}_f^{(i-1)} + \Delta\hat{\omega}_f^{(i)}) \cdot n\Delta\hat{t}^{(i)}) \\
&= \cos(\hat{\omega}_f^{(i-1)} \cdot n\Delta\hat{t}^{(i)})\cos(\Delta\hat{\omega}_f^{(i)} \cdot n\Delta\hat{t}^{(i)}) - \sin(\hat{\omega}_f^{(i-1)} \cdot n\Delta\hat{t}^{(i)})\sin(\Delta\hat{\omega}_f^{(i)} \cdot n\Delta\hat{t}^{(i)}) \\
&\cong \cos(\hat{\omega}_f^{(i-1)} \cdot n\Delta\hat{t}^{(i)}) - \sin(\hat{\omega}_f^{(i-1)} \cdot n\Delta\hat{t}^{(i)})\Delta\hat{\omega}_f^{(i)} \cdot n\Delta\hat{t}^{(i)} \\
&= \cos(\tfrac{2\pi n}{N_0}) - \sin(\tfrac{2\pi n}{N_0})\Delta\hat{\omega}_f^{(i)} \cdot n\Delta\hat{t}^{(i)}
\end{aligned}
$$

$$(8)$$

Similarly, a Taylor series expansion of the sine function in Equation (7) is taken around $\hat{\omega}_f^{(i-1)}$:

$$
\sin(\hat{\omega}_f^{(i)} \cdot n\Delta\hat{t}^{(i)}) \cong \sin(\frac{2\pi n}{N_0}) + \cos(\frac{2\pi n}{N_0})\Delta\hat{\omega}_f^{(i)} \cdot n\Delta\hat{t}^{(i)} \tag{9}
$$

Substitution of Equations (8) and (9) into Equation (7) yields:

$$
\begin{aligned}
\hat{y}_S^{(i)}(n) &\cong \hat{C}_f^{(i)}\left\{\cos(\tfrac{2\pi n}{N_0}) - \sin(\tfrac{2\pi n}{N_0})\Delta\hat{\omega}_f^{(i)} \cdot n\Delta\hat{t}^{(i)}\right\} + \hat{S}_f^{(i)}\left\{\sin(\tfrac{2\pi n}{N_0}) + \cos(\tfrac{2\pi n}{N_0})\Delta\hat{\omega}_f^{(i)} \cdot n\Delta\hat{t}^{(i)}\right\} \\
&= \hat{C}_f^{(i)}\cos(\tfrac{2\pi n}{N_0}) + \hat{S}_f^{(i)}\sin(\tfrac{2\pi n}{N_0}) + \Delta\hat{\omega}_f^{(i)} \cdot n\Delta\hat{t}^{(i)}\left\{\hat{S}_f^{(i)}\cos(\tfrac{2\pi n}{N_0}) - \hat{C}_f^{(i)}\sin(\tfrac{2\pi n}{N_0})\right\} \\
&\cong \hat{C}_f^{(i)}\cos(\tfrac{2\pi n}{N_0}) + \hat{S}_f^{(i)}\sin(\tfrac{2\pi n}{N_0}) + \Delta\hat{\omega}_f^{(i)} \cdot n\Delta\hat{t}^{(i)}\left\{\hat{S}_f^{(i-1)}\cos(\tfrac{2\pi n}{N_0}) - \hat{C}_f^{(i-1)}\sin(\tfrac{2\pi n}{N_0})\right\}
\end{aligned}
$$

$$(10)$$

The successive samples in Equation (10) can be written in matrix form:

$$
\hat{\mathbf{y}}_S^{(i)} = \hat{\mathbf{A}}^{(i)}\hat{\mathbf{x}}^{(i)} \tag{11}
$$

where:

$$
\hat{\mathbf{y}}_S^{(i)} = \begin{bmatrix} \hat{y}_S^{(i)}(N_0) & \cdots & \hat{y}_S^{(i)}(\hat{N}_M^{(i)}) \end{bmatrix}^T \qquad \hat{\mathbf{x}}^{(i)} = \begin{bmatrix} \hat{C}_f^{(i)} & \hat{S}_f^{(i)} & \Delta\hat{\omega}_f^{(i)} \end{bmatrix}^T
$$

$$
\hat{\mathbf{A}}^{(i)} = \begin{bmatrix}
\cos(\frac{2\pi N_0}{N_0}) & \sin(\frac{2\pi N_0}{N_0}) & N_0\Delta\hat{t}^{(i)}\left\{\hat{S}_f^{(i-1)}\cos(\frac{2\pi N_0}{N_0}) - \hat{C}_f^{(i-1)}\sin(\frac{2\pi N_0}{N_0})\right\} \\
\vdots & \vdots & \vdots \\
\cos(\frac{2\pi \hat{N}_M^{(i)}}{N_0}) & \sin(\frac{2\pi \hat{N}_M^{(i)}}{N_0}) & \hat{N}_M^{(i)}\Delta\hat{t}^{(i)}\left\{\hat{S}_f^{(i-1)}\cos(\frac{2\pi \hat{N}_M^{(i)}}{N_0}) - \hat{C}_f^{(i-1)}\sin(\frac{2\pi \hat{N}_M^{(i)}}{N_0})\right\}
\end{bmatrix}
$$

Since the coefficient matrix $\hat{\mathbf{A}}^{(i)}$ with dimensions $(\hat{N}_M^{(i)} - N_0 + 1) \times 3$ is not square, its pseudo inverse with dimensions 3×3 is used to find $\hat{\mathbf{x}}^{(i)}$:

$$
\hat{\mathbf{x}}^{(i)} = ((\hat{\mathbf{A}}^{(i)})^T\hat{\mathbf{A}}^{(i)})^{-1}(\hat{\mathbf{A}}^{(i)})^T\hat{\mathbf{y}}_S^{(i)} \tag{12}
$$

Finally, the frequency estimate is updated using the frequency deviation $\Delta\hat{\omega}_f^{(i)}$ in $\hat{\mathbf{x}}^{(i)}$:

$$
\hat{\omega}_f^{(i)} = \hat{\omega}_f^{(i-1)} + \Delta\hat{\omega}_f^{(i)} \tag{13}
$$

2.3. Frequency Estimation Procedure

Figure 1 shows the block diagram for the frequency estimation procedure using the two-stage algorithm. Initial estimates for $\hat{C}_f^{(0)}$, $\hat{S}_f^{(0)}$, and $\hat{\omega}_f^{(0)}$ are obtained using 3pDFT [31], a simple and efficient algorithm for estimating the fundamental frequency. At the first iteration, a new sampling time $\Delta\hat{t}^{(1)}$ is calculated corresponding to $\hat{\omega}_f^{(0)}$ and cubic spline interpolation is applied to reconstruct the power system signal at the new sampling time. The reconstructed signal passes through the tuned sine filter to eliminate harmonics. Prior to performing the modified curve fitting on the sine-filtered signal, the coefficients of $\hat{\mathbf{A}}^{(1)}$ are calculated based on $\hat{C}_f^{(0)}$, $\hat{S}_f^{(0)}$, and $\Delta\hat{t}^{(0)}$. After finding the pseudo inverse of

$\hat{\mathbf{A}}^{(1)}$ as given in Equation (12), the modified curve fitting determines the modeling parameters ($\hat{C}_f^{(1)}$, $\hat{S}_f^{(1)}$, and $\Delta\hat{\omega}_f^{(1)}$) and the frequency estimate $\hat{\omega}_f^{(1)}$ is updated using $\Delta\hat{\omega}_f^{(1)}$. The ith iteration is the same as the first iteration, but uses the most recent estimates $\hat{C}_f^{(i-1)}$, $\hat{S}_f^{(i-1)}$, and $\hat{\omega}_f^{(i-1)}$ instead of the initial estimates $\hat{C}_f^{(0)}$, $\hat{S}_f^{(0)}$, and $\hat{\omega}_f^{(0)}$. Since curve fitting algorithms with an unknown frequency double the number of significant digits in the frequency estimate at each iteration and converge very rapidly, six iterations have been considered more than adequate [30]. In this paper, six iterations are repeated to produce an accurate frequency estimate.

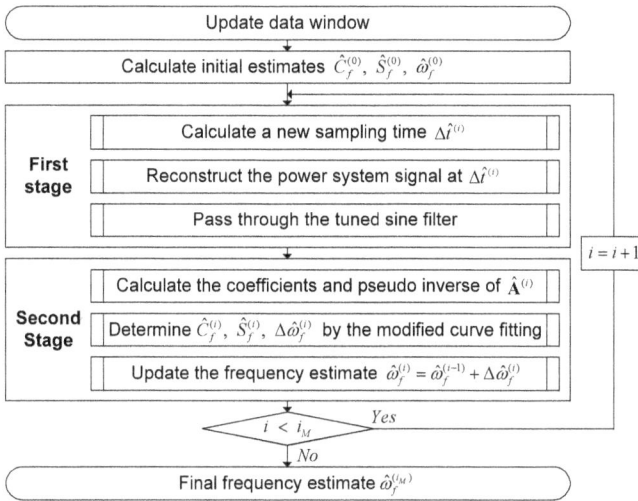

Figure 1. Frequency estimation procedure for the two-stage algorithm.

3. Performance Evaluation

To evaluate the performance of the two-stage algorithm, results were compared with those obtained using the TDIS, 4PSF, and PSFEi algorithms. All four algorithms were tested using two approaches: computer simulations and a hardware implementation.

3.1. Computer Simulations

3.1.1. Number of Cycles in the Data Window

While the sampling time and length of the data window were fixed (*i.e.*, $\Delta t = 30\,us$ and $N_M = 8192$), the number of cycles in the data window was varied by varying the fundamental frequency of the test waveform, which consisted of a single-tone sine wave and a 60 dB white noise signal. In the simulations, the number of cycles in the data window was varied between 3 and 20, which corresponds to the fundamental frequencies between 12.207 Hz and 81.380 Hz. The root-mean-square (RMS) error of frequency estimation after 1000 repetitions at each number of cycles is shown in Figure 2, together with the CRLB. The 4PSF and PSFEi algorithms perform close to the CRLB for all numbers of cycles; the TDIS algorithm performs well, but is marginally worse than the 4PSF and PSFEi algorithms for six cycles or more within the data window. It was also found that the two-stage algorithm needs at least eight cycles in the data window to perform close to the CRLB.

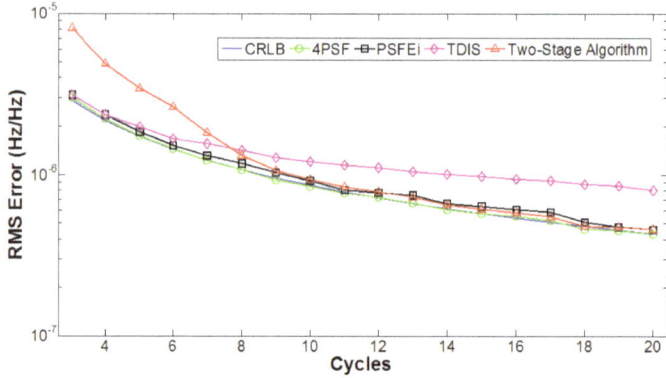

Figure 2. RMS error of frequency estimation with an increasing number of cycles in the data window.

3.1.2. Noise Level

Simulations with noise were performed using a single-tone sine waveform with a random phase and fundamental frequency of 61.2 Hz. The signal-to-noise ratio (SNR) varied from 0 dB to 100 dB with a noise increase of 10 dB. The RMS error of frequency estimation after 1000 repetitions at each SNR level is shown in Figure 3. As with previous tests, the 4PSF and PSFEi algorithms perform close to the CRLB for all noise levels, and the two-stage algorithm also performs close to the CRLB. The TDIS algorithm performs approximately 1.2-fold worse than the others.

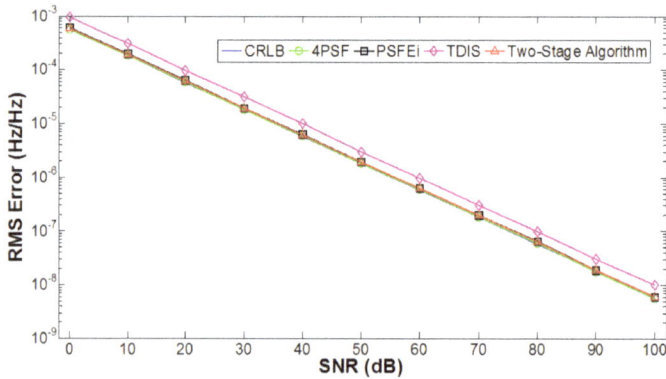

Figure 3. Root-mean-square (RMS) error of frequency estimation with increasing noise level.

3.1.3. Harmonics

The simulated test waveforms have the following form:

$$y(n) = \sum_{h=0}^{H} A_h(n) \sin(h \cdot \omega_f t_n + \theta_h) + e(n) \tag{14}$$

Tests were run with fundamental frequencies varying within 60 ± 5 Hz, with a random phase θ_h for each harmonic. The other parameters in Equation (14) were set as follows: $e(n) = 60$ dB, $H = 10$, $A_1(n) = 1$, $A_{h,\text{even}}(n) = 0.01 \times S$, $A_{h \neq 1,\text{odd}}(n) = 0.2 \times S$, where S was an integer multiplier from 0 to 4 used to increase the amplitudes of the even and odd harmonics. The maximum error of frequency estimation after 1000 repetitions at each level of THD is given in Figure 4. The performance of the 4PSF algorithm significantly degrades as the multiplier *S* and, therefore, the THD increases. The PSFEi, TDIS, and two-stage algorithms perform with greater accuracy regardless of THD level, with the two-stage algorithm exhibiting slightly improved performance over the PSFEi and TDIS algorithms.

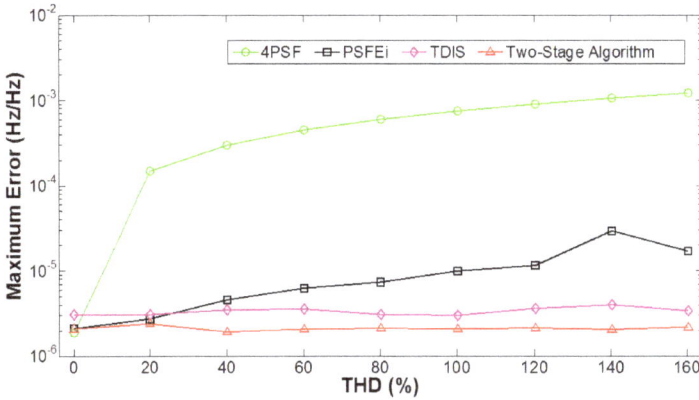

Figure 4. Maximum error of frequency estimation with increasing THD.

3.1.4. Fluctuating Harmonic Component

The simulated test waveforms with a fluctuating harmonic component of order *k* have the following form:

$$y(n) = \sum_{h=0,\neq k}^{H} A_h(n)\sin(h \cdot \omega_f t_n + \theta_h) + A_k(1 + B_k\sin(\omega_k t_n)) \cdot \sin(k \cdot \omega_f t_n + \theta_k) + e(n) \qquad (15)$$

where ω_k is the frequency of fluctuation, which was set to vary randomly within 7 ± 2 Hz; A_k is the amplitude of the fluctuating harmonic component, which was set to 0.5; and B_k is the depth of fluctuation, which was set to 0.1. The other parameters were identical to those used in Section 3.1.3, except that here the integer multiplier *S* was set to 1. The order of the fluctuating harmonic component was varied from 2 to 10, and the maximum error of frequency estimation after 1000 repetitions for each fluctuating harmonic order is given in Figure 5; comparison between Figures 4 and 5 shows that the maximum errors for most of the algorithms are significantly worse when one of the harmonics fluctuates. Performance slightly improves as the fluctuating harmonic order increases. This is most likely because spectral leakage from a fluctuating harmonic closer to the fundamental frequency has a greater effect on the estimate.

To evaluate the effect of modulation depth, the order of the fluctuating harmonic was set to 3 and B_k was varied from 0 to 1.0 in 0.1 increments. The maximum error of frequency estimation after 1000 repetitions for each modulation depth is shown in Figure 6. Similar to the effect seen when changing the fluctuating harmonic order, an increase in spectral leakage due to greater modulation depth has an adverse effect on the performance of the PSFEi, TDIS and two-stage algorithms. The 4PSF algorithm is more influenced by THD level and a fluctuating harmonic component does not have a significant effect on its performance.

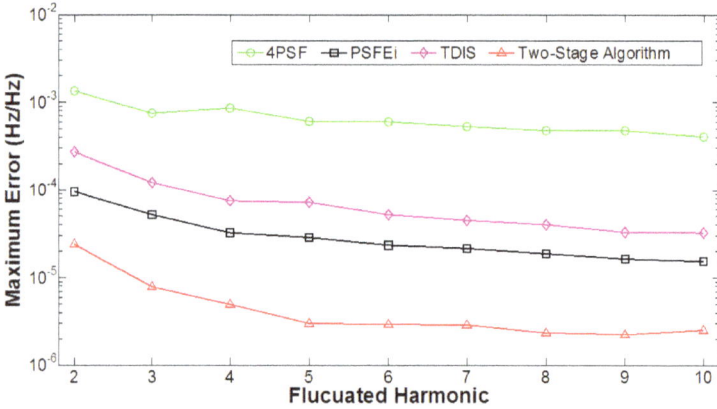

Figure 5. Maximum error of frequency estimation with increasing order of the fluctuating harmonic.

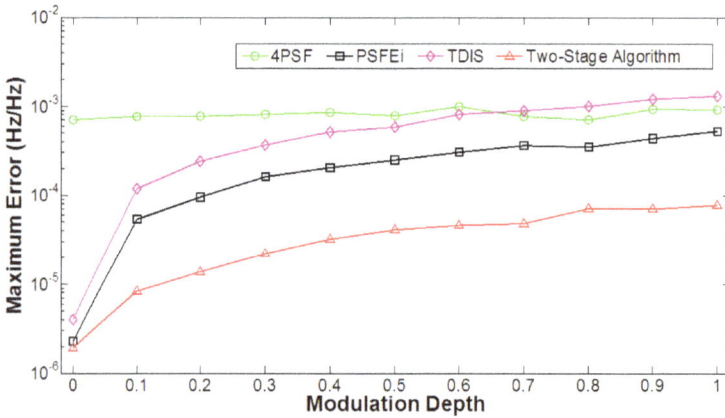

Figure 6. Maximum error of frequency estimation with increasing modulation depth of the third harmonic.

3.1.5. Fluctuating Inter-Harmonic Component

The simulated test waveforms with a fluctuating inter-harmonic component of index x have the following form:

$$y(n) = \sum_{h=0}^{H} A_h(n) \sin(h \cdot \omega_f t_n + \theta_h) + A_x(1 + B_x \sin(\omega_x t_n)) \cdot \sin(x \cdot \omega_f t_n + \theta_x) + e(n) \qquad (16)$$

where ω_x is the frequency of fluctuation, which was set to vary randomly within 7 ± 2 Hz; and B_x is the depth of fluctuation, which was set to 0.1. Other parameters were identical to those used in Section 3.1.3, except that the integer multiplier S was set to 1. The inter-harmonic amplitude A_x was varied from 0 to 0.7 in 0.1 increments. The maximum error of frequency estimation after 1000 repetitions for each of the inter-harmonic amplitude is given in Figure 7. At each repetition, index x was randomly varied from 1.5 to 10. The contamination of the fundamental component caused by the inter-harmonic component leads to increasing estimation errors for the TDIS, PSFEi, and two-stage algorithms with increasing inter-harmonic amplitude. Again, a fluctuating inter-harmonic component does not have a significant effect on the performance of the 4PSF algorithm.

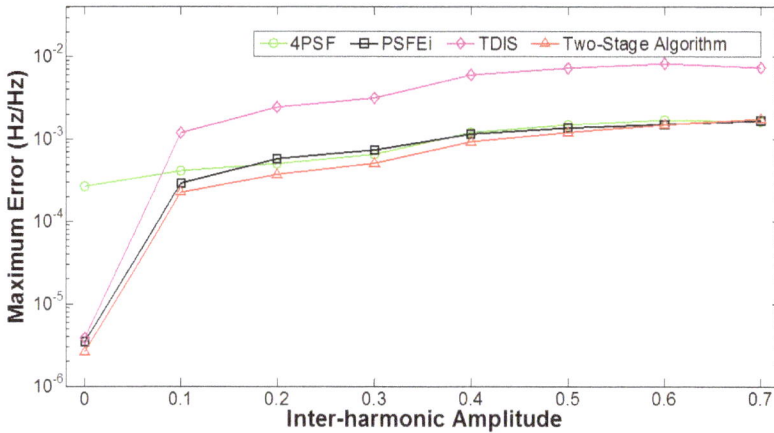

Figure 7. Maximum error of frequency estimation with increasing inter-harmonic amplitude.

3.1.6. Computational Burden

Table 1 lists the average time taken to complete a frequency estimate in the simulations for each of the algorithms. The algorithms were implemented in MATLAB, and all simulations were run on a personal computer with an Intel Core i7 processor running at 4.0 GHz. The TDIS and two-stage algorithms required significantly more time than the 4PSF and PSFEi algorithms, which is due to the time-domain interpolation.

Table 1. Average time required for one frequency estimation (ms).

Algorithm	Figure number					
	2	3	4	5	6	7
4PSF	8.3337	8.2824	8.1042	8.3044	8.2134	8.5206
PSFEi	12.388	11.279	10.444	11.102	10.710	11.762
TDIS	463.98	465.78	459.23	466.76	462.63	481.13
Two-stage	254.12	348.07	331.07	341.41	339.56	355.14

3.2. Hardware Implementation

Tests using real waveforms with known characteristics are a useful indicator of the performance of the algorithms under real conditions. A system for efficient implementation and comparison of the algorithms was developed at Korea Testing Laboratory (KTL), a representative accreditation authority in South Korea for the testing and calibration of measurement equipment. As shown in Figure 8, an electrical power quality calibrator (Fluke 6105A) generates calibration voltage and current signals, which were sampled using the data acquisition system developed for the KTL power measurement calibration service. In the data acquisition system, a current shunt (Fluke A40B) converts the calibration current signal to a shunt voltage signal. Two digitizing multi-meters (Agilent 3458A) encode the calibration voltage signal and the shunt voltage signal into digital outputs at a 16-bit resolution. During the encoding process, a waveform generator (Agilent 33500B) produces an EXT TRIG signal to synchronize the operation of the two multi-meters. Integrated software on a host computer controls the two multi-meters through a GPIB and gathers the sampled data, which are used for power measurement calibration. All four algorithms are implemented as sub-programs to estimate the fundamental frequency of the calibration voltage signal, sampled at 8.192 kS/s for 8192-point data.

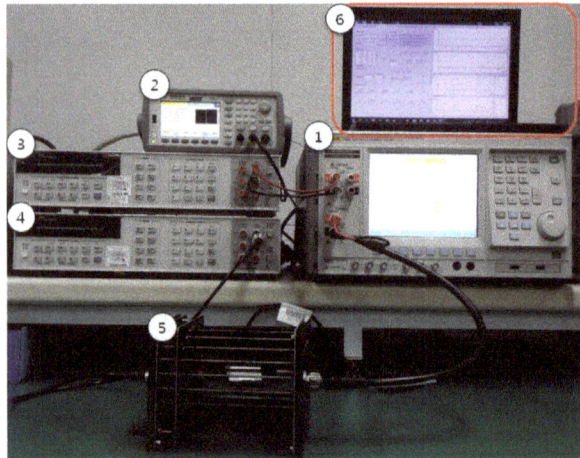

No.	Device Name
1	Electrical power quality calibrator (Fluke 6105A)
2	Waveform generator (Agilent 33500B)
3,4	Digitizing multi-meter (Agilent 3458A)
5	Current shunt (Fluke A40B)
6	Host computer

Figure 8. Hardware implementation for tests using real waveforms.

Table 2 summarizes the maximum errors of frequency estimation when tests were run at a fundamental frequency of 60 Hz. Two calibration voltage levels (110 V and 220 V) were used in the tests and all algorithms demonstrated similar performance regardless of the calibration voltage level and harmonic. In particular, the maximum errors of the two-stage algorithm are maintained stably between 9.5416×10^{-6} and 9.9564×10^{-6}.

Table 2. Maximum errors of frequency estimation with a fundamental frequency of 60 Hz ($10^{-6} \times Hz/Hz$).

Voltage	$H_1 = 110$ V, $\theta_1 = 0$				$H_1 = 220$V, $\theta_1 = 0$			
	$H_3 = 10\%$	$H_3 = 10\%$	$H_{49} = 10\%$	$H_{49} = 10\%$	$H_3 = 10\%$	$H_3 = 10\%$	$H_{49} = 10\%$	$H_{49} = 10\%$
Harmonic	$\theta_3 = 0$	$\theta_3 = \pi$	$\theta_{49} = 0$	$\theta_{49} = \pi$	$\theta_3 = 0$	$\theta_3 = \pi$	$\theta_{49} = 0$	$\theta_{49} = \pi$
4PSF	6.6318	12.492	9.9227	9.6065	8.7664	11.733	9.5059	9.7659
PSFEi	9.4811	9.9378	9.9649	9.6864	9.6831	9.4355	9.8214	9.9948
TDIS	11.623	8.4567	9.3250	9.1283	10.472	9.7417	9.0150	9.0917
Two-stage	9.5266	9.8951	9.9366	9.6903	9.5992	9.5416	9.7377	9.9564

4. Conclusions

A two-stage algorithm is proposed for estimation of the fundamental frequency of asynchronously sampled signals in power systems. In the first stage, time-domain interpolation reconstructs the power system signal at a new sampling time and the reconstructed signal passes through a tuned sine filter. The tuned sine filter retains its ability to eliminate harmonics by adjusting its reference frequency to synchronize with an estimate of the fundamental frequency. Prior to applying the tuned sine filter, time-domain interpolation renders it possible to match the filter length of the tuned sine filter to the number of samples per cycle in the reconstructed signal. In the second stage, the fundamental frequency is estimated using a modified curve fitting, which is robust to noise. Since the tuned sine filter eliminates the DC component in addition to harmonics, the DC component is removed from the modeling parameters of the modified curve fitting.

The performance of the two-stage algorithm was evaluated using computer-simulated signals, which were asynchronously sampled and contaminated with noise, harmonics, and an inter-harmonic component. The comparison showed that the two-stage algorithm required approximately eight cycles within the data window to provide a level of performance similar to CRLB, and enabled estimation of the fundamental frequency accurately under a range of conditions. In particular, the two-stage algorithm could estimate the fundamental frequency with greater accuracy than the other algorithms in the presence of high levels of THD, a fluctuating harmonic component and a fluctuating inter-harmonic component. The two-stage algorithm was then implemented on a KTL data acquisition system with 16-bit resolution. The results of this implementation demonstrate both the efficiency and validity of the two-stage algorithm, and show that it can achieve accurate estimations of the fundamental frequency in practical conditions. Therefore, the two-stage algorithm may be considered useful for high-precision applications, such as calibrations in metrology laboratories.

Acknowledgments: This research was supported in part by Basic Science Research Program through the National Research Foundation of Korea (NRF) funded by the Ministry of Education (No. NRF-2013R1A1A2062924). This research was also supported in part by the Human Resources Program in Energy Technology of the Korea Institute of Energy Technology Evaluation and Planning (KETEP), granted financial resource from the Ministry of Trade, Industry & Energy, Republic of Korea (No. 20154030200770).

Author Contributions: Joon-Hyuck prepared the manuscript and completed the simulations. Soon-Ryul Nam supervised the study and coordinated the main theme of this paper. Dong-Hun Ryu and Jae-Lim Chang developed the hardware implementation. Sang-Hee Kang discussed the results and implications, and commented on the manuscript. All of the authors read and approved the final manuscript.

Conflicts of Interest: The authors declare no conflict of interest.

References

1. Grandke, T. Interpolation algorithms for discrete Fourier transforms of weighted signals. *IEEE Trans. Instrum. Meas.* **1983**, *32*, 350–355. [CrossRef]
2. Andria, G.; Savino, M.; Trotta, A. Windows and interpolation algorithms to improve electrical measurement accuracy. *IEEE Trans. Instrum. Meas.* **1989**, *38*, 856–863. [CrossRef]
3. Xi, J.; Chicharo, J.F. A new algorithm for improving the accuracy of periodic signal analysis. *IEEE Trans. Instrum. Meas.* **1996**, *45*, 827–831.
4. Zhang, F.; Geng, Z.; Yuan, W. The algorithm of interpolating windowed FFT for harmonic analysis of electric power system. *IEEE Trans. Power Del.* **2001**, *16*, 160–164. [CrossRef]
5. Radil, T.; Ramos, P.M.; Serra, A.C. New spectrum leakage correction algorithm for frequency estimation of power system signals. *IEEE Trans. Instrum. Meas.* **2009**, *58*, 1670–1679. [CrossRef]
6. Begovic, M.M.; Djuric, P.M.; Dunlap, S.; Phadke, A.G. Frequency tracking in power network in the presence of harmonics. *IEEE Trans. Power Del.* **1993**, *8*, 480–486. [CrossRef]
7. Moore, P.J.; Carranza, R.D.; Johns, A.T. A new numeric technique for high-speed evaluation of power system frequency. *IEE Proc. Gen. Transm. Distrib.* **1994**, *141*, 529–536. [CrossRef]
8. Akke, M. Frequency estimation by demodulation of two complex signals. *IEEE Trans. Power Del.* **1997**, *12*, 157–163. [CrossRef]
9. Sidhu, T.S. Accurate measurement of power system frequency using a digital signal processing technique. *IEEE Trans. Instrum. Meas.* **1999**, *48*, 75–81. [CrossRef]
10. Yang, J.Z.; Liu, C.W. A precise calculation of power system frequency. *IEEE Trans. Power Del.* **2001**, *16*, 361–366. [CrossRef]
11. Nam, S.R.; Lee, D.G.; Kang, S.H.; Ahn, S.J.; Choi, J.H. Power system frequency Estimation in Power Systems Using Complex Prony Analysis. *Int. J. Electr. Eng. Tech.* **2011**, *6*, 154–160. [CrossRef]
12. Ren, J.; Kezunovic, M. A Hybrid method for power system frequency estimation. *IEEE Trans. Power Del.* **2012**, *27*, 1252–1259. [CrossRef]

13. Yamada, T. High-accuracy estimations of frequency, amplitude, and phase with a modified DFT for asynchronous sampling. *IEEE Trans. Instrum. Meas.* **2013**, *48*, 1428–1435. [CrossRef]

14. Nam, S.R.; Kang, S.H.; Kang, S.H. Real-time estimation of power system frequency using a three-level discrete fourier transform method. *Energies* **2015**, *8*, 79–93. [CrossRef]

15. Pintelon, R.; Schoukens, J. An improved sine-wave fitting procedure for characterizing data acquisition channels. *IEEE Trans. Instrum. Meas.* **1996**, *45*, 588–593. [CrossRef]

16. Handel, P. Properties of the IEEE-STD-1057 Four-Parameter Sine Wave Fit Algorithm. *IEEE Trans. Instrum. Meas.* **2000**, *49*, 1189–1193. [CrossRef]

17. Ramos, P.M.; da Silva, M.F.; Martins, R.C.; Cruz Serra, A.M. Simulation and experimental results of multi-harmonic least squares fitting algorithms applied to periodic signals. *IEEE Trans. Instrum. Meas.* **2006**, *55*, 646–651. [CrossRef]

18. Ramos, P.M.; Cruz Serra, A.M. Least squares multi-harmonic fitting: Convergence improvements. *IEEE Trans. Instrum. Meas.* **2007**, *56*, 1412–1418. [CrossRef]

19. Giarnetti, S.; Leccese, F.; Caciotta, M. Non recursive multi-harmonic least squares fitting for grid frequency estimation. *Measurement* **2015**, *66*, 229–237. [CrossRef]

20. Zhu, L.M.; Ding, H.; Zhu, X.Y. Extraction of periodic signal without external reference by time-domain average scanning. *IEEE Trans. Ind. Electron.* **2008**, *55*, 918–927. [CrossRef]

21. Clarkson, P.; Wright, P. Evaluation of an asynchronous sampling correction technique suitable for power quality measurements. In Proceedings of the IMEKO World Congress Fundamental and Applied Metrology, Lisbon, Portugal, 6–11 September 2009; pp. 907–912.

22. Zhou, F.; Huang, Z.; Zhao, C.; Wei, X.; Chen, D. Time-domain quasi-synchronous sampling algorithm for harmonic analysis based on Newton's interpolation. *IEEE Trans. Instrum. Meas.* **2011**, *60*, 2804–2812. [CrossRef]

23. Wang, K.; Teng, Z.; Wen, H.; Tang, Q. Fast Measurement of Dielectric Loss Angle with Time-Domain Quasi-Synchronous Algorithm. *IEEE Trans. Instrum. Meas.* **2015**, *64*, 935–942. [CrossRef]

24. Lapuh, R.; Clarkson, P.; Pogliano, U.; Hallstrom, J.K.; Wright, P.S. Comparison of asynchronous sampling correction algorithms for frequency estimation of signals of poor power quality. *IEEE Trans. Instrum. Meas.* **2011**, *60*, 2235–2241. [CrossRef]

25. Ristic, B.; Boashash, B. Comments on "The Cramer-Rao lower bounds for signals with constant amplitude and polynomial phase". *IEEE Trans. Signal. Process.* **1998**, *46*, 1708–1709. [CrossRef]

26. Lapuh, R. Phase Estimation of Asynchronously Sampled Signal Using Interpolated Three-Parameter Sinewave Fit Technique. In Proceedings of the Instrumentation and Measurement Technology Conference, Austin, TX, USA, 3–6 May 2010; pp. 82–86.

27. Lapuh, R. Phase Sensitive Frequency Estimation Algorithm for Asynchronously Sampled Harmonically Distorted Signals. In Proceedings of the Instrumentation and Measurement Technology Conference, Binjiang, China, 10–12 May 2011; pp. 1–4.

28. Negusse, S.; Handel, P.; Zetterberg, P. IEEE-STD-1057 Three Parameter Sine Wave Fit for SNR Estimation: Performance Analysis and Alternative Estimators. *IEEE Trans. Instrum. Meas.* **2014**, *63*, 1514–1523. [CrossRef]

29. Press, W.H.; Teukolsky, S.A.; Vetterling, W.T.; Flannery, B.P. *Numerical Recipes in C: The Art of Scientific Computing*; Cambridge University Press: Cambridge, UK, 1992.

30. Souders, M.; Blair, J.; Boyer, W. IEEE Std. 1057-2007. In *IEEE Standard for Digitizing Waveform Recorders*; Institute of Electrical and Electronics Engineers (IEEE): New York, NY, USA, 2008.

31. Agrez, D. Dynamics of Frequency Estimation in the Frequency Domain. *IEEE Trans. Instrum. Meas.* **2007**, *56*, 2111–2118. [CrossRef]

![energies logo] *energies*

MDPI

Article

Interval Type-II Fuzzy Rule-Based STATCOM for Voltage Regulation in the Power System

Ying-Yi Hong * and Yu-Lun Hsieh

Department of Electrical Engineering, Chung Yuan Christian University, Taoyuan 32023, Taiwan;
hsiehtabo@gmail.com

* Correspondence: yyhong@dec.ee.cycu.edu.tw; Tel.: +886-3-265-1200; Fax: +886-3-265-1299

Academic Editor: Akhtar Kalam
Received: 9 July 2015; Accepted: 17 August 2015; Published: 21 August 2015

Abstract: The static synchronous compensator (STATCOM) has recently received much attention owing to its ability to stabilize power systems and mitigate voltage variations. This paper investigates a novel interval type-II fuzzy rule-based PID (proportional-integral-derivative) controller for the STATCOM to mitigate bus voltage variations caused by large changes in load and the intermittent generation of photovoltaic (PV) arrays. The proposed interval type-II fuzzy rule base utilizes the output of the PID controller to tune the signal applied to the STATCOM. The rules involve upper and lower membership functions that ensure the stable responses of the controlled system. The proposed method is implemented using the NEPLAN software package and MATLAB/Simulink with co-simulation. A six-bus system is used to show the effectiveness of the proposed method. Comparative studies show that the proposed method is superior to traditional PID and type-I fuzzy rule-based methods.

Keywords: photovoltaic array; interval type II fuzzy rule; STATCOM; voltage variation

---

1. Introduction

A static synchronous compensator (STATCOM) is a volt-ampere-reactive (VAR)/voltage regulation device that is used in both electric transmission and distribution networks. The STATCOM utilizes a voltage- or current-source converter and can act as either a source or sink of reactive power in the power system. If the STATCOM is connected to a source of power, it can also provide the real power.

Traditionally, STATCOMs have been used to study the stability problem. Wang and Hsiung presented a STATCOM control scheme to enhance the damping of a grid-connected 80 MW offshore wind farm and 40 MW marine-current farm [1]. The damping controller was designed using modal control theory and has been examined under different operating conditions. Wang and Truong presented a damping controller of the STATCOM by using a pole-assignment approach to ensure adequate damping of the dominant modes of the studied power system [2]. Beza and Bongiorno presented an adaptive power oscillation damping controller for a STATCOM equipped with an energy storage device [3]; this was achieved using a signal estimation technique based on a modified recursive least-square algorithm, which supported the adaptive estimation of low-frequency electromechanical oscillations from locally-measured signals during power system disturbances.

Recently, STATCOMs have been also used to deal with the voltage problems in the power system. Wang and Crow developed a feedback linearization controller to regulate bus voltages [4]. Chen *et al.* investigated the use of a STATCOM for system voltage control, during peak solar irradiation, in order to increase the PV installation capacity of a distribution feeder and avoid the voltage violation problem [5]. Aziz *et al.* presented a method for VAR planning with a STATCOM for an industrial system that comprised multiple distributed generation units. The impact of the placement of STATCOM on voltage recovery and the rating of the STATCOM were determined through time-domain simulations [6]. Xu and Li proposed an adaptive PI control, which can self-adjust the control gains in response to a disturbance such that the desired response is maintained, regardless of any change in the operating conditions [7].

On the other hand, conventional fuzzy rules (also called a type-I (T1) fuzzy logic system) have been used to design controllers and have been found to be effective when the controlled plant is not complex. On the contrary, an interval type-II (IT2) fuzzy logic system outperforms T1 if the system is complex. Zadeh introduced the concept of type-II fuzzy sets, as an extension of type-I fuzzy sets, incorporating an additional dimension that represents uncertainty in degrees of membership [8]. Liang and Mendel proposed a simplified method to compute the input and antecedent operations for IT2 fuzzy logic systems [9]. They showed that all of the results that are needed to implement an IT2 fuzzy logic system can be obtained using T1 fuzzy logic mathematics [10]. Other works have elaborated IT2 fuzzy logic systems, such as [11–14]. A tutorial about IT2 fuzzy logic systems is also available in [15].

Recently, the IT2 fuzzy logic system was applied to solve power system problems. Tripathy and Mishra proposed an IT2 fuzzy logic-based thyristor-controlled series capacitor to improve power system stability. They used the concept of uncertainty bounds for type-reduction to overcome the limitation of the time-consuming iterative method generally used for type reduction [16]. Khosravi *et al.* proposed the application of IT2 fuzzy logic systems to the problem of short-term load forecasting [17]. Mikkili and Panda proposed a shunt active filter to improve the power quality of the electrical network by mitigating the harmonics with the help of T1 and IT2 fuzzy logic controllers [18]. Jafarzadeh *et al.*, proposed T1 and IT2 Takagi-Sugeno-Kang (TSK) fuzzy systems for both modeling and prediction of solar power [19]. Khosravi and Nahavandi used IT2 fuzzy logic systems for day-ahead load forecasting, which introduced an optimal type reduction algorithm to improve the approximation [20]. Murthy *et al.* addressed that the failure rate and the repair rate of any component of a power system for reliability studies are uncertain; therefore, a membership representation using precise and crisp functions seems practically unreasonable [21].

In this work, IT2 fuzzy rules are applied to the STATCOM to mitigate voltage variations caused by large load changes and intermittent photovoltaic power. The proposed IT2 fuzzy rule-based STATCOM is more robust than those described elsewhere [4–7]. The proposed method is implemented by integrating the NEPLAN software package with MATLAB/Simulink. The NEPLAN software package is utilized to perform a dynamic power flow simulation while MATLAB/Simulink is used to design the IT2 fuzzy rule-based controller and STATCOM.

The rest of this paper is organized as follows. Section 2 provides the background of IT2 fuzzy logic systems. Section 3 presents the method based on the IT2 fuzzy rule base for mitigating the voltage variations in the power system. Section 4 summarizes the simulation results of a six-bus distribution system with PV generation. The conclusions are presented in Section 5.

2. Background of IT2 Fuzzy Logic Systems

2.1. IT2 Fuzzy Sets

A traditional type 1 (T1) fuzzy set $\tilde{A}$ in X is defined as a set of ordered pairs, as follows.

$$\tilde{A} = \left\{ \left(x, \mu_{\tilde{A}}(x) \right) \middle| x \in X \right\} \tag{1}$$

where X is the universal set and $\mu_{\tilde{A}}(x)$ is the membership function (MF) of x in $\tilde{A}$. Notably, the MF represents the degree to which x belongs to $\tilde{A}$ and the value of MF is normally between zero and unity. A high value of the MF indicates that x is very likely to be in $\tilde{A}$. Figure 1a shows an example of a T1 fuzzy set, $\tilde{A}$. Any number in the x domain corresponds to a membership value. The MF $\mu_{\tilde{A}}(x)$ of a T1 fuzzy set can be chosen either based on the user's experience or using algorithms.

A traditional T1 fuzzy set is certain in the sense that its membership values are crisp. In contrast, the type II fuzzy set is characterized by MFs that are themselves fuzzy. The IT2 fuzzy set is a special case of type II fuzzy sets. Figure 1b shows an example of an IT2 fuzzy set. The membership value of an IT2 fuzzy set lies within an interval. An IT2 fuzzy set is bounded by two type-1 MFs, $\bar{A}$ and $\underline{A}$, which are referred to as upper MF (UMF) and lower MF (LMF), respectively. The area between $\bar{A}$ and $\underline{A}$ is called the footprint of uncertainty (FOU). IT2 fuzzy sets are particularly useful when specifying exact MFs is difficult.

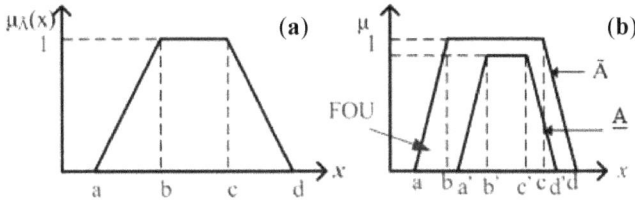

Figure 1. Examples of (**a**) a T1 membership function; (**b**) IT2 membership functions (UMF $\bar{A}$ and LMF $\underline{A}$).

2.2. IT2 Fuzzy Rules

Figure 2 shows a schematic diagram of an IT2 fuzzy logic system. The outputs of the inference engine are IT2 fuzzy sets. A type-reducer is needed to convert IT2 fuzzy sets into a T1 fuzzy set before defuzzification is carried out.

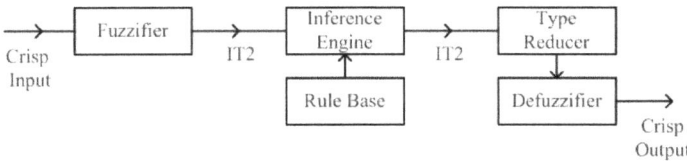

Figure 2. IT2 fuzzy logic system.

Consider N IT2 fuzzy rules. The *n*th rule can be expressed as follows.
IF x_1 is $\tilde{A}_{1n}$ and ... and x_M is $\tilde{A}_{Mn}$ then y is Y_n. $n = 1, 2, \ldots , N$.
Where $\tilde{A}_{in}$ ($i =1, 2, \ldots , M$) are IT2 fuzzy sets and $Y_n=[\underline{y}_n, \bar{y}_n]$ is an interval. Assume that the input vector $x^* = (x_1^*, x_2^*, \ldots , x_M^*)$. The IT2 fuzzy reasoning can be implemented using the following steps [22].

(1) Evaluate the membership of x^* on each $\tilde{A}_{in}$, $[\mu_{\underline{A}_{in}}(x_i^*), \mu_{\bar{A}_{in}}(x_i^*)]$, $i = 1, 2, \ldots , M$, $n = 1, 2, \ldots , N$.

(2) Evaluate the firing interval of the *n*th rule, $F_n(x^*) \equiv [\underline{f}_n, \bar{f}_n]$, $n = 1, 2, \ldots , N$.

$$[\underline{f}_n, \bar{f}_n] = [\mu_{\underline{A}_{1n}}(x_1^*) \times \ldots \times \mu_{\underline{A}_{Mn}}(x_M^*), \mu_{\bar{A}_{1n}}(x_1^*) \times \ldots \times \mu_{\bar{A}_{Mn}}(x_M^*)] \tag{2}$$

(3) Conduct type reduction using $F_n(x^*)$ and $[\underline{y}_n, \bar{y}_n]$.

$$y_\ell = \frac{\min}{j \in [1, N-1]} \frac{\sum_{n=1}^{j} \bar{f}_n \underline{y}_n + \sum_{n=j+1}^{N} \underline{f}_n \underline{y}_n}{\sum_{n=1}^{j} \bar{f}_n + \sum_{n=j+1}^{N} \underline{f}_n} \tag{3}$$

$$y_\ell = \frac{\min}{j \in [1, N-1]} \frac{\sum_{n=1}^{j} \underline{f}_n \bar{y}_n + \sum_{n=j+1}^{N} \bar{f}_n \bar{y}_n}{\sum_{n=1}^{j} \underline{f}_n + \sum_{n=j+1}^{N} \bar{f}_n} \tag{4}$$

The sets $\{\underline{y}_n\}$ and $\{\bar{y}_n\}$ are sorted in ascending order. The values of y_ℓ and y_r in Equations (3) and (4) can be evaluated using Karnik-Mendel algorithms [23].

(4) Estimate the defuzzified output.

$$y = \frac{y_\ell + y_r}{2} \tag{5}$$

3. Proposed Method

A STATCOM can be operated with either current source or voltage source converters; the latter are the more popular. Essentially, the main function of the STATCOM is to regulate voltages by reactive power control unless an additional energy storage system is connected to the DC bus of the STATCOM. Figure 3 illustrates the model of the STATCOM that is used herein [1,2].

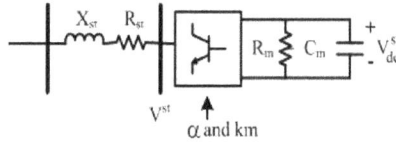

Figure 3. Model of static synchronous compensator (STATCOM).

The DC voltage V_{dc}^{st} can be transformed to be the d- and q-axis components of the output terminal of the STATCOM as follows [1,2].

$$V_d^{st} = V_{dc}^{st} \times km \times \sin(\theta_{bus} + \alpha) \tag{6}$$

$$V_q^{st} = V_{dc}^{st} \times km \times \cos(\theta_{bus} + \alpha) \tag{7}$$

where θ_{bus} and V_{dc}^{st} denote the phase angle of the controlled AC bus and the DC voltage across the DC capacitance C_m, respectively The variables km and α are the modulation index and the phase angle of the STATCOM, respectively. From Figure 3, one can obtain Equations (8) and (9) as follows [1,2].

$$C_m \frac{d}{dt} V_{dc}^{st} = \omega_b \left(I_{dc}^{st} - \frac{V_{dc}^{st}}{R_m} \right) \tag{8}$$

$$I_{dc}^{st} = i_q^{st} km \, \cos(\theta_{bus} + \alpha) + i_d^{st} km \, \sin(\theta_{bus} + \alpha) \tag{9}$$

where I_{dc}^{st} is the DC current that flows into the positive terminal of V_{dc}^{st}. R_m is the equivalent resistance of the STATCOM. i_d^{st} and i_q^{st} are the d- and q-axis currents that flow into the terminals of the STATCOM, respectively.

3.1. Dynamic Equations of STATCOM

Let V^{st} be the voltage at the STATCOM and Vd_{bus} and Vq_{bus} be the voltages of the d- and q-axes at the controlled AC bus, respectively. From the model of the STATCOM, the following dynamic equations are attained. Let the controlled bus voltage magnitude be $Vt = \sqrt{Vd_{bus}^2 + Vq_{bus}^2}$. The following dynamic equations can be obtained:

$$\frac{d}{dt} V^{st} = \frac{km}{C_m} \left(\frac{i_q^{st} Vq_{bus}}{Vt} \cos\alpha + \frac{i_q^{st} Vd_{bus}}{Vt} \sin\alpha \right) + \frac{km}{C_m} \left(\frac{i_d^{st} Vd_{bus}}{Vt} \cos\alpha + \frac{i_d^{st} Vq_{bus}}{Vt} \sin\alpha \right) - \frac{V^{st}}{C_m R_m} \tag{10}$$

$$\frac{d}{dt}\alpha = (Kc \times (V_{ref} - Vt) - \alpha) \times \frac{1}{Tc} \tag{11}$$

where Kc and Tc are parameters of the transfer function between $\Delta\alpha$ and ΔV_{dc}^{st}. V_{ref} is a given voltage reference.

$$\frac{d}{dt}km = \left(Ks \times \left(V^{st} - \sqrt{2}\right) - km\right) \times \frac{1}{Ts} \tag{12}$$

where Ks and Ts are parameters of the transfer function between Δkm and ΔVt

$$\frac{d}{dt}i_d^{st} = \left(Vd_{bus} - R_{st} \times i_d^{st} + i_q^{st} \times X_{st} - km \times V^{st}\left(\frac{Vd_{bus}}{Vt}\cos\alpha + \frac{Vq_{bus}}{Vt}\sin\alpha\right)\right) \times \frac{1}{X_{st}} \tag{13}$$

$$\frac{d}{dt}i_q^{st} = \left(Vq_{bus} - R_{st} \times i_q^{st} + i_d^{st} \times X_{st} - km \times V^{st}\left(\frac{Vq_{bus}}{Vt}\cos\alpha + \frac{Vd_{bus}}{Vt}\sin\alpha\right)\right) \times \frac{1}{X_{st}} \tag{14}$$

The dynamic phenomenon in the power system can be studied using the above dynamic equations, the swing equations of synchronous machines, and network equations.

3.2. IT2 Fuzzy Rules for Controllers

This paper proposes an IT2 fuzzy rule-based controller for mitigating the voltage variation, as shown in Figure 4. The term $V_{ref} - Vt$ in Equation (11) will be replaced by the output (modified ΔVt) of the IT2 fuzzy rule base. The proposed IT2 fuzzy rule is expressed as

IF x_1 is $\tilde{A}_{1n}$ and x_2 is $\tilde{A}_{2n}$ then y is Y_n. $n = 1, 2, \ldots, 25$.

where x_1 is the output of the PID controller (denoted as $\Delta V_t'$) and x_2 is the change rate of $\Delta V_t'$. The variable y is the modified ΔVt, which replaces the term $V_{ref} - Vt$ in Equation (11).

A total of 25 IT2 fuzzy rules are implemented in the proposed method (*i.e.*, N = 25), as shown Table 1. The symbols NB, NS, ZR, PS, PB, IB, IS, KV, DS, DB, represent "Negative Big", "Negative Small", "Zero", "Positive Small", "Positive Big", "Increasing and Big", "Increasing and Small", "Keeping Value", "Decreasing and Small", and "Decreasing and Big", respectively. All UMF and LMF of $\tilde{A}_{1n}$ and $\tilde{A}_{2n}$ are expressed as trapezoid functions. Possible conditions of x_1 and x_2 could be NB, NS, ZR, PS or PB and the consequent actions y could be IB, IS, KV, DS or DB, as shown in Table 1.

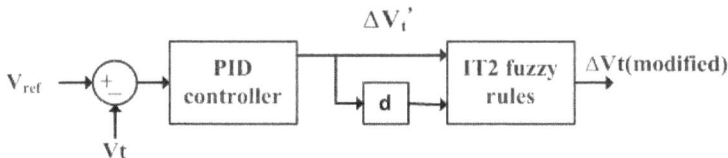

Figure 4. IT2 fuzzy rule-based controller. (Symbol "d" means "derivative").

Table 1. Twenty-five IT2 fuzzy rules.

Linguistic Variables of x_1 and x_2	x_1 = NB	x_1 = NS	x_1 = ZR	x_1 = PS	x_1 = PB
x_2 = PB	KV	IS	IB	IB	IB
x_2 = PS	DS	KV	IS	IB	IB
x_2 = ZR	DB	DS	KV	IS	IB
x_2 = NS	DB	DB	DS	KV	IS
x_2 = NB	DB	DB	DB	DS	KV

Notes: Negative Big (NB), Negative Small (NS), Zero (ZR), Positive Small (PS), Positive Big (PB), Increasing and Big (IB), Increasing and Small (IS), Keeping Value (KV), Decreasing and Small (DS), Decreasing and Big (DB).

3.3. Co-Simulation between NEPLAN and SIMULINK

Two software packages, namely NEPLAN [24] and MATLAB/Simulink (MathWorks, Natick, MA, USA), are used to implement the proposed method. The NEPLAN software package deals with the dynamic power flow simulation while the IT2 fuzzy rule-based controller and STATCOM are modeled by MATLAB/Simulink, as shown in Figure 5.

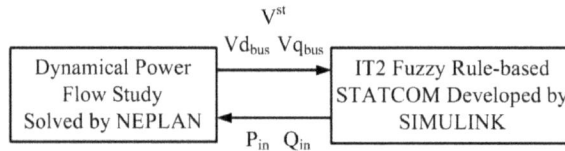

V^{st}

Vd_{bus} Vq_{bus}

| Dynamical Power Flow Study Solved by NEPLAN | IT2 Fuzzy Rule-based STATCOM Developed by SIMULINK |

P_{in} Q_{in}

Figure 5. Co-simulation conducted by integrating NEPLAN with Simulink.

The NEPLAN software package has interfaces for linking with Simulink. The dynamic power flow studies performed in the NEPLAN package are conducted by running a power flow program with varying real and reactive power injections (P_{in} and Q_{in}). A constant time step for running the power flow program is specified in the NEPLAN package. V^{st}, Vd_{bus}, and Vq_{bus}, computed by the NEPLAN package, are fed to Simulink.

3.4. Discussions of the Proposed Method

Two comments about the proposed method are discussed as follows:

(1) The proposed method utilized 25 IT2 fuzzy rules to implement the STATCOM controllers. In power system problems many facilities require controllers, which may be realized by different types of fuzzy rules, to stabilize the power system. Table 2 summarizes eight fuzzy rule-based controllers with their corresponding problems, number of fuzzy rules, types of fuzzy rules, and hardware [25–32]. It can be found that the numbers of fuzzy rules are in the range of 16–60. The state-of-the-art DSP chip or CPU is able to accommodate these numbers of fuzzy rules and provide fast calculation for controlling the kW- or MW-scaled facilities in the power system.

(2) The existing STATCOM works, such as References [4–7] mentioned in Section 1, have the following limitations and are less robust than the proposed method. (i) The work of Wang and Crow in [4] needs to linearize the nonlinear system before applying the feedback linearization transformation. The state feedback matrix needs to be determined by linear control techniques, such as pole assignment. As described in [4], when an operating point crosses the singular surface of the power system problem, the system matrix becomes ill-conditioned and the trajectory curves are more likely to oscillate dramatically near this point. This may cause unexpected uncontrollability and chaotic behavior in the STATCOM dynamics. (ii) All the control gains in Chen's work [5] are fixed and not adaptive to a change of scenarios. (iii) Aziz's work in [6] addressed the application of STATCOM at the planning stage. Aziz tuned the PID parameters by taking the step response of the open-loop plant into consideration. The results were verified in the peak load condition only. (iv) The work of Xu and Li lies in three essential parameters: allowed delayed time of response and ideal ratio of K_p over K_i for both current and voltage control loops, where K_p and K_i denote the parameters of proportional and integral gains, respectively. Actually, the "ideal ratio" of K_p over K_i is not well-defined and may depend on the operating conditions of different power systems.

Table 2. Eight fuzzy rule-based controllers and their corresponding features.

Problems	Number of Fuzzy Rules	Types of Fuzzy Rules	Hardware	References
Maximum power-point tracker for photovoltaic arrays	16	Mamdani	Infineon TriCore TC1796	[25]
Design of wide-area damping controller to damp the inter-area oscillations	30	Mamdani	-	[26]
Improvement of transient stability using FACTS devices	30	Mamdani	-	[27]
Improvement of transient stability using bang-bang controller	49	Mamdani	-	[28]
Control of the inverter for utilization of the wind energy	49	Mamdani	PC with DT2821 Data Card	[29]
Design of power system stabilizer	49	Mamdani	-	[30]
Design of power system stabilizer	49	Mamdani	-	[31]
Power management of energy of storage systems	60	Takagi-Sugeno	DSP TMS320F2812	[32]

4. Simulation Results

A six-bus power system, as shown in Figure 6, is studied. The PV farm with a generating capacity of 6.3 MW is located at bus 6. An IT2 rule-based STATCOM is located at bus 4. For comparison, the performance obtained by the traditional fuzzy rule-based (Type I) STATCOM is also provided herein. The values of K_p, K_i and K_d in Figure 4 are 10, 10 and 0, respectively. Suppose that the variation of irradiation is linear. The MVA base is 100. The Appendix A provides the parameters of IT2 fuzzy rules as well as parameters of the STATCOM model.

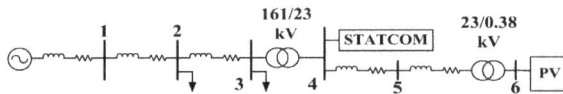

Figure 6. Studied system.

4.1. Scenario 1: Switching a Heavy Load

The demands of buses 2 and 3 are 30 MW + j20 MVAR and 100 MW + j30 MVAR, respectively. At $t = 10$ s, an additional heavy load 60 MW + j60 MVAR is switched on at bus 3. The irradiation in this scenario is zero.

Figure 7 shows the controlled voltage at bus 3 obtained using three different methods—traditional PID, T1 fuzzy rules, and IT2 fuzzy rules. As shown in Figure 7a, the voltage drops at 10 s and recovers slowly when traditional PID control is used. The steady-state voltage is still oscillating. When T1 fuzzy rules are used, the voltage recovers quickly after dropping but an overshoot occurs, as shown in Figure 7b. Again, the steady-state voltage is still oscillating. When the proposed IT2 fuzzy rules are used, the rising time and overshoot of the voltage are almost negligible, as shown in Figure 7c.

Figure 8 illustrates the reactive power injection from the STATCOM obtained using the three methods. The variations of voltages, shown in Figure 7, are consistent with those of the reactive powers shown in Figure 8.

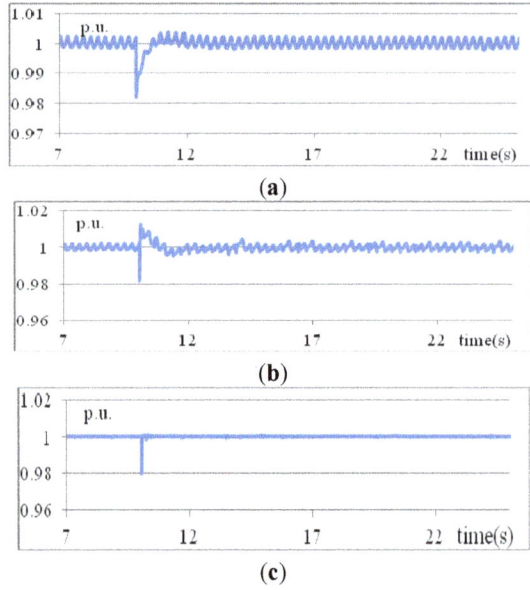

Figure 7. Voltages at bus 3 obtained using three methods: (**a**) traditional PID; (**b**) T1 fuzzy rules; (**c**) IT2 fuzzy rules (Scenario 1).

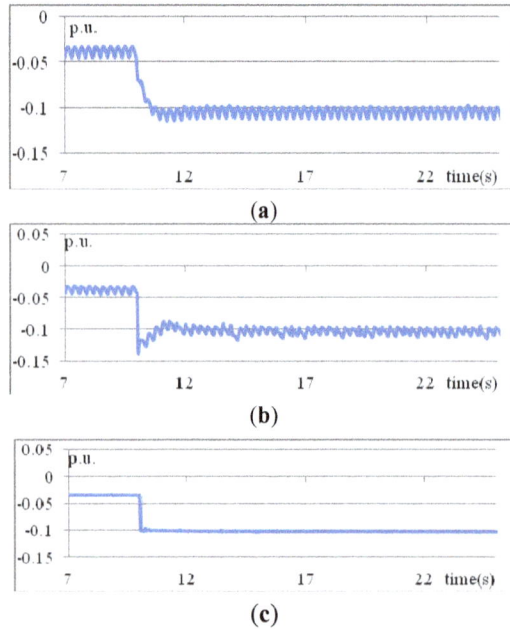

Figure 8. MVARs from STATCOM obtained using three methods: (**a**) traditional PID; (**b**) T1 fuzzy rules; (**c**) IT2 fuzzy rules (Scenario 1).

4.2. Scenario 2: Increasing and then Decreasing Irradiations

The loads at buses 2 and 3 are 30 MW + j20 MVAR and 10 MW + j3 MVAR, respectively. The irradiation changes linearly. Initially, the MW generation from the PV is 0.063 p.u., as shown in Figure 9. The MW generation starts to decrease at $t = 17$ s and becomes 0.0363 p.u. at $t = 21$ s.

Figure 10 shows the controlled voltage at bus 3 obtained using three methods—traditional PID, T1 fuzzy rules, and IT2 fuzzy rules. As shown in Figure 10a, the voltage decreases slightly near $t = 18$ s and increases back to the nominal value near $t = 21$ s. The proposed IT2 fuzzy rules yield the best voltage performance, as shown in Figure 10c.

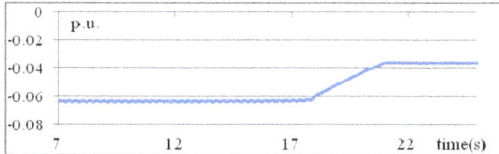

Figure 9. Variations of MW generation from PV (Scenario 2).

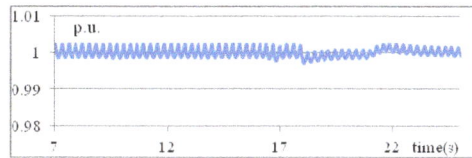

(a)

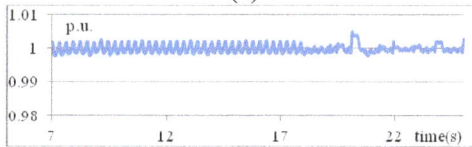

(b)

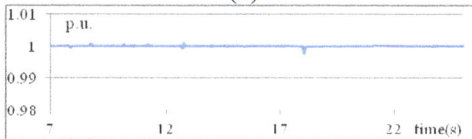

(c)

Figure 10. Voltages at bus 3 obtained using three methods: (**a**) traditional PID; (**b**) T1 fuzzy rules; and (**c**) IT2 fuzzy rules (Scenario 2).

The MVAR injections from the STATCOM increase from 0.031 p.u. close to $t = 17$ s to 0.114 p.u. near $t = 21$ s, as shown in Figure 11. Of the three methods, the proposed IT2 fuzzy rule- based method performs best.

355

(a)

(b)

(c)

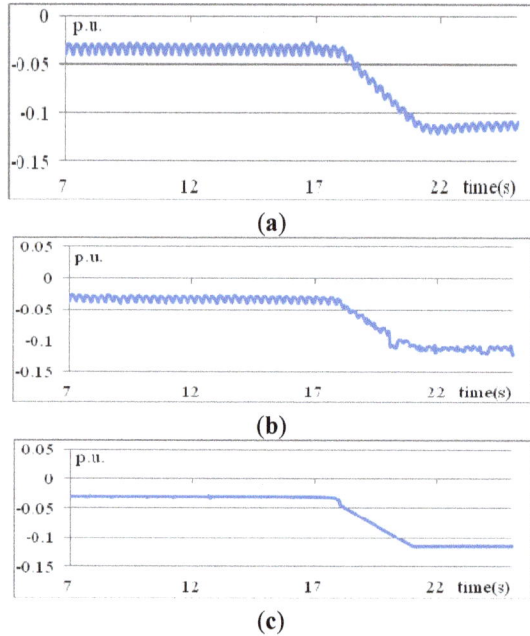

Figure 11. MVARs from STATCOM obtained using three methods: (**a**) traditional PID; (**b**) T1 fuzzy rules; and (**c**) IT2 fuzzy rules (Scenario 2).

4.3. Scenario 3: Varying Irradiations

Buses 2 and 3 have demands of 30 MW + j20 MVAR and 10 MW + j3 MVAR, respectively. The irradiation is 1000 W/m^2 at $t = 0$ s. The irradiation increases to 1400 W/m^2 at $t = 10$, and thereafter linearly changes to be 1000, 1400, and 1000 W/m^2 at $t = 12$, 15 and 18 s, respectively, in the third scenario. The variations of MW generations from PV are shown in Figure 12.

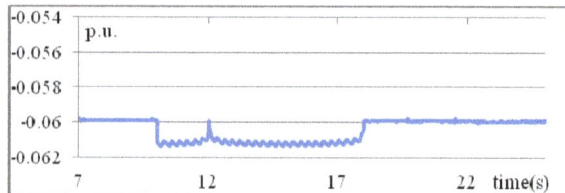

Figure 12. Variations of MW generations from PV (Scenario 3).

Figures 13 and 14 plot the controlled voltages and MVAR from the STATCOM, respectively. It can be found that the proposed IT2 fuzzy rule-based method always acquire the most stable performance than traditional PID and T1 fuzzy rule-based methods.

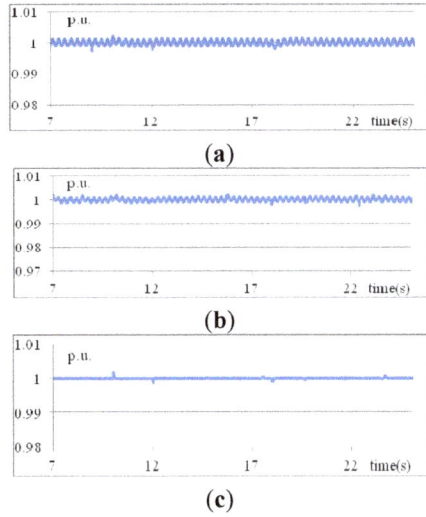

Figure 13. Voltages at bus 3 using obtained three methods: (**a**) traditional PID; (**b**) T1 fuzzy rules; (**c**) IT2 fuzzy rules (Scenario 3).

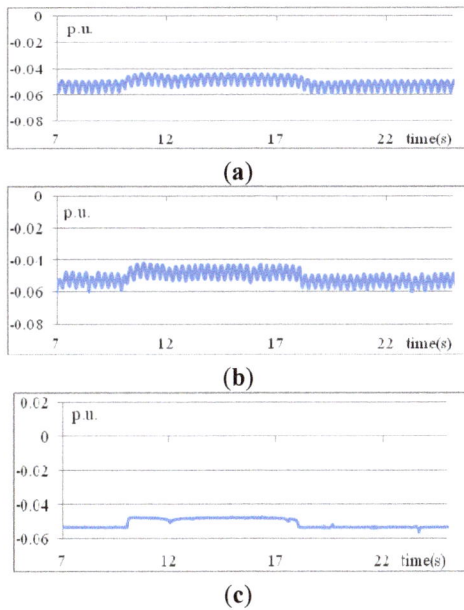

Figure 14. MVARs from STATCOM obtained using three methods: (**a**) traditional PID; (**b**) T1 fuzzy rules; (**c**) IT2 fuzzy rules (Scenario 3).

4.4. Unscheduled Photovoltaic Outage

In this scenario, an unscheduled PV outage occurs at $t = 10$ s. Figure 15 reveals that the voltage increases rapidly at $t = 10$ s. The proposed method attains quickly-damped voltage variations. The dynamic voltage obtained by the T1 fuzzy rules is slowly damped from $t = 10$ to 12 s and oscillates as t approaches infinity. To reduce the voltage at $t = 10$ s, the reactive power produced from the STATCOM must be reduced as shown in Figure 16. The proposed IT2 fuzzy rule-based method outperforms both the traditional PID and T1 fuzzy rule-based methods.

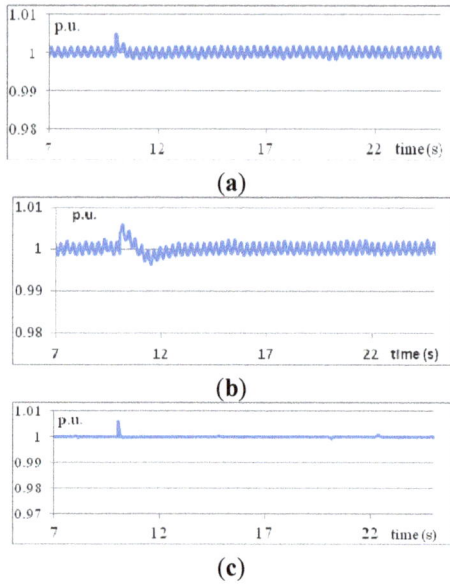

Figure 15. Voltages at bus 3 obtained using three methods: (**a**) traditional PID; (**b**) T1 fuzzy rules; (**c**) IT2 fuzzy rules (Scenario 4).

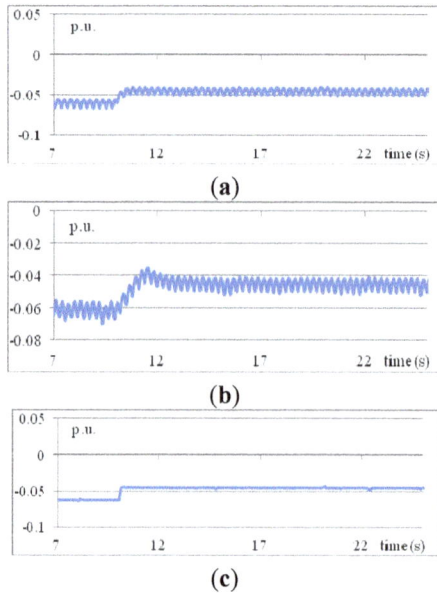

Figure 16. MVARs from STATCOM obtained using three methods: (**a**) traditional PID; (**b**) T1 fuzzy rules; and (**c**) IT2 fuzzy rules (Scenario 4).

5. Conclusions

An interval type-II (IT2) fuzzy rule-based STATCOM is proposed to mitigate voltage variations in this paper. The voltage error and rate of change of the voltage error are utilized to serve as inputs of the IT2 fuzzy rules; the action parts of IT2 fuzzy rules produce modified voltage errors. IT2 fuzzy rules are more applicable to nonlinear and time-varying systems than traditional T1 fuzzy rules because the upper and lower membership functions are implemented in an IT2 fuzzy set. A six-bus system is used to validate the proposed method. Four scenarios that involve large load changes and variations of MW generation from a PV farm are studied. Simulation results indicate that the proposed method always outperforms the traditional PID and T1 fuzzy rule-based methods.

Acknowledgments: The authors would like to thank Dinh-Nhon Truong for discussions about the modeling of STATCOM.

Author Contributions: Ying-Yi Hong designed the IT2 fuzzy rule base, wrote this manuscript and supervised the progress of research. Yu-Lun Hsieh developed the MATLAB code, and investigated the model of STATCOM and IT2 fuzzy theory.

Conflicts of Interest: The authors declare no conflict of interest.

Appendix

The parameters of STATCOM used in this paper are as follows: $R_m = 20.3$ p.u., $C_m = 0.123$ p.u., $R_{st} = 0.00813$ p.u., $X_{st} = 0.0325$ p.u., $K_c = 30$, $T_c = 0.001$ s, $K_s = 2$, $T_s = 0.01$ s. The bases of power and voltage are 100 MVA and 23 kV, respectively.

Table A1 provides the upper and lower MFs for the IT2 fuzzy rules used in Section 4. Definitions of [a, b, c, d] and [a', b', c', d'] are given in Figure 1 of Section 2.1.

Table A1. Upper and lower membership functions.

Linguistic Variables	Upper MF [a, b, c, d]	Lower MF [a', b', c', d']
NB	$[-2, -2, -1.7, -1.5]$	$[-2, -2, -1.5, -1.2]$
NS	$[-1.5, -1.2, -1, -0.5]$	$[-1.2, -1.1, -1.1, -1]$
ZR	$[-1, -0.5, 0.5, 1]$	$[-0.5, 0, 0, 0.5]$
PS	$[0.5, 1, 1.2, 1.5]$	$[1, 1.1, 1.1, 1.2]$
PB	$[1.2, 1.5, 2, 2]$	$[1.5, 1.7, 2, 2]$

References

1. Wang, L.; Hsiung, C.T. Dynamic stability improvement of an integrated grid-connected offshore wind farm and marine-current farm using a STATCOM. *IEEE Trans. Power Syst.* **2011**, *26*, 690–699. [CrossRef]
2. Wang, L.; Truong, D.N. Dynamic stability improvement of four parallel-operated PMSG-based offshore wind turbine generators fed to a power system using a STATCOM. *IEEE Trans. Power Deliv.* **2013**, *28*, 111–119.
3. Beza, M.; Bongiorno, M. An adaptive power oscillation damping controller by STATCOM with energy storage. *IEEE Trans. Power Syst.* **2015**, *26*, 484–493. [CrossRef]
4. Wang, K.Y.; Crow, M.L. Power system voltage regulation via STATCOM internal nonlinear control. *IEEE Trans. Power Syst.* **2011**, *26*, 1252–1262. [CrossRef]
5. Chen, C.S.; Lin, C.H.; Hsieh, W.L.; Hsu, C.T.; Ku, T.T. Enhancement of PV penetration with DSTATCOM in Taipower distribution system. *IEEE Trans. Power Syst.* **2013**, *28*, 1252–1262.
6. Aziz, T.; Hossain, M.J.; Saha, T.K.; Mithulananthan, N. VAR planning with tuning of STATCOM in a DG integrated industrial system. *IEEE Trans. Power Deliv.* **2013**, *28*, 875–885. [CrossRef]
7. Xu, Y.; Li, F.X. Adaptive PI control of STATCOM for voltage regulation. *IEEE Trans. Power Deliv.* **2014**, *29*, 1002–1011. [CrossRef]

8. Zadeh, L.A. The concept of a linguistic variable and its application to approximate reasoning—I. *Inf. Sci.* **1975**, *8*, 199–249. [CrossRef]
9. Liang, Q.L.; Mendel, J.M. Interval type-2 fuzzy logic systems: Theory and design. *IEEE Trans. Fuzzy Syst.* **2000**, *8*, 535–550. [CrossRef]
10. Mendel, J.M.; John, R.I.; Liu, F.L. Interval Type-2 fuzzy logic systems made simple. *IEEE Trans. Fuzzy Syst.* **2006**, *14*, 808–821.
11. Juang, C.F.; Chen, C.Y. An interval type-2 neural fuzzy chip with on-chip incremental learning ability for time-varying data sequence prediction and system control. *IEEE Trans. Neural Netw. Learn. Syst.* **2014**, *25*, 216–228. [CrossRef] [PubMed]
12. Juang, C.F.; Jang, W.S. A type-2 neural fuzzy system learned through type-1 fuzzy rules and its FPGA-based hardware implementation. *Appl. Soft Comput.* **2014**, *18*, 302–313. [CrossRef]
13. Lin, Y.Y.; Liao, S.H.; Chang, J.Y.; Lin, C.T. Simplified interval type-2 fuzzy neural networks. *IEEE Trans. Neural Netw. Learn. Syst.* **2014**, *25*, 959–969. [CrossRef] [PubMed]
14. Lin, Y.Y.; Chang, J.Y.; Lin, C.T. A TSK-type-based self-evolving compensatory interval type-2 fuzzy neural network (TSCIT2FNN) and its applications. *IEEE Trans. Ind. Electron.* **2014**, *61*, 447–459. [CrossRef]
15. Mendel, J.M. General type-2 fuzzy logic systems made simple: A tutorial. *IEEE Trans. Fuzzy Syst.* **2014**, *22*, 1162–1182. [CrossRef]
16. Tripathy, M.; Mishra, S. Interval type-2-based thyristor controlled series capacitor to improve power system stability. *IET Génér. Transm. Distrib.* **2011**, *5*, 209–222. [CrossRef]
17. Khosravi, A.; Nahavandi, S.; Creighton, D.; Srinivasan, D. Interval type-2 fuzzy logic systems for load forecasting: A comparative study. *IEEE Trans. Power Syst.* **2012**, *27*, 1274–1282.
18. Mikkili, S.; Panda, A.K. Types-1 and -2 fuzzy logic controllers-based shunt active filter Id–Iq control strategy with different fuzzy membership functions for power quality improvement using RTDS hardware. *IET Génér. Transm. Distrib.* **2013**, *6*, 818–833. [CrossRef]
19. Jafarzadeh, S.; Fadali, M.S.; Evrenosoğlu, C.Y. Solar power prediction using interval type-2 TSK modeling. *IEEE Trans. Sustain. Energy* **2013**, *4*, 333–339.
20. Khosravi, A.; Nahavandi, S. Load forecasting using interval type-2 fuzzy logic systems: Optimal type reduction. *IEEE Trans. Ind. Inf.* **2014**, *10*, 1055–1063.
21. Murthy, C.; Varma, K.A.; Roy, D.S.; Mohanta, D.K. Reliability evaluation of phasor measurement unit using type-2 fuzzy set theory. *IEEE Syst. J.* **2014**, *8*, 1302–1309. [CrossRef]
22. Wu, D.R. On the fundamental differences between interval type-2 and type-1 fuzzy logic controllers. *IEEE Trans. Fuzzy Syst.* **2012**, *20*, 832–848. [CrossRef]
23. Mendel, J.M. *Uncertain Rule-Based Fuzzy Logic Systems: Introduction and New Directions*; Prentice-Hall: Upper Saddle River, NJ, USA, 2001.
24. NEPLAN 5.5 User Manual. Available online: http://www.neplan.ch/ (accessed on 18 August 2015).
25. Alajmi, B.N.; Ahmed, K.H.; Finney, S.J.; Williams, B.W. Fuzzy-logic-control approach of a modified hill-climbing method for maximum power point in microgrid standalone photovoltaic system. *IEEE Trans. Power Electron.* **2011**, *26*, 1022–1030. [CrossRef]
26. Mokhtari, M.; Aminifar, F.; Nazarpour, D.; Golshannavaz, S. Wide-area power oscillation damping with a fuzzy controller compensating the continuous communication delays. *IEEE Trans. Power Syst.* **2013**, *28*, 1997–2005. [CrossRef]
27. Sadeghzadeh, S.M.; Ehsan, M.; Said, N.H.; Feuillet, R. Improvement of transient stability limit in power system transmission lines using fuzzy control of FACTS devices. *IEEE Trans. Power Syst.* **1998**, *13*, 917–922. [CrossRef]
28. Chang, H.C.; Wang, M.H. Neural network-based self-organizing fuzzy controller for transient stability of multimachine power systems. *IEEE Trans. Energy Convers.* **1995**, *10*, 339–347. [CrossRef]
29. Hilloowala, R.M.; Sharaf, A.M. A rule-based fuzzy logic controller for a PWM inverter in a stand alone wind energy conversion scheme. *IEEE Trans. Ind. Appl.* **1996**, *32*, 57–65. [CrossRef]
30. Hosseinzadeh, N.; Kalam, A. A rule-based fuzzy power system stabilizer tuned by a neural network. *IEEE Trans. Energy Convers.* **1999**, *14*, 773–779. [CrossRef]

Energies **2015**, *8*, 8908–8923

31. Hoang, P.; Tomsovic, K. Design and analysis of an adaptive fuzzypower system stabilizer. *IEEE Trans. Energy Convers.* **1996**, *11*, 445–461. [CrossRef]
32. Talla, J.; Streit, L.; Peroutka, Z.; Drabek, P. Position-based T-S fuzzy power management for tram with energy storage system. *IEEE Trans. Ind. Electron.* **2015**, *62*, 3061–3071. [CrossRef]

MDPI AG

St. Alban-Anlage 66

4052 Basel, Switzerland

Tel. +41 61 683 77 34

Fax +41 61 302 89 18

http://www.mdpi.com

Energies Editorial Office

E-mail: energies@mdpi.com

http://www.mdpi.com/journal/energies